Lecture Notes in Electrical Engineering

Volume 1219

The book series *Lecture Notes in Electrical Engineering* (LNEE) publishes the latest developments in Electrical Engineering—quickly, informally and in high quality. While original research reported in proceedings and monographs has traditionally formed the core of LNEE, we also encourage authors to submit books devoted to supporting student education and professional training in the various fields and applications areas of electrical engineering. The series cover classical and emerging topics concerning:

- Communication Engineering, Information Theory and Networks
- Electronics Engineering and Microelectronics
- Signal, Image and Speech Processing
- Wireless and Mobile Communication
- Circuits and Systems
- Energy Systems, Power Electronics and Electrical Machines
- Electro-optical Engineering
- Instrumentation Engineering
- Avionics Engineering
- Control Systems
- Internet-of-Things and Cybersecurity
- Biomedical Devices, MEMS and NEMS

For general information about this book series, comments or suggestions, please contact leontina.dicecco@springer.com.

To submit a proposal or request further information, please contact the Publishing Editor in your country:

China

Jasmine Dou, Editor (jasmine.dou@springer.com)

India, Japan, Rest of Asia

Swati Meherishi, Editorial Director (Swati.Meherishi@springer.com)

Southeast Asia, Australia, New Zealand

Ramesh Nath Premnath, Editor (ramesh.premnath@springernature.com)

USA, Canada

Michael Luby, Senior Editor (michael.luby@springer.com)

All other Countries

Leontina Di Cecco, Senior Editor (leontina.dicecco@springer.com)

**** This series is indexed by EI Compendex and Scopus databases. ****

Sabu M. Thampi · Patrick Siarry ·
Mohammed Atiquzzaman · Ljiljana Trajkovic ·
Jaime Lloret Mauri
Editors

Fifth International Conference on Computing and Network Communications

Proceedings of CoCoNet 2023, Volume 1

Springer

Editors
Sabu M. Thampi
Department of Computer Science
and Engineering
Digital University Kerala
Trivandrum, Kerala, India

Patrick Siarry
Laboratory of Images, Signals
and Intelligent Systems
Paris-Est Créteil University
Vitry-sur-Seine, France

Mohammed Atiquzzaman
School of Computer Science
University of Oklahoma
Norman, OK, USA

Ljiljana Trajkovic
School of Engineering Science
Simon Fraser University
Burnaby, BC, Canada

Jaime Lloret Mauri
Departamento Comunicaciones
Universidad Politecnica de Valencia
Valencia, Spain

ISSN 1876-1100 ISSN 1876-1119 (electronic)
Lecture Notes in Electrical Engineering
ISBN 978-981-97-4542-5 ISBN 978-981-97-4540-1 (eBook)
https://doi.org/10.1007/978-981-97-4540-1

This Springer imprint is published by the registered company Springer Nature Singapore Pte Ltd.
The registered company address is: 152 Beach Road, #21-01/04 Gateway East, Singapore 189721, Singapore

Organized By

PES University, Bengaluru, India

Conference Organization

Chief Patron

M. R. Doreswamy, Chancellor, PES University

Patrons

D. Jawahar, Pro-Chancellor, PES University
Ajoy Kumar, COO, PES Institutions
J. Surya Prasad, Vice Chancellor, PES University
K. S. Sridhar, Registrar, PES University

General Chairs

Ljiljana Trajkovic, Simon Fraser University, Canada
Jaime Lloret, Universitat Politecnica de Valencia, Spain
Shikha Tripathi, PES University, Bengaluru

General Executive Chair

Sabu M. Thampi, Digital University Kerala, India

Steering Committee

Sudarshan T. S. B., Dean of Research, PES University (Chair)
San Murugesan, BRITE Professional Services and University of Western Sydney, Australia
Kuan Ching Li, Providence University, Taiwan

TPC Chairs

Patrick Siarry, University Paris-Est Créteil Val de Marne, France
Mohammed Atiquzzaman, School of Computer Science, University of Oklahoma, USA
Dilip Krishnaswamy, Quantum Walks Technologies, USA
Vipin Chaudhary, Case Western Reserve University, USA

Workshop and Symposium Chairs

Peter Mueller, IBM Zurich Research Laboratory, Switzerland
Axel Sikora, University of Applied Sciences Offenburg, Germany

Tutorial Chairs

Sameep Mehta, IBM Research AI, India
Al-Sakib Khan Pathan, United International University (UIU), Dhaka, Bangladesh

Organizing Chair

Anuradha M., PES University

Organizing Secretaries

M. S. Sunitha, PES University

Event Management Chair

M. Rajasekar, PES University

TPC Members

https://coconet-conference.org/2023/?q=committee

Preface

The Fifth International Conference on Computing and Network Communications (CoCoNet'23) took place in Bengaluru (Bangalore), India, during December 18–20, 2023. The conference provided a platform for experts in the emerging areas of computing and communications to share original research outcomes and practical development experiences. The event was organized by PES University, Bengaluru, with technical support from the IEEE Signal Processing Society Bangalore Chapter and the IEEE Communications Society Bangalore Chapter.

The program included keynote talks, a Start-up Expo, Panel Discussions, lightning talks, tutorials, symposiums, parallel sessions, workshops, Hands-on Sessions, and Hot off the Press. CoCoNet'23 convened a tailored and selected collection of eminent speakers from renowned universities and industries in and outside India. The keynote speakers included Dr. C. Mohan, Dr. Ranga Rao Venkatesha Prasad, Dr. Sanjay K. Madria, Dr. S. R. Mahadeva Prasanna, and Dr. Sri Krishnan. An industrial keynote was delivered by Dr. Vishwas Lakkundi. The tutorial speakers were Dr. Abdullah Uz Tansel, Dr. Nithin Nagaraj, Dr. Amitava Mukherjee, Dr. Vishnu Srivastava, Dr. H. L. Shashirekha, Asha Hegde, Karthik Sivarama Krishnan, and Dr. Jaya Dofe.

The conference received 212 submissions, with 91 papers accepted after a rigorous review process evaluating their significance, novelty, and technical quality. The accepted papers were presented in regular paper sessions, symposiums, workshops, and short paper sessions.

The proceedings of the conference are organized into two volumes. This volume comprises 44 papers. The topical sections include Networking and Communication, IoT and Edge Computing, Privacy, Security, Forensics and Signal Processing, VLSI Design, and Embedded Computing.

The success of the conference ultimately depends on the numerous individuals to plan and organize the technical program and local arrangements. In particular, we extend our gratitude for the insightful advice and innovative ideas provided by the Program Chairs, Workshop and Symposium Chairs, and Tutorial Chairs. Our heartfelt thanks also go to the organizing committee and steering committee members.

The accomplishment of the conference is attributed to the tireless efforts of many individuals. We extend our sincere gratitude to the TPC members and reviewers who generously shared their technical expertise and assisted us in the thorough review of all submitted papers. Heartfelt thanks go to the general chairs, organizing committee members, steering committee, keynote speakers, tutorial speakers, session chairs, and conference attendees. We appreciate the authors for choosing this conference as the platform to present their research works. Our wholehearted appreciation goes to all those who dedicated their valuable time to ensure the success of CoCoNet'23.

We are thankful to PES University for hosting the conference. Recognition is due to the Local Organizing Committee members who worked diligently on the crucial aspects of the conference programs. We acknowledge the contributions of the faculty and staff of PES University, as well as the student volunteers who selflessly dedicated their time to contribute to the success of the conference. We would like to thank Kamiya Khatter, Associate Editor at Springer Nature, for her valuable assistance and cooperation.

We sincerely believe that CoCoNet'23 served as a platform for excellent discussions that facilitated the emergence of new ideas and the promotion of collaborative research. We are confident that this proceedings will serve as a significant source of research references and knowledge, contributing to scientific and engineering findings as well as the development of new products and technologies.

Trivandrum, India — Sabu M. Thampi
Vitry-sur-Seine, France — Patrick Siarry
Norman, USA — Mohammed Atiquzzaman
Burnaby, Canada — Ljiljana Trajkovic
Valencia, Spain — Jaime Lloret Mauri
December 2023

Contents

IoT and Edge Computing

About the Editors

Sabu M. Thampi is a Professor at the School of Computer Science and Engineering, Digital University Kerala. His research interests include the Internet of Things (IoT), Cognitive Security, Endpoint Security, Cognitive Surveillance, and Smart Cyber-Physical Systems. He also coordinates the Connected Systems and Intelligence (CSI) Lab. With a Ph.D. in Computer Engineering from the National Institute of Technology Karnataka, Sabu has actively contributed to funded research projects and has a notable record of publications in books, journals, and conferences. He has authored and edited books and conference proceedings published by reputed publishers such as Springer, IEEE, ACM, and Elsevier. Sabu's contributions extend to serving as a Guest Editor for special journal issues and participating as a program committee member in international conferences and workshops. He has taken on leadership roles as a co-chair for international workshops and conferences and has initiated and organized multiple annual conferences and symposiums. Dr. Sabu actively contributes as a reviewer for esteemed international journals and holds an Editorial Board membership in the Springer Applied Machine Learning book series. Dr. Sabu is a Senior Member of IEEE and ACM and serves as the founding Chair of the ACM Trivandrum Professional Chapter.

Patrick Siarry received the Ph.D. degree from the University Paris 6, in 1986 and the Doctorate of Sciences (Habilitation) from the University Paris 11, in 1994. He was first involved in the development of analog and digital models of nuclear power plants at Electricité de France (E.D.F.). Since 1995 he is a professor in automatics and informatics. His main research interests are computer-aided design of electronic circuits, and the applications of new stochastic global optimization heuristics to various engineering fields. He is also interested in the fitting of process models to experimental data, the learning of fuzzy rule bases, and of neural networks.

Mohammed Atiquzzaman obtained his M.S. and Ph.D. in Electrical Engineering and Electronics from the University of Manchester (UK) in 1984 and 1987, respectively. He currently holds the Edith J. Kinney Gaylord Presidential professorship in the School of Computer Science at the University of Oklahoma. Dr. Atiquzzaman is the Editor-in-Chief of *Journal of Networks and Computer Applications*, the founding Editor-in-Chief of *Vehicular Communications*, and serves/served on the editorial boards of many journals including *IEEE Communications Magazine, Real Time Imaging Journal, International Journal of Communication Networks and Distributed Systems* and *Journal of Sensor Networks* and *International Journal of Communication Systems*. He co-chaired the IEEE High Performance Switching and Routing Symposium (2003, 2011), IEEE Globecom and ICC (2014, 2012, 2010, 2009, 2007, 2006), IEEE VTC (2013) and the SPIE Quality of Service over Next Generation Data Networks conferences (2001, 2002, 2003). He was the panels co-chair of INFOCOM'05, and is/has been in the program committee of many conferences such as INFOCOM, Globecom, ICCCN, ICCIT, Local Computer Networks, and serves on the review panels at the National Science Foundation. Dr. Atiquzzaman received IEEE Communication Society's Fred W. Ellersick Prize, IEEE Distinguished Technical Achievement Award for contributions in switching and routing, IEEE Satellite Communications Technical Contribution Award, and NASA Group Achievement Award for "outstanding work to further NASA Glenn Research Center's effort in the area of Advanced Communications/Air Traffic Management's Fiber Optic Signal Distribution for Aeronautical Communications" project. He is the co-author of the book *Performance of TCP/IP over ATM Networks* and has over 300 refereed publications. His current research interests are in areas of Internet of Things, wireless and mobile networks, ad hoc networks, satellite networks, vehicular communications, and optical communications.

Ljiljana Trajkovic received the Dipl. Ing. degree from University of Pristina, Yugoslavia, in 1974, the M.Sc. degrees in electrical engineering and computer engineering from Syracuse University, Syracuse, NY, in 1979 and 1981, respectively, and the Ph.D. degree in electrical engineering from University of California at Los Angeles, in 1986. She is currently a Professor in the School of Engineering Science at Simon Fraser University, Burnaby, British Columbia, Canada. From 1995 to 1997, she was a National Science Foundation (NSF) Visiting Professor in the Electrical Engineering and Computer Sciences Department, University of California, Berkeley. She was a Research Scientist at Bell Communications Research, Morristown, NJ, from 1990 to 1997, and a Member of the Technical Staff at AT&T Bell Laboratories, Murray Hill, NJ, from 1988 to 1990. Her research interests include high-performance communication networks, control of communication systems, computer-aided circuit analysis and design, and theory of nonlinear circuits and dynamical systems.

Jaime Lloret Mauri received his B.Sc. + M.Sc. in Physics in 1997, his B.Sc. + M.Sc. in electronic Engineering in 2003 and his Ph.D. in telecommunication engineering (Dr. Ing.) in 2006. He is a Cisco Certified Network Professional Instructor and he has 7 Cisco Networking Academy Certifications. He also has the Hewlett-Packard IT Architect Certification. He worked as a network designer and administrator in several enterprises. He is Full Professor at the Polytechnic University of Valencia. He is the Chair of the Integrated Management Coastal Research Institute (IGIC) since January 2017. He was the founder of the "Communications and Networks" research group of the IGIC and he is the head (and founder) of the "Active and collaborative techniques and use of technologic resources in the education (EITACURTE)" Innovation Group. He has authored 15 books and has more than 800 research papers published in national and international conferences, international journals (more than 400 with Clarivate Analytics JCR). He has been the co-editor of 54 conference proceedings and guest editor of several international books and journals. He is editor-in-chief of the *Ad Hoc and Sensor Wireless Networks* (with Clarivate Analytics Impact Factor), the international journal *Networks Protocols and Algorithms*, and the *International Journal of Multimedia Communications*. Moreover, he is Associate editor of *Sensors* in the Section Sensor Networks, and in Wireless Communications and Mobile Computing, he is advisory board member of the *International Journal of Distributed Sensor Networks* (all with Clarivate Analytics Impact factor), and he is *IARIA* Journals Board Chair (8 Journals). Furthermore, he is (or has been) associate editor of 46 international journals (16 of them with Clarivate Analytics Impact Factor). He has led many local, regional, national and European projects.

Networking and Communication

A Novel Approach for Bridging the Gap Between SDN and MITRE for Agile Incident Response

Anjana Jimmington, P. Alfas Hakeem, Preetam Mukherjee, and Sabu M. Thampi

Abstract The growing complexity of potential attacks in network infrastructure demands advanced incident response (IR) techniques. With organization adopting the Software-Defined Networks the response capabilities can be improved to a greater extend. SDN's centralized control is utilized in this process. However the lacking of a list of possible defense mechanisms in the SDN, effects the swift respondence against the occurrence of a security incident. To address this challenge, we studied the MITRE D3FEND knowledge base, which provides various standard defense mechanisms. Later, within the knowledge base the defense mechanisms that can be implemented in SDN were identified. A valid list of possible implementable defense mechanisms in SDN was created. With the help of the developed implementer tool ANTq, which runs on top of the Ryu controller the suggested MITRE defense mechanisms were automatically implemented.

Keywords Incident response · Software-defined network · Cyber defense · MITRE D3FEND

1 Introduction

In the ever-evolving landscape of cybersecurity, organizations face various threats, ranging from common malware infections to sophisticated targeted attacks. While prevention is always the primary goal, security incidents will inevitably occur. This

A. Jimmington (✉) · P. Alfas Hakeem · P. Mukherjee · S. M. Thampi
Connected Systems and Intelligence Lab, School of Computer Science and Engineering, Digital University Kerala, Technopark Phase IV, Thiruvananthapuram, Kerala, India
e-mail: anjana.res21@duk.ac.in

P. Alfas Hakeem
e-mail: alfas.hakeem@duk.ac.in

P. Mukherjee
e-mail: preetam.mukherjee@duk.ac.in

S. M. Thampi
e-mail: sabu.thampi@duk.ac.in

S. M. Thampi et al. (eds.), *Fifth International Conference on Computing and Network Communications*, Lecture Notes in Electrical Engineering 1219,
https://doi.org/10.1007/978-981-97-4540-1_1

is where incident response (IR) plays a crucial role. Incident response empowers organizations to contain and mitigate the impact of security incidents [1–3]. With the increasing complexities of organization networks and the dynamic nature of threats, traditional incident response faces many challenges. The Automated Incident Response (AIR) systems have emerged as a practical solution. Characterized by its predefined actions triggered by security tools, AIR eliminates manual intervention, significantly improving efficiency and enhancing response time in addressing cybersecurity threats [4–6]. Imagine an AIR system in a network identifying unusual patterns indicating a potential Distributed Denial of Service (DDoS) attack. Without manual intervention, the AIR system dynamically adjusts network configurations to divert and filter malicious traffic, mitigating the attack's impact. This timely approach aligns with the challenges posed by evolving cyber threats.

With the emergence of Software-Defined Networks (SDN), organizations can seamlessly integrate these automated incident response capabilities into their organization [7]. SDN decouples the control plane from the data plane in network infrastructure, enabling centralized and programmatic control over network resources. The ability to provide a centralized global view of the network and granular control over individual network devices makes it an ideal platform for AIR systems [8]. By utilizing SDN's capabilities, AIR systems can dynamically change network configurations, isolate affected areas, and implement real-time mitigation strategies in response to security incidents. For example, upon detecting suspicious network activity, an AIR system communicates with the SDN controller to dynamically isolate the affected segment, showcasing the effectiveness of this integrated approach in responding swiftly to potential security threats. This dynamic capability enhances the agility of incident response strategies, ensuring adaptability to changing network conditions and emerging threats.

Researchers have utilized this capability of SDN and proposed diverse frameworks for incident response. SDN4S [9] prioritizes automated countermeasures for occurred incidents through predefined playbooks. Patzer et al. [10] introduced an SDN-based incident response framework for Industrial Control Systems. Additionally, diverse approaches, such as an SDN-based ship communication security framework (CyberShip), the scalable distributed monitoring methodology in TENNISON, the adaptable FLOWGUARD for firewall policy violations, and Murillo et al.'s exploration of NFV and SDN for ICS incident response, contribute to the incident response landscape of SDN environments [11–14].

While these frameworks have effectively addressed security incidents, they face challenges from uncertainties regarding the efficacy of suggested defense strategies, the difficulty of determining optimal deployment points within the SDN environment and assessing the aptness of applied defenses. These challenges underscore the pressing need for a comprehensive and validated list of defense suggestions in SDN automated incident response.

Hence, a gap exists in a well-defined, readily accessible, valid list of defense suggestions in SDN automated incident response. We aim to fill this gap with our exhaustive list of defense suggestions. This list we provide can act as practical guidance for network administrators and security professionals in safeguarding their SDN infrastructure.

Recognizing the need for a valid list, our research compelled a thorough exploration of diverse resources to address this substantial gap. We studied the reputable framework MITRE D3FEND [15] and selected it for our study. Its systematic approach to mapping defense tactics and techniques against a broad spectrum of threat scenarios provided a comprehensive foundation for constructing a robust defense strategy. From the D3FEND, through a detailed study, we identified those defenses that can be implemented in the SDN environment. We practically tested each defense suggestion derived to prove the validity of our research and also the exhaustive list we provide. The implementation of the MITRE defense mechanism in SDN was tested using the ANTq tool. By presenting a comprehensive list of defense suggestions, this paper aims to empower network administrators and security professionals with the necessary knowledge and guidance to effectively secure their SDN environments. This will contribute to a more resilient and secure SDN ecosystem.

In short, our paper has the following contributions:

1. A valid list of implementable defense suggestions in SDN. As far as we are aware, this is the first work that provides an exhaustive list of all possible defense suggestions in SDN.
2. We have also mapped the implementable defense suggestions in SDN to the MITRE D3FEND knowledge base. This can prove the validity of the list provided through this work.
3. The SDN flow rule for each defense suggestion listed. This makes the list a practical guide for implementing all defenses in SDN.
4. For implementing the suggested defense technique an implementer tool was developed.

The rest of the paper is organized as follows. Section 2 discusses the related works and their lackings. MITRE D3FEND is discussed in detail in Sect. 3. The implementation of defense mechanisms in SDN is discussed in Sect. 4. The developed threat intelligence message format Automated Defense Message Format (ADMF) is introduced in Sect. 5. Section 6 introduces ANTq, a tool designed for the implementation of our defenses. The results and evaluation are discussed in Sect. 7. We conclude this paper with a discussion of future plans Sect. 8.

2 Literature Review

This section explores the integration of automated incident response within Software-Defined Networking. It examines existing frameworks, methodologies, and their effectiveness. This section also identifies research gaps and lays the groundwork for proposing comprehensive defense strategies for SDN environments.

In [9], the authors introduced SDN4S, a framework designed to expedite the resolution of events in the context of SDN. Addressing the challenge of a growing number of network threats and limited resources for analysis and resolution, SDN4S prioritizes automating countermeasures. The system employs playbooks predefined response plans to customize incident responses for specific business settings. While SDN4S interfaces with the BD4S tool for real-time alerts, the absence of evaluation findings on threat mitigation and incident response in a physical test introduces uncertainties about the tool's practical impact.

Patzer et al. [10] introduced an incident response framework for Industrial Control Systems using SDN. Utilizing SDN's centralized control and programmability, the framework automates incident response in complex ICS environments through three key components: anomaly detection, incident classification, and automated response. The automated response component enforces actions, such as isolating affected devices and blocking malicious traffic, based on the classified incidents. The authors demonstrated the framework's effectiveness in a simulated ICS environment, highlighting its potential to enhance security by streamlining incident response processes.

In [11], the authors propose an SDN-based framework to secure a ship's communication system from cyberattacks, simplifying policy management for non-experts through human-readable Controlled Natural Language (CNL). Utilizing SDN-enabled switches to connect ship components like the Autonomous Engine Monitoring Controller (AEMC) and Integrated Bridge Controller (IBC), the framework employs the Event-Condition-Action (ECA) paradigm to translate high-level policies into low-level OpenFlow rules for flexible deployment. The incorporation of multi-path routing enhances defenses against online threats. However, challenges include a simpler implementation topology due to AEMC's alert generation and potential latency issues in larger networks.

TENNISON, as proposed in [12], introduces a distributed monitoring and remedial methodology featuring multi-level monitoring, multiple controls, and tunneling for effective attack detection and response. In an example evaluation, the authors demonstrated scalable network monitoring, low-latency detection, and robust protective capabilities. The methodology suggests strategically placing monitoring rules and controllers, maximizing network protection. The authors propose that adding more controllers can effectively handle higher network loads, emphasizing the potential for multiple-tier scaling within the TENNISON architecture and providing an automated scaling procedure.

FLOWGUARD [13] stands out as one of the earliest works enabling precise detection and adaptable resolution of firewall policy violations. Implemented as a prototype in Floodlight, experimental results show manageable performance overhead, enabling real-time monitoring in SDNs. Unlike traditional methods focusing solely on packet-in behaviors for flow violations, FLOWGUARD adopts a more effective approach. It checks violations based on packet behavior and addresses flow policies that breach firewall policies established proactively or due to network state and configuration changes.

Murillo et al. [14] explain how Network Function Virtualization (NFV) and SDN may aid in creating automated incident response systems for an ICS. In an industrial context, the authors are utilizing SDN to assist in managing networks and generate a faster incident response time. A Programmable Logic Controller (PLC), sensors, actuators, and other components were used to simulate a nonlinear control system that simulated three water tanks. The "Unknown Input Observer" (UIO) is a novel method developed to improve security. The objective of this technique was to recognize prospective assaults on sensors, pinpoint the source of fraudulent data originating from sensors, and eventually lessen the impact of those attacks.

Brew [16] proposes a system that uses run-time decision-making for conflict resolution. It discusses flow rule conflicts and methods for automatically resolving them in a distributed SDN-based cloud. The authors offered a number of strategies for resolving flow rule conflicts and examined the advantages and disadvantages of each. Two limitations were that it is inappropriate for use in a highly dynamic environment and that the environment's topology is not considered when detecting conflicts. The processing time is also linear, meaning that it might be further optimized for increased efficiency.

Within the context of our research, the current state of automated incident response in Software-Defined Networking (SDN) lacks a comprehensive and validated list of defense suggestions. While existing works contribute valuable insights, the effectiveness and reliability of the suggested defense strategies remain uncertain. This gap in the literature underscores the need for a meticulous exploration and categorization of defense mechanisms, leading to the developing of a robust and validated list tailored specifically for SDN environments. By addressing this gap, our work aims to provide cybersecurity professionals with a definitive and practical resource for enhancing the security of SDN networks through a well-founded set of defense suggestions.

3 MITRE D3FEND

The MITRE Corporation plays an important role in preparing knowledge bases related to cybersecurity, like the ATT&CK and D3FEND framework. MITRE D3FEND[1] is a comprehensive knowledge base of defensive techniques to help organizations systematically fortify their cyber infrastructure. It supplements the MITRE ATT&CK,[2] which focuses on characterizing adversaries, offensive tactics, techniques, and procedures (TTPs). D3FEND framework is organized hierarchically and includes tactics, techniques, and sub-techniques. Tactics such as Harden, Detect, Isolate, Evict, Deceive, and Restore address different defense aspects with specific goals.

- Harden: Strengthens the environment, making attacks more challenging through measures like network segmentation, software patching, etc.

[1] https://d3fend.mitre.org/.
[2] https://attack.mitre.org/.

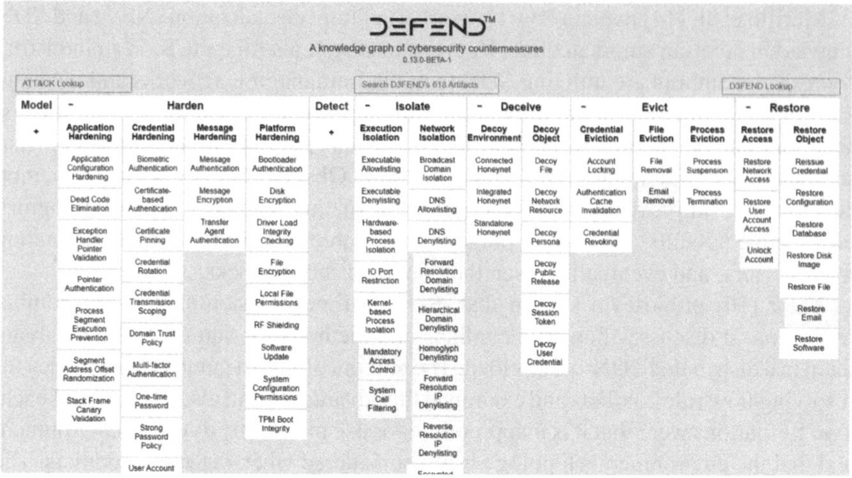

Fig. 1 MITRE D3FEND tactics [15]

- Detect: Identifies ongoing attacks using tools such as intrusion detection systems.
- Isolate: Separates compromised networks or systems to prevent the spread of attacks, employing techniques like network segmentation.
- Evict: Remediate infected systems by removing malware, resetting passwords, and terminating network connections.
- Deceive: Deflects attacks and gathers intelligence about adversaries using tactics like decoys and honeypots.
- Restore: Returns the system to a better state by restoring network access, user access, etc.

3.1 D3FEND Tactics-Techniques and SDN

The MITRE D3FEND framework offers a wide range of defensive tactics and techniques, each intended to accomplish certain goals within the detection, denial, and disruption framework. Each tactic, as shown in Fig. 1, has a variety of techniques, and sub-techniques with unique MITRE IDs, that are in line with its overall objective. Although the MITRE D3FEND architecture includes a wide range of techniques, our discussion focuses on a few that are particularly relevant to the scope of SDN, as listed below.

- D3-NI: Network Isolation
  Network isolation enhances cybersecurity by segmenting networks, preventing malware spread, and limiting unauthorized access through techniques like allow listing and filtering.

- D3-NTF: Network Traffic Filtering
  Network Traffic Filtering (NTF) is a technique used to restrict and monitor network traffic to prevent unauthorized access, malicious activity, and data ex-filtration. Network Traffic Filtering can be achieved through:
  - D3-OTF—Outbound Traffic Filtering.
  - D3-ITF—Inbound Traffic Filtering.
  - D3-EF—Email Filtering.
- D3-DE: Decoy Environment
  A Decoy Environment (Honeypot) strategically misleads attackers through deceptive hosts and networks. It observes and documents activities, aiding in detecting signs of compromise and potential cyber attacks on other systems. The techniques for creating a decoy environment are listed below:
  - D3-CHN—Connected Honeynet.
  - D3-IHN—Integrated Honeynet.
  - D3-SHN—Standalone Honeynet.
- D3-RNA: Restore Network Access
  Restoring network access for an entity entails adjusting configurations or permissions to ensure seamless connectivity, allowing the entity to regain the necessary system access.
- D3-NTA: Network Traffic Analysis
  Network Traffic Analysis (NTA) is a cybersecurity practice that involves monitoring and analyzing network traffic patterns to identify anomalies and potential threats.

4 Implementing Defenses in SDN Environments

At the onset of a security incident, analyzers promptly detect incidents and provide immediate suggestions for implementing effective defense measures. In Software-Defined Networking, translating defense suggestions into configurable rules is pivotal for automating incident response. Flow rules, acting as fundamental instructions for SDN, define how SDN-enabled devices should handle network traffic [17]. Each rule comprises matching criteria (e.g., source/destination IP addresses, port numbers) and actions (e.g., forwarding, dropping). This section explores SDN flow rules corresponding to various defense suggestions [18]. The syntax of an SDN flow rule typically includes the following elements [19]:

- Match criteria: Specifies the conditions that must be met for the rule to apply. Common match criteria include source and destination IP addresses, port numbers, etc.
- Actions: Defines the actions to be taken when the match criteria are met, such as forwarding, dropping, or modifying traffic.

- Priority: Determines the order in which flow rules are evaluated. Flow rules with higher priority are evaluated first.

The structure of a sample SDN flow rule is shown as follows:

```
{
    ''dpid'': 0000000000000001,
    ''hard_timeout'': 30,
    ''priority'': 1267,
    ''match'': {
        in_port: 5,
        eth_src: ''00:11:22:33:44:55'',
        eth_dst: ''66:77:88:99:AA:BB'';
        ip_proto: 6;
        tcp_dst: 80;
   },
    ''actions'': [
        {
            ''type'': ''OUTPUT'',
            ''port'': 2
        }
    ]
}
```

In the flow rule inserted into the switch with dpid '0000000000000001', the incoming packet undergoes matching based on attributes within the match field. These attributes include the ingress port, source and destination MAC addresses, and the IP protocol of the packet. The output is then directed toward port 2 of the switch. The rule is configured with a hard timeout, resulting in the deletion of the rule from the flow table after 30 s. Additionally, it is assigned a priority value of 1267, determining the instructions to be applied to similar incoming packets by matching them against rules in the flow table.

Table 1 shows the implementation of the discussed MITRE D3FEND techniques in SDN. It also shows a mapping of the MITRE D3FEND ID to the corresponding implementable flow rules. This acts as the repository for enforcing the given defense suggestions to the target network.

Apart from those discussed in Table 1, a few more defense suggestions are possible in the SDN. MITRE D3FEND, a daily updating list, hasn't incorporated these in its knowledge base, but we expect them to be added to the list in the future. These defense techniques are:

- Time-based communication.
- Quality-of-service.
- VLAN segmentation.

Table 2 discusses these techniques along with the flow rules needed to implement them.

Table 1 MITRE D3FEND ID mapped to ANTq ID

MITRE ID	ANTq ID	Flow rule	MITRE ID	ANTq ID	Flow rule
1. D3-NI	host-iso	{ “dpid”: 0x2, “match”: {“in_port”: 3, “eth_src”: “00:00:00:00:00:05” }, “actions”: [{“type”: “CLEAR_ACTIONS”}] }	3. D3-DE	honeypot-iso	{ “match”: { “in_port”: 5, “eth_src”: “00:11:22:33:44:55” }, “actions”: [{“type”: “OUTPUT”, “port”: 1000}] }
		{ “dpid”: 0x2, “match”: {“in_port”: 3, “eth_dst”: “00:00:00:00:00:05” }, “actions”: [{“type”: “CLEAR_ACTIONS”}] }		traffic-red	{ “match”: { “in_port”: 3, “tcp_dst”: 80 }, “actions”: [{“type”: “OUTPUT”, “port”: 1000 }] }
	switch-iso	{ “dpid”: 0x2, “priority”: 32768, “match”: { }, “actions”: [{“type”: “CLEAR_ACTIONS”}] }	4. D3-RNA	host-res	{ “match”: { “in_port”: 10, “eth_src”: “00:11:22:33:44:55” }, “actions”: [{“type”: “OUTPUT”, “port”: 8088}] }
2. D3-NTF	host-ntf	{ “dpid”: 0x2, “match”: { “in_port”: 3, “ip_proto”: 6, “eth_type: 2048; “eth_src: 00:00:00:00:00:02 }, “actions”: [{“type”: “CLEAR_ACTIONS”}] }		switch-res	{ “match”: { “in_port”: 10, “eth_src”: “00:11:22:33:44:55”}, “actions”: [{“type”: “OUTPUT”, “port”: 8088}] }
	switch-ntf	{ “dpid”: 0x2, “priority”: 32768, “match”: {“ip_proto”: 17, “eth_type”: 2048 }, “actions”: [{“type”: “CLEAR_ACTIONS”}] }			
	subnet-ntf	{ “dpid”: 0x4, “priority”: 32768, “match”: {“in_port”: 5, “ipv4_src”: “192.168.1.2”, “ipv4_dst”: “10.0.0.0/24” }, “actions”: [ ]}			

– The configuration messages in the table are in the OpenFlow format, these will be pushed by the application using Ryu API to configure the flow tables
– Fictitious values and addresses are used in the table, ‘*’ denotes the wild card entry
– dpid is used to identify the targeted SDN device

5 Automated Defense Message Format (ADMF)

Prompt implementation of defense suggestions in the target network depends on adopting a standardized exchange format for efficient communication between analyzers and defense implementers. The section below discusses the various existing formats for exchanging messages.

Table 2 SDN defense techniques

Defense suggestion	ANTq ID	Flow rule	Defense suggestion	ANTq ID	Flow rule
1. Time based communication	time-com	{ "match": { "in_port": 1, "eth_src": "00:11:22:33:44:55", "time": "16:00:00–20:00:00" }, "actions": [{"type": "OUTPUT", "port": 2 }] }	3. VLAN segmentation	vlan-seg	{ "match": {"in_port", "ip_src": "192.168.1.10", "ip_dst": "192.168.1.20", "vlan_id": 10 }, "actions": [{"type": "OUTPUT", "port": 2, "vlan_id": 10}]}
2. QoS	dba-qos	{ "match": {"in_port": 1}, "bandwidth-range": {"min-rate": "00000", "max-rate": "500000"}}			
	lb-qos	{ "match": {"in_port": 1, "eth_type": 0x0800, "ip_proto": 6}, "actions": [{"type": "OUTPUT", "port": [2, 3, 4]}]}			

– *QoS* quality of service

5.1 Existing Defense Message Format

The Incident Object Description Exchange Format (IODEF) [20] facilitates structured representation for security incidents, streamlining incident response. Structured Threat Information eXpression (STIX) [21] is a language for describing cyber threat intelligence, supporting information sharing on threat actors, tactics, techniques, and indicators of compromise. Trusted Automated eXchange of Indicator Information (TAXII) [22] is a protocol for real-time cyber threat intelligence exchange, enabling automated sharing and integration with security systems. Cyber Observable eXpression (CybOX) [23] provides a standardized language for representing and sharing cyber observables, aiding threat analysis. Open Indicators of Compromise (OpenIOC) [24] defines a standard for describing indicators of compromise, facilitating detection, and response to security incidents. The Malware Information Sharing Platform (MISP) [25] is a threat intelligence platform supporting various formats for collaborative defense. Intrusion Detection Message Exchange Format (IDMEF) [26] is an XML-based format for exchanging information on intrusion detection and prevention, enabling standardized representation for real-time incident response.

After evaluating these threat intelligence exchange formats, the decision was made to create an exchange format called Automated Defense Message Format (ADMF). This choice was driven by considerations such as user-friendly design and simple syntax, which makes it easy to comprehend. The need for handling complicated data structures and customized formats required to meet different applications' specific needs was also considered.

5.2 *Message Format*

Automated Defense Message Format (ADMF) adopts the JavaScript Object Notation (JSON) data interchange format for efficient message transfer, driven by several advantages. JSON's readability is crucial, providing operational accessibility for individuals with limited networking or SDN expertise. Its platform independence ensures the framework's versatility, allowing seamless operation across various operating systems, devices, and applications.

JSON's adaptability enables ADMF to handle various data types and structures, ensuring flexibility for diverse application needs.

An example of Host network traffic filtering ADMF message is given below.

```
{
        ''mitre-code'': ''D3-NTF'',
        ''target_device'': ''H2'',
        ''match'': {
            ''ip_proto'': ''TCP'',
        }
}
```

The above ADMF message, is a suggestion for TCP traffic filtering for host H2.

Table 3 outlines attribute and supported structures for defensive proposals in ADMF. Each attribute's data type and brief description is detailed in the table.

5.3 *ADMF to Implementable Rule Conversion*

This subsection tells us how ADMF can be converted to implementable rules in SDN. The incoming ADMF message, along with the defense suggestion, contains the MITRE ID of the corresponding defense suggestion. This MITRE ID is mapped to the corresponding implementation rules which can be enforced on the network via the SDN controller. An example of conversion of ADMF to implementable rule is shown in Fig. 2.

Table 3 Automated defense message format structure

Attribute	Type	Description	Example
mitre_code	String	MITRE DEFEND code for identifying and mapping defense suggestions	D3-NTF
antq_code	String	Custom ANTQ code for identifying and mapping defense suggestion	host-ntf
target device	String/list of strings	Network asset identification	H3
src device	String		H4
dst device	String		H5
match	Object	All attributes allowed are in the support version of OpenFlow with exact name	{"ip proto": 5, "eth type": 34525}
forwarding device	String/list of strings	Datapath identifier can be a string or list of identifiers depending on ANTq code	S7
time-info	Nested object (string)	Information for setting time based segmentation	"time-range": {"start time": "14:00:00", "end time": "17:00:00"}, "duration": [1, 2, 3, 4, 5]
bandwidth-range	Object	Information for dynamic bandwidth allocation (in bits)	{"min-rate": "20", 'max-rate": "50"}
out port	String	Port into which the switch forwards the packet	2
action	String	Instruction to deal with incoming packet to switch (default action = "deny")	allow

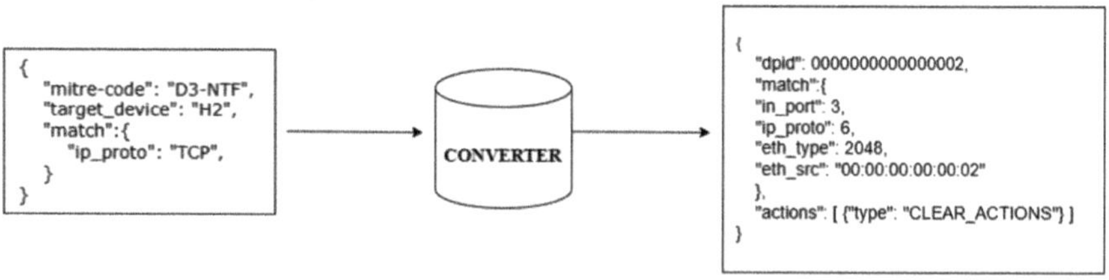

Fig. 2 ADMF to OpenFlow configuration

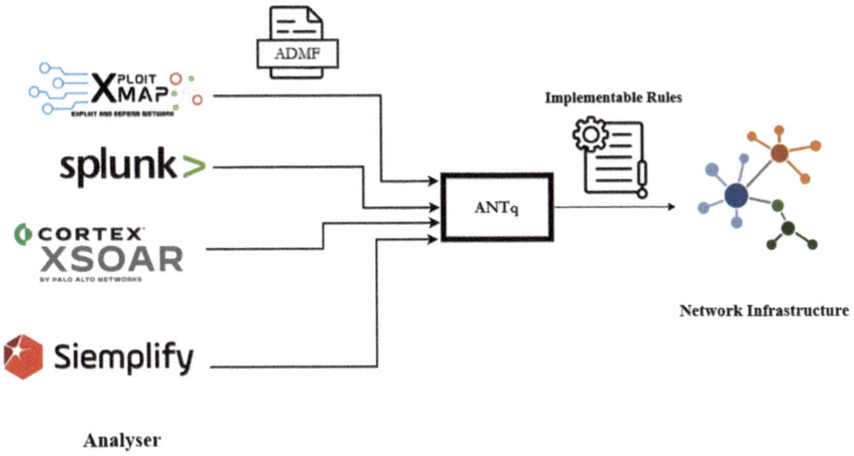

Fig. 3 Overview of ANTq

6 ANTq: Automated Network Defense Implementation Tool

ANTq is a software application running on top of the Ryu SDN controller in the application plane in the SDN [27]. It aims at translating cyber defense suggestions to implementable defense measures. Any network analyzer such as XploitMAP [28], Splunk [29], etc. can utilize this tool to implement their defense suggestions on the target network.

Figure 3 shows the overview of how the ANTq tool helps in implementing suggested cyber defenses. The analyzers analyze the network to detect security incidents and can suggest defenses that are contained in the MITRE D3FEND knowledge base. The defense suggestions are fed to the ANTq in the specially structured Automated Defense Message Format (ADMF). The ANTq on reception of these suggestions converts them to the corresponding implementable rules and enforces them on the target network infrastructure.

6.1 Working of ANTq

Upon identifying any malicious event within the network, security analyzers (including log analyzers, threat analyzers, etc.) can generate defense suggestions and transmit them to the ANTq tool. Those messages that are not in the ADMF format are sent to the ADMF converter module. On reception of the defense suggestions, they are first validated in ANTq. During the validation process ANTq checks if the attributes in the message are meeting the requirements of the defense suggested. After validation,

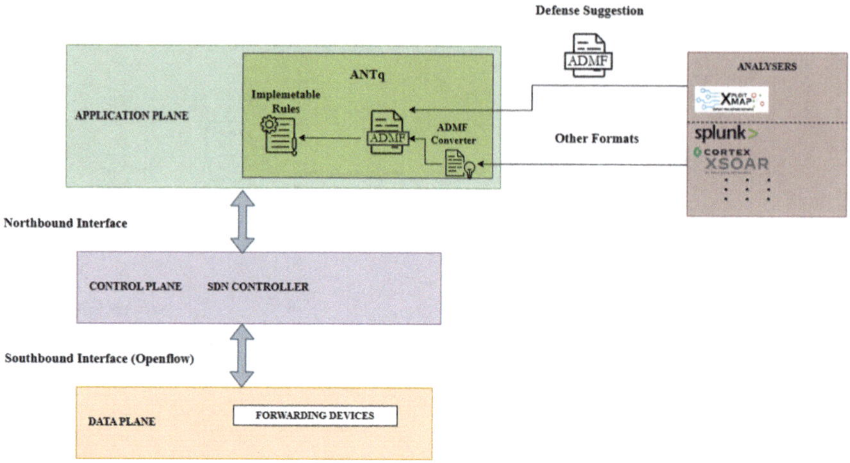

Fig. 4 ANTq work flow

they are passed to the device identification module, where the target network devices are identified. Finally, the appropriate rules will be implemented on the network via the SDN controller. Figure 4 shows the working of the ANTq in detail.

7 Results and Evaluation

To check the reliability of ANTq, we have created a virtual scenario network. The network was created in the Mininet simulator, with seven hosts H1, H2, H3, H4, H5, H6, and H7, three switches S1, S2, and S3, and a controller C0. In our setup here, we have used the Ryu Controller. Figure 5 shows the complete topology created.

Let us assume due to some detected incident, an analyzer is suggesting Network Traffic Filtering (host-ntf) for host H2 and TCP protocol. ANTq accepts the suggestion, and enforces the flow rules automatically. The tool will examine the received message and will choose the host-ntf to enforce the defense suggestion. To achieve the host-ntf, TCP packet transfer between the host2 and host7 are blocked. Figure 6 shows the ANTq dashboard with open terminals before enforcing the rules. We can observe that ICMP and TCP packets are successfully sent/received between the hosts H2 and H7. The red highlighted box shows the successful transfer of packets via TCP with 0% packet loss.

Figure 7 shows part of the ANTq dashboard showing how the flow rules are updated in the SDN device flow table. This flow rule is implemented on the switch connected to the host. It was verified that TCP packets were blocked to/from host H2, but the ICMP packets were transferred between all seven hosts using the ping and

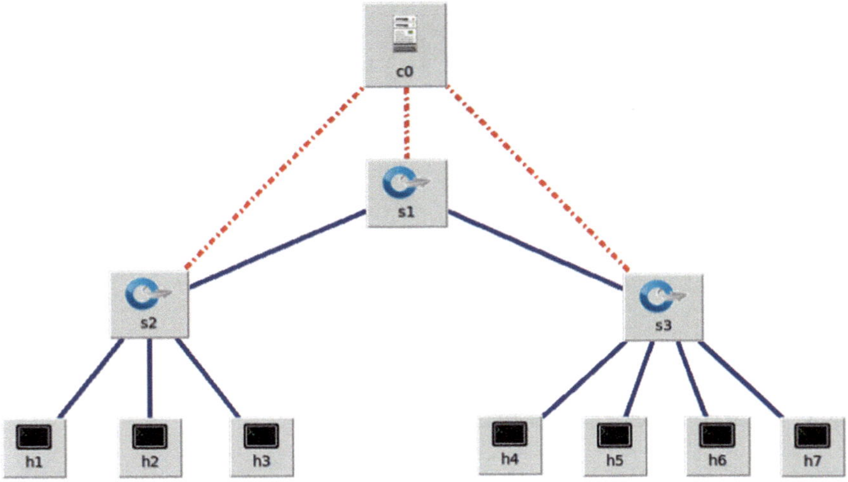

Fig. 5 Network topology

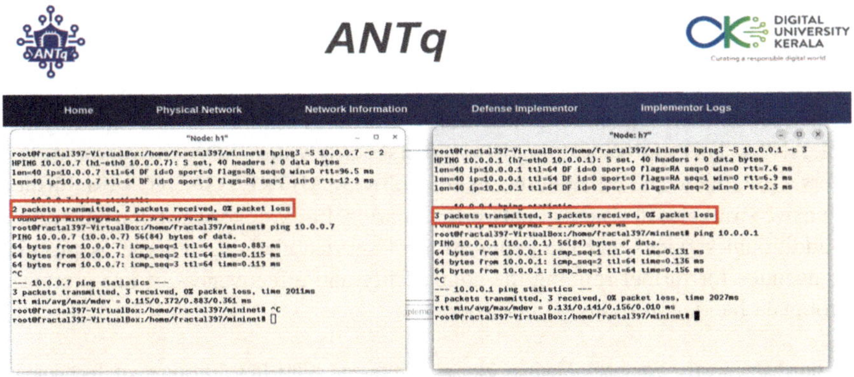

Fig. 6 Before implementing host NTF defense suggestion

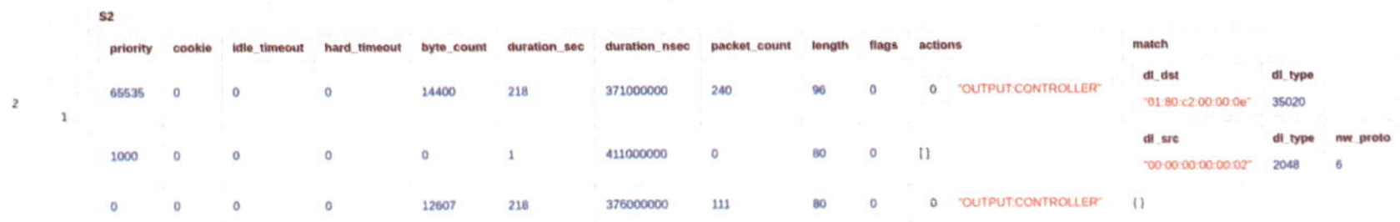

Fig. 7 Flow rule updated in switch flow table

hping3 commands, respectively, as shown in Fig. 8. The red highlighted box shows the successful blocking of the TCP traffic with 100% packet loss, thereby proving the reliability of our proposed framework.

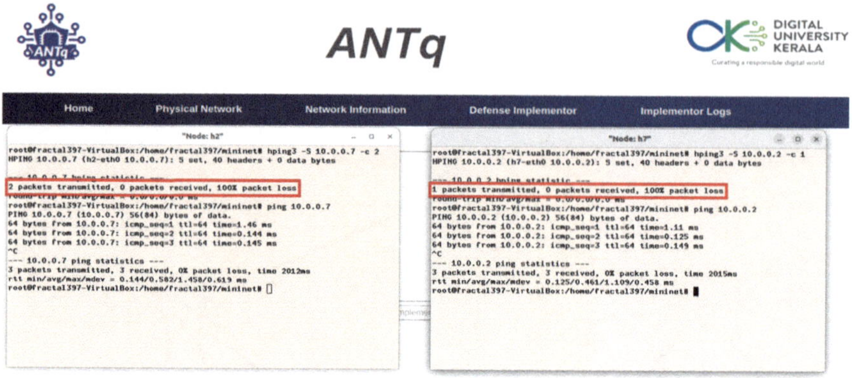

Fig. 8 TCP packets blocked after implementation of host NTF

8 Conclusion

Our research is dedicated to enhancing incident response capabilities within the context of Software-Defined Networking (SDN) through the strategic implementation of MITRE D3FEND techniques. The core of our methodology lies in translating these techniques into actionable SDN flow rules, a process seamlessly facilitated by our tool, ANTq. This approach strengthens cybersecurity within SDN environments and offers a pragmatic and efficient means of deploying MITRE D3FEND's established defensive strategies in practical network scenarios. Looking forward, the integration of additional services and the exploration of service chaining emerge as promising avenues for further refining the adaptability and effectiveness of our proposed automated incident response framework.

Acknowledgements We thank Mathias Ekstedt, Professor, KTH Royal Institute of Technology, and Chandan Mazumdar, Professor, Jadavpur University, for their helpful and perceptive comments during this project.

References

1. Cichonski P, Millar T, Grance T, Scarfone K et al (2012) Computer security incident handling guide, vol 800, pp 1–147
2. Schlette D, Caselli M, Pernul G (2021) A comparative study on cyber threat intelligence: the security incident response perspective. IEEE Commun Surv Tutor 23(4):2525–2556. https://doi.org/10.1109/COMST.2021.3117338
3. Bada M, Creese S, Goldsmith M, Mitchell C, Phillips E (2014) Computer security incident response teams (CSIRTs): an overview. The Global Cyber Security Capacity Centre
4. Anastopoulos V, Giovannelli D (2022) Automated/autonomous incident response

5. Ossenbühl S, Steinberger J, Baier H (2015) Towards automated incident handling: how to select an appropriate response against a network-based attack? In: 2015 ninth international conference on IT security incident management & IT forensics. IEEE, pp 51–67
6. Leite C, Hartog J, Santos D, Costante E (2022) Actionable cyber threat intelligence for automated incident response. In: Nordic conference on secure IT systems, pp 368–385. https://doi.org/10.1007/978-3-031-22295-5_20
7. Kreutz D, Ramos FM, Verissimo P (2013) Towards secure and dependable software-defined networks. In: Proceedings of the second ACM SIGCOMM workshop on hot topics in software defined networking, pp 55–60
8. McKeown N (2009) Software-defined networking, vol 17, pp 30–32
9. Koulouris T, Mont MC, Arnell S (2017) SDN4S: software defined networking for security. Technical report. Hewlett Packard Labs, Palo Alto, CA, USA
10. Patzer F, Meshram A, Heß M (2019) Automated incident response for industrial control systems leveraging software-defined networking. In: ICISSP, pp 319–327
11. Sahay R, Meng W, Estay DS, Jensen CD, Barfod MB (2019) CyberShip-IoT: a dynamic and adaptive SDN-based security policy enforcement framework for ships, vol 100, pp 736–750
12. Fawcett L, Scott-Hayward S, Broadbent M, Wright A, Race N (2018) TENNISON: a distributed SDN framework for scalable network security, vol 36, pp 2805–2818. https://doi.org/10.1109/JSAC.2018.2871313
13. Hu H, Han W, Ahn G-J, Zhao Z (2014) FLOWGUARD: building robust firewalls for software-defined networks. In: Proceedings of the third workshop on hot topics in software defined networking, pp 97–102
14. Murillo Piedrahita AF, Gaur V, Giraldo J, Cárdenas A, Rueda SJ (2018) Leveraging software-defined networking for incident response in industrial control systems. IEEE Softw 35(1):44–50. https://doi.org/10.1109/MS.2017.4541054
15. MITRE Corporation. MITRE D3FEND. https://d3fend.mitre.org
16. Pisharody S, Natarajan J, Chowdhary A, Alshalan A, Huang D (2017) Brew: a security policy analysis framework for distributed SDN-based cloud environments, vol 16, pp 1011–1025
17. Shalimov A, Zuikov D, Zimarina D, Pashkov V, Smeliansky R (2013) Advanced study of SDN/OpenFlow controllers. In: Proceedings of the 9th central & eastern European software engineering conference in Russia, pp 1–6
18. Kuźniar M, Perešíni P, Kostić D (2015) What you need to know about SDN flow tables. In: Passive and active measurement: 16th international conference, PAM 2015, New York, NY, USA, 19–20 Mar 2015, proceedings 16. Springer, pp 347–359
19. OpenFlow switch specification version 1.3.0. Accessed 25 June 2012. https://opennetworking.org/wp-content/uploads/2014/10/openflow-spec-v1.3.0.pdf. Version 1.3.0 (wire protocol 0x04)
20. Takahashi T, Landfield K, Kadobayashi Y (2014) An incident object description exchange format IODEF extension for structured cybersecurity information. Technical report
21. Cyber Threat Intelligence Technical Committee (2017) Structured threat information eXpression (STIX)
22. Connolly J, Davidson M, Schmidt C (2016) Trusted automated eXchange of indicator information (TAXII™), 2 May 2014
23. Barnum S, Martin R, Worrell B, Kirillov I (2012) The cybox language specification. The MITRE Corporation
24. Mavroeidis V, Bromander S (2017) Cyber threat intelligence model: an evaluation of taxonomies, sharing standards, and ontologies within cyber threat intelligence. In: 2017 European intelligence and security informatics conference (EISIC). IEEE, pp 91–98. https://doi.org/10.1109/EISIC.2017.20
25. Wagner C, Dulaunoy A, Wagener G, Iklody A (2016) MISP: the design and implementation of a collaborative threat intelligence sharing platform. In: Proceedings of the 2016 ACM on workshop on information sharing and collaborative security, pp 49–56
26. Debar H, Curry D, Feinstein B. The intrusion detection message exchange format IDMEF. https://doi.org/10.17487/RFC4765. IETF. https://tools.ietf.org/rfc/rfc4765.txt

27. Hakeem A (2023) Automated translation and deployment of suggested cyber defenses in an SDN environment. M.Tech. thesis, Kerala University of Digital Sciences Innovation and Technology
28. Mukherjee P, Thampi SM, Rohith N, Poddar BK, Sen I (2023) Detection and hardening strategies to secure an enterprise network. In: Proceedings of the 19th ICISS, pp 91–108. https://doi.org/10.1007/978-3-031-49099-6_6
29. Splunk Inc. Splunk. https://www.splunk.com

Evaluation of Compressed Sensing and Recovery of Sound Signals Using Sparse Bayesian Learning Methods

Ebin M. Manuel and M. P. Ananya

Abstract Compressed sensing is a signal processing technique that is used for the efficient acquisition and reconstruction of signals by finding solutions to under-determined linear systems. In most cases, the compressed signal is recovered using the conventional l_1 norm minimization technique, which is a convex optimization procedure. However, the global minimum is not necessarily the sparsest solution. Therefore, another method called sparse Bayesian learning (SBL) is introduced for the reconstruction of compressed audio signals. Using the different SBL algorithms, a better reconstruction for the compressed audio signal is achieved. The SBL algorithm for the multiple measurement vector (MMV) model is also implemented for the audio signal. The results of different SBL techniques (SBL, T-SBL, MSBL, T-MSBL) are compared. We show via the experimental results that the time-varying model outperforms the other techniques. Further, we show that the T-MSBL algorithm reconstructed the original audio signal from the compressed signal more efficiently than the other algorithms.

1 Introduction

Compressed sensing (CS) refers to an efficient signal acquisition and reconstruction technique, using properties of under-determined linear system of equations. The researchers provide Theoretical guarantees for recovery of the signal from lesser samples than that specified by the Nyquist sampling theorem.

The quantum of characteristic features embodied in huge volumes of data is much less in many applications compared to the number of coefficients in the actual data

E. M. Manuel (✉)
Department of Electronics and Communication, Rajiv Gandhi Institute of Technology Kottayam, Kottayam, India
e-mail: ebin@rit.ac.in

M. P. Ananya
Department of Electronics and Communication, College of Engineering Trivandrum, Trivandrum, Kerala, India

S. M. Thampi et al. (eds.), *Fifth International Conference on Computing and Network Communications*, Lecture Notes in Electrical Engineering 1219,
https://doi.org/10.1007/978-981-97-4540-1_2

representation, resulting in a huge scope for data compression. In conventional data compression, the full-length data is compressed before storage or transmission, and the data recovery is straightforward. However, the (CS) is a paradigm shift where the encoding part is very light, and the approach is efficient when the data vector is either sparse or can be represented as a sparse signal in any of the transform domains. However, the complexity lies in the decoding part for the data recovery [1]. Therefore, the CS can be preferred for wireless sensor network (WSN) applications. However, because of its high compaction, the CS is widely applied in many wireless sensor network applications for data aggregation [2].

Compressed sensing is an efficient method of data compression. The recovery of the compressed data is a challenge. There are a lot of algorithms that exist for the recovery of CS data. However, little research has been done on recovering compressed audio signals using the sparse Bayesian learning (SBL) algorithm. This work evaluates the performance of the CS technique under different SBL approaches. We implement the SBL algorithms to recover the audio signal and compare its results to find which algorithm is better for recovering the CS audio signal.

This paper is organized as follows: Section 2 describes the related works. Sections 3 and introduce Compressed Sensing and SBL algorithms, respectively. Section 3 discusses the methodology used. Section 4 provides the results and their analysis, followed by the conclusion in Sect. 5.

2 Motivation and Related Works

A detailed study on compressed sensing and its application in a wireless sensor network is given in [3]. Xu et al. [4] and He et al. [5] explains the learning of the dictionary by using thresholds on the cosine similarity matrix from the input signal instead of using a random Gaussian matrix for source recovery and classification. The problem of obtaining sparse solutions to an undetermined line system under an unknown sparsity template is addressed in [6]. Fu et al. [7] and Song [8] addresses the problem of data recovery under multiple measurement vectors (MMV) when elements in each non-zero row of the solution matrix are temporally correlated, and also different SBL (T-SBL [9], T-MSBL [10]) algorithms are investigated.

The CS is a technique leveraging the sparsity before reducing the data quantum required to specify a signal. The CS is guaranteed to be efficient with structured and correlated signals. The principle of the CS is based on the concept that instead of transmitting the data vector values x_i for $k = 0, \dots, N-1$ as such, only certain linear combinations are transmitted from which we recover the data vector

$$y_j = \sum_{j=0}^{N-1} A_{i,j} x_i + v_j \tag{1}$$

where $j = 0, \dots, M-1$ with $M < N$ and v_j denote the noise component. The vector $\mathbf{y}$ is the measurement of the data vector $\mathbf{x}$. The matrix $\mathbf{A}$ is referred to as the measurement matrix. If $\mathbf{x}$ is known to be sparse concerning the basis $\mathbf{D}$, then $\mathbf{y}$ can be represented as

$$\mathbf{y} = \mathbf{AD}\xi + \upsilon, \tag{2}$$

where ξ is a sparse vector representation of $\mathbf{x}$.

3 Proposed Method: Sparse Bayesian Learning and MAP Estimation

Sparse signal recovery, or compressed sensing, is an emerging field in signal processing. The basic mathematical model is

$$\mathbf{y} = \mathbf{Ax} + \mathbf{v} \tag{3}$$

where $\mathbf{A} \in \mathbb{R}^{N \times M}$ $(N \ll M)$ is commonly referred to as a dictionary matrix that abides by the restricted isometric property. The columns of $\mathbf{A}$ are orthogonal. The measurement vector $\curvearrowright \in \mathbb{R}^{N \times 1}$ is derived such that $M < N$ and $\mathbf{v}$ is the vector containing noise values. The recovery problem is to find an estimate of the data vector $\mathbf{x}$ that minimizes the error $\mathbf{y} = \mathbf{Ax}$. Inspired by applications that have a sequence of data vectors, the CS model can be extended to the multiple measurement vector (MMV) model as follows:

$$Y = \mathbf{AX} + \mathbf{V} \tag{4}$$

where $\mathbf{Y} = [Y_1, \ldots, Y_L] \in R^{N \times L}$ is the extended measurement a matrix consisting of $\mathbf{L}$ measurement vectors, $\mathbf{X} = [X_1, \ldots, X_L] \in \mathbb{R}^{M \times L}$ is the multiple data matrix.

For the recovery of the data matrix, Bayesian base approaches are proven to be efficient and demonstrate superior recovery performance. Sparse Bayesian learning refers to one important class of Bayesian algorithms that are used for recovery purposes. The sparse Bayesian learning was first tested in a single measurement vector model and later applied to multiple vector cases. The solution obtained by the common sparse signal recovery algorithms based on the ell_1 minimization approaches is not necessarily the globally optimal solution. However, SBL algorithms guarantee the global optimum solution and thus provide superior data recovery performance. Further, the Bayesian approaches provide much fewer local minima compared to the conventional ell_1 minimization algorithms. However, little research has been done on recovering compressed audio signals using the sparse Bayesian learning algorithms. So, in this work, we implement the SBL algorithms to recover audio signals and compare the results to find which algorithm is better for recovering CS audio signals.

Maximum aposteriori estimate of an unknown quantity can be used to estimate the unknown quantity based on empirical data. It employs an apriori distribution over the unknown quantity to be estimated. Assume the need to estimate an unknown quantity θ based on observation vector $\mathbf{x}$. Let f be the sampling distribution of $\mathbf{x}$, so $f(\mathbf{x}|\theta)$ is the probability of $\mathbf{x}$ given parameter θ. Then the function:

$\theta \longmapsto f(\mathbf{x}|\theta)$ is known as the likelihood function and the estimate:

$$\hat{\theta}_{\mathrm{MLE}}(\mathbf{x}) = \arg_{\theta}\max f(\mathbf{x}|\theta) \tag{5}$$

is known as the maximum likelihood estimate of θ.

To estimate the hyperparameters, we can use evidence maximization or Type-II maximum likelihood. This involves marginalizing over the weights **x** and then performing maximum likelihood estimation. Refer to the whole framework, including the solution and the hyperparameter estimation, as the block sparse Bayesian learning (bSBL) framework. Note that in contrast to the original SBL framework, the bSBL framework models the temporal structures of sources in the prior density via the matrices $B_i (i = 1, \ldots, M)$. Different ways to learn the matrices result in different algorithms. We will discuss the learning of these matrices and other hyperparameters in the following sections.

4 Results and Discussion

We have a time-domain signal x of length N generated by the first-order Autoregressive(AR) process. We commonly take N measurements, but we can only take M measurements ($M < N$). Use vector y to represent our M measurements which is the compressed signal, i.e., $y = \mathbf{A}x$, where **A** is the sensing matrix that allows us to get from x to y (via random measurements, transformations, or a combination of the two). The Gaussian matrix is used as the sensing matrix since it satisfies the URP and RIP properties. The compressed audio signal is reconstructed using sparse Bayesian learning algorithms. The idea is to find posterior probability $p(x|y; \theta)$ via the Bayesian rule where θ set of all the hyperparameters. The solution x is given by the Maximum-A-posterior (MAP) estimate. The hyperparameters are estimated from data by marginalizing over x and then performing type-II Maximum Likelihood Basic model is extended to multiple measurement vector (MMV). Here, the sequence of measurement vectors is made available. To solve the MMV problem, the MSBL algorithm is used. Most existing work does not deal with the temporal correlation of sources. Another SBL framework is used to exploit temporal correlation called block sparse Bayesian learning (bSBL). MMV model is transformed to block SMV model, i.e., $y = Dx + v$. The T-SBL algorithm, which is derived from this concept, is used for the reconstruction of the compressed audio signal, which is very effective but is slow due to its operation in a higher dimensional parameter space. A fast version, called T-MSBL, which operates in the original parameter space, is also used for the reconstruction. The performance of all the above algorithms is compared.

A sparse audio signal is generated randomly. It is compressed using a Gaussian matrix. The compressed signal is reconstructed using various sparse Bayesian learning algorithms such as MSBL, T-SBL, and T-MSBL. Also, reconstruction is done using FOCUSS algorithm (a classical algorithm). The results of these algorithms have been compared. For that, two performance measures have been used. One is

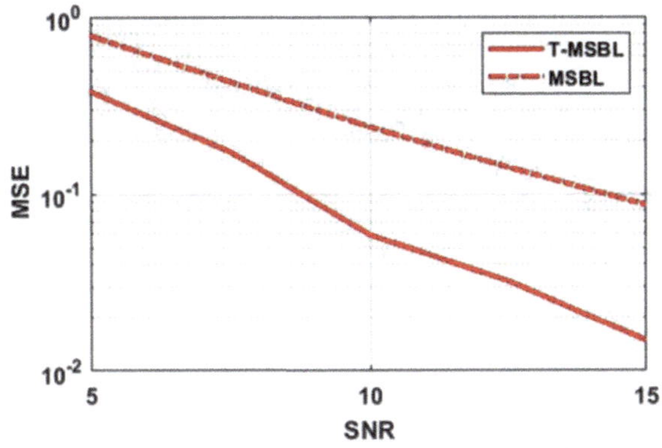

Fig. 1 MSE of MSBL and T-MSBL algorithms at different noise levels

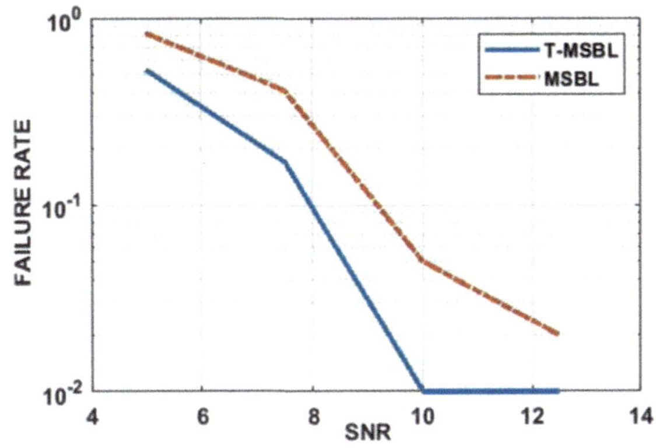

Fig. 2 Failure rate of MSBL and T-MSBL algorithms at different noise levels

the Failure Rate, which indicates the percentage of failed trials in the total trials. In noiseless cases, the failed trial was recognized in such a way that if the indexes of the estimated source are not the same as that of the true index of the corresponding source, then that particular trial is considered to be a failed trial. In noisy cases, since any algorithm is not able to recover X_{gen} exactly in these cases, a failed trial was recognized in such a way that if the indexes of estimated sources with the K. The largest l_2 norms differed from the true indexes. The second one is a mean square error. In noisy cases, the mean square error (MSE) is used as a performance measure, defined by $||\ \hat{X}-X_{\text{gen}}||_f^2/||X_{\text{gen}}||_f^2$, where $\hat{X}$ was the estimated source matrix.

4.1 MSE and Failure Rates at Different Noise Levels

The Signal to Noise ratio varies from 5 to 15 dB, and the corresponding mean square error is evaluated. A graph is plotted between the SNR and MSE (Fig. 1). In this figure, it can be seen that T-MSBL outperformed MSBL in all the noise levels. This

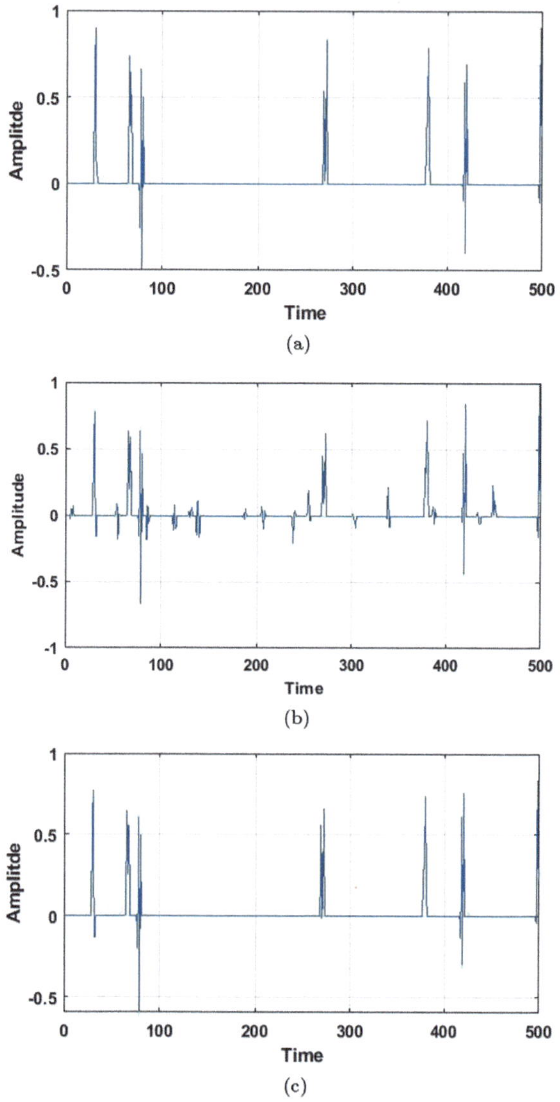

Fig. 3 Comparison of **a** The Input audio signal and reconstructed signal using **b** T-MSBL and **c** MSBL algorithms

implies that even in low SNR cases, exploiting the temporal correlation of sources is beneficial. Figure 2 plots a graph between the SNR and Failure rate. Here also it can be seen that T-MSBL outperformed MSBL in all the noise levels, i.e., the failure rate of the T-MSBL algorithm is much lower than the failure rate of MSBL algorithm even at low SNR cases.

In Fig. 3a the randomly generated audio signal is plotted. Figure 3b shows the reconstructed outputs using the MSBL algorithm, and Fig. 3c shows the reconstructed

Fig. 4 Failure rates of all the algorithms when K varied from 10 to 18 at different temporal correlation levels

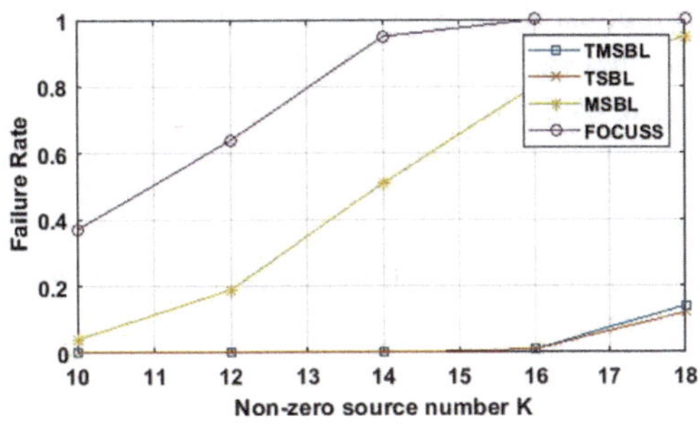

(a)

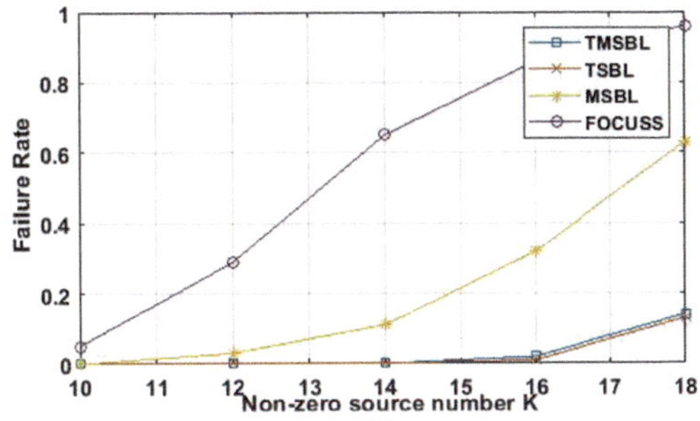

(b)

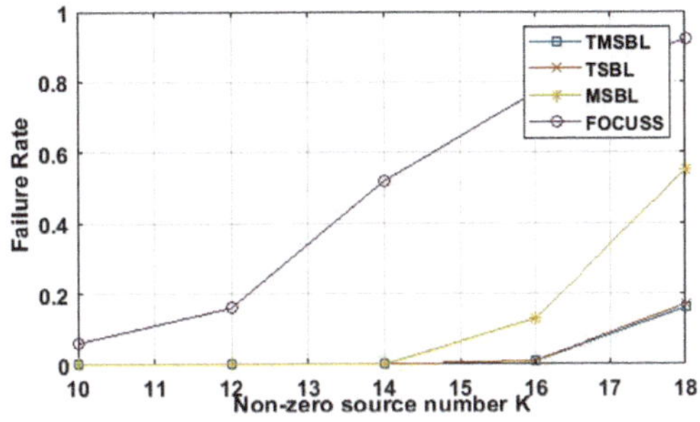

(c)

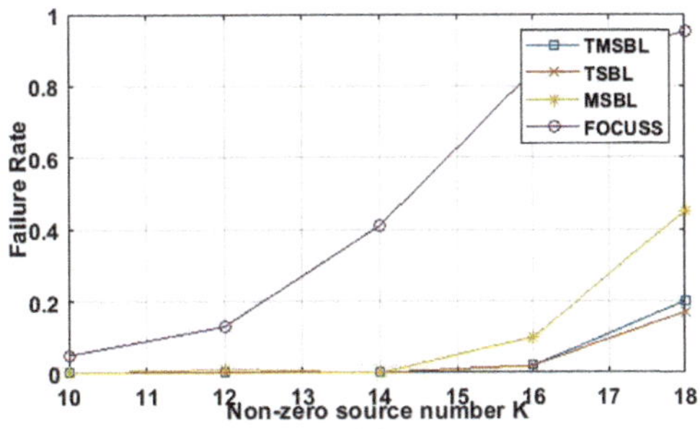

(d)

output using the T-MSBL algorithm. From the figure, it can be seen that the reconstructed signal using T-MSBL is closer to the original signal than the reconstructed signal using MSBL.

4.2 Failure Rate at Different Temporal Correlation

This result studied the effects of temporal correlation on the number of accurately recovered sources in a noiseless case. The four algorithms (MSBL, T-MSBL, T-SBL, and FOCUSS) were compared at four different temporal correlation levels. Figure 4a–d represent the variation of failure rate with the number of non-zero sources at a temporal correlation of 0, 0.5, 0.9, and 0.99, respectively. Results show that T-MSBL and T-SBL accurately recovered many more sources than other algorithms, especially at high temporal correlation levels.

5 Conclusion

The CS is efficient in aggregating the readings of a local number of sensors into a compact data representation to be transmitted to a control center where the original data is reconstructed back from the compact representation via exploiting the sparsity prior. The work presented in the paper investigates the CS-based multiple measurement models for audio applications. The simulation results demonstrate the superior performance of the Bayesian-based approaches in recovering the original data vector from its measurement samples. The simulation results show that the T-SBL and T-MSBL algorithms outperform their conventional peers.

References

1. Zhang X, He J, Zhou Y, Chi Y (2022) A subspace approach to sparse sampling based multi-attribute data aggregation in IoT. IEEE Internet Things J
2. Manuel EM, Pankajakshan V, Mohan MT (2023) Energy-efficient data aggregation in low-power wireless networks with sensors of discrete transmission ranges: a mathematical framework for network design. IEEE Trans Netw Sci Eng
3. Leinonen M, Codreanu M, Giannakis G (2019) Compressed sensing with applications in wireless networks. Found Trends Signal Process 13(1–2)
4. Xu Y, Sun G, Geng T, Li Z (2019) Compressive multi-timeslots data gathering with total variation regularization for wireless sensor networks. IEEE Commun Lett 23(4):648–651
5. He J, Sun G, Li Z, Zhang Y (2017) Compressive data gathering with low-rank constraints for wireless sensor networks. Signal Process 131:73–76
6. Qin Z, Fan J, Liu Y, Gao Y, Li GY (2018) Sparse representation for wireless communications: a compressive sensing approach. IEEE Signal Process Mag 35(3):40–58
7. Fu H, Dai F, Hong L (2023) Two-dimensional off-grid DOA estimation with metasurface aperture based on MMV sparse Bayesian learning. IEEE Trans Instrum Meas

8. Song Y (2023) Inversion-free sparse Bayesian learning for temporally correlated signal recovery
9. Guo Q, Xin Z, Zhou T, Xu S (2023) Off-grid space alternating sparse Bayesian learning. IEEE Trans Instrum Meas 72:1–10
10. Shao Y, Ma H, Liu H (2022) A study and comparison of different sparse Bayesian learning algorithms in DOA estimation. In: 2022 5th international conference on information communication and signal processing (ICICSP). IEEE, pp 1–5

Maximizing Efficiency: Unveiling the Potential of Kubernetes Metrics

Josephine Eskaline Joyce and Shoney Sebastian

Abstract In the realm of Kubernetes cluster management, the importance of metrics cannot be overstated. Metrics serve as a powerful lens, providing a quantitative perspective into a cluster's performance, behavior, and resource utilization. In the ever-evolving landscape of cloud-native computing, metrics are the key to informed decision-making. They empower administrators to navigate scaling, resource allocation, and the holistic optimization of Kubernetes clusters with a data-driven confidence. This paper stands as a vital contribution, placing metrics at the forefront of the discussion. It underscores their transformative potential by shedding light on how they drive administrators' decisions, enable the identification of performance bottlenecks, and enhance application responsiveness. Moreover, metrics play a pivotal role in proactive capacity planning, ensuring resources are allocated with precision to meet both current and future workload demands. In essence, this paper's core contribution lies in providing a comprehensive overview of Kubernetes metrics and highlighting their profound impact on Autoscaling strategies. By revealing the constraints that metrics may impose on the efficient scaling of application resources, it equips administrators with a navigational tool for building dynamic and resilient computing environments within Kubernetes clusters.

Keywords kubernetes · Metrics · Scaling · Custom metrics · Metrics pipeline

1 Introduction

Kubernetes [1] helps to manage, deploy, and scale containerized applications in a highly available, automated, and scalable manner. Kubernetes has become one of the most popular container orchestration platforms in recent years [2] and is widely used in production environments by organizations of all sizes. Kubernetes monitoring is a

J. E. Joyce (✉) · S. Sebastian
CHRIST (Deemed to be University), Bengaluru, India
e-mail: josephine.j@res.christuniversity.in

S. Sebastian
e-mail: shoney.sebastian@christuniversity.in

S. M. Thampi et al. (eds.), *Fifth International Conference on Computing and Network Communications*, Lecture Notes in Electrical Engineering 1219,
https://doi.org/10.1007/978-981-97-4540-1_3

critical practice for maintaining a reliable, high-performing cluster. Kubernetes monitoring involves systematically tracking and evaluating the performance, health, and resource utilization of a Kubernetes cluster. It's essential for ensuring the reliability and efficiency of applications running in the cluster.

1.1 Our Contribution

Kubernetes monitoring provides real-time visibility into the health, performance, and resource utilization of containerized applications within a Kubernetes cluster. Without effective monitoring, it's challenging to detect and respond promptly to issues like resource constraints, application failures, or network problems. In essence, it's a cornerstone in ensuring the reliability, performance, and cost-effectiveness of containerized applications in Kubernetes environments.

Despite its robust capabilities, Kubernetes monitoring does have some gaps that organizations should be aware of [3]. One significant challenge is the complexity of correlating metrics across the various layers of the Kubernetes stack. Understanding the relationships between nodes, pods, containers, and services can be intricate, especially in large and dynamic environments. Additionally, Kubernetes doesn't provide out-of-the-box support [4] for monitoring application-level metrics or specific business KPIs.

In this research, we investigate a variety of metrics and the tools that can be used to monitor Kubernetes cluster. This study also focuses on how these can be used by autoscaler to scale Kubernetes effectively.

2 Background

2.1 Kubernetes Metrics Architecture Pipeline

A Kubernetes monitoring pipeline is a structured set of processes, tools, and components used to collect, process, store, and analyze metrics and logs from a Kubernetes cluster. The primary goal of a monitoring pipeline is to ensure the health, performance, and reliability of applications and infrastructure running on Kubernetes. These pipelines are critical for maintaining the health, performance, and reliability of applications and infrastructure running on Kubernetes clusters.

Kubernetes Core Metrics Pipeline

Metrics Collection: In the core metrics pipeline, metrics are collected directly from Kubernetes components and resources. This includes data from nodes, pods, containers, services, and other Kubernetes objects.

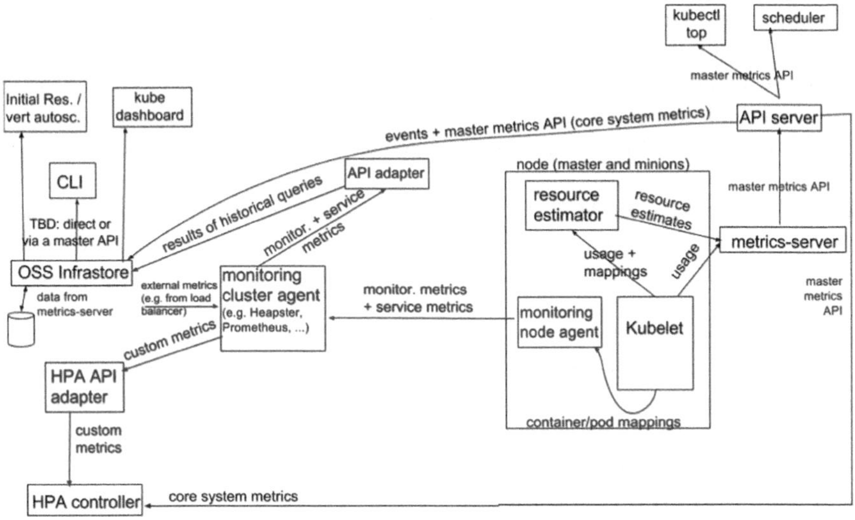

Fig. 1 Kubernetes monitoring pipeline

- kube-state-metrics [5]: This component provides valuable metrics about the state of various Kubernetes objects, allowing for insights into their status and health.
- cAdvisor [6]: This agent runs on each node and collects detailed information about container resource usage, providing crucial data for understanding container behavior.
- Kubelet [7]: It provides metrics related to nodes, pods, and containers, giving insights into resource utilization and health.

Aggregation and Storage: Metrics are typically aggregated using tools like Prometheus [8]. Prometheus is purpose-built for monitoring Kubernetes and can efficiently scrape, store, and query metrics. It's a popular choice for collecting and aggregating core Kubernetes metrics [9].
Visualization and Dashboarding: Visualization tools like Grafana [10] are commonly used in conjunction with Prometheus to create informative and customizable dashboards. These dashboards provide a visual representation of the collected metrics. *Alerting and Notifications*: Alerting systems like Alertmanager, integrated with Prometheus, are used to define and manage alerts based on predefined conditions or thresholds. Notifications can be sent to appropriate channels (e.g., email, Slack) when alerts are triggered (Fig. 1).

Monitoring Pipeline
The broader "monitoring pipeline" encompasses not only core Kubernetes metrics but also includes additional components for monitoring application-specific metrics, logs, and potentially even traces.

- **Logs Collection and Aggregation**: Tools like Fluentd or Fluent Bit are commonly used for collecting, processing, and forwarding logs from containers and system components. The ELK Stack (Elasticsearch, Logstash, Kibana) is another popular choice for log aggregation.

- **Logs Collection and Aggregation**: Tools like Fluentd or Fluent Bit are commonly used for collecting, processing, and forwarding logs from containers and system components. The ELK Stack (Elasticsearch, Logstash, Kibana) is another popular choice for log aggregation.
- **Application-Specific Metrics**: In addition to core Kubernetes metrics, organizations often monitor application-specific metrics. These are metrics that are specific to the behavior and performance of the applications or services running on Kubernetes.
- **Tracing**: Distributed tracing tools like Jaeger or Zipkin may be integrated for monitoring and understanding the flow of requests across microservices.
- **Long-Term Storage and Analysis**: Specialized time-series databases (TSDB) like InfluxDB or other TSDBs can be employed to store metrics over longer periods for trend analysis.
- **Kubernetes Metrics Categories**: Kubernetes metrics can be categorized into multiple types. Figure 2 shows multiple types of metrics supported in Kubernetes.

Core metrics

- **Role in Monitoring Pipeline**: Core metrics are fundamental metrics related to the resource usage and health of Kubernetes objects such as nodes, pods, and containers. They are typically provided by the Kubernetes API server or kubelet.
- **Collection**: These metrics are collected directly from the Kubernetes components and resources, providing a basic level of monitoring for the cluster.
- **Examples**: CPU Usage, Memory Usage, Network Metrics.
- **Use in Monitoring Pipeline**: Core metrics serve as the foundational data points for understanding the overall health and performance of the Kubernetes cluster. They are often used for basic monitoring and scaling decisions.

Non-core Metrics

- **Role in Monitoring Pipeline**: Non-core metrics, also known as custom or application-specific metrics, are specific to a particular application or service running on Kubernetes. They are often defined by users or developers and may not be directly related to Kubernetes resources.
- **Collection**: Non-core metrics are collected from within the application or service itself, often by exposing a custom metric endpoint or API.
- **Examples** :Number of Pending Jobs in a Queue, Response Time of an API Endpoint.
- **Use in Monitoring Pipeline**: Non-core metrics provide application-specific insights and allow for fine-grained control over monitoring and scaling decisions. They are crucial for understanding and optimizing the behavior of individual applications.

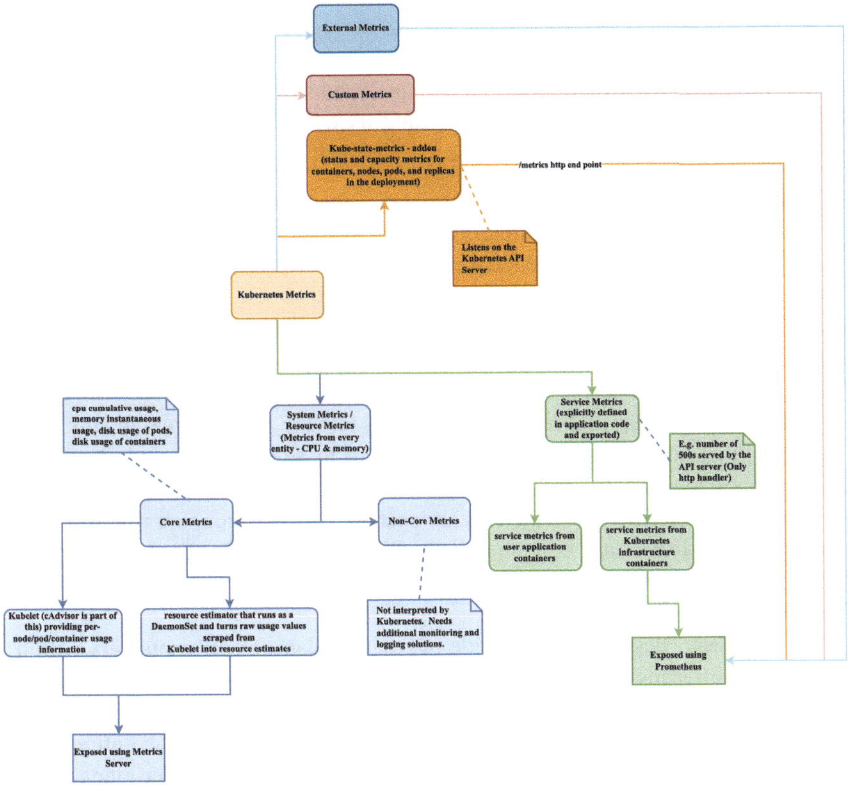

Fig. 2 Kubernetes metrics categories

Service Metrics

- **Role in Monitoring Pipeline**: Service metrics pertain to the performance and utilization data related to services within a Kubernetes cluster. They provide insights into how well services are handling incoming requests from clients or other parts of the application.
- **Collection**: Service metrics are collected from the services themselves and are focused on measuring the behavior and performance of the services.
- **Examples**: Request Rate, Error Rate, Latency.
- **Use in Monitoring Pipeline**: Service metrics are essential for assessing the performance, availability, and reliability of services within a Kubernetes cluster. They enable operators and developers to identify and address any issues affecting the user experience.

2.2 Kubernetes Metrics Collection

Various types of metrics are collected and analyzed to gain insights into the health, performance, and behavior of the cluster and its workloads. Here are the different types of metrics commonly observed in the pipeline:

- **Resource Metrics**: These metrics are related to the utilization of physical resources such as CPU, memory, and storage. They help understand how efficiently resources are being used. Examples of these metrics include CPU usage, memory usage, disk I/O, network traffic.
- **Workload Metrics**: These metrics focus on the behavior and performance of applications and services running on the Kubernetes cluster. Examples of these metrics include Request rates, error rates, latency, concurrency.
- **Node Metrics**: These metrics provide insights into the performance and utilization of individual nodes in the cluster. Examples of these metrics include Node CPU usage, node memory usage, node network metrics.
- **Container Metrics**: These metrics are specific to the containers running within pods. They help monitor container resource usage and performance. Examples of these metrics include Container CPU usage, container memory usage, container network metrics.
- **Pod Metrics**: These metrics provide aggregated data for all containers within a pod. They offer insights into the performance and behavior of pods. Examples of these metrics include Pod resource usage, restart count, status, network metrics.
- **Cluster-Wide Metrics**: These metrics offer a holistic view of the entire Kubernetes cluster, including components like the API server, controller manager, and scheduler. Examples of these metrics include API server performance, controller manager metrics, scheduler metrics.
- **Service Metrics**: These metrics focus on the performance and utilization of services within the Kubernetes cluster, including request rates and response times. Examples of these metrics include Request rates, error rates, availability.
- **Custom Metrics**: These metrics are specific to individual applications or services and are defined by users or developers. They may reflect unique performance indicators. Examples of these metrics include Application-specific metrics, business-specific KPIs.
- **External Metrics**: These metrics are sourced from external systems or services that may interact with applications running in the Kubernetes cluster. Examples of these metrics include Metrics from external APIs, third-party service performance.

2.3 *Metrics-Based Dynamic Workload Adjustment*

Autoscaling Metrics

Kubernetes autoscaler [11] uses metrics to dynamically adjust the number of resources, such as pods or nodes, in a cluster based on observed workload demands [12]. It constantly monitors specified metrics, like CPU usage or custom application-specific metrics, to understand the resource needs of deployed applications. When certain predefined thresholds or conditions are met, the autoscaler can automatically scale the resources up or down to ensure optimal performance and resource utilization. For instance, if CPU usage exceeds a specified threshold, indicating high demand, the autoscaler can add more pods to distribute the load. Conversely, if resource usage decreases, the autoscaler can scale down, conserving resources and reducing costs. This dynamic adjustment based on real-time metrics allows Kubernetes to efficiently allocate resources, ensuring that applications can handle varying workloads effectively and providing a more responsive and cost-effective environment for running workloads. Following are the three types of metrics that cover a wide range of scenarios for autoscaling, including both internal resource usage and external factors that influence workload demands. By utilizing these metrics, Kubernetes autoscaler can dynamically adjust resources to meet the needs of applications and services running on the cluster.

- **Resource Metrics**: This category includes metrics related to the utilization of physical resources like CPU and memory. For instance, the autoscaler may monitor CPU usage to determine if pods or nodes need to be scaled up or down based on demand.
- **Custom Metrics**: Custom metrics are application-specific metrics defined by users or developers. These metrics reflect unique performance indicators that are specific to a particular application or service. They allow for fine-tuned autoscaling decisions based on the behavior of the application.
- **External Metrics**: External metrics are sourced from systems or services external to the Kubernetes cluster. These metrics may reflect the behavior and performance of dependencies outside the cluster. For example, the autoscaler might monitor the response time of an external API and adjust resources accordingly.

2.4 *Autoscaling Metrics Aggregator*

Custom metrics and external metrics are important for autoscaling in Kubernetes because they allow for more fine-grained and context-specific scaling decisions. While core metrics like CPU and memory utilization are essential, they may not always capture the full picture of an application's performance or resource needs. Custom metrics, which are application-specific metrics defined by users or developers, provide insights into unique performance indicators that are crucial for efficient scaling. For example, the number of pending jobs in a queue or response time of a critical API endpoint can be critical factors in determining when to scale an application. External metrics, on the other hand, allow Kubernetes to scale based on factors external to the cluster, such as the load on an external service or API. This is particularly important when applications rely on third-party services, as scaling decisions may need to take into account the availability and performance of those services. By incorporating custom and external metrics into autoscaling strategies, Kubernetes can dynamically adjust resources to meet the specific demands of each application, ensuring optimal performance and resource utilization.

A Kubernetes metrics aggregator is a crucial component in a Kubernetes cluster responsible for collecting, processing, and centralizing metrics data from various sources within the cluster. It acts as a centralized hub that aggregates data on resource utilization, performance, and behavior of nodes, pods, containers, and other Kubernetes objects. By consolidating this information, the metrics aggregator provides a unified view of the cluster's health and performance. This aggregated data can be queried and analyzed by users and monitoring systems, enabling them to make informed decisions about resource allocation, scaling, and optimization. Additionally, Kubernetes metrics aggregators often integrate with visualization tools and alerting mechanisms, allowing for the creation of informative dashboards and timely notifications in case of critical events or threshold breaches. Overall, a Kubernetes metrics aggregator plays a vital role in ensuring the smooth operation and efficient utilization of resources within a Kubernetes environment. Figure 3 shows the architecture of Kubernetes metrics aggregator using Prometheus. Custom metrics and external metrics can be aggregated using Prometheus. Solutions like NewRelic <todo ref>, Sysdig <todo ref>, Amazon cloudwatch <todo ref>, Datadog are various solutions that uses aggregator to send metrics to Kubernetes.

2.5 *Impact of Custom Metrics in Autoscaling*

Kubernetes autoscaling based on custom metrics empowers organizations to create a more responsive, efficient, and cost-effective environment for their applications. By dynamically adjusting resources in real-time, Kubernetes ensures that applications can meet the demands of their users and adapt to changing conditions effectively.

Custom metrics parameters are the key indicators or measurements that are defined by users or developers to be used for autoscaling purposes. These metrics are specific to individual applications or services and are not part of the built-in, core metrics provided by Kubernetes. Custom metrics parameters play a crucial role in making fine-grained scaling decisions based on the behavior and requirements of the application. The specific custom metrics parameters that can be used for autoscaling depend on how they are defined and exposed by the application. They can cover a wide range of performance indicators, which may include:

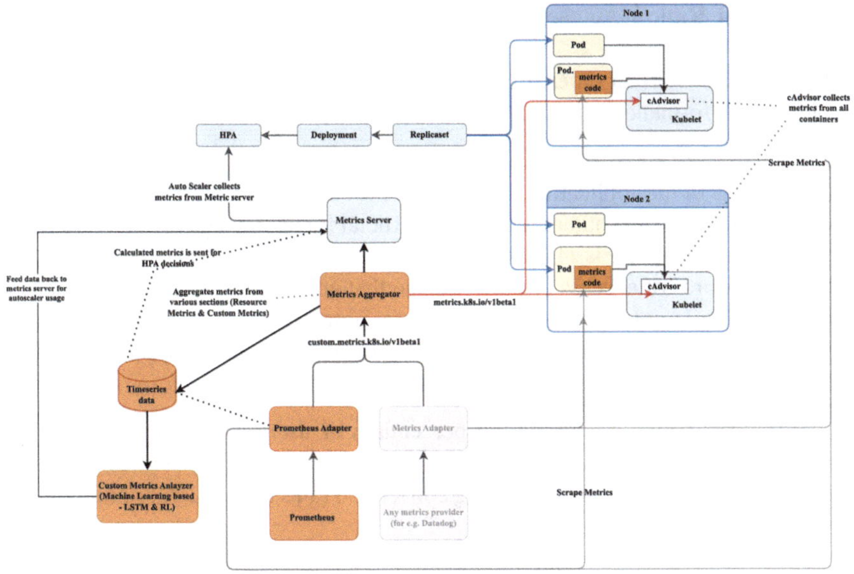

Fig. 3 Kubernetes metrics aggregator

Application-Specific Metrics [13]: These are metrics that are directly relevant to the behavior and performance of the application. For example, it could be the number of pending tasks in a queue, response time of a specific API endpoint, or any other metric that is critical for the application's performance.

Metrics used for autoscaling in Kubernetes are custom measurements that are specific to the behavior and performance of a particular application or service. These metrics are defined by users or developers based on the unique requirements and characteristics of the application. The choice of application-specific metrics depends on the nature of the application and the key performance indicators (KPIs) that are critical for its operation. Here are some examples of application-specific metrics that can be used for autoscaling

- **Number of Pending Tasks or Jobs**—Autoscaling based on this metric can help ensure that the application can handle high demand.
- **Response Time of Critical Endpoints**—Autoscaling based on response times can help maintain a responsive user experience.
- **Queue Length or Depth**—Autoscaling based on queue length can help manage workloads with varying processing times.
- **Throughput or Transactions per Second (TPS)**—Autoscaling based on throughput helps ensure the application can handle increased transaction volumes.
- **Error Rates**—Autoscaling based on error rates helps maintain application reliability.
- **Concurrency or Active Sessions**—Autoscaling based on concurrency helps manage concurrent user loads.

- **Cache Hit Ratio**—Autoscaling based on cache hit ratio can optimize resource usage.
- **Latency for Critical Operations**—Autoscaling based on latency helps maintain optimal performance.

Business-Specific KPIs: Key Performance Indicators (KPIs) that are specific to the business logic of the application. These metrics could be related to user engagement, conversion rates, or any other metric that is crucial for achieving business objectives.

These Kubernetes KPIs [14] are unique to each business or industry and reflect the specific measures of success or performance that are critical for achieving business objectives. While not all KPIs are suitable for autoscaling in Kubernetes, they can still be valuable for monitoring and optimizing the performance of applications. Here are some examples of Business-Specific KPIs that may be relevant for monitoring:

- **Conversion Rate**—Measures the percentage of users who complete a desired action, such as making a purchase or signing up for a service. While not directly related to autoscaling, understanding conversion rates is crucial for optimizing the user experience.
- **Customer Acquisition Cost (CAC)**—Calculates the cost of acquiring a new customer. Monitoring CAC can help ensure that marketing efforts are efficient and cost-effective.
- **Customer Lifetime Value (CLTV)**—Estimates the total revenue a business expects to earn from a customer throughout their entire relationship with the company. Understanding CLTV helps guide customer acquisition and retention strategies.
- **Churn Rate**—Measures the percentage of customers who stop using a product or service over a specific period. While not used for autoscaling, monitoring churn rate is important for customer retention efforts.
- **Average Revenue Per User (ARPU)**—Calculates the average revenue generated by each user or customer. Understanding ARPU is vital for optimizing pricing strategies.
- **Net Promoter Score (NPS)**—Gauges customer satisfaction and loyalty by asking how likely customers are to recommend a product or service. While not an autoscaling metric, NPS provides valuable insights into customer sentiment.
- **User Engagement Metrics (e.g., Daily Active Users, Monthly Active Users)**—Measures the level of user interaction and engagement with a product or service. While not used for autoscaling, these metrics are crucial for understanding user behavior.
- **Retention Rate**—Measures the percentage of customers who continue to use a product or service over a specific period. Monitoring retention rate helps assess customer satisfaction and product stickiness.
- **Average Transaction Value (ATV)**—Calculates the average value of transactions made by customers. Understanding ATV helps optimize sales and marketing strategies.

- **Customer Satisfaction (CSAT)**—Measures customer satisfaction with a product or service through surveys or feedback. While not directly related to autoscaling, CSAT provides valuable insights into customer sentiment.

Custom Application Health Metrics: Metrics that provide insights into the health and status of the application. These reflect the overall health and status of an application. These metrics are defined by users or developers based on the unique requirements and characteristics of the application. While they may not be directly used for autoscaling in Kubernetes, monitoring these metrics is crucial for ensuring the application is operating smoothly and efficiently. Here are some examples of Custom Application Health Metrics that can be monitored:

- **Error Rates**—A low error rate indicates a healthy application.
- **Latency or Response Time**—Consistently low response times are indicative of a healthy and responsive application.
- **Uptime and Availability**—High uptime and availability are key indicators of a healthy application.
- **Transaction Success Rates**—A high success rate is indicative of a well-functioning application.
- **Database Connection Pooling Metrics**—Monitors the utilization and performance of database connections, including metrics like connection pool size, idle connections, and connection timeouts.
- **Request Rate and Throughput**—Consistent request rates indicate a healthy application.
- **Resource Utilization**—Well-balanced resource utilization is crucial for application health.
- **Concurrency Levels**—Consistent concurrency levels indicate a well-scaled application.
- **Component-Specific Metrics**—Tracks metrics specific to individual components or services within the application, providing insights into their health and performance.

Unique Workload Characteristics: Metrics that capture unique workload characteristics that are specific to the application. Unique Workload Characteristics-based Metrics are specific measurements that capture the distinctive behavior and performance patterns of an application's workload. These metrics are defined by users or developers based on the specialized requirements and characteristics of the workload. While not directly used for autoscaling in Kubernetes, monitoring these metrics is crucial for ensuring that the application can handle its workload effectively. Here are some examples of Unique Workload Characteristics-based Metrics that can be monitored:

- **Specialized Processing Tasks**—Measures the performance of specific tasks or operations that are unique to the workload, such as complex computations, data processing, or specialized algorithms.

- **Batch Processing Metrics**—Tracks metrics related to batch processing jobs, such as job completion times, batch sizes, and processing rates. These metrics are crucial for optimizing batch job execution.
- **File System I/O Performance**—Monitors metrics related to file read and write operations, including latency, throughput, and IOPS. These metrics are relevant for applications that heavily rely on file operations.
- **Database Query Performance**—Measures the performance of database queries, including metrics like query response times, transaction rates, and database connection utilization.
- **Custom Message Queue Metrics**—Tracks metrics related to message queue operations, such as message processing times, queue depth, and message delivery rates.
- **Workflow Execution Times**—Measures the time it takes to complete specific workflows or processes within the workload. Monitoring workflow execution times helps ensure efficient process flows.
- **Job Queue Length**—Reflects the number of jobs or tasks waiting to be processed. Monitoring job queue length is crucial for optimizing workload processing.
- **Algorithm-Specific Metrics**—Metrics that capture the behavior and performance of specific algorithms or computational methods used within the workload.
- **Specialized Data Transformation Metrics**—Measures the performance of data transformation operations or data manipulation tasks unique to the workload.
- **Specialized Network Traffic Patterns**—Monitors network traffic patterns that are specific to the workload, such as specialized communication protocols or data transfer patterns.

2.6 Analysis

Kubernetes autoscaler, while a powerful tool for managing resource allocation, does have certain limitations and considerations with respect to metrics. Two of the major limitations include the below -

- **Complexity of Custom Metrics**—Kubernetes autoscaler with custom metrics offers a powerful mechanism for fine-tuning resource allocation based on specific application behavior. Implementing custom metrics for autoscaling can be complex. It requires users to define, expose, and instrument their applications to provide the necessary metrics, which may necessitate code modifications. Additionally, exposing custom metrics demands a deep understanding of the application's inner workings. Moreover, monitoring custom metrics can be resource-intensive, potentially imposing an overhead on the application's performance. Scaling based on custom metrics also demands a deep understanding of the application's inner workings, making it crucial for developers and operators to collaborate closely. Finally, scaling decisions based on custom metrics may not always align perfectly with the application's actual needs, and fine-tuning these parameters may require iterative adjustments. Despite these challenges, leveraging custom metrics in Kubernetes autoscaling can lead to highly tailored and efficient resource allocation for specific workloads.

- **Dependency on External Metrics**—Kubernetes autoscaler with external metrics introduces several potential gaps and considerations. Utilizing external metrics introduces a dependency on external monitoring systems or APIs. If there are issues with these external sources, it could potentially lead to inaccurate or delayed autoscaling decisions. The accuracy and timeliness of autoscaling decisions are contingent on the availability and reliability of these external sources. Security and authentication measures must also be carefully managed to ensure the integrity of the metrics data. Moreover, external metrics may not always perfectly align with the specific needs and behavior of the application, potentially leading to suboptimal scaling decisions. This approach also requires a robust monitoring infrastructure, as any failures or disruptions in the external metrics source can impact the autoscaling process. Overall, while leveraging external metrics provides valuable insights, it introduces additional complexity and dependencies that must be carefully managed for effective autoscaling in Kubernetes.

The other limitations and challenges include:

- **Limited Native Metric Types**—Kubernetes autoscaler primarily relies on core metrics like CPU and memory utilization for scaling decisions. This may not always capture the complete picture of an application's performance, potentially leading to suboptimal scaling decisions.
- **Resource Intensive Monitoring**—Monitoring custom or external metrics can be resource-intensive, potentially adding an overhead to the application's performance. Careful consideration of the impact on the application's resources is necessary.
- **Fine-Tuning Challenges**—Achieving optimal scaling decisions based on metrics may require iterative adjustments and fine-tuning. It can be challenging to find the right balance and thresholds for scaling.
- **Potential Overlooking of Critical Metrics**—Depending on the implementation, there is a possibility of overlooking important metrics or failing to capture all relevant aspects of an application's behavior, potentially leading to suboptimal scaling decisions.
- **Limited Visibility into Application Behavior**—Depending on the complexity of the application, it may be challenging to identify and expose all the metrics that truly reflect its behavior and performance.
- **External Dependency Risks**—Relying on external metrics introduces a potential point of failure. Disruptions or inaccuracies in the external metrics source can impact the autoscaling process.

It's important for users to carefully evaluate their specific application and workload requirements when deciding which metrics to use for autoscaling in Kubernetes. Additionally, regular monitoring and fine-tuning of autoscaling parameters are essential to ensure optimal performance.

3 Conclusion

This review paper presents a summary of Kubernetes metrics that can are available in the cluster. A literature review was conducted to perform a comprehensive study, which helped identify different metrics that can be utilized for Kubernetes Autoscaling.

The analysis of this study exposes the need for custom metrics and external metrics which impacts autoscaling strategies in a cluster. This study also delves the challenges and limitations with the custom metrics and external metrics that can be used for autoscaling Kubernetes. Furthermore, this research shed light on the architecture of metrics aggregator to aggregate custom and external metrics to define an efficient autoscaling policy.

References

1. Kubernetes documentation. http://kubernetes.io
2. CNCF Annual Survey 2022. https://www.cncf.io/reports/cncf-annual-survey-2022/
3. Rossi F, Valeria C, Francesco Lo P, Matteo N (2020) Geo-distributed efficient deployment of containers with Kubernetes. J Comput Commun 159(1):161–174
4. Nguyen T-T, Yu-Jin Y, Taehong K, Dae-Heon P, Sehan K (2020) Horizontal pod autoscaling in kubernetes for elastic container orchestration. J Sens 20(16):4621
5. Kube-state-metrics github documentation. https://github.com/kubernetes/kube-state-metrics
6. cAdvisor github documentation. https://github.com/google/cadvisor
7. kubelet documentation. https://kubernetes.io/docs/reference/command-line-tools-reference/kubelet/
8. Prometheus documentation. https://prometheus.io/
9. Greneche N, Cerin C (2022) Autoscaling of containerized HPC clusters in the cloud. In: 2022 IEEE/ACM International workshop on interoperability of supercomputing and cloud technologies (SuperCompCloud)
10. Grafana documentation. https://grafana.com/
11. Kubernetes Autoscaler documentation. https://github.com/kubernetes/autoscaler
12. Duan R, Zhang F, Kha SU (2021) A case study on five maturity levels of a kubernetes operator. In: 2021 IEEE cloud summit (cloud summit)
13. Dittakavi RSS (2023) Achieving the delicate balance: resource optimization and cost efficiency in kubernetes. Eduzone: Int Peer Revi/Refereed Multidiscip J (EIPRMJ) 12(2)
14. Klaise J, Van Looveren A, Cox C, Vacanti G, Coca A (2020) Monitoring and explainability of models in production. In: Workshop on challenges in deploying and monitoring machine learning systems (ICML 2020)

Resource-Aware Learning Automata for Low-Latency Task Scheduling in Fog Computing

S. Sheela, G. M. Vishnupriya, and S. M. Dilip Kumar

Abstract Cloud computing capabilities are extended to the edge of the network with fog computing in the modern computing landscape. This proximity to data sources and end-users offers several advantages, including reduced latency, enhanced real-time processing, and improved privacy. In this context, task scheduling becomes crucial to orchestrate computational tasks efficiently across fog nodes. Effective task scheduling ensures optimal resource utilization, minimizes latency, and enhances the overall performance of fog computing systems. This proposed work introduces a resource-aware task scheduling approach in a fog computing environment to improve latency and responsiveness for latency-sensitive applications. The model considers the residual energy along with memory constraint of the fog node as the basis for assigning the fog node for task execution. Tournament Selection algorithm is used by the task scheduler for action selection. The results show that the proposed approach performs better in reducing the task response time compared to existing approaches.

Keywords Energy consumption · Fog computing · Learning automata · Memory · Task scheduling

1 Introduction

A fog network is a group of peer nodes that share computing and storage resources and spare spectrum for node-to-node and device-to-device communications and task scheduling [1]. Fog computing helps to overcome the limitations of cloud computing in terms of mobile support, delay, and location awareness by moving tasks from the cloud to the edge device [2, 3]. The explosive growth of number of devices connected to the network and an increase in the demand for real-time, reliable application execution has necessitated the need to reduce the response time of the fog nodes [4,

S. Sheela (✉) · G. M. Vishnupriya · S. M. D. Kumar
Department of Computer Science and Engineering, University Visvesvaraya College of Engineering (UVCE), Bangalore University, Bengaluru, India
e-mail: sheelasresearch@gmail.com

S. M. Thampi et al. (eds.), *Fifth International Conference on Computing and Network Communications*, Lecture Notes in Electrical Engineering 1219,
https://doi.org/10.1007/978-981-97-4540-1_4

5]. In addition, with the vast amount of data generated by IoT devices and the need for quick processing, fog computing serves as a decentralized computing structure that complements cloud computing [6]. Though the fog computing has served in improving the quality of services in the computational landscape, the fog nodes come with the constrained capacity in terms of memory, energy and computational capabilities [7].

Hence to effectively utilize the capabilities of the fog nodes, efficient application placement, resource allocation, task scheduling, and load balancing techniques needs to be in place [8]. Managing resource allocation and minimizing latency in fog environments presents challenges owing to the decentralized and heterogeneous nature of their hardware capabilities. In this context, task scheduling emerges as a pivotal determinant of system performance [9]. Users send in requests, called jobs, which involve inputting information into the system, processing data, gaining access to necessary software or infrastructure, or storing data in designated data centers. The output is then returned to the user. When assigning a task to a specific resource, several factors are considered, including the task's deadline, waiting period, input duration, job cost, and resource energy consumption while performing the task. The primary aim of a task scheduler is to allocate tasks to resources to maximize resource usage [10].

In this work, a resource-aware task scheduling approach leveraging the learning automata is proposed. The proposed approach makes the mapping of the incoming task based on the pre-estimated values of memory required, energy consumption and the task run time. The learning automata on the fog node decides the best action, that is the best fog node for assignation based on its learning from the environment.

1.1 Motivation

The growing demand for real-time applications and the proliferation of IoT has elevated the significance of fog computing. Fog computing significantly reduces the latency by empowering the edge devices to process data locally, hence assisting in minimizing latency and improving the response time of the fog devices. The applications ranging from autonomous vehicles, tele-medicine, and industrial automation require quick data processing and decision-making at the edge. Inefficient task scheduling can compromise critical system reliability and performance due to increased latency. Our work seeks to bridge this gap by proposing a resource-aware task scheduling approach leveraging stochastic automata.

1.2 Contributions

The contributions of the proposed work are as follows:

1. The resource-aware task scheduling approach, leveraging the stochastic automata considering the dynamic nature of the task execution time is proposed.
2. A normal distribution-based dynamic task run time model in fog environment to estimate task execution time is presented.
3. An energy, memory-aware decision-making mechanism for task scheduling to achieve resource optimization is introduced. This mechanism takes into account the residual energy and memory of fog nodes before assigning new tasks.
4. We present simulation experiment results to evaluate the proposed algorithm's reliability and robustness in minimizing energy consumption while ensuring task QoS.

1.3 Paper Organization

The rest of the paper is organized as follows. The related work is discussed in Sect. 2. The problem statement and objectives of the proposed work are listed in Sect. 3. Section 4 provides an overview of the scheduling strategy proposed for the fog model. The empirical evaluation of the obtained simulation results is presented in Sect. 5. Finally, concluding remarks and future research directions are presented in Sect. 6.

2 Related Work

Many static and dynamic approaches for task scheduling can be found in the literature. Some of the works in terms of references, title, work proposed and limitations of the relevant works, are presented in Table 1. The authors in work [1] propose an energy-efficient task scheduling approach for homogeneous fog network in which the resources of fog nodes are shared among the neighboring nodes for optimal resource utilization, the trade-off between performance gains and energy costs is investigated. However only energy consumption is considered for optimizing the performance. The work proposed in [12] proposes a contract establishment scheme based on sealed-bid bilateral auction scheme for resource sharing between fog nodes. Fog nodes are selected based on the betweenness centrality, computing performance and communication delay.

Table 1 Summary of related work on task scheduling in fog computing

Ref	Paper title	Work proposed	Limitations
[1]	MEETS: Maximal energy efficient task scheduling in homogeneous fog networks	An optimal scheduling decision for a task node and multiple neighboring helper nodes under feasible modulation schemes and time allocations is proposed	Energy consumption is only metric used to measure the system performance
[11]	Contract-based resource sharing for time effective task scheduling in fog-cloud environment	A scheduling approach based on the construction of functional domain to mitigate the influence of mobility of fog nodes is proposed	Betweenness centrality does not always yield optimal results
[12]	Green fog planning for optimal internet-of thing task scheduling	Integer Linear Programming models to solve the fog planning issue under the integrated Cloud–Fog framework are proposed	Power consumption is only metric used to measure the system performance
[13]	A method based on the combination of laxity and ant colony system for cloud-fog task scheduling	A task scheduling problem into a constrained optimization problem in cloud-fog environment is proposed	Results are not optimal when tasks are with similar deadlines
[14]	Automata-based dynamic fault tolerant task scheduling approach in fog computing	A learning automata-based task scheduling approach for fault tolerant fog environment is proposed	Fog nodes for task assignment are chose only on the basis of residual energy of the node

Two Integer Linear Programming (ILP) models for reducing the cost and power consumption of the fog nodes while maximizing the number of successfully provisioned IoT tasks are proposed by the authors in their work [13], however only power consumption is considered as the metric for improving the performance and reducing cost of execution. In the work [14], authors propose a laxity and ant colony system algorithm-based task scheduling for cloud-fog integrated infrastructure considering the priority and task deadline. The work reduces the energy consumption and failure rate of associated tasks scheduling with mixed deadlines, however does not yield optimal results when tasks are similar deadlines. The authors in work [15] discusses a learning automata-based task scheduling approach considering the residual energy of the fog nodes for making the system fault tolerant. However, the work only considers the residual energy as the metric for task scheduling.

From the works reviewed in this section, it is observed that most of the works consider on energy of the fog node as the constraint for task execution, while running a task on fog node require many other additional resources like memory, storage and network. In this work, a task scheduling approach based on residual energy along with the memory constraint of the fog node is considered as the criteria for selecting the fog node for task execution.

3 Problem Statement and Objectives

The problem of this work is to dynamically and efficiently schedule the set of incoming tasks to the set of fog nodes while improving the system utilization and the performance.

The objectives of the proposed work as follows:

1. To efficiently and dynamically schedule the incoming tasks to the set of fog nodes available in the system.
2. To avoid failure of tasks by pre-calculating the memory and the energy required for the task execution.
3. To reduce the latency of the fog system by efficient task scheduling.
4. To simulate the proposed model to evaluate the performance of the system.

4 Proposed System

This section presents the proposed approach for memory-aware learning automata-based task scheduling for a fog computing paradigm. The proposed architecture in Fig. 1 depicts fog computing as a distributed computing scheme that acts as an intermediate layer between cloud data centers, sensors, and IoT devices. Fog Layer consists of m fog nodes, $F = \{f_1, f_2, \ldots, f_m\}$ and a task scheduler. User layer consists of n user devices $U = \{u_1, u_2, \ldots, u_n\}$. The fog nodes are assumed to be close to the end-users, typically a hop away and are connected through high-speed wired/ wireless broadband connections. The task scheduler is responsible to provisioning the resources of fog nodes to end-user requests. We assume service requests can be split into multiple tasks that are independent, non-preemptive and deadline bound. Hence any task J from user $u_i \in U$ can be defined by a five-tuple is shown in Eq. 1.

$$t_i = \langle s_i, p_i, a_i, l_i, d_i \rangle \quad (1)$$

where t_i is the task id, s_i is the size of the task in *bits*, p_i is the priority, a_i is the task arrival time, l_i is the length of the tasks in number of CPU cycles and d_i is the deadline defined for the task.

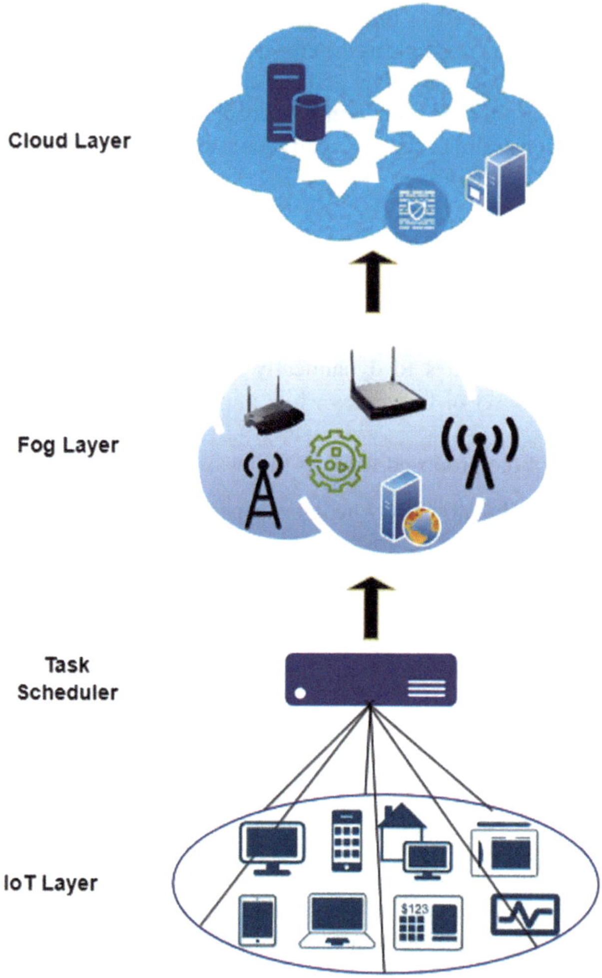

Fig. 1 Architecture of the proposed fog system

These tasks are generated randomly to simulate the real world scenarios. The task priority is determined based on two main factors: inherent priority (p^{ρ}) and user-defined priority (p^{υ}). The inherent priority of the task t_i is calculated based on d_i and l_i of the task, while the total priority p_1 is calculated as follows:

$$p_i = k_1 \cdot p_i^{\rho} + k_2 \cdot p_i^{\upsilon} \tag{2}$$

where k_1 and k_2 represent the weights of the inherent and user-defined priorities respectively.

In contrast to the static run time estimation done in many works from the literature, this work uses the dynamic run time. The memory required and the energy required by the task is estimated in prior. The proposed approach of learning automata-based task scheduling uses a learning automata that can learn from the environment and perform actions based on the learning. Each action selected by the automata is evaluated by the environment to provide a response as feedback to the learning automata. The feedback from the environment assists the automata in selecting the action for next task.

When a task from user in IoT layer is assigned to a fog node in fog layer for execution is chosen based on the following conditions:

1. The processor node is not in an error state.
2. The remaining energy level in the processor node is $\geq 20\%$.
3. The residual memory of the processor node is $\geq$ the memory required for task execution.

The parameters `Earliest Start Time (EFT)` i.e., the earliest time the task t_i can start execution and the `Earliest Finish Time (EFT)` i.e., the earliest execution time for task t_i on node f_i, the available energy on the node A_e and the available memory on the node A_m decides whether the system is rewarded or penalized based on the following conditions:

- Case 1: **Deadline of task t_i is < EFT or t_i["E_r"] > A_e or t_i["M_r"] > A_m**: The task cannot be completed and hence the action is penalized.
- Case 2: **Deadline of task $t_i \geq L_{EFT}$ and t_i["dl"] < U_{EFT} and t_i["E_r"] > A_e or** t_i["M_r"] > A_m: The task may be completed but may not be processed completely, hence the action is probably penalized.
- Case 3: **Deadline of task t_i is $\geq$ *[EFT]* and $\geq$ *[EFT]* and t_i["E_R"] $\leq A_e$ and** t_i["M_r"] $\leq A_m$: Task will be completed before its deadline, hence the action is rewarded.

In this work, every fog node is equipped with automata with equal probability of being selected by the task scheduler. Based on the above conditions, when a user submits a request, the task scheduler selects the automatons using Roulette Wheel method [16] and Tournament Selection method [17] is used to select an action in an automaton. Based on the action selected, the fog node corresponding to the action is selected. If the selected fog node successfully processes the tasks, it sends a Confirmation Message to the scheduler based on which the action is rewarded or penalized. The Algorithm 1 presents the scheduling algorithm proposed in this work.

Algorithm 1 Resource Aware Task Scheduling Algorithm

$\mathcal{F} = \{f_1, f_2, ..., f_n\}$ (Set of fog nodes) , $\mathcal{J} = \{t_1, t_2, ..., t_{||t||}\}$ (Set of received tasks) , $\mathcal{AL} = \{al_i | \forall f_i \in \mathcal{F}\}$ (Set of automatons with the initial value of $1/|\mathcal{F}|$), δ (Threshold value for task priority)
BEGIN The task scheduler is running the scheduling algorithm
Initialisation: Initialize the probabilities for each fog node
Choose an α_i using the Tournament Selection method
Calculate the task priority based on $p_i = k_1.p_i^{\rho} + k_2.p_i^{\upsilon}$ (where p_i is the probability of fog node i, k_1 and k_2 are constants, and ρ and υ are parameters)
if $p_i \geq \delta$ **then**
    The source node stores the task until the end of processing
    Check current energy level and memory level of the fog node
    if $E_{current} < E_{required}$ or $M_{current} < M_{required}$ **then**
        Apply penalty function $P(j)$
        if *task* is completed **then**
            $P(j) = (1 - b_1) * j$
        else
            $P(j) = b_2/(r - 1) * (1 - b_2) * j$
        end if
    end if
    Calculate reward function $R(j)$ based on performance
    $R = r/t$
    Update probabilities using both penalty and reward values: $p_i(n) = p_i(n-1) + R - P$
else
    Case 1: Penalised Action (PA)
    if $t_i[dl] < EFT$ or $t_i[E_r] > A_e$ or $task[M_r] > A_m$ **then**
        $P_{new}[i] = (1 - b_1) * P[i]$ for $j = i$
        $P_{new}[k] = b_1/(n - 1) * (1 - b_1) * P[k]$ for $j \neq i$
    end if
    Case 2: Probability Penalised Action (PPA)
    if $t_i[dl] >= L_{EFT}$ and $t_i[dl] < U_{EFT}$ and if $t_i[E_r] > A_e$ or $task[M_r] > A_m$ **then**
        $P_{new}[i] = (1 - b_2) * P[i]$ for $j = i$
        $P_{new}[k] = b_2/(r - 1) * (1 - b_2) * P[k]$ for $j \neq i$
    end if
    Case 3: Reward Action (RA)
    if $t_I[dl] >= U_{EFT}$ and $task[E_R] <= A_e$ and $task[M_r] <= A_m$ **then**
        reward $= c * (E_{available} - E_{estimated}) + c * (M_{available} - M_{estimated})$ for $j = i$
        P_{new}[Selected]$+= reward$ for $j = i$
    end if
    Update probabilities Task Scheduler
end if

5 Experimental Results

This section presents the results obtained from simulating the proposed memory-aware learning automata-based task scheduling approach for fog computing. The system was simulated for different scales in terms of number of fog nodes and number of tasks. The results are presented for 10 fog nodes and 20 tasks. Figure 2 depicts the frequency of the arrival times (in ms) of the tasks. Figure 3 depicts the assignment of the task to fog nodes. Figures 4 and 5 depicts the memory consumption and the energy consumption of the tasks obtained using the proposed algorithm. Finally the Fig. 6 depicts the response time of the tasks when deployed over the fog nodes selected using the proposed approach. In comparison to the baseline method [15], the proposed approach is observed to be more efficient in terms of memory utilization and energy consumption.

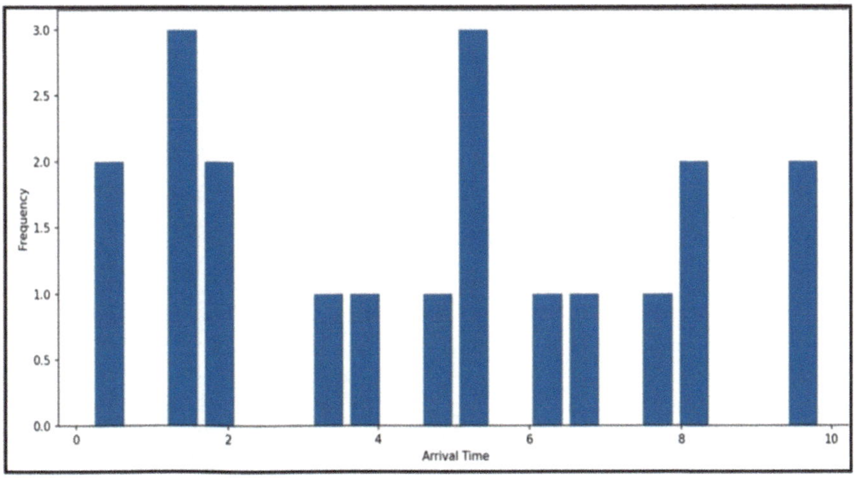

Fig. 2 Frequency distribution of the task arrival time

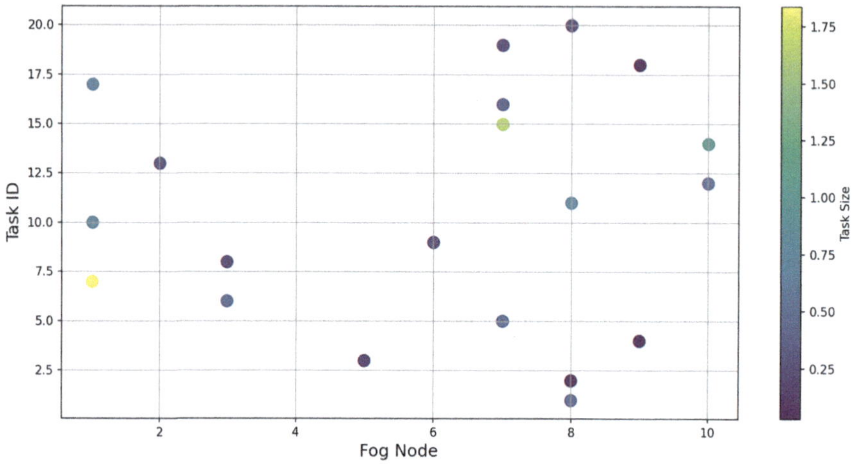

Fig. 3 Task assignment versus fog node

Fig. 4 Energy consumption by the tasks

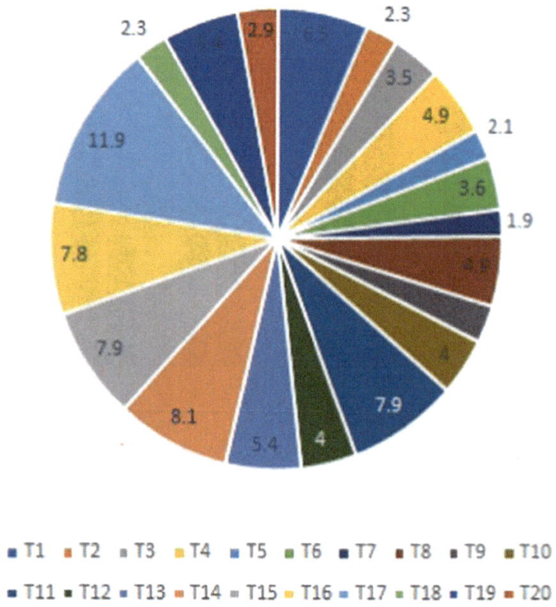

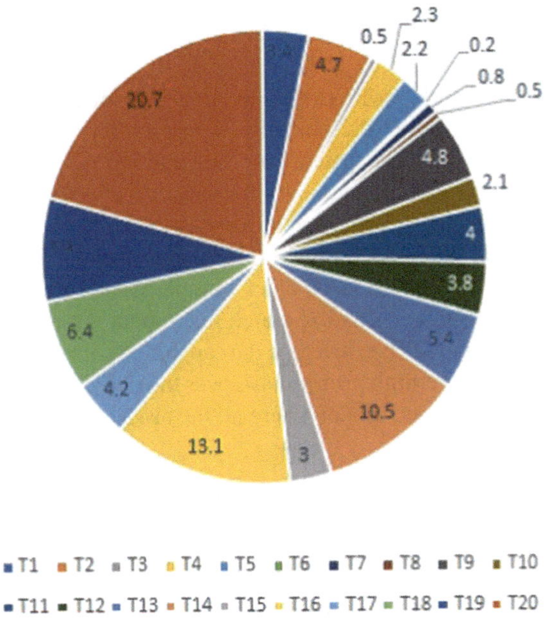

Fig. 5 Memory usage by the task

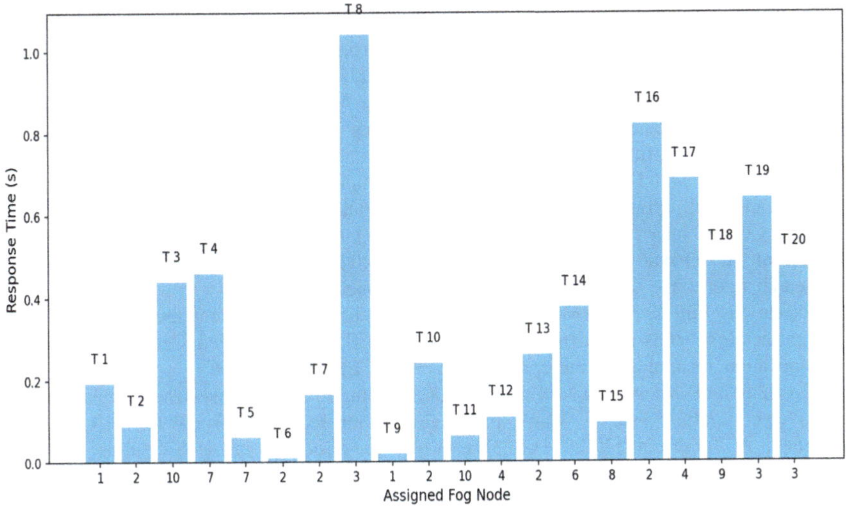

Fig. 6 Response time of tasks

6 Conclusion

In this work, to leverage the advantage of the learning automata, we propose a memory aware task scheduling approach for reducing the latency in fog computing environment. The system proposed is tested and verified in a simulation environment. The allocation of fog nodes for task execution is determined by the residual energy and memory constraint of the processing node. The proposed algorithm uses variable learning automata for making decisions in choosing the actions. The obtained results show that the proposed algorithm performs better than the other systems in terms of response time and resource utilization. As part of the future work, we plan to extend this work to leverage the capabilities of cloud-fog collaborative infrastructure. Deploying a high priority task over the cloud can lead to better results as compared to the proposed model where all the tasks are executed on fog nodes.

References

1. Yang Y, Wang K, Zhang G, Chen X, Luo X, Zhou M-T (2018) Meets: maximal energy efficient task scheduling in homogeneous fog networks. IEEE Internet Things J 5(5):4076–4087
2. Wang S, Zhao T, Pang S (2020) Task scheduling algorithm based on improved firework algorithm in fog computing. IEEE Access 8:32385–32394
3. Movahedi Z, Defude B, Hosseininia AM (2021) An efficient population-based multi-objective task scheduling approach in fog computing systems. J Cloud Comput 10:1–31
4. Abdel-Basset M, Mohamed R, Elhoseny M, Bashir AK, Jolfaei A, Kumar N (2020) Energy-aware marine predators algorithm for task scheduling in iot-based fog computing applications. IEEE Trans Indus Inform 17(7):5068–5076
5. Sheela S, Chouhan D, Dilip Kumar SM (2022) Computational framework for node placement and bandwidth optimization in dynamic fog computing environments. In: 2022 IEEE 19th India Council International Conference (INDICON). IEEE, pp 1–6
6. Harkut DG, Thakar P, Mutneja L (2022) An adaptive task scheduling in fog computing. In: Search algorithm-essence of optimization. IntechOpen
7. Sheela S, Dilip Kumar SM (2022) Dynamic application placement technique for intelligent heterogeneous fog environment. In: 2022 IEEE International conference for women in innovation, technology & entrepreneurship (ICWITE). IEEE, pp 1–6
8. Hosseinioun P, Kheirabadi M, Tabbakh SRK, Ghaemi R (2022) A task scheduling approaches in fog computing: a survey. Trans Emerg Telecommun Technol 33(3):e3792
9. Enami A, Torkestani JA (2021) Intelligent resource allocation in fog computing: a learning automata approach. J Adv Comput Eng Technol 7(1):19–34
10. Alizadeh MR, Khajehvand V, Rahmani AM, Akbari E (2020) Task scheduling approaches in fog computing: a systematic review. Int J Commun Syst 33(16):e4583
11. Rezvanian A, Moradabadi B, Ghavipour M, Khomami MMD, Meybodi MR, Rezvanian A, Moradabadi B, Ghavipour M (2019) Introduction to learning automata models. In: Learning automata approach for social networks, pp 1–49
12. Sun H, Yu H, Fan G (2020) Contract-based resource sharing for time effective task scheduling in fog-cloud environment. IEEE Trans Network Service Manage 17(2):1040–1053
13. He Z, Zhang Y, Tak B, Peng L (2019) Green fog planning for optimal internet-of-thing task scheduling. IEEE Access 8:1224–1234
14. Xu J, Hao Z, Zhang R, Sun X (2019) A method based on the combination of laxity and ant colony system for cloud-fog task scheduling. IEEE Access 7:116218–116226

15. Ghanavati S, Abawajy J, Izadi D (2020) Automata-based dynamic fault tolerant task scheduling approach in fog computing. IEEE Trans Emerg Top Comput 10(1):488–499
16. Yu F, Fu X, Li H, Dong G (2016) Improved roulette wheel selection-based genetic algorithm for TSP. In: 2016 International conference on network and information systems for computers (ICNISC). IEEE, pp 151–154
17. Blickle T (2000) Tournament selection. Evol Comput 1:181–186

Design of High Gain Wideband Petal-Shaped Magneto-Electric Dipole Metal Antenna for 5G Sub-6 GHz and Wi-Fi Applications

B. Praveen Kumar, K. Usha Kiran, and R. Ramesh

Abstract This paper proposes a novel petal-shaped Magneto-Electric (ME) dipole antenna configuration loaded with slot, optimized for 5G sub-6 GHz and Wi-Fi applications. The antenna is designed in a compact size of $1.18\lambda_0 \times 1.18\lambda_0 \times 0.27\lambda_0$ (λ_0, wavelength at 3.5 GHz), in conjunction with a Γ-shaped feeding structure and box-shaped reflector, providing a notable range of wide bandwidth and high gain. Furthermore, Γ-shaped feeding strip incorporation also helps to improve the impedance matching of the proposed design. The antenna demonstrates a broad frequency coverage from 2.02 to 5.26 GHz, resulting in an impedance bandwidth of 88.7%. The wide range of frequencies this antenna covers enables it to effectively accommodate the diverse spectrum demands of 5G networks (mobile and Wi-Fi). The antenna has a significant broadside gain of 10.4 dB with excellent stability throughout the frequency band. The optimization using Ansys High-Frequency Structure Simulator (HFSS) shows better antenna performance in terms of return loss, gain, and radiation pattern. The innovative design of this compact antenna represents a notable advancement in the development of 5G antenna technology, laying the foundation for improved wireless communication capabilities. As it is a full metal antenna, it has the advantage of providing high gain without any dielectric loss.

Keywords ME dipole antenna · Compact size · High gain · Wide bandwidth · 5G applications

---

B. Praveen Kumar · K. Usha Kiran (✉) · R. Ramesh
School of Electronics Engineering (SENSE), Vellore Institute of Technology, Chennai, India
e-mail: usha.kiran@vit.ac.in

B. Praveen Kumar
e-mail: praveenkumar.b2020a@vitstudent.ac.in

R. Ramesh
e-mail: ramesh.r@vit.ac.in

S. M. Thampi et al. (eds.), *Fifth International Conference on Computing and Network Communications*, Lecture Notes in Electrical Engineering 1219,
https://doi.org/10.1007/978-981-97-4540-1_5

1 Introduction

The demand for high-speed communication networks has increased rapidly due to data traffic and increased users. With improvements over earlier generations including increased data capacity, decreased latency, and great device connection capabilities, 5G stands out as the best option for seamless communication [1, 2]. Moreover, the increasing development of technologies demands a 5G antenna with high gain, wide bandwidth, compact size, and stable radiation performance.

There is always a challenge in obtaining a high gain in compact design structure, because of its tradeoff between size and radiation performance [3]. To address the needs and challenges of 5G communication systems, ME dipole antennas have been suggested recently. In 2006, there was the first mention of ME dipole, the antenna was constructed to have both electric and magnetic dipole, which are concurrently excited. Multiple ME dipole antenna designs have been proposed recently to satisfy the need for 5G applications [4, 5]. For instance, in [6] an ME dipole antenna is presented with T-shaped coupling stub and achieves an impedance bandwidth of 49.4% (3.2–5.3 GHz) with a unidirectional pattern. An antenna with a unique bent feeding structure has been proposed in [7] and obtained an impedance bandwidth of 53.2% (3.06–5.28 GHz) with a gain of 8 dBi and stable radiation. In 2019, a broadband ME dipole antenna presented in a planar form [8] and obtained a bandwidth of 27.6% (3.3–4.2 GHz). The antenna further maintained a compact design with good radiation characteristics but not enough operating bandwidth to meet the needs of network demand. For obtaining a wide impedance bandwidth, an ME dipole antenna presented in [9], the antenna has achieved a bandwidth of 80%, but fabrication of the antenna structure is complex due to the staircase magneto dipole design and it also has a disadvantage in applications where space is limited because of its physical dimensions. Besides, the ME dipole antennas have become more popular in 5G applications due to their unidirectional radiation pattern with minimal back lobe radiation, cross-polarization levels, and stable gain over the antenna's operating frequency band [10].

The present work introduces a novel petal-shaped ME dipole antenna that operates in 5G sub 6 GHz frequency, making it ideal for 5G Wi-Fi and mobile base station applications. The suggested antenna has achieved better compactness compared to other previously proposed ME dipole antennas [4, 9, 11–13]. It also achieved an impedance bandwidth of about 88.7% because of Γ-shaped feeding strip excitation and the inclusion of a slot in the radiating element. The operating frequency ranges from 2.02 to 5.26, with wide bandwidth, compact size, and high gain, making it useful for 5G sub 6 GHz applications.

2 Antenna Design

2.1 *Proposed Antenna Geometry*

The geometric configuration of the proposed Petal-shaped ME dipole antenna comprises many elements, as seen in Figs. 1 and 2. This design approach is intended to maintain consistent gain and radiation performance over the whole frequency range. The electric dipole is constructed as a petal-shaped planar patch with a slot and the vertically oriented shorted patches form a magneto dipole, which together forms the ME dipole antenna. The feeding technique is comprised of a transmission line and a Γ-shaped coupling strip. The complete antenna structure is built on copper material with a thickness of 1 mm, and the proposed design has been modeled and tested using Ansys HFSS.

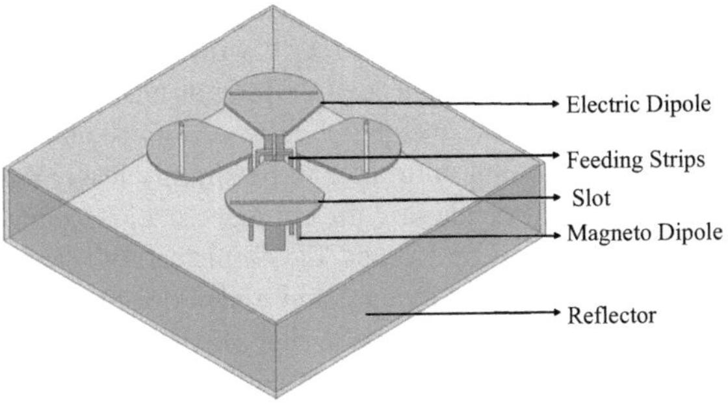

Fig. 1 Proposed petal-shaped ME dipole antenna design

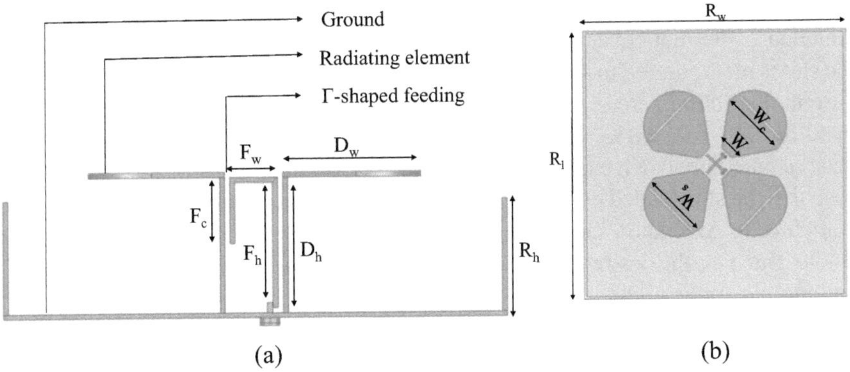

Fig. 2 **a** Side view and **b** top view of the proposed petal-shaped ME dipole antenna

The horizontal component of the coupling strips deals with the transfer of electrical energy to the antenna, while the vertical component adds capacitance to counterbalance the effects of inductance. The impedance bandwidth of the transmission line is enhanced by using a linear tapering metallic strip that is folded, therefore functioning as an air microstrip. In addition, the boxy rectangular reflector obtains gain that is both high and steady over the whole usable frequency.

The antenna's dimensions were modeled and optimized using the HFSS software, with careful attention to detail. The optimized antenna design parameters have been presented: $R_l = 101$, $R_w = 101$, $W_c = 26$, $W_s = 23$, $W = 5$, $R_h = 23$, $D_h = 26.5$, $F_h = 27.5$, $F_c = 13$, $F_w = 9.5$, $D_w = 27.6$ (units = mm).

2.2 Design Approach

Table 1 consists of the design analysis from the initial design to the final proposed design. The antenna base model (Antenna 1) starts with the design aspect of a general ME dipole antenna but with a sharp edge petal shape as an electric dipole and a shorted patch as a magnetic dipole as shown in Table 1. The petal shape planar structure which acts as an electric dipole has been optimized to be slightly lesser than $\lambda/4$ to obtain a minimal electrical length for achieving the compact size. Antenna 1 has obtained a dual band with a bandwidth of 65.4% and 17.8% instead of wideband, because of the minimal width of the radiating element, and it also does not achieve stable gain over the required frequency band as shown in Fig. 3b, due to the radiation distortion because of its sharp edges of the radiating element [14].

From the design analysis of Antenna 1, Antenna 2 is designed with a curved edge radiating element and the width of the electric dipole has been increased to the same as the length, which is slightly lesser than $\lambda/4$. It has obtained a better bandwidth and stable gain compared to Antenna 1 since the current path has been maximized. Yet, Antenna 2 does not cover the desired frequency band of interest. So, in Antenna 3, the slot is incorporated in a petal-shaped radiating dipole by analyzing the maximum current distribution area which introduces a new resonance at the higher frequency band of 4.78–5.26 GHz. As a result, the final design Antenna 3 has obtained the desired outcome which can be seen from the comparison graphs of Fig. 3a. It can be seen that the slot-loaded petal-shaped antenna gives better bandwidth and return loss compared to Antenna 1 and Antenna 2. Fig. 3b shows that Antenna 3 has obtained a stable gain over the desired frequency band which is much more needed in 5G Wi-Fi and mobile applications.

Table 1 Antenna design process with results

Antenna design	Frequency range (GHz)	Bandwidth (%)	Gain (dB)
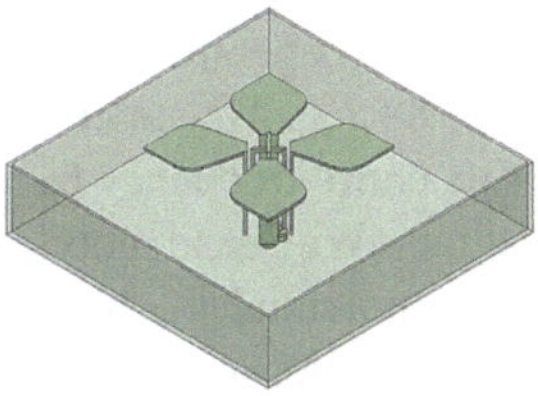 Antenna 1	2.1–4.1	64.5	11 @ 3.5 GHz (Not stable gain)
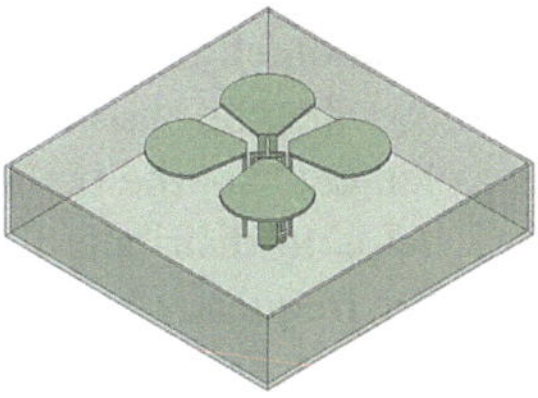 Antenna 2	2–4.7	80.6	10.2 @ 3.5 GHz
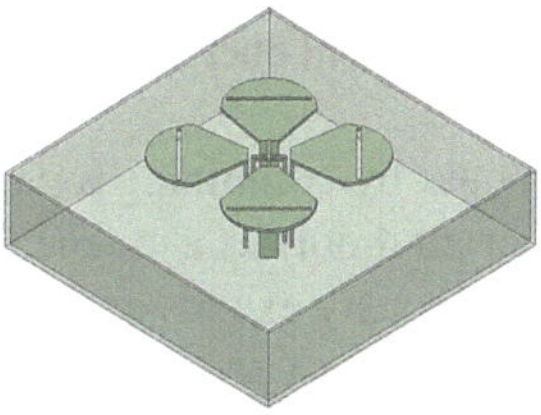 Antenna 3	2.02–5.26	88.7	10.4 @ 3.5 GHz

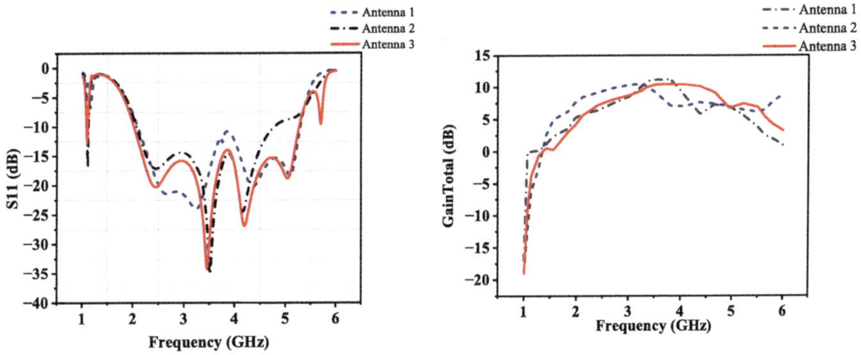

(a) (b)

Fig. 3 **a** Comparison graph for return loss and **b** comparison graph for gain

3 Results and Discussion

The proposed Antenna 3 is built and simulated using Ansys HFSS as shown in Fig. 1. From Fig. 4. It has been evident that the proposed antenna is operational from 2.02 GHz to 5.26 GHz with a wide impedance bandwidth of 88.7%. The antenna's broadside gain of 10.4 dB was remarkably consistent over its entire frequency band which makes the antenna ideal for use in several 5G applications (Wi-Fi and mobile). The incorporation of a slot in the antenna structure has the potential to enhance the bandwidth characteristics via modification of current distribution path and the introduction of additional resonant frequencies.

Figure 5 shows the surface current distribution of the proposed ME dipole with and without a slot, and it exhibits variations that affect both radiation efficiency and bandwidth. The surface current distribution without the slot shows that the maximum current is not distributed to the major part of the radiating element. So, the incorporation of the slot with a length slightly higher than $\lambda/2$ helps to distribute the current further and resonate an additional frequency at a higher frequency band of 4.78–5.26 GHz compared to the slot of $\lambda/2$ length.

The slot functions as a resonator, generating an E-field distribution as shown in Fig. 6, and perimeter current that enhance stable radiation as shown in Fig. 3b, thereby antenna's bandwidth has been enhanced by enabling efficient energy transfer at distinct frequencies.

As discussed earlier, the ME dipole antenna is constructed by combining a pair of two planar electric dipoles with a vertically oriented shorted dipole as a magnetic dipole that is fed by a Γ-shaped feeding strip which simultaneously excites both the dipoles. These electric and magnetic dipoles together produce time-varying electric and magnetic fields, leading to a complementary effect, characterizing electric and magnetic currents that are orthogonal and in phase differences of 90°. As a result, there is a minimal back lobe and symmetrical radiation patterns in the E- and H-planes as shown in Fig. 7.

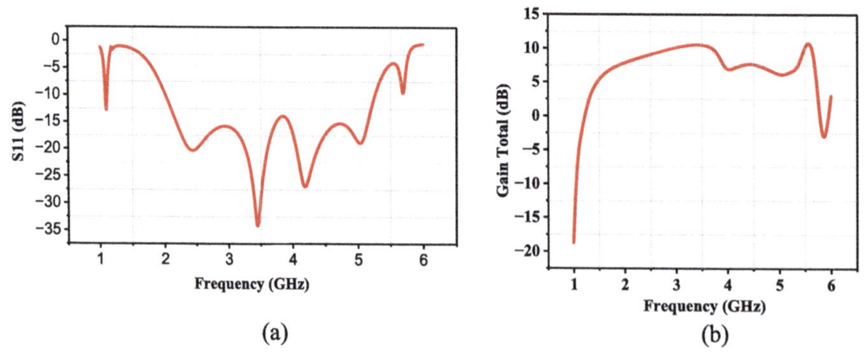

Fig. 4 **a** Return loss and **b** gain versus frequency plot

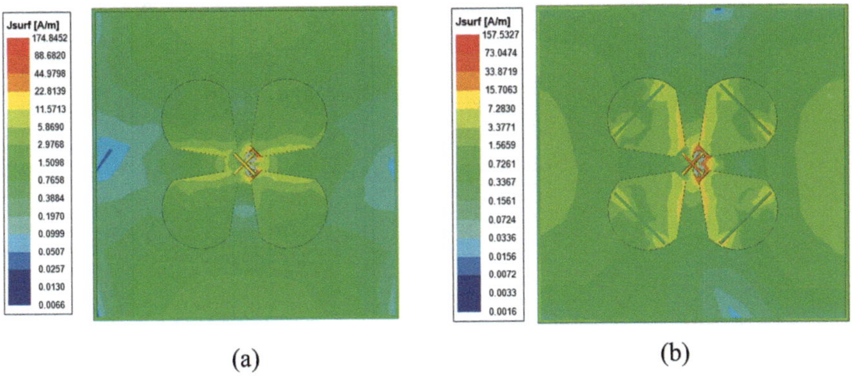

Fig. 5 Surface current distribution **a** without slot **b** with slot

Fig. 6 E-field distribution of the proposed antenna

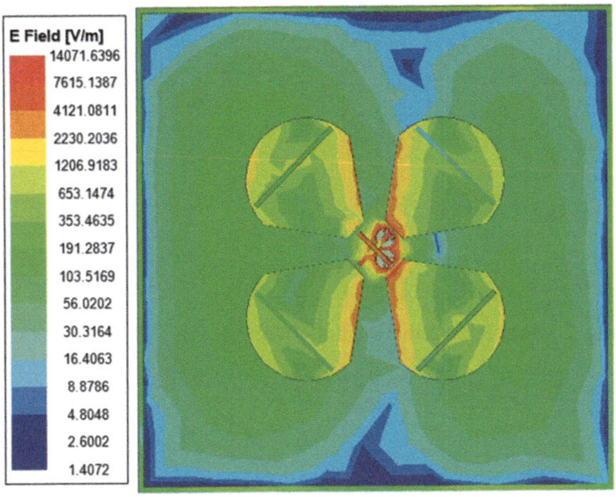

The antenna's radiation pattern was measured unidirectional over its whole bandwidth. The reflector used in this structure helps to obtain stable radiation for the proposed antenna. The electric dipole length has been chosen slightly lesser than λ/4 which strongly impacted the impedance matching, bandwidth, and stability of the antenna. The suggested antenna design is optimal for several 5G sub-6 GHz communication applications due to its wide bandwidth, high gain, and stable radiation. From Fig. 7, the simulated radiation pattern for both the E plane and H plane for multiple frequencies can be analyzed.

Table 2 gives a comparison of the dimensions, frequency range, bandwidth, and gain of the several antenna designs proposed recently. From the comparison, the proposed petal-shaped ME dipole antenna gives better results when compared to others, especially in terms of the compactness of the design and also with the gain and bandwidth performance.

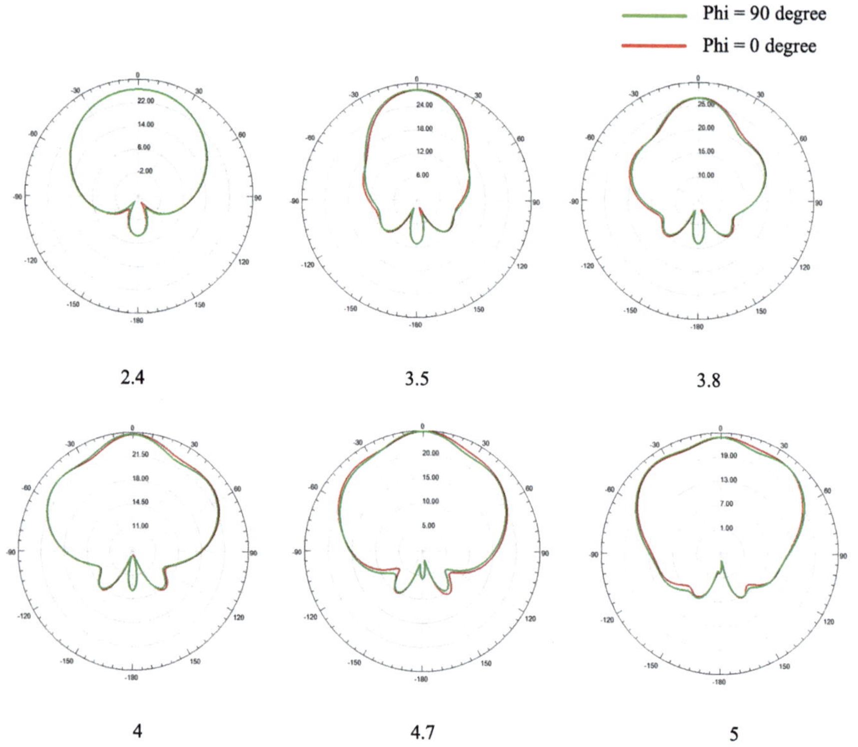

Fig. 7 Simulated radiation pattern at 2.4, 3.5, 3.8, 4, 4.7, and 5 GHz

Table 2 Comparison table on 5G sub 6 GHz application antennas

Ref	Antenna design	Dimensions (λ_0)	Frequency range (GHz)	Bandwidth (%)	Gain (dB)
Sun et al. [4]	Dual polarized ME dipole antenna	1.85 × 1.85 × 0.26	3.05–4.42	37	9.3
Li et al. [9]	ME dipole antenna	1.21 × 1.21 × 0.24	1.68–3.92	80	9.2
Zhu et al. [11]	ME dipole antenna with parasitic element	1.63 × 1.63 × 0.28	2.14–3.66 and 4.68–5.04	52.4 and 7.4	9.5
Yang et al. [12]	Dual layer ME dipole antenna	1.40 × 1.40 × 0.30	1.70–2.76 and 4.50–6.10	47.5 and 30.2	8.5
Zhang et al. [13]	Claw-shaped ME dipole antenna	1.62 × 1.62 × 0.17	1.18–3.72	103.6	7.6
Proposed work	Petal-shaped ME dipole antenna	1.18 × 1.18 × 0.27	2.02–5.26	88.7	10.4

4 Conclusion

In this paper, a compact ME dipole antenna has been presented in the form of a petal shape and improved its performance, especially for emerging 5G sub-6 GHz applications. When the Γ-shaped feeding structure is combined with the novel slot-loaded petal-shaped radiating element, the resulting impedance bandwidth and high gain are better compared to other designs. The antenna covers an operating frequency range of 2.02–5.26 GHz, and the impedance bandwidth resulted at 88.7%. The modeled antenna supports the wide range of frequencies required by 5G networks such as Wi-Fi and 5G mobile base stations. A broadside gain of 10.4 dB further demonstrates the antenna's stable radiation characteristics. This high gain is an essential factor for delivering sufficient signal strength and coverage in challenging environments.

References

1. Attaran M (2023) The impact of 5G on the evolution of intelligent automation and industry digitization. J Ambient Intell Humaniz Comput 14(5):5977–5993. https://doi.org/10.1007/s12652-020-02521-x
2. Salahdine F, Han T, Zhang N (2023) 5G, 6G, and Beyond: recent advances and future challenges. Ann Telecommun 78(9–10):525–549. https://doi.org/10.1007/s12243-022-00938-3

3. Xu Y, Luk K-M, Li A, Sun J (2021) A novel compact magneto-electric dipole antenna for millimeter-wave beam steering applications. IEEE Trans Veh Technol 70(11):11772–11783. https://doi.org/10.1109/TVT.2021.3114217
4. Sun JN, Li JL, Xia L (2019) A dual-polarized magneto-electric dipole antenna for application to N77/N78 band. IEEE Access 7:161708–161715. https://doi.org/10.1109/ACCESS.2019.2951414
5. Liu T, Zhu WJ, Hu XJ, Ye LH, Wu D-L (2023) Dual-wideband dual-polarized magnetoelectric dipole antenna for sub-6 GHz applications. IEEE Antennas Wirel Propag Lett 1–5. https://doi.org/10.1109/LAWP.2023.3243572
6. Li Y, Zhao Z, Tang Z, Yin Y (2020) Differentially fed, dual-band dual-polarized filtering antenna with high selectivity for 5G sub-6 GHz base station applications. IEEE Trans Antennas Propag 68(4):3231–3236. https://doi.org/10.1109/TAP.2019.2957720
7. Wu S, Shang F (2023) Broadband dual-polarized magnetoelectric dipole antenna with compact structure for 5G base station. IEEE Access 11:20806–20813. https://doi.org/10.1109/ACCESS.2023.3247821
8. Yang SJ, Pan YM, Zhang Y, Gao Y, Zhang XY (2019) Low-profile dual-polarized filtering magneto-electric dipole antenna for 5G applications. IEEE Trans Antennas Propag 67(10):6232–6243. https://doi.org/10.1109/TAP.2019.2925151
9. Li Z, Sun Y, Yang M, Wu Z, Tang P (2017) A broadband dual-polarized magneto-electric dipole antenna for 2G/3G/LTE/WiMAX applications. Progr Electromagnetics Res C 73:127–136. https://doi.org/10.2528/PIERC17022005
10. Fakhte S, Zahir Joozdani M, Zarifi D (2022) Magneto-electric dipole antenna for 5G. Print Antennas 5G Netw 109–151. https://doi.org/10.1007/978-3-030-87605-0_5
11. Zhu C, Wang B, Luo W, Hao H, Wang P (2020) Dual-wideband dual-polarised magnetoelectric dipole antenna for 4G/5G microcell base station. Electron Lett 56(6):269–271. https://doi.org/10.1049/el.2019.3535
12. Yang M, Zhou J, Lian W, Chen B (2020) Dual-band dual-polarized magneto-electric dipole antenna with dual-layer structure. Prog Electromagn Res M 92 193–202
13. Zhang X, Jiao Y, Weng Z, Zhang Y, Feng S (2019) Wideband magneto-electric dipole antenna with a claw-shaped reflector for 5G communication systems. Microw Opt Technol Lett 61(9):2098–2104. https://doi.org/10.1002/mop.31869
14. Mishra B, Verma RK, Yashwanth N, Singh RK (2022) A review on microstrip patch antenna parameters of different geometry and bandwidth enhancement techniques. Int J Microw Wirel Technol 14(5):652–673. https://doi.org/10.1017/S1759078721001148

A Novel Billhook-Shaped 2-Element MIMO Array for Sub-6 GHz Millimeter Wave Application

K. A. Ansal, Mercy Mathew, G. K. Ragesh, Jaimon Yohannan, and C. G. Jagadish

Abstract The effectuation of multiple input multiple output (MIMO) systems is severely hampered by mutual interaction between the various antenna configurations. This work depicts a novel billhook-shaped 2-element MIMO array with self-isolation for sub-6 GHz millimeter wave applications. A CPW fed-antenna uses FR4 substrate material with a thickness of 1.6-mm. This MIMO antenna uses defective coplanar ground plane, and spacing between two-patch elements is 3-mm. The two elements are positioned perfectly out of phase to mitigate mutual entanglement. The suggested MIMO antenna of $32 \times 67 \times 1.6\text{-mm}^3$ size has shown wide 10-dB impedance bandwidth (4–5.71 GHz), maximal gain of 5.63 dB, and a minimal self-isolation of $-$ 27 dB. Additionally, the MIMO diversity metrics like diversity gain (DG $\cong$ 10), mean effective gain (MEG $< -$ 3 dB), channel capacity loss (CCL $<$ 0.4 bits/s/Hz), and envelope correlation coefficient (ECC $\ll$ 0.01) are obtained and addressed. In comparison with mm wave, the proposed antenna for sub-6 GHz frequency waves suffers from lower penetration losses. Consequently, the higher version of it is preferable to install 5G networks in major cities.

Keywords MIMO · ECC · DG · CCL

K. A. Ansal (✉) · G. K. Ragesh
Indian Institute of Information Technology, Kottayam 686635, India
e-mail: ansal.pdf20221@iiitkottayam.ac.in

M. Mathew · J. Yohannan
Amaljyothi College of Engineering, APJKTU, Koovappally 686518, India

C. G. Jagadish
KMEA Engineering College, Cochin 683561, India

S. M. Thampi et al. (eds.), *Fifth International Conference on Computing and Network Communications*, Lecture Notes in Electrical Engineering 1219,
https://doi.org/10.1007/978-981-97-4540-1_6

1 Introduction

The data rate, range, and reliability of the communication link can all be enhanced using MIMO technology, which enables the use of several antennas for transmitting and receiving data by exploring transmit and receive diversity simultaneously. For most wireless communication applications like Wi-Fi, WiMAX, 3G, 4G, and the most potential 5G and beyond and future 6G, sub-6 GHz bands assigned were having pivotal role in the transmission and reception. MIMO antenna design development and characterization for multiuser connection are still difficult. The antenna isolation by reducing mutual coupling effects and miniaturization are crucial considerations.

The various state-of-the-art technologies have reported for enhancing isolation between antennas. Due to their ground plane and radiating patch structures, some antennas offer self-isolation. However, other antennas use a defective ground design approach, slots or slits in the radiating patch, or a combination of the two to accomplish isolation. The research on MIMO antennas mostly focuses on techniques for reducing mutual coupling. The diversity effect may be used to provide better isolation between antenna elements. Utilizing an external network or techniques like parasitic or slot elements, electronic band gap structures (EBG), metamaterial structures, complementary split ring resonators (CSRR), neutralization lines, defected ground structures (DGS), neutralization lines (NL), and dielectric resonators can reduce mutual coupling. [1–17].

Without the use of any additional isolation components or decoupling methods, an 8-port MIMO array system was designed. The system substrate has dimensions of $150 \times 75 \times 0.8\text{-mm}^3$. The self-isolated antenna provides maximum isolation characteristics of about − 20 dB with acceptable correlation performance and efficiency [2]. Four conducting radiators that are coupled to a common shared ground and printed on a flexible substrate make up the planar antenna structure [3]. The optimized array elements emit in the 2.37–5.85 GHz range with radiation efficiency of 84.64%. Self-isolation for the antenna is approximately − 17.5 dB with dimension of $145 \times 70 \times 0.2\text{-mm}^3$. A double dual-band monopole structure loaded with meandered ground plane with decoupling resonator makes up a dual-band antenna, which operates in the 2.4–2.48 GHz and 5.15–5.825 GHz bands. The meandering line offers isolation [4] on a FR4 with dimensions of $50 \times 50 \times 1.6\text{-mm}^3$. A 4-port MIMO antenna is discussed in [5] have a metallic disc with a circular shape loaded in the ground plane which is further extended by combining four single wideband elements to make it as 4-port MIMO antenna. This disc functions as a 180°-phase different pool of current that isolates separate ports. The proposed structure employs the disc for achieving port-to-port isolation which is less than – 20 dBi. To achieve a strong isolation, a rectangular micro-strip-based MIMO antenna utilizes a stub with a defect on the ground plane demonstrated in [6]. The ground planes are cross-connected which acts as a neutralization line for better isolation combined with a new decoupling element of the antenna made of I-shaped and II-shaped metal strips. The proposed antenna is designed in $55 \times 55 \times 1.6\text{-mm}^3$ dimension. The antenna's impedance bandwidth ranges from 3.28 to 3.80 GHz [7]. An eight-element MIMO antenna array with ground plane which is loaded by inverted L-shape strips gives strong isolation (> 18 dB) demonstrated for the frequency range of (3.82–7.5 GHz) [8]. An appreciable value of self-isolation is achieved for a triband antenna with dimension 60 × 60 × 0.79-mm^3 [9]. To obtain the desired gain and necessary impedance bandwidth characteristics, the suggested method uses a substrate loaded with four circularly polarized pentagonal-shaped patch elements and loaded slits. Antennas are placed orthogonal to achieve good isolation [10].

A mirrored F-shaped defected ground and three grounded stubs are used for achieving isolation [11]. The combined effects of stubs defective structure provide good isolation and various MIMO performance metrics for the demonstrated structure. Another design which employs decoupling with X-shaped strips is discussed in [14]. The strips give good isolation in the operating band. The H-shaped strips for achieving isolation are elaborated in [15], which also gives the enough mutual coupling reduction in the desired frequency range. Compact MIMO antenna for wireless applications is reported in [2, 12, 13] which gives appreciable performance metrics for the fifth-generation MIMO operation. A dual-port multiuser antenna array with wide band and good isolation is achieved in [16, 17]. For current wireless applications, a miniaturized multi-band multiuser antenna array with polarization diversity approach is reported [1, 9, 18–21]. In terms of several performance metrics, such as envelope correlation coefficient (ECC), mean effective gain (MEG), diversity gain (DG), and channel capacity loss (CCL), the antenna's diversity performance for multiuser communication is studied and reported.

A novel two-port billhook-shaped MIMO with the dimensions 32 × 67 × 1.6-mm^3 is designed in the present study. The proposed structure provides good self-isolation without the aid of additional components. This antenna exhibits a good reflections and radiation characteristics with two resonance frequencies with wide bandwidth and successfully achieves mutual coupling reduction without any isolation technique. The analysis of the diversity performance ECC, CCL, DG, and MEG in the Ansys electronic desktop 2022 solver shows results that are comparable to standard values.

2 Billhook-Shaped Antenna Design

The following section discusses the antenna design evolution, geometry, various parametric results, and implications based on the S parameter, gain, radiation efficiency, radiation patterns, and various performance metrics with reference to the MIMO requirements.

2.1 Single-Element Antenna Geometry

The suggested coplanar waveguide-fed antenna is constructed on a 1.6-mm-thick FR4 substrate. This antenna utilizes substrate which has a dielectric permittivity

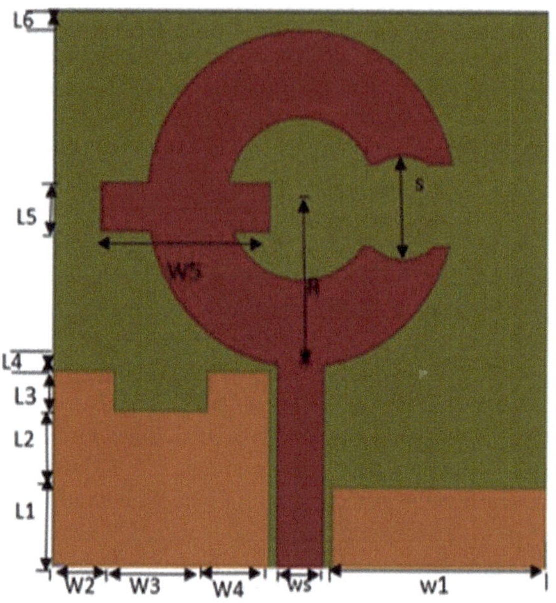

Fig. 1 Geometry of proposed one-element antenna

value of 4.4 and a loss tangent factor of 0.02. Overall size of the proposed billhook structure is of 32 × 33 × 1.6-mm^3 shown in Fig. 1. For obtaining bandwidth and good gain, a defective asymmetric coplanar ground and a circular patch with stubs and slits are inserted on the design.

2.2 *Design Evolution of Single-Element Billhook Antenna*

A circular-shaped patch antenna with asymmetrical ground and a coplanar waveguide (CPW) feed initially designed, and the various design iterations for the evolution of the demonstrated structure depicted in Fig. 2 are used to maximize bandwidth and gain characteristics of the proposed structure. A defected coplanar ground which is different from conventional ground plane results in bandwidth and gains improvement. The bandwidth response is good in step 1 and step 2, but it exhibits poor gain performances, and in step 3 the structure gives good bandwidth and reflection characteristics but fail to obtain the desirable gain. But additional stub loaded in center portion of billhook structure is made to boost gain. The inclusion of a stub and the curve of the slit mentioned in step 4, represents the proposed antenna structure provides bandwidth of 3.93–5.73 GHz with two prominent resonances. The reflection characteristics, $S_{11,}$ of the proposed prototype in their design iteration phases are given in Fig. 3a. The various parametric attributes of the suggested structure are given in Table 1. From the table its clear that the suggested single-element antenna has dimension of 32 × 33-mm^2 with substrate thickness of 1.6-mm.

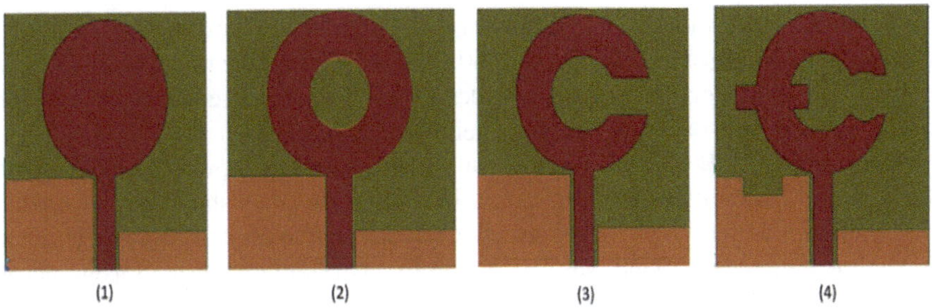

Fig. 2 Design iterations of the proposed antenna

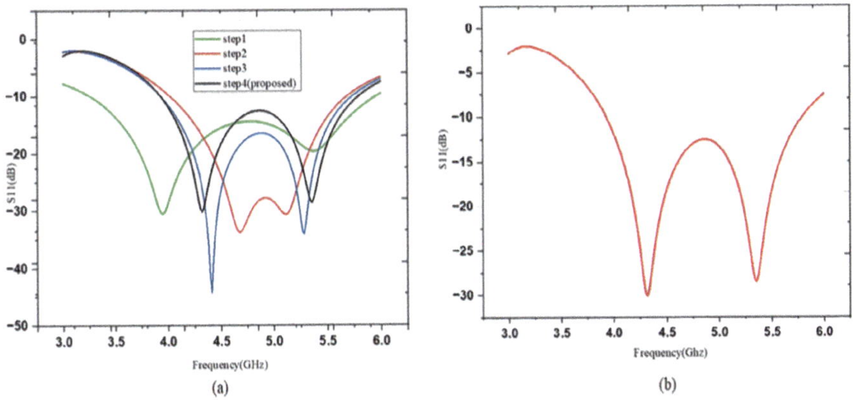

Fig. 3 Reflection characteristics for **a** design evolution of proposed antenna; **b** proposed antenna

Table 1 Dimensional parameters in mm

Parameter	Dim (mm)	Parameter	Dim (mm)
$L1$	5	$W1$	14
$L2$	5	$W2$	4
$L3$	2.3	$W3$	6
$L4$	0.3	$W4$	4
$L5$	3	$W5$	6.25
$L6$	1	$W6$	3
S	4.5	R	10

2.3 Results and Discussion of Single-Element Billhook-Shaped Antenna

The simulated reflection attribute of single-element billhook-shaped antenna is shown in Fig. 3b. There are two resonant frequencies present, one at 4.3 GHz and the other at 5.3 GHz with a good reflection of − 29 dB and − 28 dB, correspondingly. The antenna exhibits a bandwidth of 1.860 GHz in its operating range of sub-6 GHz band. The proposed coplanar waveguide design provides wide frequency band which allows the desirable date rate of fifth-generation deployments.

The simulated 2D and 3D patterns on the antennas are shown in Fig. 4. From the figure, it's evident that the pattern in azimuth and elevation shows the acceptable characteristics with little tilt on left direction which is approximately omni in H-plane and figure of eight in E-plane. The asymmetricity in the pattern due to its asymmetric ground plane is little different from the conventional coplanar waveguide structure. The defected asymmetric ground plane at the left side provides the good impedance bandwidth for the proposed billhook-shaped antenna configuration. The 3D patterns of the projected structure for dual resonances at 4.3 GHz and 5.3 GHz. demonstrate that the gain of the antenna for lower resonance is about 1.6dBi and the higher resonance is about 4.49dBi. The legends show the electric filed strength of proposed antenna when it radiates. The tuned stub which is in the billhook-shaped radiating element provides the enhanced gain during the upper resonant frequency which is highly acceptable for fifth-generation base station equipment.

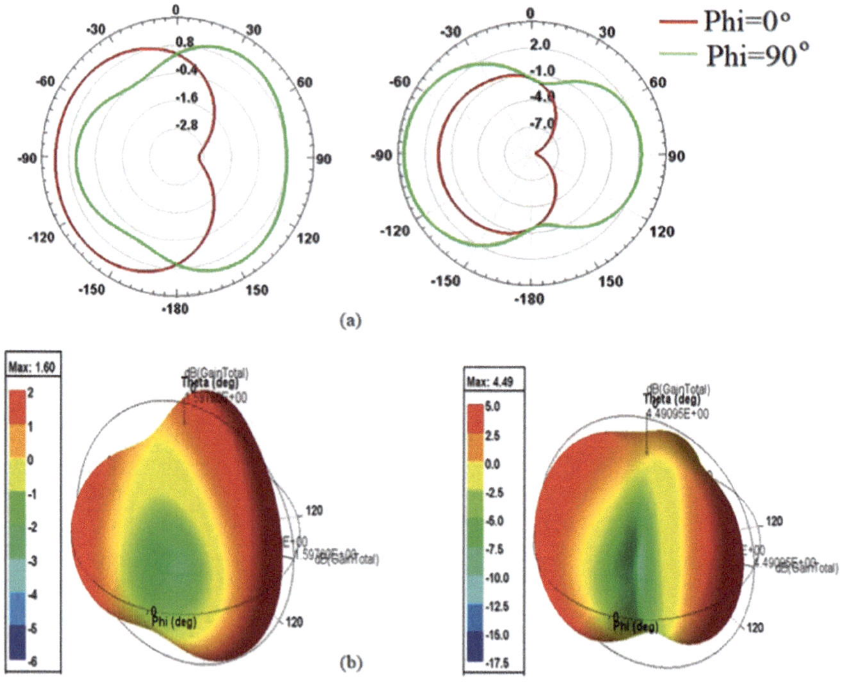

Fig. 4 2D and 3D patterns of 1 element suggested antenna **a** @ 4.3 GHz, 5.3 GHz; **b** @ 4.3 GHz, 5.3 GHz

3 Two-Element MIMO Antenna Geometry

3.1 Results and Discussions of Two-Element Antenna

This section elucidates the configuration of 2-element multiuser antenna which is aligned by two single-element antennas that are placed exactly out of phase with one another to the radiating patch elements placed by a 3-mm gap. The two-element array's dimensions are $32 \times 67 \times 1.6\text{-mm}^3$. Figure 5a depicts the two-element MIMO array structure. The proposed two-element structure gives better self-isolation for the discussed sub-6 GHz frequencies. The various performance metrics for MIMO antennas along with reflection radiation characteristics and diversity performance were analyzed in the following sections.

The impedance bandwidth obtained for proposed MIMO antenna is from 4 to 5.7 GHz and is illustrated in Fig. 5b. It is evident from the S [11] reflection characteristics that the first resonance frequency is at 4.3 GHz with a reflection of − 34 dB and the second is at 5.25 GHz with a return loss of − 32 dB. According to the transmission characteristics S [1], the isolation is at least − 20 dB at 4.4 GHz and − 27 dB at 5.3 GHz. The self-isolation values are higher than that of similar literatures reported in Table 2. The S21 characteristics prove that the proposed structure has good self-isolation characteristics, which is approximately doubles the theoretical value of isolation needed for MIMO performances.

The two-dimensional and three-dimnesional radiation pattern of the two-element array plotted in Fig. 6. At dual resonance frequencies, the gain about 2.72 dBi and 5.63 dBi, respectively, is well improved from single-element structure. This gain is

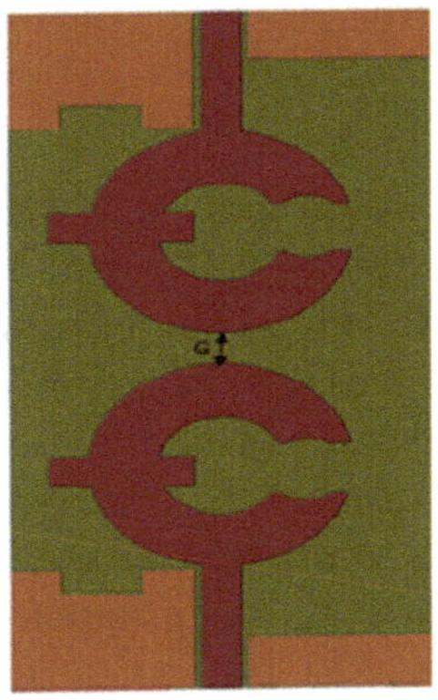

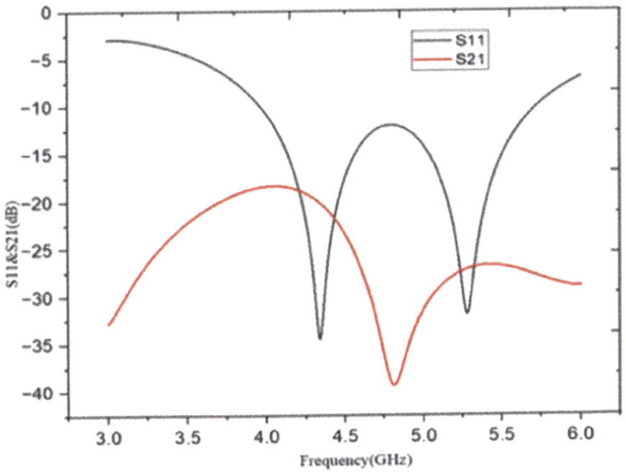

Fig. 5 **a** 2 element billhook antenna arrays; **b** reflection characteristics of two-element billhook MIMO antenna

Table 2 Comparison of suggested prototype with existing configuration

Ref No.	Size (mm)	Bands (GHz)	Gain (dBi)	Isolation (dB)	ECC	DG (dB)	MEG (dB)
[2]	80 × 80 × 1.6	2.2/3.3	3	− 15	0.016	9.6	–
[3]	70 × 145 × 0.2	2.3–5.8	4.5	− 17.5	0.05	9.8	–
[5]	50 × 50 × 1.6	3.4–3.8	–	− 11	0.009	–	–
[12]	120 × 65 × 1.6	3.3–5	4.71	− 18.8	0.018	–	–
[2]	80 × 80 × 1.6	1.5–5	3	− 15	0.016	~10	–
[13]	61 × 61 × 1.6	3.44–5.57	–	− 20	0.001	9.85	–
[14]	70 × 60 × 1.6	2.40/3.35	1.2	–	0.01	9.6	–
[18]	70 × 70 × 1.6	2.54/5.26	3.98/4.13	− 21/9	Low	–	< − 3
[19]	30 × 66 × 1.6	3.3–5	–	− 18	0.05	~10	< − 3
[1]	59 × 55 × 1.6	3–7	5.6	− 18		9.64	< − 3
[9]	60 × 60 × 0.79	2.2–3.45	4.25	− 20	0.0001	9.998	< − 5
		4.9–6.1	5				
		7.6–9.8	5.5				
Proposed work	32 × 67 × 1.6	3.9–5.7	5.63	− 27	0.0014	9.98	< − 3

comparatively better in proposed 2 element arrays at these frequencies for MIMO performance. The azimuthal and elevation 2D pattern is also plotted in Fig. 5a.

The gain and efficiency versus frequency of both single-element and dual-element antennas is plotted in Fig. 7a, b, respectively, from it can be seen that the proposed single structure exhibits the maximal gain of 4.49 dBi and maximal radiation efficacy of 94%. In the operating band, proposed structure provides more than 90% of average efficiency. The 2-element structure provides peak gain of 5.63 dB, and the entire operating range its average radiation efficiency is greater than 90%. The mentioned values are acceptable for the desired efficacy.

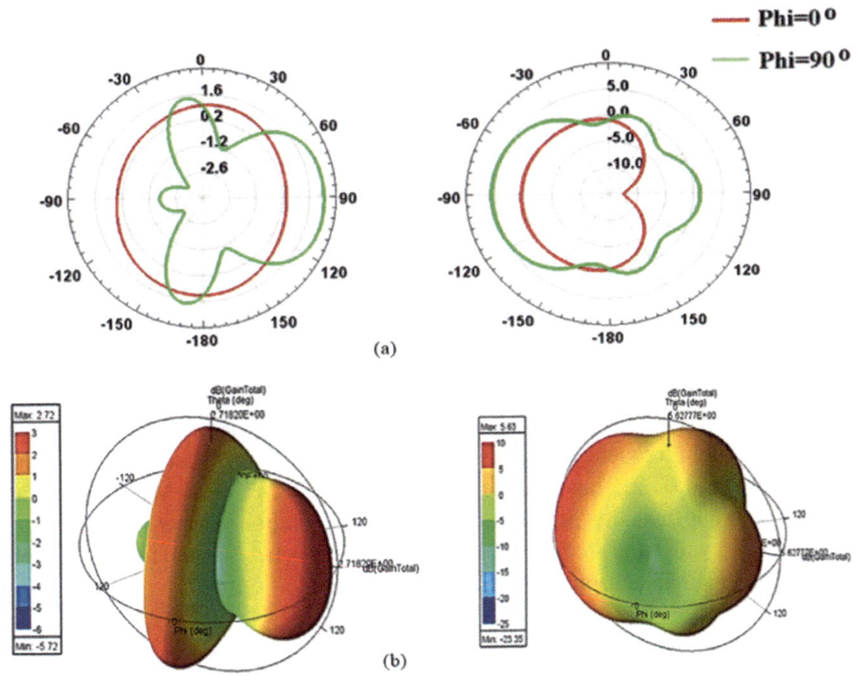

Fig. 6 2D and 3D pattern of 2 element suggested antenna **a** @ 4.3 GHz and 5.3 GHz; **b** @ 4.3 GHz and 5.3 GHz

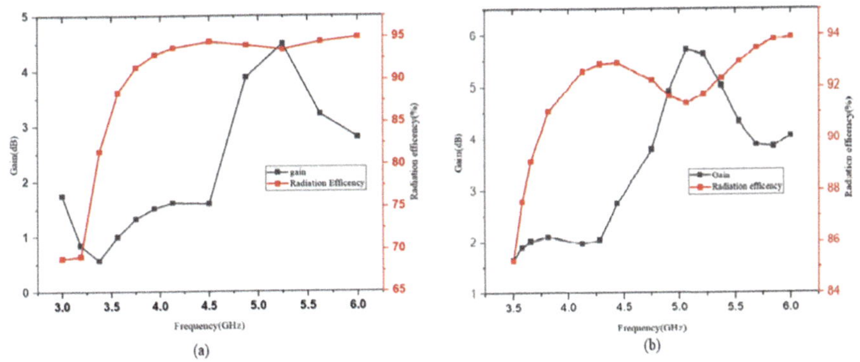

Fig. 7 Gain and radiation efficiency versus frequency: **a** single element; **b** dual element billhook

3.2 Diversity Performance Characteristics

Envelope Correlation Coefficient (ECC)

This parameter shows how much the communication channel is isolated when transmit simultaneously. ECC has a practical acceptable value of 0.5 in a free space. ECC for a two-port antenna in terms of port-to-port isolation can be calculated from S parameter in Eq. (1), [1, 9, 18–21].

$$\text{ECC} = \frac{\left|s_{11}^{*}s_{12} + s_{21}^{*}s_{22}\right|^{2}}{\left(1 - |s_{11}|^{2} + |s_{21}|^{2}\right)\left(1 - (|s_{22}|^{2} + |s_{12}|^{2}\right)}. \tag{1}$$

where S_{11} and S_{22} are reflection coefficients and S_{21} and S_{12} are transmission coefficients. The whole frequency spectrum for this billhook-shaped antenna displays an ECC of less than 0.02. This indicate that the proposed 2-element MIMO structure provides enough diversity performance for fifth-generation devices. When calculating ECC for wideband antennas, the far-field pattern is favored and for this antenna the value is less than acceptable case.

Diversity Gain (DG)

For good quality and dependability, wireless systems require a high DG of the multiple antenna design, with a value close to 10 dB in the operational bandwidth. Calculating the DG based on the ECC value is given in Eq. (2). The ECC and DG responses are plotted against operating frequency in Fig. 8, and the diversity gain of this antenna in the frequency band is close to 10 dB and the ECC characteristics are well below 0.02 which is appreciable for the desired operation

$$\text{DG} = 10\sqrt{1 - |\text{ECC}|^{2}} \tag{2}$$

Mean Effective Gain (MEG)

The power reception of any wireless system in a fading environment can be measured using mean effective gain (MEG). The MEG is the difference between the power received by a diversity antenna and an isotropic antenna, and it is a critical component in the analysis of multiple antenna designs [1, 9, 18–21]. According to Eq. (3) the difference in MEGs in dB should be close to unity and should not be larger than 3 dB for increased diversity performance [1, 9, 18–21]. In Fig. 8a, MEG 1 and MEG 2 are clearly below − 3 dB range (Fig. 9).

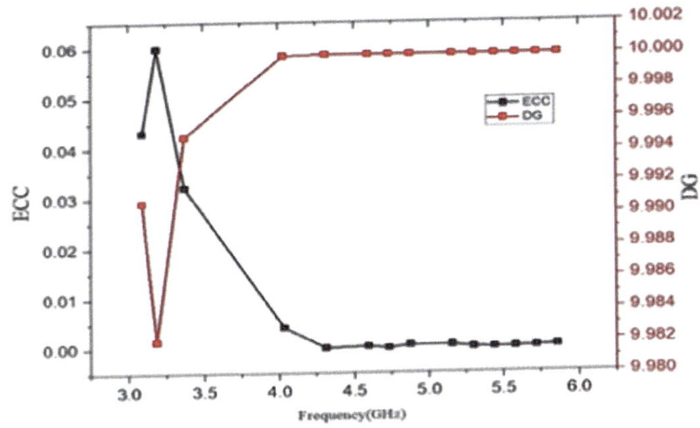

Fig. 8 ECC and DG response of billhook antenna

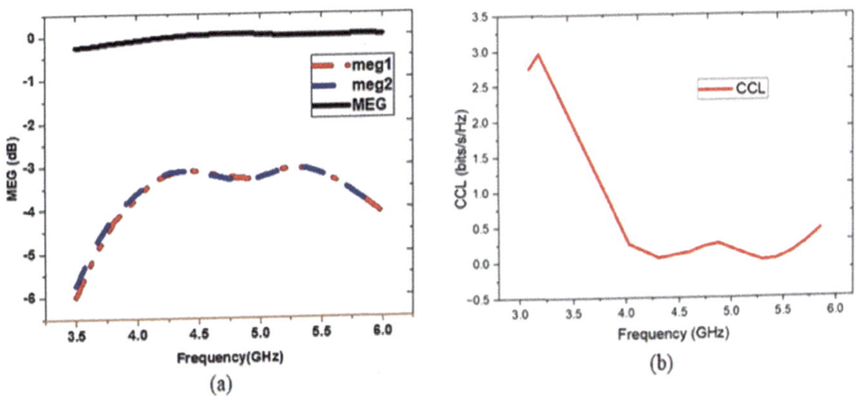

Fig.9 **a** MEG versus frequency; **b** CCL versus frequency

$$\mathrm{MEG}_i = 0.5\left[1 - \sum_{j=1}^{M} |S_{ij}|^2\right]$$

$$\mathrm{MEG}_1 = 0.5\left(1 - |S_{11}|^2 - |S_{12}|^2\right)$$
$$\mathrm{MEG}_2 = 0.5\left(1 - |S_{21}|^2 - |S_{22}|^2\right) \quad (3)$$

Channel Capacity Losses (CCL)

Channel capacity losses (CCL) in MIMO antenna systems arise by correlation losses. The CCL parameter specifies the quantity of bits lost during transmission. Using S-parameters, it is computed. Acceptable CCL values are those smaller than 0.4 bits/

s/Hz [1, 9, 18–21]. CCL can be calculated by Eq. (4). In Fig. 8b, the computed CCL is displayed. It is considerably below than 0.4 bits/s/Hz, assuring that the suggested multiuser array may transmit data at higher speeds

$$\mathrm{CCL} = -\log_2\left|\varphi^R\right| \tag{4}$$

$$\varphi^R = \begin{bmatrix} \varphi_{ii} \ \varphi_{ij} \\ \varphi_{ji} \ \varphi_{jj} \end{bmatrix}$$

$$\varphi_{ii} = 1 - \left(|s_{ii}|^2 + |s_{ij}|^2\right)$$

$$\varphi_{jj} = 1 - \left(|s_{jj}|^2 + |s_{ji}|^2\right)$$

$$\varphi_{ij} = -\left(s_{ii}^* s_{ij} + s_{ji}^* s_{jj}\right)$$

$$\varphi_{ji} = -\left(s_{jj}^* s_{ji} + s_{ij}^* s_{ii}\right)$$

3.3 Comparison of Suggested Structure with Similar Literatures

Table 2 provides the contrast between size, operating band of frequencies, gain, isolation, and various performances metrics for the MIMO characterization of proposed two-element arrays with existing similar literatures. The proposed antenna outperforms well in contrast with various analogous described in the table. When analyzing with reference values avilable in the liturature, the parameters like antenna dimension, gain, ECC, DG and MEG, make the proposed structure is suitable for sub-6GHz wireless communication. In comparison also seen that the proposed antenna utilizes a FR4 substrate with minimal dimensions gained the MIMO performance metrics and miniaturization trade off required for 5G and beyond.

4 Conclusion

A novel billhook-shaped two-element multiuser array for mid-band millimeter wave application was designed. This structure has an impedance operating bandwidth range of 4–5.7 GHz frequency range. The two-element antenna provides a maximal gain of 5.63 dB and self-sequestration of − 27 dB at that frequency range. The diversity performances are good for this antenna, its ECC value is very less, 0.0014, DG value is close to 10 dB, MEG1 and MEG2 value shows − 3 dB performance, and CLL is also less than acceptable limit. The mutual coupling of the antenna can be downgraded by loading metamaterials or slots in suggested antenna. In future, this antenna can be measured and result can be analyzed with simulation results. Wi-Fi 5 and the ns77/78 band work well with this structure.

Acknowledgements Authors greatly acknowledged the Kerala State Higher Education Council and Abdul Kalam Technological University, Kerala, for the fellowship and technical support provided for the work presented.

References

1. Kumari T et al (2019) Design approach for dual element hybrid MIMO antenna arrangement for wideband applications. Int J RF Microwave Comput Aided Eng 29(1):e21486
2. Zhao A, Ren Z (2018) Size reduction of self-isolated MIMO antenna system for 5G mobile phone applications. IEEE Antennas Wirel Propag Lett 18(1):152–156
3. Kulkarni J, Alharbi AG, Desai A, Sim C-Y-D, Poddar A (2021) Design and analysis of wideband flexible self-isolating MIMO antennas for sub-6 GHz 5G and WLAN smartphone terminals. Electronics 10:3031
4. Deng J, Wang Z, Li J, Guo L (2018) A dual-band MIMO antenna decoupled by a meandering line resonator for WLAN applications. Microw Opt Technol Lett 60:759–765
5. Saxena S, Kanaujia BK, Dwari S, Kumar S, Tiwari R (2018) MIMO antenna with built-in circular shaped isolator for sub-6 GHz 5G applications. Electron Lett 54:478–480
6. Yang R, Xi S, Cai Q, Chen Z, Wang X, Liu G (2021) A compact planar dual-band multiple-input and multiple-output antenna with high isolation for 5G and 4G applications. Micromachines 12:544
7. Jamal MY, Li M, Yeung KL (2020) Isolation enhancement of closely packed dual circularly polarized MIMO antenna using hybrid technique. IEEE Access 8:11241–11247 (2020)
8. Li Q, Wei Y, Tan M, Lei X, Wu G, Huang M, Gong Y (2019) Flexibly extensible planar self-isolated wideband MIMO antenna for 5G communications. Electronics 8:994
9. Hussain M, Awan WA, Alzaidi MS, Elkamchouchi DH (2023) Self-decoupled tri band MIMO antenna operating over ISM, WLAN and C-band for 5G applications. Heliyon 9(7):e17404
10. Bhadade RS, Mahajan SP (2019) Circularly polarized 4 × 4 MIMO antenna for WLAN applications. Electromagnetics 39(5):325–342
11. Iqbal A, Saraereh OA, Ahmad AW, Bashir S (2018) Mutual coupling reduction using F-shaped stubs in UWB-MIMO antenna. IEEE Access 6:2755–2759
12. Biswas A, Gupta VR (2020) Design and development of low-profile MIMO antenna for 5G new radio smartphone applications. Wirel Pers Commun 111:1695–1706
13. Ashar D, Yadav AKS, Sharma M (2022) Design and simulation of compact mimo antenna for the 5G communication in C band. Phys: Conf Ser 2312:012032
14. Zou X, Wang G, Wang Y, Li H (2019) Decoupling antenna array with X-shaped strip. Int J RF Microw Comput Aided Eng 29:1–7
15. Li Y, Zhao Z, Tang Z, Yin Y (2019) Differentially fed, dual-band filtering antenna with H-shaped slot. Microw Opt Technol Lett 62:448–452
16. Urimubenshi F, Konditi DB, de Dieu Iyakaremye J, Moukala Mpele P, Munyaneza A (2021) A novel approach for low mutual coupling and ultra-compact Two Port MIMO antenna development for UWB wireless application. Heliyon 8
17. Kulkarni JS, Desai AH, Sim C (2021) Two port CPW-fed MIMO antenna with wide bandwidth and high isolation for future wireless applications. Int J RF Microw Comput-Aided Eng 31
18. Malviya L, Panigrahi RK, Kartikeyan MV (2016) A 2 × 2 dual-band mimo antenna with polarization diversity for wireless applications. Progr Electromagn Res C 61:91–103
19. Gollamudi NK, Narayana YV, Prasad AM (2022) A novel cow-head shaped multiple input multiple output antenna for 5G sub: 6 GHz N77/N78 & N79 bands applications. Progr Electromagn Res C 122:83–93

20. Manteghi MM, Rahmat-Samii Y (2005) Multiport characteristics of a wideband cavity backed annular patch antenna for multipolarization operations. IEEE Trans Antennas Propag 53(1):466–474
21. Chae SH, Oh S-K, Park SO (2007) Analysis of mutual coupling, correlations, and TARC in WiBro MIMO array antenna. IEEE Antennas Wirel Propag Lett 6:122–125

Energy-Efficient Transmission in Wireless Sensor Network Using Compressive Sensing

Ebin M. Manuel and A. P. Darshana

Abstract Minimizing energy consumption and maximizing network longevity through efficient data transmission are open research problems in practical wireless sensor networks. The proposed work focuses on data transmission based on two methods: Compressive Sensing (CS) and Partial Canonical Identity matrix-based Compressive Sensing (PCI-CS). The performance of these two methods on different energy-efficient clustering and routing protocols, such as Low Energy Adaptive Clustering Hierarchy (LEACH), Stable Election Protocol (SEP), and Threshold sensitive Energy-Efficient Sensor Network (TEEN) is compared. The combination of CS or PCI-CS, along with these energy-efficient clustering and routing protocols, can immensely reduce energy consumption. Simulation results show that the PCI-CS method outperforms other methods in terms of energy efficiency as well as network longevity.

1 Introduction

Wireless Sensor Networks (WSNs) contribute to the rapid advancement in the Internet of Things (IoT) worldwide. With the evolution of the next-generation IoT networks, numerous sensors are being deployed for various applications such as monitoring environments, tracking objects, and controlling industrial operations. The data collected from all these sensor units has to be fetched to the base station (BS), where the aggregation and extraction of information are performed. In many applications, these sensors are battery-operated, and replenishing or replacing the batteries is not viable. Therefore, energy-efficient methods are designed to reduce their energy consumption and extend the network's longevity.

E. M. Manuel (✉)
Department of Electronics and Communication, Rajiv Gandhi Institute of Technology Kottayam, Kottayam, India
e-mail: ebin@rit.ac.in

A. P. Darshana
Department of Electronics and Communication, College of Engineering Trivandrum, Trivandrum, Kerala, India

S. M. Thampi et al. (eds.), *Fifth International Conference on Computing and Network Communications*, Lecture Notes in Electrical Engineering 1219,
https://doi.org/10.1007/978-981-97-4540-1_7

Past studies focused on developing different clustering and routing protocols that are energy-efficient [1]. In many such approaches, the geographical region of interest was divided into several clusters, and cluster-based techniques were designed to reduce energy consumption. A single node within each cluster acts as a cluster head (CH). All the nodes within a cluster send the sensed data to their respective CH, which then sends these data to the BS. Alternatively, some sensor devices can be designated to transmit data to the BS while other sensors transmit the data to their nearby designated sensor. As data transmission is the major cause of energy depletion of the sensors, efficient data transmission schemes are to be deployed in the next-generation IoT networks to reduce energy consumption and extend the network life. To further reduce energy consumption, another method, namely PCI-CS [2], which is a modification of CS, is used.

In this work, we focus on Compressive Sensing (CS) to achieve the objective of minimizing the data transmission to reduce energy depletion. Further, we design a variant of the Partial Canonical Identity matrix-based Compressive Sensing (PCI-CS) to reduce the data transmission. The performance of these two methods is compared with established routing protocols such as LEACH [3], SEP [4], and TEEN [5]. Simulation results show that PCI-CS could significantly reduce energy consumption and thus prolong the network longevity.

Minimizing energy consumption and maximizing network longevity through efficient data transfer are active problems in next-generation IoT networks. This paper relies on data transfer based on two methods: Compressive Sensing (CS) and Partial Canonical Identity matrix-based Compressive Sensing (PCI-CS) [6]. The performance of these two methods is compared with established routing protocols such as LEACH, SEP, and TEEN. The combination of CS or PCI-CS, along with these energy-efficient clustering and routing protocols, can immensely reduce energy consumption. Simulation results show that the PCI-CS technique performs more energy-efficient compared to other methods leading to an improved network lifetime.

This paper is organized as follows. Section 2 describes the related works on CS, PCI-CS, and different clustering and routing protocols. Sections 2.1, 2.2 and 2.3 briefly describe the network model, the clustering and routing protocols, and CS. Section 3 describes the PCI-CS scheme. Section 4 describes the proposed method. Section 5 gives the simulation results and discussions related to this work.

2 Preliminaries

2.1 Network Model

Let N number of sensors be randomly deployed over a square area. BS is also deployed within this area itself. The network is divided into several clusters where each cluster contains a CH and its corresponding members of the cluster. CHs receive the data sent by the cluster members and forward the aggregated data to the BS.

The energy consumption model used in this paper is adopted from [7], and the energy consumed while transmitting an l-bit data over a distance d is given by:

$$E_{\mathrm{Tx}}(l,d) = \begin{cases} lE_{\mathrm{elec}} + l\epsilon_{\mathrm{fs}}d^2, & d < d_0 \\ lE_{\mathrm{elec}} + l\epsilon_{\mathrm{mp}}d^4, & d \geq d_0 \end{cases} \tag{1}$$

and the energy consumed while receiving an l-bit data is given by:

$$E_{\mathrm{Rx}}(l) = lE_{\mathrm{elec}} \tag{2}$$

where d is the distance between the transmitter node and the receiver node, d_0 is the threshold.

The proposed method uses the following set of values: Electronics energy, E_{elec} = 50 nJ/bit, ϵ_{fs} = 10 pJ/bit/m^2, ϵ_{mp} = 0.0013 pJ/bit/m^4 and the energy for data aggregation, E_{DA} = 5 nJ/bit/signal.

2.2 *Clustering and Routing Protocols*

Some of the energy-efficient clustering and routing protocols are briefly described in this section.

LEACH protocol

LEACH is a cluster formation technique that enables self-organization of the sensor nodes deployed and allows CH rotation to distribute the energy load evenly among all the nodes. The operation of the LEACH protocol is explained in rounds and can be divided into two steps: the set-up step and the steady-state step. During the set-up step, the formation of the clusters take place, while during the steady-state step, the CHs receive data from their corresponding cluster members and send the data to the BS.

Let each node become a CH with probability p_{opt} and k denotes the optimal number of clusters per round, in the CH designation phase of the set-up step. Then, every node becomes a CH once in every $1/p_{\mathrm{opt}}$ rounds [8]. Each node n in the network chooses a random number between 0 and 1, which is compared with a threshold given by:

$$T(n) = \begin{cases} \frac{p_{\mathrm{opt}}}{1-p_{\mathrm{opt}}(k \mod \frac{1}{p_{\mathrm{opt}}})}, & n \in G \\ 0, & \text{otherwise} \end{cases} \tag{3}$$

where the present round is represented by k, and the sensors that were not CHs in the previous rounds are included in the set G. The sensor n is to be designated as a CH if $T(n)$ is lower than the threshold. Each CH then sends an advertisement message to other nodes, which contains the sensor's header and identifier. The non-CH nodes find their nearest CH and send a joint request message back to the CH. In the steady-state phase, data transmission occurs, and the time slots are set frame-wise. It is only in frames the data is sent from the cluster members to their CHs.

SEP

SEP is also known as heterogeneous aware protocol. SEP used the concept of a WSN with nodes having different energies. Here, two types of nodes exist: normal nodes and advanced nodes. Advanced nodes are those having higher energy compared to normal nodes. Let m denote the fraction of advanced nodes in the network, and a denote the additional energy factor. Out of the N sensor nodes in the network, Nm denotes the cardinality of forward sensors, and $N(1-m)$ denotes the cardinality of usual sensors. If E_I denotes the initial energy of each usual sensor node, $E_\text{I}(1+f)$ denotes the initial energy of the forward sensors. Thus, the total energy of the network is given as:

$$E_\text{total} = N(1-m)E_\text{I} + NmE_\text{I}(1+u) = NE_\text{I}(1+fm) \tag{4}$$

The SEP protocol is designed such that the forward sensors are given a better chance to act CHs compared to the usual nodes. The probabilities with which the sensors are chosen as the usual and forward sensors are as follow:

$$p_\text{nrm} = \frac{p_\text{opt}}{1+fm} \tag{5}$$

$$p_\text{adv} = \frac{p_\text{opt}}{1+fm} \times (1+f) \tag{6}$$

Both normal and forward nodes generate random numbers, which are compared with their respective thresholds given by:

$$T(S_\text{nrm}) = \begin{cases} \frac{p_\text{nrm}}{1-p_\text{nrm}(r \mod \frac{1}{p_\text{nrm}})}, & n \in G' \\ 0, & \text{otherwise} \end{cases} \tag{7}$$

$$T(S_\text{adv}) = \begin{cases} \frac{p_\text{adv}}{1-p_\text{adv}(r \mod \frac{1}{p_\text{adv}})}, & n \in G'' \\ 0, & \text{otherwise} \end{cases} \tag{8}$$

where G' and G'' denote usual and forward sensors that are yet to be theCHs in the previous rounds, respectively. If the generated seed for a sensor is lower than the respective threshold value, that sensor will become CH.

TEEN

TEEN is the first protocol developed for reactive networks where the sensors respond to the variations of the acquired sensor readings, and therefore appropriate for real-time applications for which the time is critical. In this protocol, at every cluster change time, the CHs broadcast to its members two different threshold values, one acts as a hard while the other is a soft.

The acquired sensor readings are governed by a hard threshold. when the threshold value is exceeded, the matter is transmitted to the corresponding CH. The sensor transmits the sensor reading to its CH if the reading exceeds the soft threshold.

The sensors are sensing the geographical region continuously. When a sensor node senses a value greater than H_T for the first time, that particular node transmits the sensed data stored in an internal variable called sensed value (SV). Suppose the sensor node has sensed another value. The node will transmit this current sensed value only if the current value of the sensed attribute exceeds H_T and differs from the SV by a value greater than or equal to S_T. The current value of the sensed attribute will be stored in SV only after the node transmits the current value.

As mentioned above, the data transfer in the network by fixing a range over which only the data is transmitted is reduced by the hard threshold. Similarly, the non-CH to CH transmissions are reduced by appropriately setting the soft threshold, so that when the change in the previous and current sensed value of the attribute falls below the value of the soft threshold, the transmissions are aborted.

2.3 *Compressive Sensing*

Compressive Sensing means compressing at the time of sensing itself. In CS, the data is sampled much below the Nyquist sampling frequency.

Suppose the WSN consists of N sensor nodes. So, the data sensed by the sensor nodes x is of length N and can be represented as $x = [x_1, x_2 \dots x_N]^{\mathrm{T}}$. The ith sensor node data is represented as x_i. According to CS, a k-sparse signal x, can be retrieved from a few linear combinations of x upon satisfying certain conditions [9]. A signal x is said to be k-sparse if there exist only $k(k \ll N)$ non-zero entries in x. Since x is k-sparse, instead of sending N data to the BS, only some projections of x whose cardinality is lesser can be sent to the BS, which is given by [9]:

$$y = \phi x \tag{9}$$

where ϕ is called the sensing matrix and the vector y consists of M measurements which are projections of x. The sensing matrix can be any random matrix, such as the Gaussian or Bernoulli matrix. After y is collected at the BS, the real data x can be retrieved by exploring the minimization of the l1-norm approaches or via other sparse signal recovery algorithms.

Suppose the signal x is sparse in the ψ domain. Then, x can be represented as $x = \psi s$, where ψ is an $N \times N$ transform matrix and s is a vector of coefficients. Therefore, y can be represented as:

$$y = \phi\psi s = As \tag{10}$$

where A is the measurement matrix.If the measurement matrix satisfies the mutual coherence or isometric property, the sparse signal x can be retrieved from y without any loss.

Thus, the network traffic could be reduced using CS as the sensor transmits only M symbols in place of the actual N symbols. But, computing the projections of the sensed data requires MN multiplications and $M(N-1)$ additions for a Gaussian matrix. Thus, CS increases the computational complexity.

3 Partial Canonical Identity Matrix-Based Compressive Sensing

PCI-CS is a sparsest random sampling scheme, which is a modification of CS. Here, the sensing or the measurement matrix is the sparsest one. Out of the N sensor nodes, only M is used to send data to the BS.

In PCI-CS, $y = [x_1, x_2 \dots x_M]^T$ denotes the sensor nodes transmitting data to the CH and can be represented as:

$$y = \begin{bmatrix} 0\,0\,1\,0 \dots 0 \\ 1\,0\,0\,0 \dots 0 \\ \vdots\,\vdots\,\vdots\,\vdots\;\vdots\;\vdots \\ 0\,0\,0\,0 \dots 1 \end{bmatrix}_{M \times N} \times \begin{bmatrix} x_1 \\ x_2 \\ \vdots \\ x_N \end{bmatrix}_{N \times 1} = \phi x \tag{11}$$

where ϕ is the sensing matrix of size $M \times N$ of which the elements are represented as:

$$\phi(i, j) = \begin{cases} 1, & 1 \le i \le M, \text{for distinct}\, m \\ 0, & \text{otherwise} \end{cases} \tag{12}$$

The PCI-CS selects only M elements of x, i.e., and thus M projections of x are captured [6]. The number of multiplications and additions is reduced to zero as the linear combinations of the sensed data are no longer required to recover the original signal x from y. Thus, the PCI sensing matrix could reduce the computational complexity compared to other random sensing matrices.

4 Results and Discussion

In this work, we focus on Compressive Sensing (CS) to achieve the objective of minimizing the data transmission to reduce energy depletion. Further, we design a variant of the Partial Canonical Identity matrix-based Compressive Sensing (PCI-CS) to reduce the data transmission. The performance of these two methods is compared with established routing protocols described in the literature part.

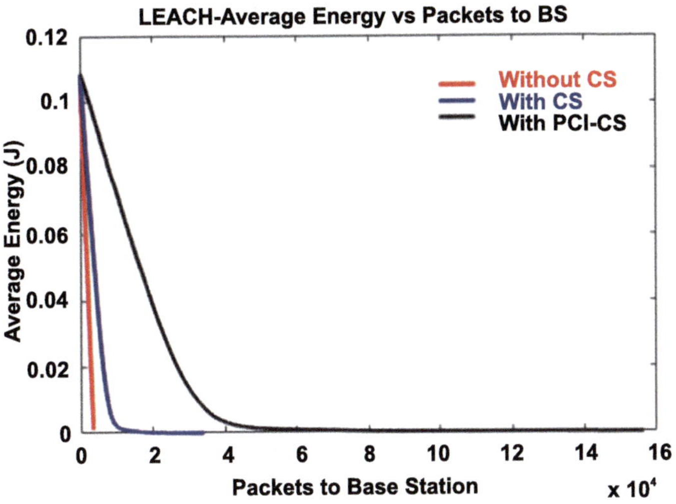

Fig. 1 Average energy of nodes versus packets to BS using LEACH protocol

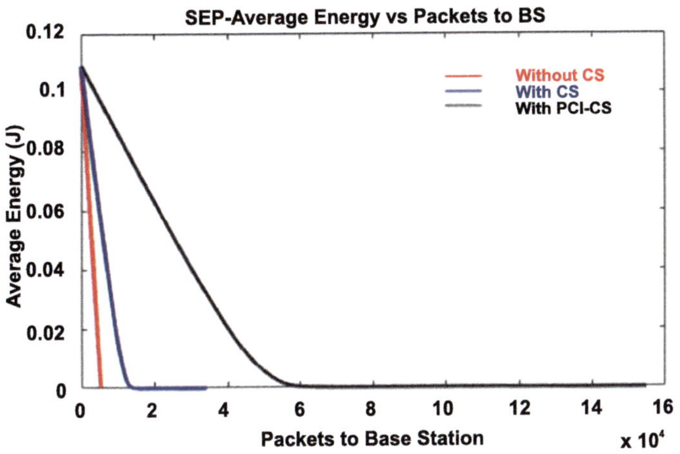

Fig. 2 Average energy versus packets to BS using SEP protocol

Figure 1 describes the LEACH's energy efficiency, i.e., the mean energy spent for transmitting data to the BS. Simulation results show that by using PCI-CS, more information can be sent to the base station compared to using CS only or without using CS. The normal method without CS requires transmission of data from all the N sensors, whereas in CS and PCI-CS, only M measurements are required. This reduces the number of transmissions and, thus, the energy consumed. Thus, the network longevity is also increased, as shown in Fig. 4, which shows that the sensors can send more information to the base station before the battery drains out.

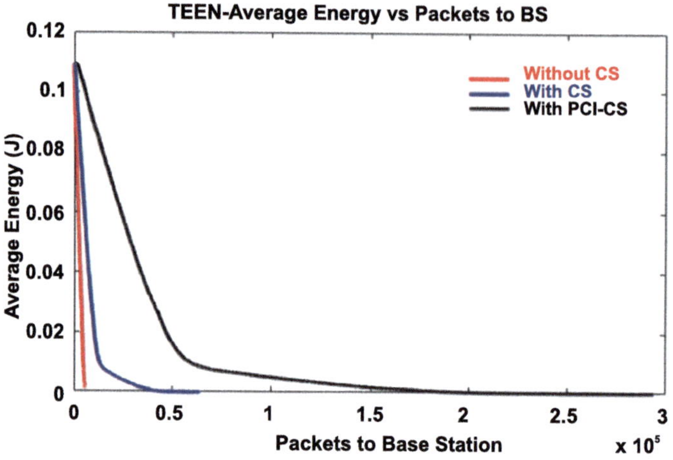

Fig. 3 Average energy versus packets to BS using TEEN protocol

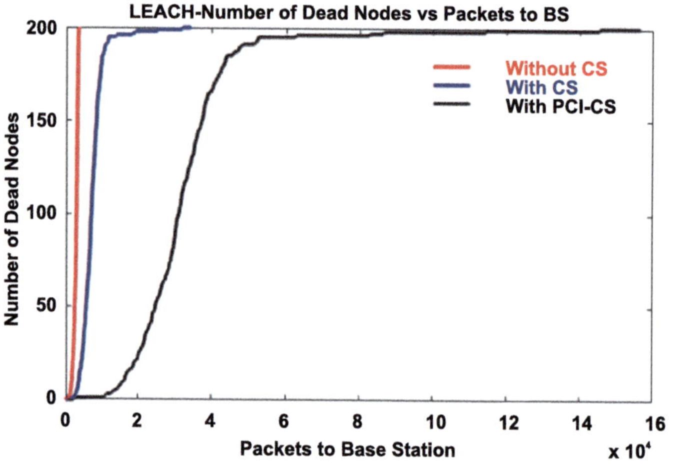

Fig. 4 Number of dead nodes versus packets to BS using LEACH protocol

Similar is the case for SEP and TEEN protocols. Simulation results show that the PCI-CS method is much more energy efficient than that with and without CS, as shown in Figs. 2 and 3. Also, the network longevity is improved using PCI-CS as shown in Figs. 4, 5 and 6.

Simulation results have shown that Partial Canonical Identity matrix-based Compressive Sensing has shown better performance in improving energy efficiency as well as prolonging the network longevity. Also, it can be seen that the TEEN protocol, along with PCI-CS, has shown better performance compared to the other methods used here regarding energy efficiency and network longevity.

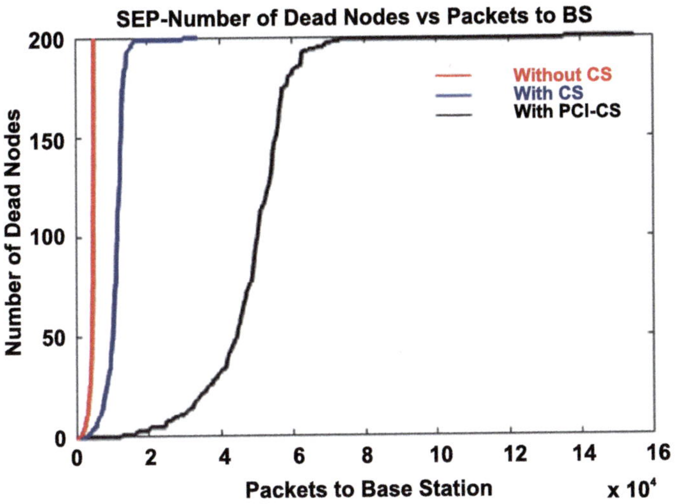

Fig. 5 Number of dead nodes versus packets to BS using SEP protocol

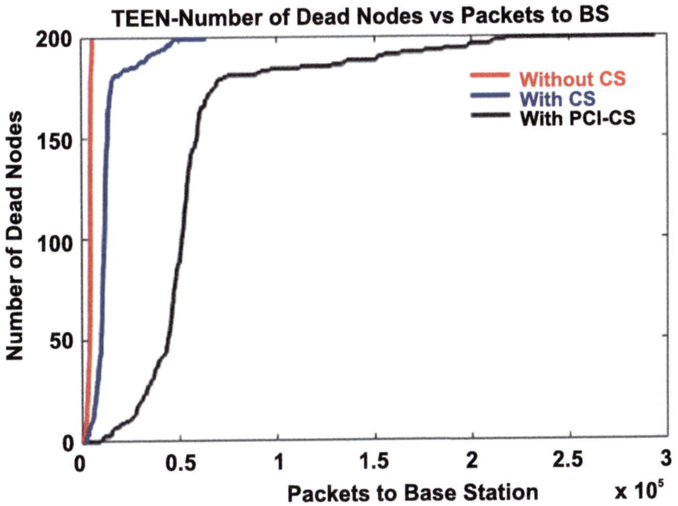

Fig. 6 Number of dead nodes versus packets to BS using TEEN protocol

5 Conclusion

Sensor networks are widely used for many real-life applications. Regardless of the sensor network's application, the data generated in the network eventually must be delivered to the BS. The transmission and reception of the sensor readings cause the major energy spending of the sensor nodes. In low-power applications, the batteries are considered to be neither replaceable nor reachable. To achieve this aim of reducing energy consumption and thereby prolonging the network longevity, several protocols have been used along with CS and PCI-CS. The proposed technique could achieve minimum energy consumption and increase network longevity. Experimental results demonstrate that the proposed Partial Canonical Identity matrix-based Compressive Sensing outperforms other methods in terms of energy efficiency and network longevity.

References

1. Huo J, Deng X, Al-Neshmi HMM (2020) Design and improvement of routing protocol for field observation instrument networking based on leach protocol. J Electr Comput Eng 2020
2. Ansari N, Gupta A (2017) Image reconstruction using matched wavelet estimated from data sensed compressively using partial canonical identity matrix. IEEE Trans Image Process 26(8):3680–3695
3. Daanoune I, Abdennaceur B, Ballouk A (2021) A comprehensive survey on leach-based clustering routing protocols in wireless sensor networks. Ad Hoc Netw 114:102409
4. Jeevanantham S, Rebekka B (2022) Hierarchical stable election protocol for WSN-based IoT inhabitant and environmental monitoring applications. Int J Commun Syst 35(15):e5301
5. Nagaraju R, Goyal S, Verma C, Safirescu CO, Mihaltan TC et al (2022) Secure routing-based energy optimization for IoT application with heterogeneous wireless sensor networks. Energies 15(13):1–16
6. Zhang X, He J, Zhou Y, Chi Y (2022) A subspace approach to sparse sampling based multi-attribute data aggregation in IoT. IEEE Internet Things J
7. Saxena M, Joshi A, Dutta S, Mishra KC, Giri A, Neogy S (2021) Comparison of different multi-hop algorithms to improve the efficiency of leach protocol. Wirel Pers Commun 118(4):2505–2518
8. Lee J-S, Kao T-Y (2016) An improved three-layer low-energy adaptive clustering hierarchy for wireless sensor networks. IEEE Internet Things J 3(6):951–958
9. Leinonen M, Codreanu M, Giannakis G (2019) Compressed sensing with applications in wireless networks. Found Trends Signal Process 13(1–2)

Smart City Solutions for Traffic Management

Maitreya Niranjan, K. S. Nachiketh, Naman Kashyap, Abhinand B. Balakrishna, and Anand M. Subbarao

Abstract Traffic congestion is a major problem in urban areas of India affecting the quality of life for the citizens. Traditional methods of traffic monitoring and management have limitations in terms of accuracy and scalability. There is a need for a more efficient and reliable system that can collect real-time traffic data and provide direct actionable insights to traffic authorities. In this paper, we propose to develop a cloud-based system consisting of a mobile application and a website, supported by a cloud-based server, that will collect real-time traffic data from collaborative users anonymously and use analytics to generate traffic density reports. The mobile application will be used to collect the real-time data from the user. These data will be sent to the cloud server where it will be used for report generation. This report will be displayed to the authorities using a website, which would assist them in taking the right decisions.

Keywords Real-time traffic data · Cloud-based system · Mobile application · Website · Report generation · Data analysis · Anonymity

1 Introduction

Urban areas across India face an ever-growing challenge: traffic congestion. The exponential increase in vehicular traffic, driven by rapid urbanization and economic growth, has left citizens grappling with deteriorating quality of life. Commuting times have become unpredictable, and the frustration of being stuck in endless traffic jams has become an unfortunate hallmark of daily life. This is especially true in Bengaluru, the Silicon Valley of India. According to the TomTom traffic index [1], the capital city of Karnataka recorded a travel time of nearly 29 min and 10 s to drive 10 km in the year 2022. This statistic alone underscores the urgent need for innovative and effective traffic management solutions in metropolitan cities like Bengaluru.

M. Niranjan · K. S. Nachiketh · N. Kashyap · A. B. Balakrishna · A. M. Subbarao (✉)
Department of Computer Science, PES University, Bengaluru, India
e-mail: anand.subbarao@gmail.com

S. M. Thampi et al. (eds.), *Fifth International Conference on Computing and Network Communications*, Lecture Notes in Electrical Engineering 1219,
https://doi.org/10.1007/978-981-97-4540-1_8

Traditional methods of traffic monitoring and management have long been in place, yet they increasingly exhibit limitations in terms of accuracy and scalability. Static infrastructures and periodic surveys struggle to keep pace with the dynamic nature of traffic patterns in a modern city. Consequently, congestion persists, leading to lost productivity, increased fuel consumption, and environmental degradation. Moreover, the effectiveness of decision-making by traffic authorities is hampered by the lack of real-time, granular data.

To address this multifaceted problem, we propose the development of an innovative cloud-based traffic monitoring and management system tailored for Bengaluru. This system, comprising a user-friendly mobile application, a dedicated website for authorities, and a robust cloud-based server, represents a paradigm shift in how traffic data are collected, analyzed, and utilized.

1.1 Objectives and Significance

- The primary objective of this research is to design, implement, and evaluate a cloud-based system that empowers both citizens and traffic authorities to collaboratively combat traffic congestion.
- Collect real-time location data from the commuters and visualize it as a dynamic heatmap, presenting live traffic density to authorities for informed traffic management decisions.
- Implement a density-based scanning mechanism to analyze real-time traffic data, identifying and ranking the most congested areas in Bangalore, and provide authorities with a prioritized list for targeted intervention and planning.
- Provide commuters with a direct channel to the authorities where they can voice their concerns, grievances, and issues. Also, provide them an interface to report accidents to the authorities.
- Implement a user-friendly SOS feature, enabling commuters to swiftly alert authorities during emergency situations through a dedicated button on the application interface.
- Implement real-time tracking of emergency vehicles, enabling authorities to swiftly respond to emergencies and manage traffic effectively for faster commute times of emergency vehicles and overall traffic alleviation.

2 Review of Literature

In this literature review, we explore a selection of research papers illuminating various facets of traffic data collection and traffic management technologies:

Paper [2] talks about GPS/GPRS/MM-based traffic data collection systems, employing floating cars equipped with GPS and GPRS modules for real-time traffic insights. Paper [3] introduces On-Board Units (OBU) in modern cars, with smartphones aiding data visualization. It introduces "floating cellular data" (FCD) and emphasizes the use of custom libraries and non-relational databases for efficient data handling. Paper [4] outlines a methodology fusing wireless communication, real-time traffic monitoring, and green wave coordination to craft a smart traffic control system. Wireless tech connects traffic lights, real-time monitoring provides accurate traffic data, and green wave coordination synchronizes light timings. Paper [5] presents a real-time adaptive traffic control system using sensor data for dynamic signal adjustments. Paper [6] proposes automatic number plate recognition (ANPR) technology for traffic enforcement and emergency response improvement. Paper [7] introduces a real-time method using passive Wi-Fi and Bluetooth sensors, along with algorithms for correcting biased speed estimates and identifying modes of transport. Paper [8] discusses the development of a smart traffic cloud as an infrastructure to support various traffic applications essentially integrating cloud computing with traffic management. This paper also goes a step further in using the map-reduce technology to process data. Paper [9] aims to present an Internet of Things cloud system designed for the prevention of vehicular accidents by processing of mobile sensor data, which are collected using a tracker device installed on a vehicle.

Collectively, these papers reveal significant gaps in the existing traffic monitoring and management infrastructure. Currently, the government lacks a comprehensive system for real-time traffic data collection, storage, monitoring, and user-friendly presentation. Such a system would involve continuous automated data gathering from mobile devices, server-based data storage, pattern monitoring, and data visualization.

While various traffic-related applications exist, none offer a unified set of features within a single system, differentiating our proposed solution:

- Real-time traffic visualization: Authorities have access to live traffic density heatmaps.
- Data privacy: Prioritizing user privacy, our application refrains from storing location data.
- Voice alerts: Real-time voice alerts enhance safety by warning users near accident-prone areas.
- SOS feature: Immediate communication with first responders and location transmission during emergencies.
- Accident reporting: Users can report accidents directly to authorities.
- Grievance support: A direct channel for commuters to communicate concerns and grievances with authorities.
- Historical map visualization: The system offers a unique feature allowing users to engage in historical map simulations by selecting specific start and end dates.
- Congested area ranking: Grant authorities access to the ranked list of congested areas through the implementation of a density-based scanning mechanism, facilitating targeted intervention and planning.

Our proposed cloud-based system, comprising a mobile application and a website, aims to fill this gap, providing a holistic solution for real-time traffic data collection, analysis, and user engagement. Beyond mitigating the limitations of existing traffic apps, our solution bolsters privacy and safety features, making it an asset for urban traffic management. These insights will be pivotal in guiding our strategy for developing a comprehensive mobile app and centralized backend server.

3 Proposed Methodology

3.1 System Design and Development

Mobile Application Development

- Design and develop a user-friendly mobile application for Android, adhering to native development guidelines and best practices.
- Implement features for user registration and login authentication using secure and industry-standard protocols.
- Integrate features for adding vehicle details, emergency information, submitting grievances, and reporting accidents ensuring user privacy and data protection.
- Develop functionalities for providing user support and initiating data collection during active journeys for real-time traffic monitoring.
- Implement a robust SOS feature that, upon activation, promptly transmits user details from the local system to authorities, providing crucial information about the user's location and user details.
- Activate audio alerts for when the commuter approaches an accident-prone area and when an emergency vehicle is in the vicinity of the commuter.
- Develop a dedicated view of the app for emergency vehicle drivers. Their location must be shown as a priority to the authorities on the website.

Server Development

- Establish a robust and scalable server infrastructure capable of handling real-time data ingestion, processing, and storage.
- Implement APIs for receiving sensor data, accident reports, and grievances from the mobile application, adhering to RESTful architecture principles.
- Establish a database system capable of storing and managing the collected traffic data, accident reports, and grievances, ensuring data integrity and security.
- Establish a WebSocket connection between the server and the website to facilitate real-time transmission of SOS alerts received from the application to the website for immediate response.

Website Development

- Design and develop a responsive and user-friendly website for authorized personnel to access and visualize traffic data.
- Integrate a mapping platform to display real-time traffic density maps, utilizing geospatial data manipulation techniques.
- Implement user authentication mechanisms to restrict unauthorized access to the website and its functionalities.
- Create a dashboard to provide an overview of traffic conditions, including real-time updates and historical trends.
- Develop separate pages for displaying accidents, grievances, and SOS alerts, enabling informed decision-making by traffic authorities.
- Introduce a dedicated feature to display and prioritize ranked congested areas, enabling informed decision-making by traffic authorities.

3.2 *Data Collection, Processing, and Visualization*

Mobile Application Data Collection

- Collect real-time sensor data from user's smartphones during active journeys.
- Transmit collected data to the server via secure and efficient protocols, ensuring data integrity and minimizing latency.
- Store vehicle details, emergency information, and grievance data locally on the user's smartphone for offline access and potential usage in emergencies.

Server-side Data Processing

- Receive and store sensor data, accident reports, and grievances from the mobile application, ensuring data validation and consistency.
- Process sensor data to generate real-time traffic density maps, employing distributed computing frameworks for scalability and performance.
- Analyze accident reports and grievances to identify trends and patterns to extract valuable insights.
- Using density-based scanning, the system processes traffic records every 15 min with DBSCAN, generating high-density clusters stored on the server for authorities' prompt access and informed decision-making.

Website Data Visualization

- Retrieve real-time traffic density data from the server, ensuring data synchronization and accuracy.
- Visualize real-time traffic density on a map, using effective color-coding and heatmaps to represent congestion levels.
- Visualize historical traffic data by generating traffic heatmaps chosen according to date and time.
- Display accidents, grievances, and SOS alerts.

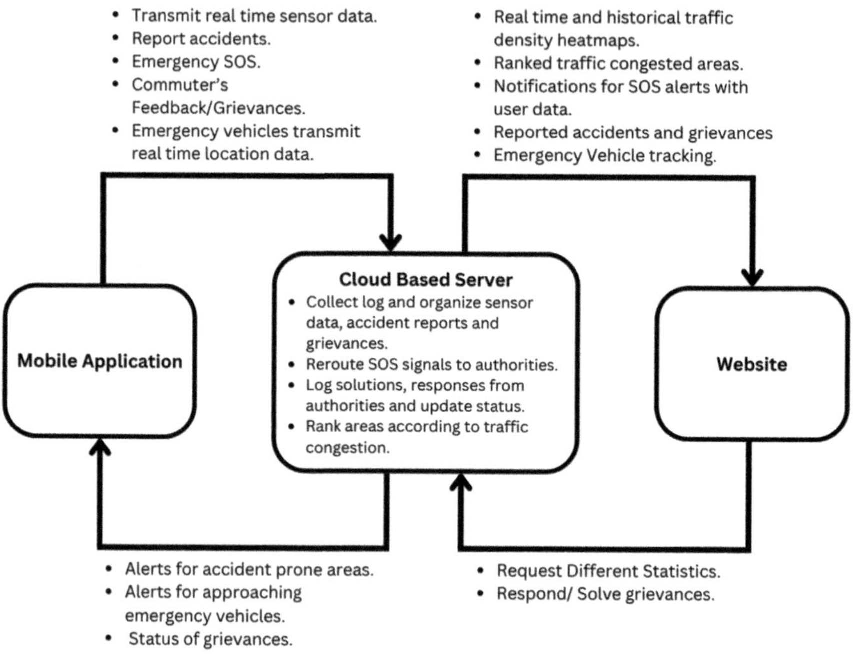

Fig. 1 System operation diagram

- Visualize the ranked areas based on traffic congestion by fetching the prepared report from the server, complemented by an additional display on the map for enhanced presentation (Fig. 1).

4 Implementation

The mobile application, developed in Kotlin within the Android Studio IDE, integrates user registration and login authentication through Google Firebase. Using the mobile device's GPS sensor, the application collects location and speed data. For user privacy, sensitive SOS data are stored locally. SOS data transmission and other relevant information occur through HTTPS. Gunicorn serves as the WSGI server for handling HTTPS requests, and Uvicorn operates as the ASGI server for managing WebSocket connections. For testing, SUMO (simulation of urban mobility) simulates real-time location and speed data, mimicking commuter information. Simulated data undergo preprocessing and error-checking on the server before being inserted into the database. The server, hosted on Google Cloud Platform, uses a MySQL database for storing comprehensive details, including grievances, location coordinates, and accident data. Data transfer between the server and website is managed through Gunicorn and secure REST APIs. Web socket technology ensures instant transmission of user details during emergencies. Docker containers host Website Backend, Website Frontend, MySQL database, Gunicorn server, Uvicorn server, and MongoDB, with Nginx serving as a reverse proxy.

The front-end website, developed using the MERN stack, leverages the Bing Maps API to visually present the traffic heatmap. Secure REST APIs fetch data from the server, and password hashing secures data storage within MongoDB. This integrated technology stack ensures a robust and secure system, offering a seamless user experience and efficient data transfer between the mobile application and server. Additionally, the system fetches the last 15 min of data from the MySQL database, applies the DBSCAN algorithm to identify clusters, ranks these clusters based on the number of vehicles, and sends the ranked clusters are stored on the server. These ranked clusters will be sent to the website upon request. The areas are then displayed on the map for further analysis.

5 Results and Discussion

Figure 2 shows the images for the detailed emergency information of the user which will only be displayed to the authorities in the event of an emergency as well as the grievance page in which the user can submit any grievance they encounter during their journey and the accidents page where the user can report an accident which has occurred in their vicinity.

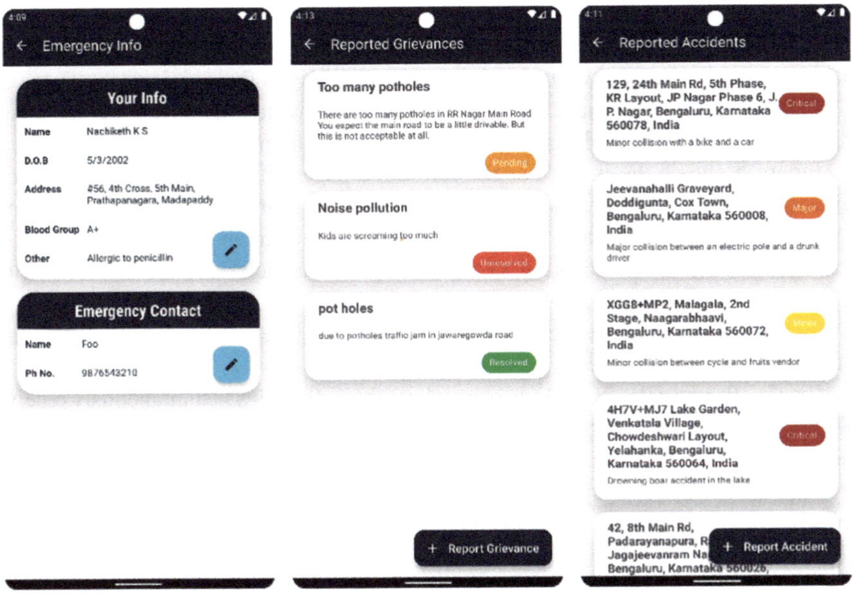

Fig. 2 Emergency information, reported accidents, and submitted grievances

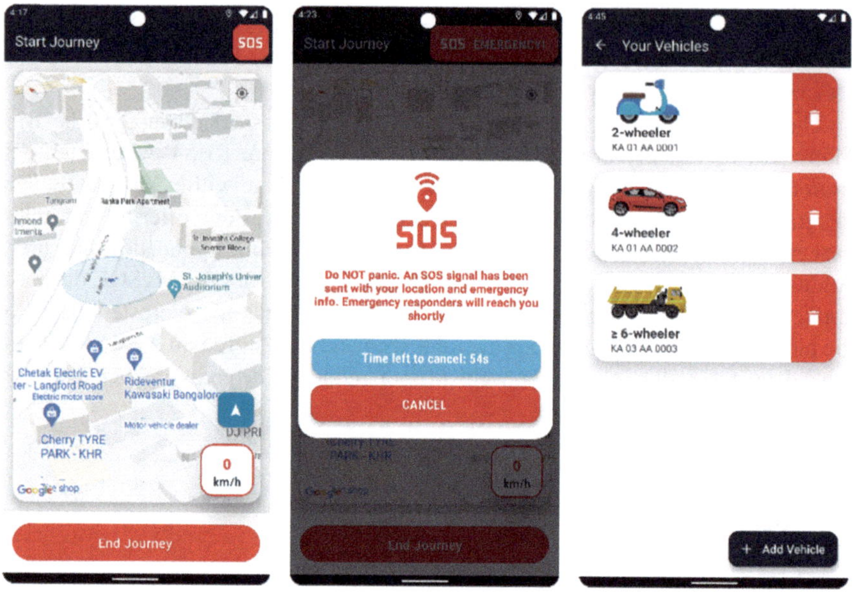

Fig. 3 Start journey, SOS alert box, and your vehicles page

Figure 3 shows the images for start journey page which displays the live location of the user as well as the SOS Alert Box with an option to cancel is displayed to the user when SOS signal is activated. Your vehicles page displays all the vehicles that the user has added to the app.

Figure 4 shows the alerts that the commuters receive when they approach an accident-prone area and when an emergency vehicle is in their vicinity.

Figure 5 shows the image for the accidents page which displayed the accidents reported by users across the city in the form of a heatmap and in a tabular format.

Figure 6 shows the image for the grievance page which displays the collected grievances in a single place where the authorities can mark them as resolved or not.

Figure 7 shows the image for the real-time traffic heatmap.

Figure 8 shows the tracking of the emergency vehicles for the authorities to efficiently move the emergency vehicle.

Figure 9 shows the historical map simulations which can be activated by selecting specific start and end dates.

Figure 10 shows the SOS alerts page where the authorities instantaneously receive SOS signals transmitted by any of the commuters. It also displays their emergency information and their live location.

Figure 11 shows the reports page which shows the top congested areas by rank in both a graphical representation and a map view.

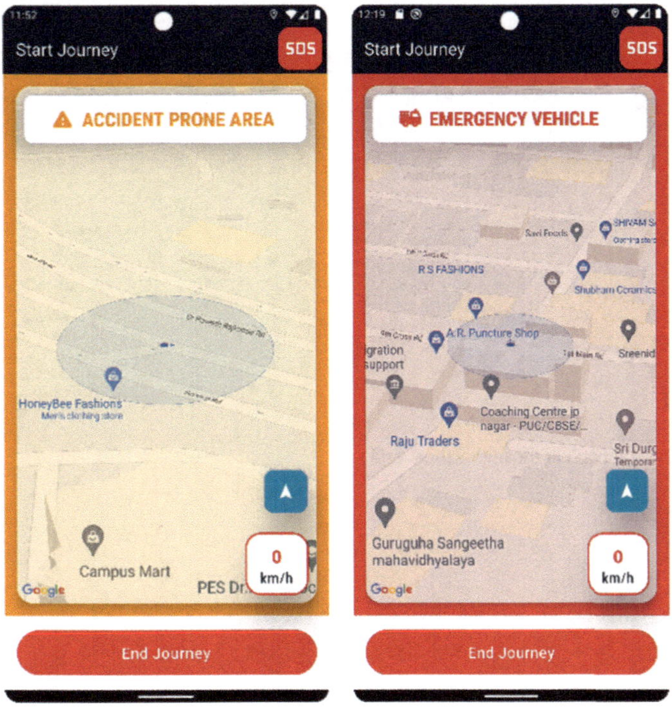

Fig. 4 Accident-prone area and emergency vehicle alert

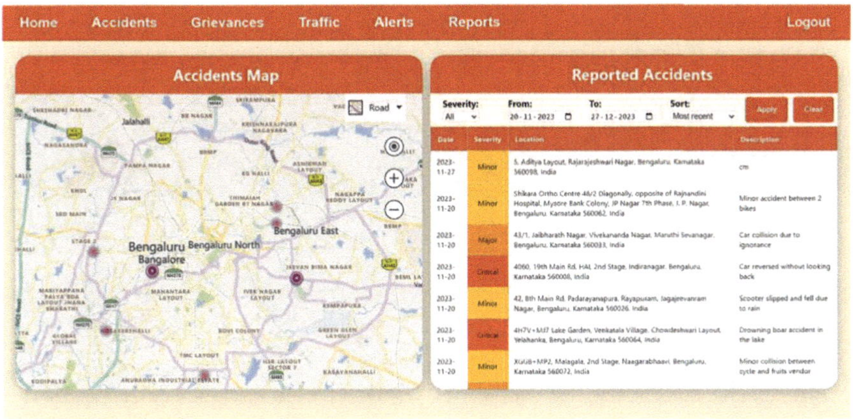

Fig. 5 Accidents page at the authorities' end

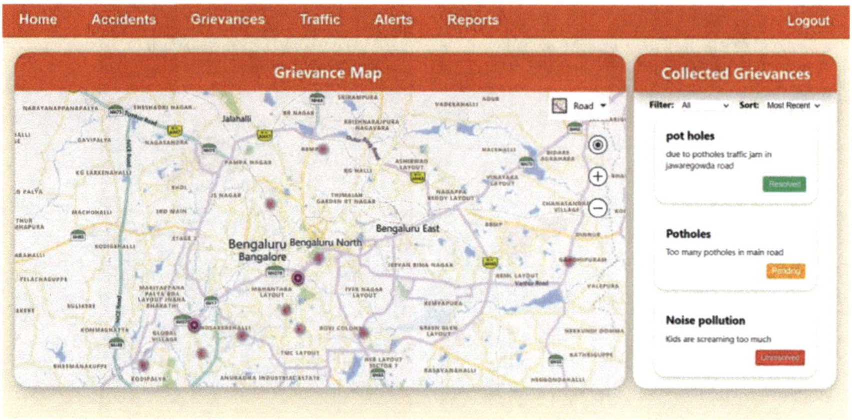

Fig. 6 Grievance page at the authorities' end

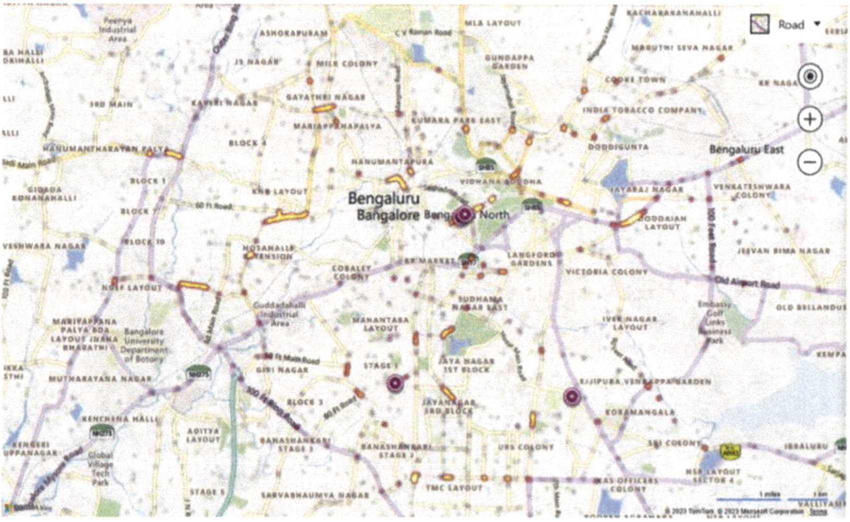

Fig. 7 Real-time traffic heatmap

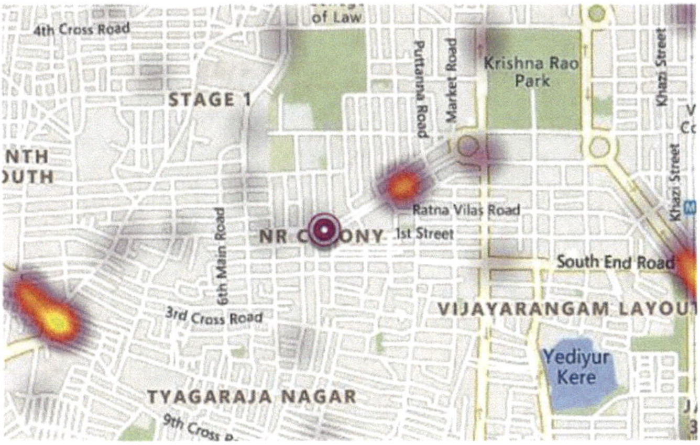

Fig. 8 Emergency vehicle tracking

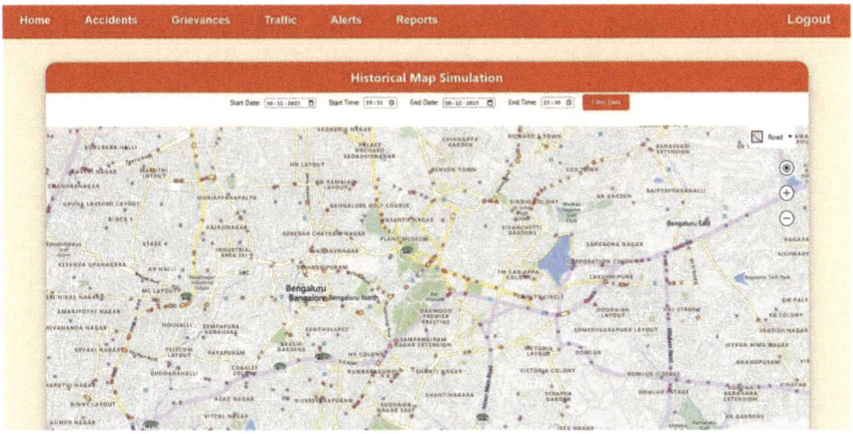

Fig. 9 Historical heatmap

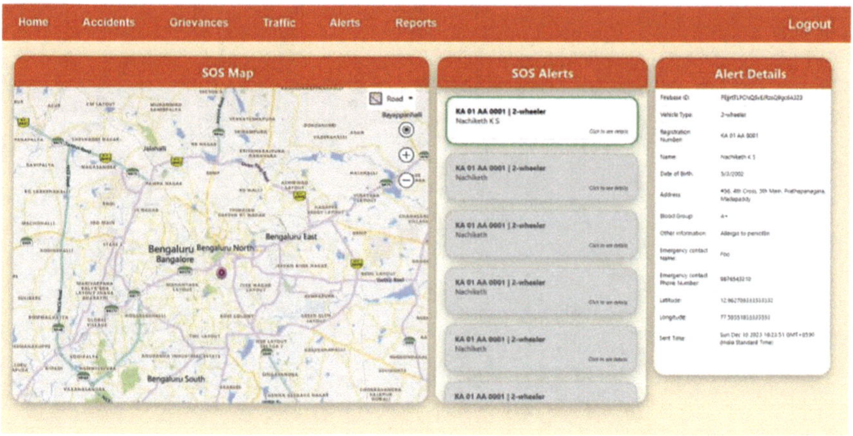

Fig. 10 SOS alert page

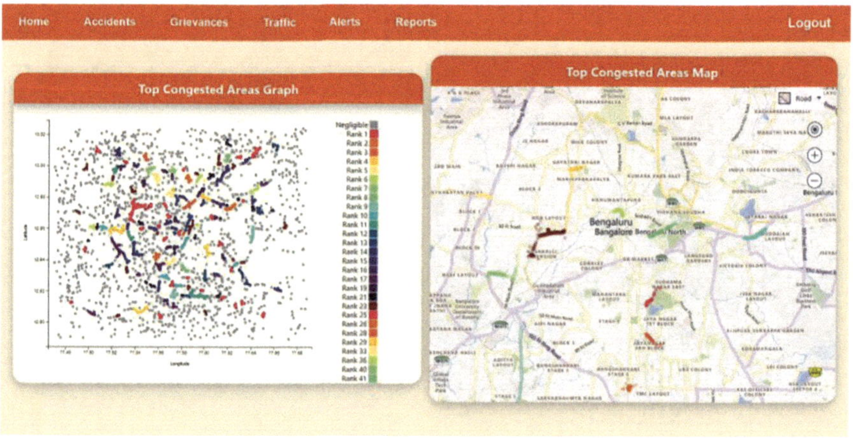

Fig. 11 Reports page

6 Conclusion

In conclusion, the project has successfully delivered a practical and user-centric solution for real-time traffic data collection and management. Through diligent research and implementation, critical deficiencies in existing traffic monitoring systems have been effectively addressed. The mobile application, in conjunction with the centralized server, provides an efficient means of gathering, processing, and analyzing traffic data. The inclusion of safety features such as voice alerts, SOS functionality, and accident reporting significantly enhances the system's utility and user safety. Furthermore, the implementation of the DBSCAN algorithm enriches the system by identifying and ranking congested areas, offering valuable insights for targeted interventions. The system's role as a direct communication channel allows commuters to voice their grievances and concerns to the government, enhancing its standing as a comprehensive and user-centric solution for modern traffic management. The dedicated website empowers authorities with easy access to reports generated within the server, thereby enhancing their decision-making capabilities.

As the project moves forward, several potential avenues for future work come into focus. Efforts can be directed toward further data refinement, with a continuous commitment to enhancing data accuracy and completeness. Scalability remains an essential consideration, and the system can be optimized to accommodate a growing user base and expanding data volumes. User experience improvements, including enhancements to the mobile application's interface and customization options, can lead to a more seamless and engaging experience for users.

References

1. TomTom (2023) Bengaluru traffic index. TomTom. https://www.tomtom.com/traffic-index/bengaluru-traffic/ https://www.tomtom.com/traffic-index/ranking/
2. Bi J, Wang L, Liu Y (2010) Traffic data collection system for floating car based on GPS/GPRS/MM. In: 2010 second WRI global congress on intelligent systems, Wuhan, China, pp 66–69 (2010). https://doi.org/10.1109/GCIS.2010.68
3. Reininger M, Miller S, Zhuang Y, Cappos J (2015) A first look at vehicle data collection via smartphone sensors. In: 2015 IEEE sensors applications symposium (SAS), Zadar, Croatia, pp 1–6. https://doi.org/10.1109/SAS.2015.7133607
4. Oliveira LFP, Luz PDG, Manera LT (2020) Development of a wireless traffic light controller system, with real-time traffic monitoring and green wave coordination applied to smart cities. In: 2020 IEEE Congreso Bienal de Argentina (ARGENCON), Resistencia, Argentina. https://doi.org/10.1109/ARGENCON49523.2020.9505576
5. Shyam Shankaran R, Rajendran L (2021) Real-time adaptive traffic control system for smart cities. In: 2021 international conference on computer communication and informatics (ICCCI), Coimbatore, India, pp 1–6. https://doi.org/10.1109/ICCCI50826.2021.9402597
6. Rachana KP, Aravind R, Ranjitha M, Jwanita S, Soumya K (2021) IOT based smart traffic management system. Int J Eng Res Technol (IJERT) 09(12)
7. Pu Z, Cui Z, Tang J, Wang S, Wang Y (2022) Multimodal traffic speed monitoring: a real-time system based on passive Wi-Fi and bluetooth sensing technology. IEEE Internet Things J 9(14):12413–12424. https://doi.org/10.1109/JIOT.2021.3136031
8. Wang WQ, Zhang X, Zhang J, Lim HB (2012) Smart traffic cloud: an infrastructure for traffic applications. In: Proceedings of the 2012 IEEE 18th international conference on parallel and distributed systems (ICPADS '12). IEEE Computer Society, USA, pp 822–827
9. Celesti A, Galletta A, Carnevale L, Fazio M, Ĺay-Ekuakille A, Villari M (2018) An IoT cloud system for traffic monitoring and vehicular accidents prevention based on mobile sensor data processing. IEEE Sens J 18(12):4795–4802. https://doi.org/10.1109/JSEN.2017.2777786
10. Top 10 Accident Prone Places in Bengaluru. https://bengaluru.citizenmatters.in/top-10-places-where-most-accidents-in-bengaluru-take-place-7260

IoT and Edge Computing

An IoT-Assisted Real-Time Urban Air Pollution Data Collection System

Banani Ghose, Sk. Minsar, Udayan Ghosh, Sk. Arif Mohammad, Adil Hossain Mallick, and Zeenat Rehena

Abstract Air pollution is one of the innumerable menaces an individual comes across every moment. Countless means to make urban lives smoother are indebted to witnessing air pollution and other types of pollution. It is contributed by unrestricted urbanization, traffic, industrial emissions, etc., causing life-threatening diseases; even death. Monitoring air pollution should be the primary objective of any society. For an effective air pollution monitoring system to run and deliver a real-time Air Quality Index (AQI) a prompt IoT-enabled air pollution sensor-based data collection system is a prerequisite. In this letter, a real-time hardware-based IoT-enabled air pollution-related data collection module is designed. This will be portable and will be recording both pollutant concentrations and meteorological factors. It consists of a number of sensors and a Wi-Fi module. Collected data will be directly sent to the cloud for further analysis. Thus, as a part of the air pollution monitoring system, this module will be capable of recording real-time data, that will lead to real-time air health monitoring. It will assist urban residents and policymakers in taking decisions that will guide them to have healthier lives.

Keywords Air pollution monitoring · Sensor data collection · IoT · Air quality sensors · Real-time air pollution data

1 Introduction

Environmental pollution is one of the challenges yet to be solved to offer healthy lives to the inhabitants of this beautiful world. Environmental pollution includes air pollution, noise pollution, water pollution, soil pollution, and many more. Air

B. Ghose (✉)
Department of Information Technology, Haldia Institute of Technology, Haldia, West Bengal, India
e-mail: bananighose@hithaldia.in

B. Ghose · Sk. Minsar · U. Ghosh · Sk. Arif Mohammad · A. H. Mallick · Z. Rehena
Department of Computer Science and Engineering, Aliah University, Kolkata, West Bengal, India

S. M. Thampi et al. (eds.), *Fifth International Conference on Computing and Network Communications*, Lecture Notes in Electrical Engineering 1219,
https://doi.org/10.1007/978-981-97-4540-1_9

pollution plays a major role in polluting the environment and makes the lives of people hilarious. 7 million people die every year due to inhalation of the harmful finest particles in the air [20]. Air pollution includes not only the pollution outside but indoor air pollution too. Air pollution contributes to several life-threatening diseases like chronic obstructive pulmonary disease (COPD), lower respiratory infection, stroke, ischemic heart disease, cancer, and many more. Air pollution monitoring is a major task nowadays. It has much applicability in both policy-making and making people aware of air pollution. Most of the monitoring stations are equipped with high-cost sensors [17] and technological constructs which make the monitoring stations hard to establish in every place in both the urban and rural environments. But if there is a possibility of having a source of air pollution in between two major air pollution monitoring stations then instituting the costly monitoring system seems to be impractical many times. To address this issue and to take these places into consideration some sort of portable air pollution monitoring system may be of great help. In both cases, the monitoring stations are established with a true understanding of several IoT devices, clouds, and a strong network as the backbone. This type of IoT-based data collection unit plays an important role in collection the raw data from the environment which is beneficial for the well-established air pollution monitoring unit. To achieve that level of accuracy the monitoring station and the inherent IoT devices need a large amount of real-time data. The data are gathered and analyzed with the right technical construct and knowledge to come to a real-time conclusion. This conclusion may be in the form of alerts or some sort of guidelines that will further help people to escape from the harmful effects of air pollution.

For the successful execution of the monitoring assignments, the monitoring stations largely need a good amount of data. The raw environmental data is the prerequisite of these monitoring stations. To collect these data from different places the IoT devices have to depend largely upon the sensors installed in the environment. The sensors are clubbed with minimum memory and limited powered batteries. Different sensors are capable of recording different kinds of pollutants and meteorological parameters.

Like every conscious country in the world, India, being a developing country is also one of the worst sufferers of air pollution. The standard parameter Air Quality Index (AQI) is evaluated as the marker of a place's AQI. Different countries ascertained different air pollution agents responsible for contaminating the ambient air quality. According to the National Air Quality Index guidelines eight pollutants are marked as responsible for air pollution in India [4]. They are PM10, PM2.5, SO_2, NO_2, O_3, CO, NH_3, Pb. In [6] the authors show a way to minimize the number of air pollutants for calculating AQI considering the correlation among them. From that knowledge, one of the highly correlated pairs of pollutants is chosen for calculating the AQI. Similarly, the same thing is done with meteorological factors too. So, by optimizing the number of inputs and making the monitoring unit simple yet functional, the proposed work considers only PM2.5, CO, and NH_3 as the pollutants and TEMP (temperature), and HUMID (humidity) as the meteorological factors for calculating the AQI of a place.

In this paper, a real-time air pollution data collection system is proposed which is accompanied by a number of low-cost gas sensors and a Wi-Fi module for recording both the pollutant concentration and meteorological factors present in the environment at any given time. The collected data is sent to the cloud and can be used for further analysis as well as can be shown on a web application for the rational use of the people.

The rest of this paper is organized as follows. Section 2 illustrates the literature review of the work. Section 3 describes the proposed system for air pollution monitoring. Section 4 explains real-time simulation and data collection. Finally, we conclude the work in Sect. 5.

2 Related Work

The modern facilities provided by trendy gadgets mostly depend on a strong dataset. So, data collection and the efficient use of the collected data are both important for the timely response of any gadget having real-time response capacity. Thus, data collection is the basic necessity of each smart gadget. The IoT-aided data collection system can be treated as an important part of the air pollution monitoring system too. Different works are done in this field which took the help of various technological advances for collecting data and further processing it. These works played a driving factor to get motivation for designing this real-time data accumulation system. A real-time air pollution monitoring system is elaborated in [11]. The system is capable of collecting pollutant concentration and meteorological factors from various stations. There lies a Wi-Fi module and an M2M communication. In [17], the authors captured the data for the pollutants like CO, CO_2, SO_2, and NO_2 with the gas sensors and used a Raspberry pi module, after collecting the data they also displayed the data through proper visuals. But they have not considered the meteorological factors as a part of the monitoring system.

CO_2, NO_2, NO_x, CO, and some other gases are measured through a club of gas sensors in [13]. A Bluetooth interface and a GPS module are used to record the temporal and spatial information of the monitoring system, as it is a portable one and is installed on a bicycle. Air pollution takes place not only outdoors but indoors too. An Indoor Air Quality Monitoring system is proposed by the authors in [3]. The system is able to capture the presence for CO_2, CO, SO_2, NO_2, O_3, Cl_2, ambient temperature, and relative humidity. In this letter, the authors gave emphasis on the storage and recovery system of the collected data for further usage in analysis purposes. A Wireless sensor-based air pollution monitoring mechanism is proposed in [18], which is capable of handling a number of gas sensors and meteorological parameter sensors. The system is capable of calculating the AQI of Bari (Italy). The calculated result is also compared with the official data issued by AROA-Puglia. It has a club of gas sensors to measure the concentrations of CO, NO_2, O_3, SO_2, PM1.0, PM2.5, and PM10 along with temperature and humidity sensors. One-step and multi-step ahead predictions of O_3, NO_2, and SO_2 are done using machine learning algorithms in [19]. The best algorithms are found by evaluating the RMSE of various machine learning algorithms. In [16], the authors proposed a portable low-cost sensor-based air pollution monitoring system that can sense the concentrations of O_2, C_6H_6, NH_3, and CO. Primary agents of air pollution in Brisbane, Australia in [12]. In [5], the authors tried to overcome the challenges of sensor-driven air pollution data-capturing systems by proposing a three-phase air pollution monitoring solution. The kit comprises various gas sensors, Arduino IDE, and an Android-based application termed as IoT-Mobair. The system made was portable and easy to install anywhere needed. In [14], the authors evaluated the performances of two types of PM2.5 sensors by installing them physically in the city of Pittsburgh, Pennsylvania and its surroundings. In the same, the authors also investigated the influence of humidity and temperature on sensor readings.

Irrespective of the fashion of recording and collecting the environmental air pollution-related parameters, the result is further used for air pollution prediction systems as described in [7, 9] or in some monitoring systems [8]. The ultimate outcome of the work is used by the residents and the policy-making institutions for the betterment of their livelihood.

By reviewing all these works and considering the air pollution guidelines provided by the Government of India, the key requirement of any air pollution monitoring station is identified as the real-time data accumulation system. The need for meteorological factors like temperature and humidity in air pollution is also taken into focus. In the proposed data collection system, there are rooms to record both pollutant concentrations and meteorological factors. Taking a minimum number of sensors the data collection module is designed. The system is designed in such proximity that it is efficient and portable. It is designed in such a fashion that the recorded data will be directly delivered to the cloud for further processing.

3 Proposed Real-Time Data Collection and Air Pollution Monitoring System

The proposed IoT-based air pollution monitoring system comprises a club of gas sensors for sensing both the important factors of air pollution present in the air, pollutant concentrations, and meteorological factors. The overall structure of the proposed system involves some layers for its smooth operations as shown in Fig. 1.

1. **Data Collection Layer**:
   The data collection layer is the lowermost layer of the proposed monitoring system. This layer is responsible for collecting pollution-related parameters from the environment in real-time. This layer consists of an array of sensors for sensing both the meteorological and pollutant concentrations. The sensors are equipped with limited storage capacity as well as limited lifetime as they are clubbed with very low-powered batteries. So whatever data is sensed by the sensors those are

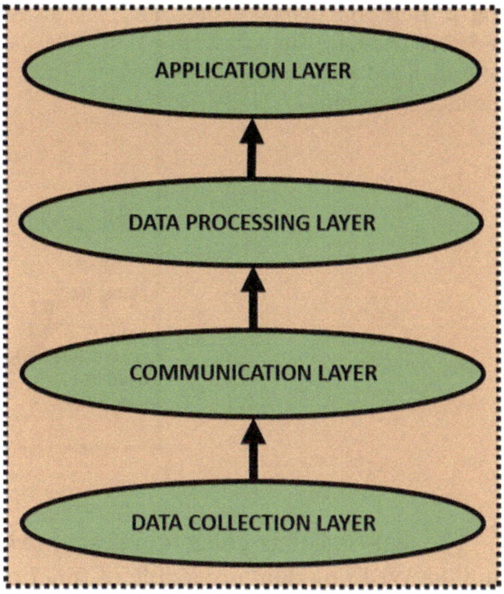

Fig. 1 Layered structure of IoT-enabled air pollution monitoring system

to be stored from time to time to ensure that there is no loss of data. So the data recorded by the sensors are to be communicated the safe storage for further usage.

2. **Communication Layer**:
   As the sensors are paired with low-powered batteries, the data captured by them are to be sent for further analysis. For this, the data gathered are to be safely sent to the next higher layer in the proposed framework, i.e., the data processing layer. The communication layer may be deployed with the help of any transmission system. During the communication, it has to be kept in mind that no data should be lost in transit. In addition to that, one more thing is to be remembered that, the communication should not cost much, if so then the applicability of the system may be disrupted.
3. **Data Processing Layer**:
   In this work, the data collection is done by an array of sensors. As the sensors are having limitations in terms of lifetime as well as storage perspective collected data are to be stored for further processing. For this purpose, a fog or a cloud layer may be deployed between the application layer and the communication layer. The actual data analysis and any other processing needed for meeting the objective of the entire work are done in this layer.
4. **Application Layer**:
   The application layer is one of the most important layers through which the users can issue their query and get the response back. The monitoring system will record the data sent by the real-time data collection module. When a user wants to know the concentrations of the major polluting agents in the air, then he or she can have them by some web applications developed as a part of the monitoring work. The entire data collection and air pollution monitoring system proposed in this section looks like the illustration depicted in Fig. 2.

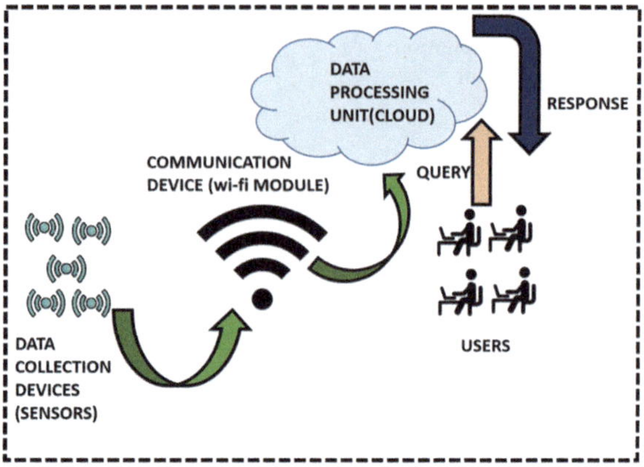

Fig. 2 Blue print of the proposed real-time data collection and air pollution monitoring system

The real-time data collection system when connected to all the required sensors and the Wi-Fi module is capable of functioning indoors as well as outdoors.

4 System Design and Implementation

In this section, the actual design of the proposed framework is illustrated. In materializing the work in practice a number of sensors and some other functioning units are used in the layers already described in Sect. 3.

4.1 Design

At the time of designing the system the blueprint of the system is prepared and the list of sensors and communication modules to be used which will make the system capable of delivering the desired output is made ready. As per the framework planned in Sect. 3 the layers are developed one after another to meet up the objective of each of the layers. The data collection layer consists of three sensors for sensing and recording the existence of the pollutants which are majorly considered for calculating AQI along with some sensors for recording the values of meteorological parameters. For recording NH_3, MQ135 is used. MQ7 is used as the sensing unit for CO, and the PM2.5 sensor (popularly known as dust sensor) is used for recording the concentration of PM2.5 in the air. On the other hand, for recording the concentrations of the various meteorological factors some other sensors are used, such as DHT11, for sensing temperature and humidity in the environment, and a PM2.5 sensor is already there to sense the presence of smoke/dust. To deploy the communication layer ESP01 is used as the Wi-Fi module for sending the sensed data to proper storage space. From where the data will be further used for analysis or for some other purposes. Here in

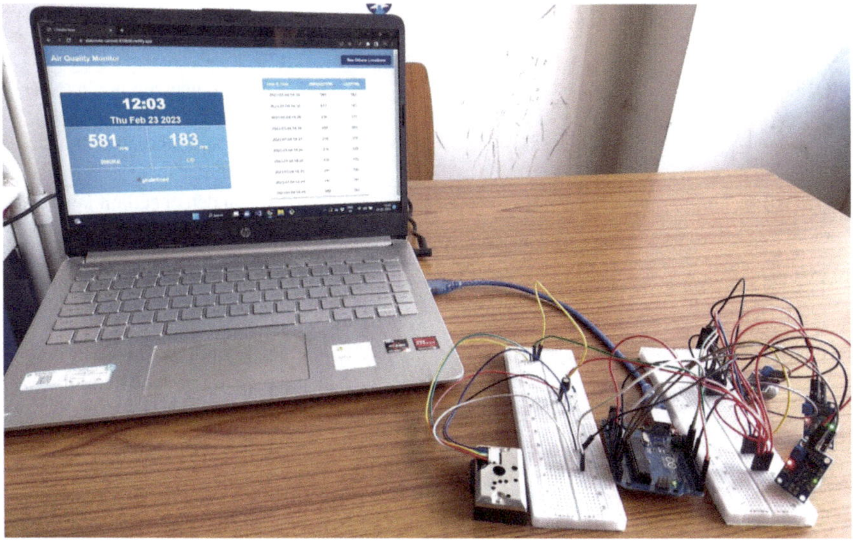

Fig. 3 The real-time air pollution data collection module with sensors and Wi-Fi module connected

this work, the data is dumped into the cloud from time to time as the cloud is an expert in assisting modern livelihood in real-time. For the data to be stored and accessed in an efficient way the ThingSpeak server is used here for a safe repository of the collected data. The collected data will be used for further monitoring purposes. The dataset generated there can be downloaded for any further necessity. The application layer is a Graphical User Interface built with HTML, CSS, and Javascript. The fully designed system looks as it is shown in Fig. 3. The system can show the response to the user's query in the form of a web page, which will further be developed in terms of some messaging service or other alert-generating services. The full list of sensors and Wi-Fi modules used in designing this proposed system is shown in Table 1.

4.2 Results

The data collection module of the proposed monitoring system is fully able to collect and record the pollutant concentration as well as the meteorological factors. After sensing the parameters the system generates a dataset that looks like Fig. 4. The users of the system at any given time can generate queries to see the recorded data. The requests from the users are handled by the ThingSpeak server. The data stored in the ThingSpeak server will be displayed on users' requests in an organized way through the Application Layer as shown in Fig. 5. Figures 6, 7, 8, 9, and 10 are the graphs plotted while recording the parameters like temperature, PM2.5, NH_3, HUMIDITY, and CO form the environment in real-time.

Table 1 Brief description of the sensors and the Wi-Fi module used

Sensor	Image	Purpose
MQ135 [21]		It can detect ammonia (NH_3), sulfur (S), benzene (C_6H_6), CO_2, and other harmful gases and smoke
MQ7 [10]		It can detect carbon monoxide (CO) gas
PM2.5 [1]		It detects particulate matter, specifically PM2.5
DHT11 [15]	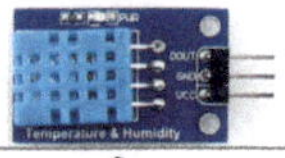	It is efficient in sensing temperature and humidity
ESP01 [2]		It is a Wi-Fi module, that allows the microcontroller access the Wi-Fi network

19	created_at	entry_id	CO	NH3	TEMP	HUMIDITY	PM2.5
20	2023-06-13 06:59:37 UTC	262	416	448	27	58	255
21	2023-06-13 07:00:51 UTC	263	416	457	26	60	251
22	2023-06-13 07:06:39 UTC	264	402	448	25	61	257
23	2023-06-13 07:07:37 UTC	265	405	448	25	60	259
24	2023-06-13 07:09:52 UTC	266	404	447	25	57	261
25	2023-06-13 07:11:52 UTC	267	403	446	25	58	262
26	2023-06-13 07:14:53 UTC	268	398	443	25	62	256
27	2023-06-13 07:19:26 UTC	269	394	440	26	63	244
28	2023-06-13 07:21:53 UTC	270	393	434	26	64	247
29	2023-06-13 07:29:54 UTC	271	386	426	24	60	395
30	2023-06-13 07:31:25 UTC	272	392	434	25	62	246
31	2023-06-13 07:33:31 UTC	273	388	433	25	63	252
32	2023-06-13 07:34:27 UTC	274	387	432	25	64	252
33	2023-06-13 07:37:57 UTC	275	387	431	25	65	249
34	2023-06-13 07:39:22 UTC	276	387	430	25	66	244
35	2023-06-13 07:42:02 UTC	277	384	425	25	59	251
36	2023-06-13 07:43:08 UTC	278	384	426	24	58	256
37	2023-06-13 07:44:14 UTC	279	383	425	24	57	255
38	2023-06-13 07:48:58 UTC	280	382	422	24	62	249
39	2023-06-13 07:50:59 UTC	281	384	413	25	64	246
40	2023-06-13 07:54:38 UTC	282	383	419	25	65	244
41	2023-06-13 07:58:13 UTC	283	379	407	25	63	249
42	2023-06-13 07:59:53 UTC	284	375	415	25	61	249
43	2023-06-13 08:02:42 UTC	285	376	408	24	58	251
44	2023-06-13 08:04:52 UTC	286	395	422	26	58	600
45	2023-06-13 08:09:57 UTC	287	392	418	26	61	601
46	2023-06-13 08:11:31 UTC	288	391	416	26	62	599
47	2023-06-13 08:15:19 UTC	289	377	415	26	64	613
48	2023-06-13 08:16:45 UTC	290	384	420	26	63	611

Fig. 4 The screen-shot of the dataset generated in the cloud

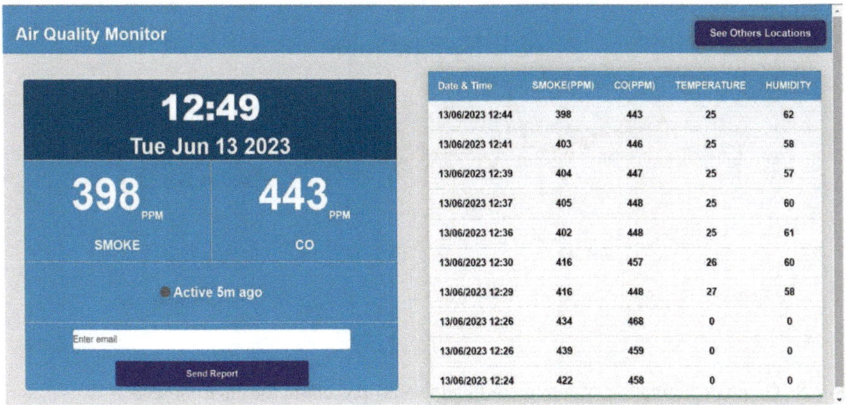

Fig. 5 The screen-shot of the response to a user's query

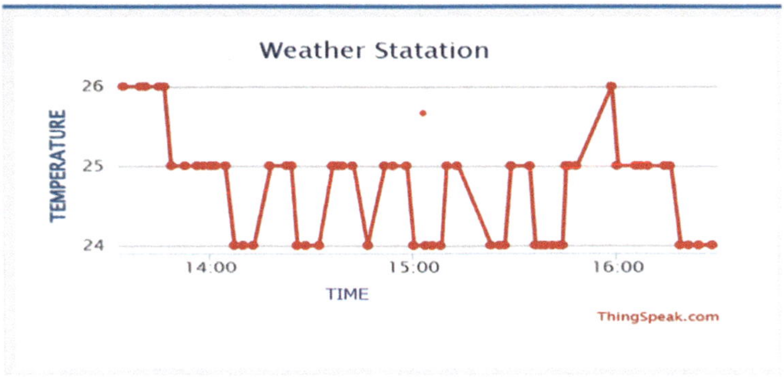

Fig. 6 Graph depicting the system recording the temperature of the environment

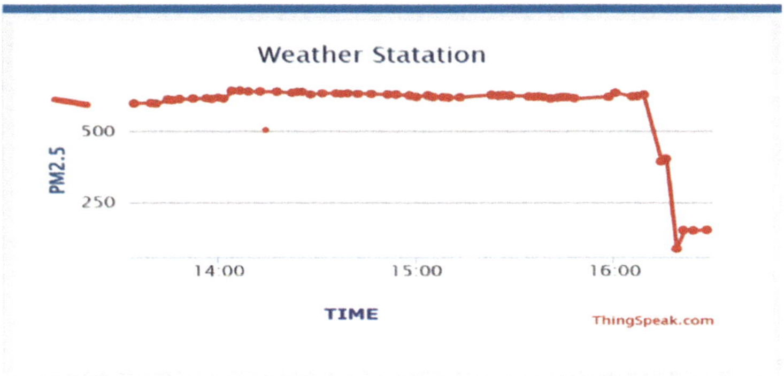

Fig. 7 Graph depicting the system recording the PM2.5 present in the environment

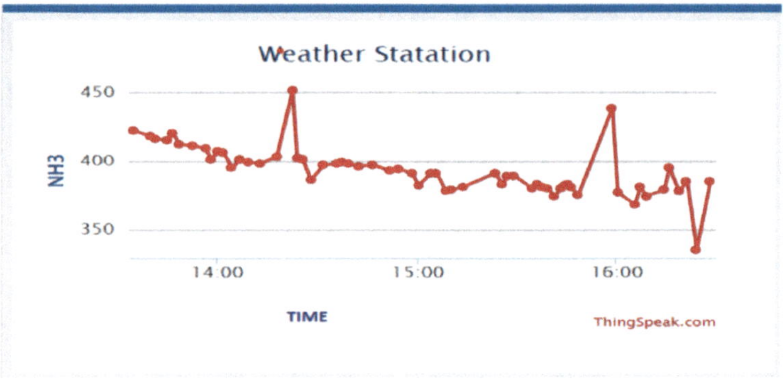

Fig. 8 Graph depicting the system recording the NH_3 present in the environment

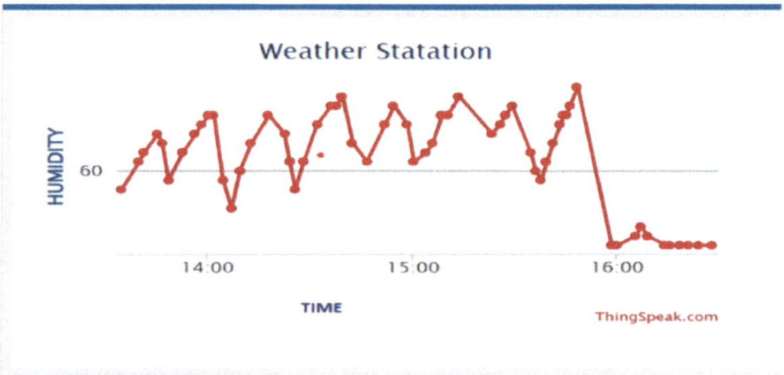

Fig. 9 Graph depicting the system recording the humidity of the environment

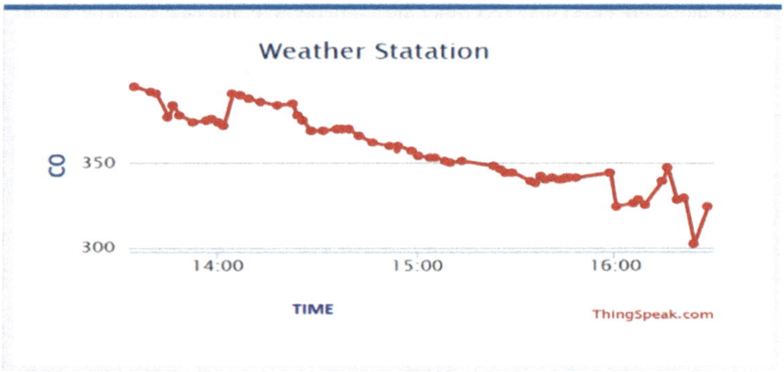

Fig. 10 Graph depicting the system recording the CO present in the environment

5 Conclusion

Air pollution monitoring and control is one of the vital issues that are to be addressed and resolved as soon as possible due to its unpropitious effects on the health of the citizens, overall environmental conditions, global warming, etc. To deploy a real-time air pollution monitoring system a real-time dataset is essential. Many researchers have already come up with solutions having many advantages as well as disadvantages, but they contributed a rich dataset for further analysis. The IoT-driven data collection system proposed in this paper is portable, efficient, and most importantly provides the collected data to the cloud where it will be analyzed for a rational outcome. This is a part of work which has a major contribution in preparing the dataset on which the monitoring system will function. After the processing and necessary analysis of the dataset, the monitoring system will reach a meaningful conclusion. Finally, the outcome of the monitoring system will be able to guide urban inhabitants to plan for healthier lives and the policymakers will be helped with this by diverting the traffic and adopting new policies, which may result in a healthier urban environment. For real-time decision making this real-time air pollution data collection system will be a vital part. The developed data collection system has certain limitations. As the proposed system comprises of low-cost sensor devices, and they are prone to failure due to several reasons, there are possibilities of missing some data or some series of data due to certain sensor failures. This missing data may lead to another challenge of dealing those missing values with utmost care and rationality. The system lacks in another perspective, there is no sensor for recording meteorological parameters like wind speed, and wind direction. In future research works, the resultant dataset will be analyzed through a monitoring system. Some more pollutants and meteorological parameters influencing the air health of a place can also be incorporated to increase the applicability of the work. The result of the monitoring system will be circulated among interested people through web applications or in the form of alerts. Finally, the entire air pollution monitoring system will definitely be helpful in monitoring and managing air pollution.

References

1. Adafruit Industries (2023) PM2.5 air quality sensor: manual. https://cdn-learn.adafruit.com/downloads/pdf/pm25-air-quality-sensor.pdf
2. AI-Thinker Team (2023) ESP-01 WiFi module: manual. https://www.microchip.ua/wireless/esp01.pdf
3. Benammar M, Abdaoui A, Ahmad SH, Touati F, Kadri A (2018) A modular IoT platform for real-time indoor air quality monitoring. Sensors 18(2):581
4. Central Pollution Control Board (2014) Air quality index. http://www.indiaenvironmentportal.org.in/files/file/Air
5. Dhingra S, Madda RB, Gandomi AH, Patan R, Daneshmand M (2019) Internet of things mobile-air pollution monitoring system (IoT-Mobair). IEEE Internet Things J 6(3):5577–5584
6. Ghose B, Rehena Z (2020) A mechanism for air health monitoring in smart city using context aware computing. Procedia Comput Sci 171:2512–2521

7. Ghose B, Rehena Z (2021) A deep learning approach for predicting air pollution in smart cities. In: Computational intelligence and machine learning: proceedings of the 7th international conference on advanced computing, networking, and informatics (ICACNI 2019). Springer, pp 29–38
8. Ghose B, Rehena Z (2022) Fog abetted early alert system for monitoring air pollution in smart cities. In: 2022 international conference on IoT and blockchain technology (ICIBT). IEEE, pp 1–6
9. Ghose B, Rehena Z, Anthopoulos L (2022) A deep learning based air quality prediction technique using influencing pollutants of neighboring locations in smart city. JUCS J Univers Comput Sci 28(8)
10. Hanwei Electronics Co., Ltd. (2023) Air quality gas sensor MQ-7: manual. https://www.sparkfun.com/datasheets/Sensors/Biometric/MQ-7.pdf
11. Kadri A, Yaacoub E, Mushtaha M, Abu-Dayya A (2013) Wireless sensor network for real-time air pollution monitoring. In: 2013 1st international conference on communications, signal processing, and their applications (ICCSPA). IEEE, pp 1–5
12. Liu X, Jayaratne R, Thai P, Kuhn T, Zing I, Christensen B, Lamont R, Dunbabin M, Zhu S, Gao J et al (2020) Low-cost sensors as an alternative for long-term air quality monitoring. Environ Res 185:109438
13. Liu X, Li B, Jiang A, Qi S, Xiang C, Xu N (2015) A bicycle-borne sensor for monitoring air pollution near roadways. In: 2015 IEEE international conference on consumer electronics-Taiwan. IEEE, pp 166–167
14. Malings C, Tanzer R, Hauryliuk A, Saha PK, Robinson AL, Presto AA, Subramanian R (2020) Fine particle mass monitoring with low-cost sensors: corrections and long-term performance evaluation. Aerosol Sci Technol 54(2):160–174
15. mouser.com (2023) DHT11 humidity & temperature sensor: manual. https://www.mouser.com/datasheet/2/758/DHT11-Technical-Data-Sheet-Translated-Version-1143054.pdf
16. Muthukumar S, Mary WS, Jayanthi S, Kiruthiga R, Mahalakshmi M (2018) IoT based air pollution monitoring and control system. In: 2018 international conference on inventive research in computing applications (ICIRCA). IEEE, pp 1286–1288
17. Parmar G, Lakhani S, Chattopadhyay MK (2017) An IoT based low cost air pollution monitoring system. In: 2017 international conference on recent innovations in signal processing and embedded systems (RISE). IEEE, pp 524–528
18. Penza M, Suriano D, Pfister V, Prato M, Cassano G (2017) Urban air quality monitoring with networked low-cost sensor-systems. In: Proceedings, vol 1. MDPI, p 573
19. Shaban KB, Kadri A, Rezk E (2016) Urban air pollution monitoring system with forecasting models. IEEE Sens J 16(8):2598–2606
20. WHO (2019) WHO: air pollution. https://www.who.int/airpollution/en/
21. Zhengzhou Winsen Electronics Technology Co., Ltd. (2023) Air quality gas sensor MQ-135: manual. https://www.winsen-sensor.com/d/files/PDF/Semiconductor

Real-Time Stock Price Prediction Using Apache Kafka

R. Deepalakshmi, N. Harshini, and S. Nivedha

Abstract Stocks play a vital role in the economy and serve as an indicator of business value and investor sentiment. To leverage improved decision-making aligned with investment strategies and market trends, real-time stock predictions enhance outcomes by offering immediate insights and enabling investors to make more informed decisions. Various machine learning techniques have been applied to this problem, but most of them rely on historical data and do not account for the dynamic and volatile nature of the stock market. In this project, we propose a novel approach for stock price prediction using real-time data streamed from Yahoo Finance with Apache Kafka. A simple linear regression model, trained on the historical data, enables to predict the closing price of the stock for every fifth minute based on the opening price. The machine learning algorithm provides a better and more efficient way to predict the prices and facilitate decision-making.

Keywords NSE · Regression analysis · Apache Kafka · Machine learning techniques

1 Introduction

The Indian economy stands as one of the largest in purchasing power parity, which will continue to grow in the foreseeable future. The fluctuations in stock prices tend to influence numerous economic factors, including consumption and business investment. As per the June 2023, India's GDP has touched $3.75 trillion marking

R. Deepalakshmi · N. Harshini (✉) · S. Nivedha
Department of Data Science, NMKRV College for Women, Bengaluru, India
e-mail: nmkc21mds001.nmkrv@rvei.edu.in

R. Deepalakshmi
e-mail: deepalakshmir.nmkrv@rvei.edu.in

S. Nivedha
e-mail: nmkc21mds004.nmkrv@rvei.edu.in

S. M. Thampi et al. (eds.), *Fifth International Conference on Computing and Network Communications*, Lecture Notes in Electrical Engineering 1219,
https://doi.org/10.1007/978-981-97-4540-1_10

India as the "bright spot" [1] in the global economy. Stock markets play a vital role in the Indian economy. Investors from diverse backgrounds gain the opportunity to achieve financial gains through stock market investments. The key point in stock market investments is that we get high returns in the long term. These stocks provide ownership to the investors through their share, and thus, the investors have a stake in the company's future success. Stock investments can help diversify the portfolio of the investors as they can invest in various kinds of securities such as stocks, bonds, real estate, gold, etc., and thereby reduce the risk associated with each investment. The demand and supply factors affect the increase and decrease of share prices to a high extent. When there is an expansion in the count of people investing in a particular share, its demand will get high and, in the meanwhile, the price also increases. When a large number of people start selling their shares, the supply increases, and its prices decrease. Many factors influence the supply and demand chain. Any kind of positive or negative developments in a company will directly affect the share price of the company. The current market trends also impact the prices of shares.

To respond more quickly to changing market conditions, real-time predictions provide investors with the most accurate and up-to-date information. Real-time prediction is of high importance during periods of high market volatility and when significant changes occur. Most algorithmic trading methodologies rely on real-time data and predictions to execute a trade at high speeds. Machine learning's computational algorithms are highly adaptable and can change as per the market conditions and adjust the predictions based on new data, helping traders stay relevant and effective in dynamic markets. Incremental learning helps models adjust to non-stationary by incorporating new data and continuously updating their parameters. So, every time new data is acquired, it is not required to retrain the model from scratch thereby reducing the training time. The central focus of this project is to predict stock prices in real time and contribute to a more informed and efficient stock market. The driving force for this project was to help the investors make well-informed decisions when purchasing and selling of stocks.

2 Literature Review

Selvamuthu et al. [2] examined the prediction accuracy of three neural networks on tick data and 15-min data from Reliance Private Limited. The study compared the performance of Bayesian Regularization, Levenberg–Marquardt (LM), and Scaled Conjugate Gradient (SCG). Bayesian Regularization outperformed the LM and SCG incomplete datasets. But when only the regression plots on the test dataset are compared, the Scaled Conjugate Gradient gives the best results. The results show that prediction using tick data delivers higher accuracy than using 15-min tick data. The study highlighted the importance of using accurate data for accurate predictions.

In order to assess the fitting of the data in linear regression and support vector machines, Gururaj et al. [3] made use of the stock data to measure the performance of the two algorithms of machine learning. The proposed system evaluated the model's performance by utilizing metrics such as root mean squared error, mean absolute error, mean absolute percentage error, mean squared error, correlation coefficient, the nonlinear regression multiple correlation coefficient. The findings indicated that support vector machines (SVM) performed better than linear regression.

Wu et al. [4] proposed a method called S_I_LSTM that integrates data from various sources and investor sentiment. To calculate the investors' sentiment index, they used the sentiment analysis method based on convolutional neural networks (CNN) for the non-traditional data. They then combined the sentiment index, technical indicators, and historical transaction data with the feature set of stock price prediction. The results obtained were closer to the true value of the close price.

To estimate the future stock movements using ensemble deep neural networks, Yang Li and Yi Pan proposed a deep learning framework that employs sentiment analysis to extract valuable insights from various textual data sources. They proposed two levels of ensemble models in which the first base level contained two RNNs and submodels contained LSTM and GRU, respectively. The outcomes achieved through this method closely approximated the real stock [5].

Xia and Su [6] proposed ARIMA-LSTM model, to model the S&P 500 stock index and its component stocks for predicting the future movement of the S&P 500. To develop the most effective investment strategy for investors, they predicted the coefficient for correlation among its component stocks and the stock index. ARIMA model was efficient in filtering out the linearity and nonlinear trends predicted in the LSTM model. The loss function test results show that the mean squared error (MSE), mean absolute error (MAE), and root mean squared error (RMSE) of the ARIMA-LSTM hybrid model are relatively less than other control models, making it possible to estimate the correlation coefficient for optimization of the portfolio.

Aldhyani and Alzahrani [7] proposed a deep neural network model to forecast the stock prices of Tesla and Apple Stocks. The results of the CNN-LSTM model were significantly greater than the LSTM model. According to the outcomes of the experiments, the CNN-LSTM and LSTM stock prediction models were superior to the existing models in terms of their ability to make accurate forecasts.

To focus on the dynamic nature of the NIFTY 50 index, Mahajan et al. [8] utilized the family of generalized autoregressive conditional heteroskedasticity (GARCH) models for forecasting and reveal asymmetric characteristics existing in the index volatility. The results showed that asymmetric models, that is, exponential generalized autoregressive conditional heteroskedasticity (EGARCH) (1, 1) and threshold ARCH (TARCH) (1, 1), performed better than the GARCH (1, 1) model on mean absolute percentage error (MAPE) and root mean squared error (RMSE), whereas the mean directional accuracy (MDA) metric showed that all the models performed equally.

Liu et al. [9] proposed a model that combines convolutional autoencoders (CAE) and a long short-term memory (LSTM) neural network, using CAE to extract stock price image feature data and combining technical data to predict short-term stock prices. To illustrate the advantages of introducing image feature data for the prediction of stock prices, the outcomes of the new model were evaluated with the LSTM model. Comparing the root mean square error (RMSE) of the LSTM model prediction using only technical data as valid input data, the RMSE of the CAE-LSTM model combined with image feature data decreased by about 4%, and the forecast performance was slightly improved.

Various factors influence the closing price of the stock. Rishi [10] investigated the closing price of the stock in S&P 500 using linear regression. The closing price was chosen as the response variable and the variables open, high, and low were used as the independent variables. The data underwent testing for various regression assumptions and obtained an R squared value of 1. Thus, they concluded that the entire variation in the closing price of the stock could be accounted by the chosen independent variables.

Kanwal et al. [11] focused on a hybrid deep neural network model for stock price prediction. They combined the Bidirectional Cuda Deep Neural Network Long Short-Term Memory (BiCuDNNLSTM) and a one-dimensional convolutional neural network. Subsequently, the suggested model was contrasted with deep learning models and state-of-the-art models. The findings, evaluated using root mean squared error (RMSE) and mean absolute error (MAE), indicated that the proposed hybrid model is efficient for accurate prediction.

3 Methodology

Most previously proposed methodologies focused on using historical data collected on a day-to-day basis for predicting the stock prices. At present, due to the rapid changes in the global economy, the volatility of the market is very high and the majority of researchers have not placed their emphasis on incremental learning. In accordance with the efficient market hypothesis (EMH), stock prices incorporate all available information, meaning that price alterations unrelated to newly disclosed information cannot be predicted [6]. In our proposed methodology, we are focusing on Intraday Stock Prices as it helps us to examine the short-term price movements and helps many intraday traders make a large number of trading decisions within a short time frame. Thus, a secondary dataset obtained from Kaggle that comprises of intraday stock prices collected for every 5 min for over a period of 5 years is incorporated in training our model. Real-time predictions can guide dynamic adjustments to investment portfolios. Real-time predictions indicate shifts in market sentiment or trends, and investors can make timely changes to their asset allocation. As a result, this paper focuses on predicting the stock prices in real time. To achieve this we have used the Yahoo finance module and Apache Kafka for live-streaming stock prices (Fig. 1). Once the model is trained and the live-streaming data are generated, it is then forwarded to the Kafka consumer through Kafka producer which predicts the stock price for the next fifth minute and the findings are displayed in a dashboard with the help of Streamlit.

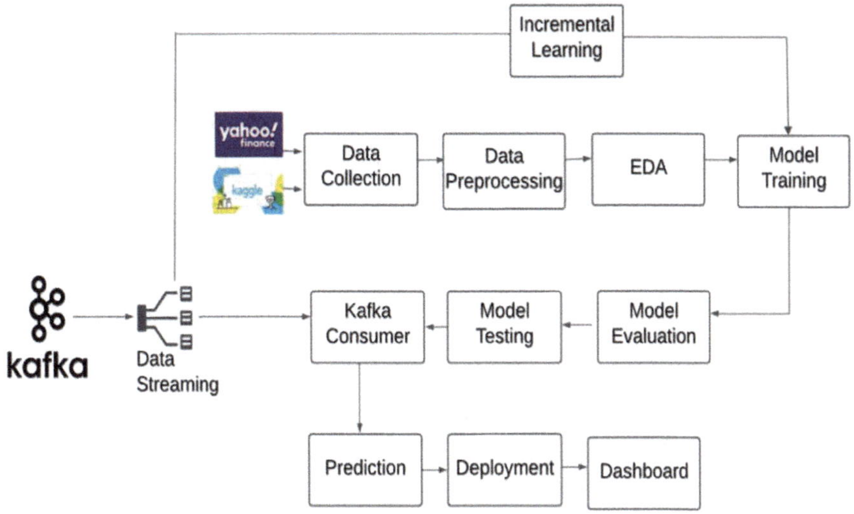

Fig. 1 Flowchart depicting the proposed methodology

4 Dataset

To train the model with historical data, we chose four stocks listed on the National Stock Exchange (NSE): ADANIENT, ITC, RELIANCE, and BIOCON. Additionally, a secondary dataset sourced from Kaggle, comprising stock prices recorded at five-minute intervals, was utilized for training purposes [12]. The dataset consists of 135,048 data points of stock prices from 2nd February 2015 to 21st October 2022. The dataframe consisted of date, open, high, low, and close variables. The yfinance python module was used to generate the stock price data from 1st June 2023 to 25th August 2023.

5 Implementation

5.1 Data Interpolation

The secondary dataset collected from Kaggle contained the stock prices from 2nd February 2015 to 21st October 2022 and the data generated from the yfinance module contained the stock prices from 1st June 2023 to 25th August 2023 (Fig. 2). To fill in the missing values from 22nd October 2022 to 21st May 2023, we have used the interpolation method.

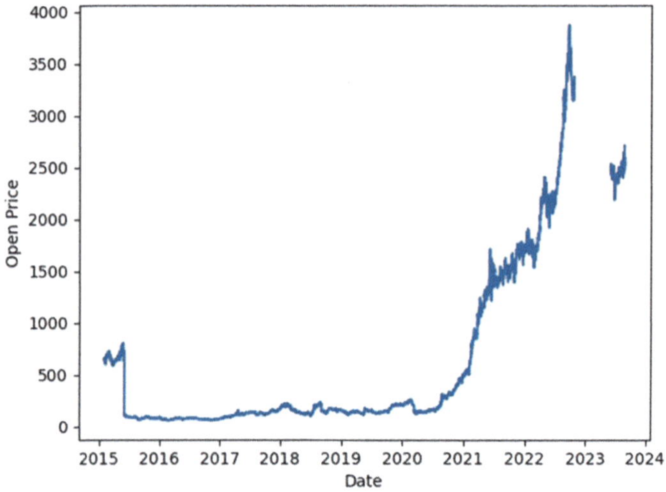

Fig. 2 Graph depicting the missing values in stock price of the open variable

Interpolation is the procedure for ascertaining unknown values that exist between two known data points. Numerous interpolation techniques are available like linear, cubic, polynomial, nearest etc. [13]. Nearest Interpolation operates on the principle of selecting the value of the nearest known point to the new point and assigning the new point the value of the nearest known point. This method results in a piecewise-constant interpolant.

From Fig. 3 it is evident that cubic and quadratic techniques have generated values beyond the scale of the actual open and close values. The linearly interpolated data lacks any differences or fluctuations within it. In our research, we have chosen the Nearest Interpolation method as it has generated values that closely approximate the remaining data points. After interpolation, the overall data points are 151,303.

5.2 Model Building

Simple linear regression proved to be the best model to provide real-time predictions for our study. Initially, we trained our data using artificial neural network (ANN), XGBoost regressor, support vector regressor, and random forest regressor.

The test loss in the artificial neural network model was very high, and there were huge variations in the train and validation losses, resulting in an overfit model (Fig. 4). The training time of the support vector regressor was very long. The predictions of the XGBoost regressor and the random forest regressor resulted in a huge difference between the actual and the predicted values and were not very accurate and resulted in a very high mean squared error (Table 1).

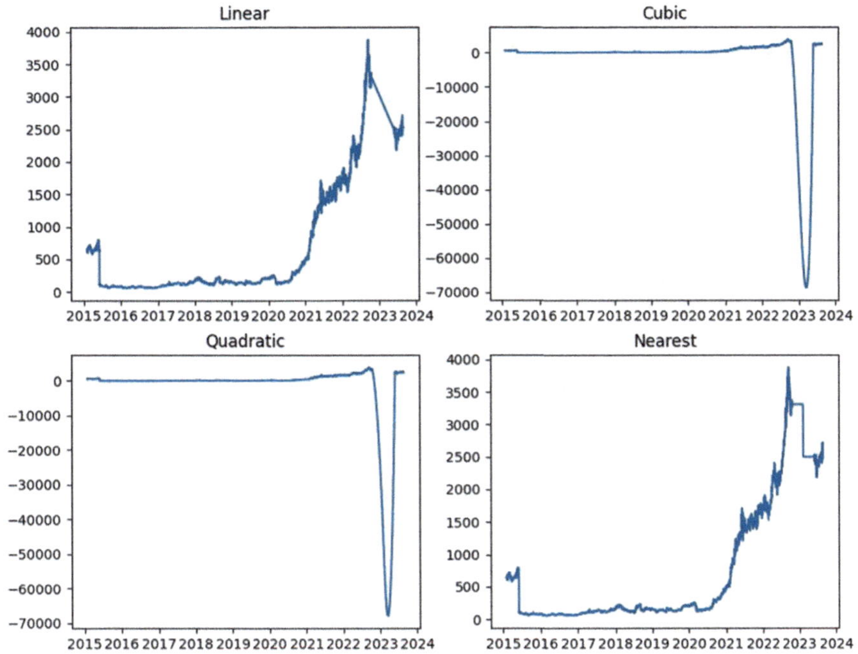

Fig. 3 Graph depicting the interpolated data for the open price variable

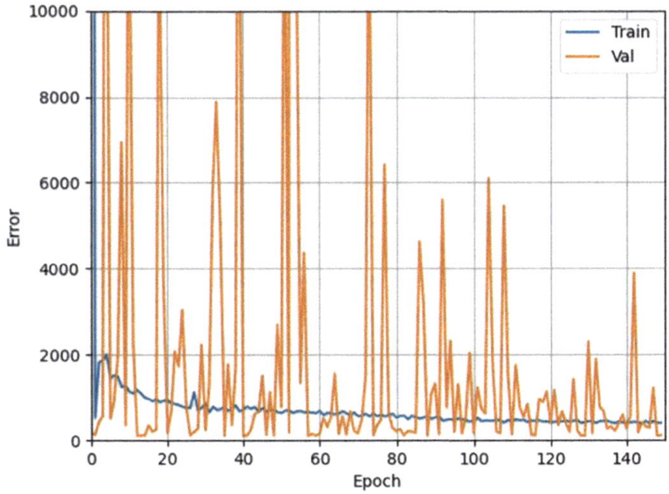

Fig. 4 Graph depicting the variations in the train and validation loss of the artificial neural network model

Table 1 Mean squared error of the models trained

Model	MSE
ANN	401.8860
XGBoost Regressor	140.86921
Random Forest Regressor	155.06058
Support Vector Regressor	36,023.07125

Due to the drawbacks of the previously mentioned models, our paper focused on predicting the stock prices using simple linear regression. The simple linear regression as seen in Eq. (1) models a linear relationship between the variables. It predicts changes in the dependent variable for each unit change in the independent variable. This method involves a single independent variable and a single dependent variable. The linear regression equation is given by [14]:

$$Y = \beta_o + \beta_1 X + e \tag{1}$$

where

- Y Dependent Variable
- β_o Intercept
- β_1 Regression Coefficient
- X Independent Variable
- e Error Term.

5.3 *Feature Selection*

As our dataset consists of five variables and the objective of the project was to predict the closing price, we eliminated the high and low variables and retained only the open and close variables. The open and close variables had a high linear relationship with a correlation coefficient of 1. The connection between close and other factors was also linear exhibiting a correlation coefficient of 1 but for intraday predictions, the open and close variables are more prominent in predicting the prices than the other variables. Thus, we chose two variables: open and close for the estimation of price.

5.4 *Model Evaluation*

Once the chosen data are fit to the model, the Slope and Intercept are obtained with a regression equation Close $= \beta_0 + \beta_1(\text{open})$. The residuals are obtained after subtracting the actual Close values from the predicted Close values that are obtained from the test dataset. The model's performance is assessed using mean squared error (MSE) and root mean squared error (RMSE), which are standard evaluation metrics for regression analysis [15]. These metrics are employed to measure how closely the predicted response value for a specific observation aligns with the true response value for that observation. The formula for mean squared error is given by Eq. (2):

$$\text{MSE} = \frac{1}{n}\sum_{i=1}^{n}\left(y_i - \hat{y_i}\right)^2 \quad (2)$$

where

MSE is the Mean Squared Error
n is the number of observations
y_i is the ith Actual Close Value
$\hat{y_i}$ is the predicted close value obtained for each open price through the regression equation.

When the mean squared error is small, it indicates that the predicted values closely match the actual values, while a large value suggests substantial differences between the predicted and actual values. The root mean squared error (RMSE) given in Eq. (3) is the square root of the mean squared error between the predicted and actual values. RMSE results are presented in the same units as the predicted variable, making it straightforward to gauge the model's performance.

$$\text{RMSE} = \sqrt{\frac{1}{n}\sum_{i=1}^{n}\left(y_i - \hat{y_i}\right)^2} \quad (3)$$

5.5 *Real-Time Streaming Using Apache Kafka*

Apache Kafka is a distributed streaming platform, which can process vast quantities of data in real-time. Initially developed by LinkedIn and now owned by Apache Software Foundation, Kafka has broad applications in constructing data pipelines, leveraging real-time data streams, enabling operational metrics, and data integration across countless sources [16]. For live streaming the prices along with the predicted close prices, we have set up two Kafka Producers and two Kafka Consumers. The Python yahoo_fin module is used to generate the live prices every five minutes. The functionalities of each Producer and Consumer are given below.

Producer: The Producer is responsible for sending the Live Open Prices every five minutes from the start of the session. Each stock ticker is stored in the form of a list.
Consumer: The simple linear regression model trained on each of the stock data was stored in the form of a pickle file. These model files are stored in the form of a dictionary with the stock symbols as the key and the path to the model files as its value.

5.6 Streamlit for Visualization

To visualize the actual and the predicted close prices, we have built an interactive dashboard. We have used the open-source Python library—Streamlit for building the dashboard. Streamlit allows us to develop beautiful web apps and provides an easy-to-use interface for model deployment [17]. Both the CSV files from Consumer 1 and Consumer 2 are loaded in a Python file, and with the help of the plotly module, we have built interactive scatter plots that plot the prices as and when the Excel files are updated.

5.7 Incremental Learning

As the stock prices get updated in the Excel sheets every day, the historical data become less accurate as the data distribution evolves over time owing to the market volatility. After adding the data pertaining to a particular day to the CSV file, a corresponding number of data points is removed from the beginning, depending on the number of new data points that were added. Then the model is trained using the new set of data points, and then, this new model is saved and loaded on the existing saved model.

6 Results

To assess the performance of the proposed model, the residual analysis was performed. Four main assumptions were tested [18]:

1. **Error Terms Are Statistically Independent**

If the error terms have to be statistically independent then, we assume that there exists no correlation between the error terms. To test this, we have used the Durbin-Watson test. The Durbin-Watson test (DW Test) is used to detect the presence of autocorrelation in the errors. The hypothesis is framed as:

- Null Hypothesis: There is no correlation among the errors.
- Alternative Hypothesis: The errors are autocorrelated.

The Durbin–Watson coefficient lies between 0 and 4. The test statistic value of 2 indicates that there is no serial correlation. The test statistic value closer to 0, indicates evidence of positive serial correlation. A value closer to 4 indicates evidence of negative serial correlation.

The results of the Durbin-Watson test for all four stocks are presented in (Table 2). The obtained test statistic value is approximately equal to 2 for stocks of ADANIENT.NS, BIOCON.NS, and, ITC.NS. For RELIANCE.NS, there is a perfect value of 2. Thus, in the model, there is no correlation among the errors in all four stocks and the error terms are statistically independent.

2. **Error Terms Have Constant Variance**

To check for the assumption of constant variance in the error terms, we use the residual plot. This assumption is also known as homoscedasticity. The residual plot displays the fitted values against the residual values. The fitted values will be plotted on the X-axis and the residuals on the Y-axis. If the residuals appear to be randomly scattered around the zeroth axis, then it indicates that the error terms have constant variance. From Fig. 5, it is evident that the errors are scattered around the zeroth axis and there is no heteroscedasticity present in the residuals.

3. **Error Terms Are Normally Distributed**

To ensure the reliability of the results, the error terms should be normally distributed. To test the error terms for normal distribution, we have used the density plot. The residuals are plotted on the X-axis and its corresponding frequencies on the Y-axis. Figure 6 depicts the density plot for the stock.

4. **The Relationship Between the Independent and the Dependent Variables is Linear and Additive**

From Fig. 7, it is evident that there exists a perfect linear relationship between the open and closing price of the stocks.

Table 2 Table depicting the values of the slope and intercept of the regression equation, Durbin–Watson coefficient for normality, and the mean of the residuals

Results of residual analysis				
Stocks	Slope	Intercept	DW test	Residual mean
ADANIENT	1.00	0.27	1.99	0.021
BIOCON	0.99	0.02	1.99	0.002
ITC	0.99	0.24	1.99	− 0.004
RELIANCE	0.99	0.32	2.00	0.02

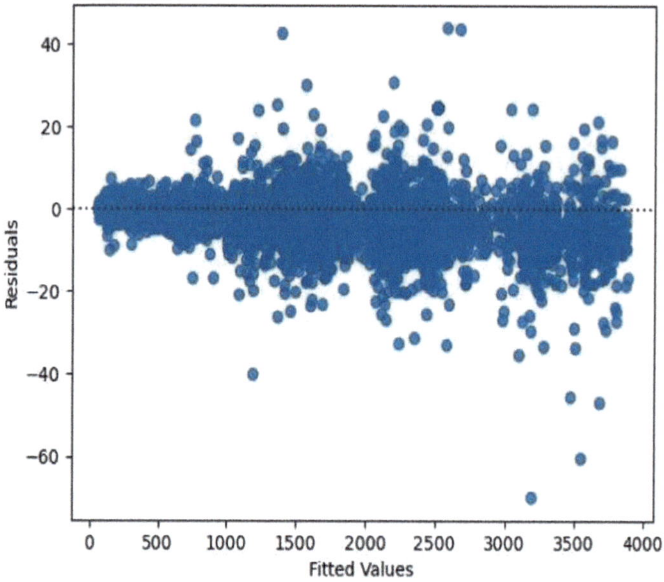

Fig. 5 Graph depicting the residuals versus fitted values

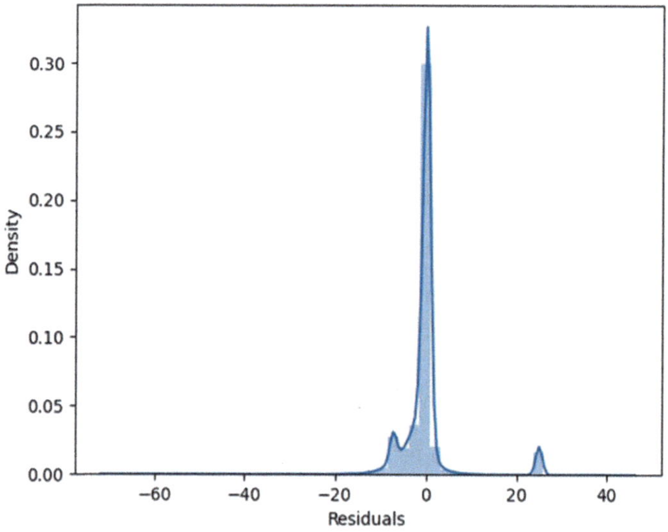

Fig. 6 Density plot depicting the normality of residuals

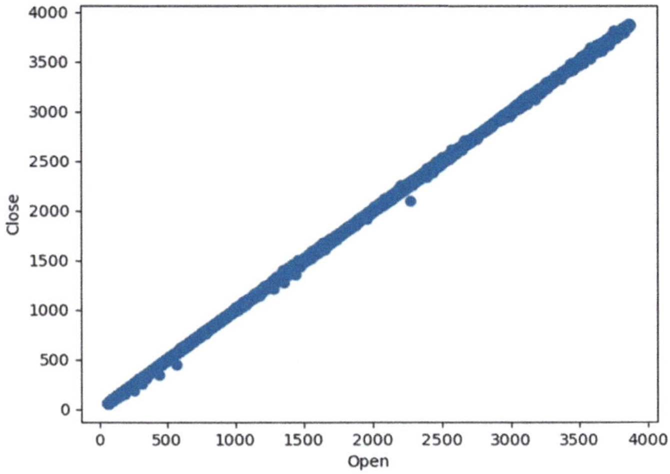

Fig. 7 Scatter plot of open and closing price

Table 3 Table depicting the mean squared error (MSE) and the root mean squared error (RMSE)

Stocks	MSE	RMSE
ADANIENT.NS	34.78650	5.89800
BIOCON.NS	0.38944	0.62405
ITC.NS	0.167084	0.40875
RELIANCE.NS	8.32761	2.88576

Table 2 displays the slope and intercept values of the regression equation for the four stocks. Another important characteristic of the residuals is that their mean should be zero. In our proposed model, the mean of the residuals is indeed zero.

The models are evaluated using the mean squared error and the root mean squared error. From Table 3, it is prominent that the mean squared error and the root mean squared error of the BIOCON.NS and ITC.NS are very low and are closer to zero. The mean squared error and the root mean squared error of ADANIENT.NS and RELIANCE.NS are very negligible compared to the scale of the open and close values in the data. And, thus, our proposed model is a good fit for the given data. Figure 8 shows the scatter plot of the actual and the predicted values with the best-fit line.

The predicted Close Price from the Kafka Consumer is stored in a comma-separated file, which is then read by the streamlit for the creation of the dashboard (Fig. 9).

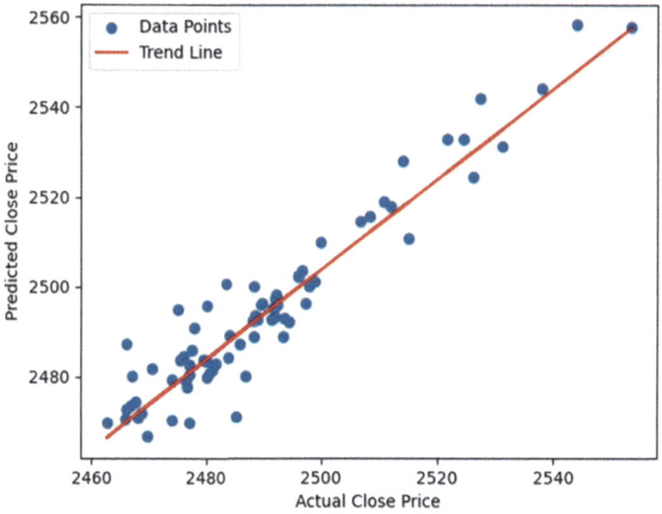

Fig. 8 Scatter plot of actual and the predicted close price

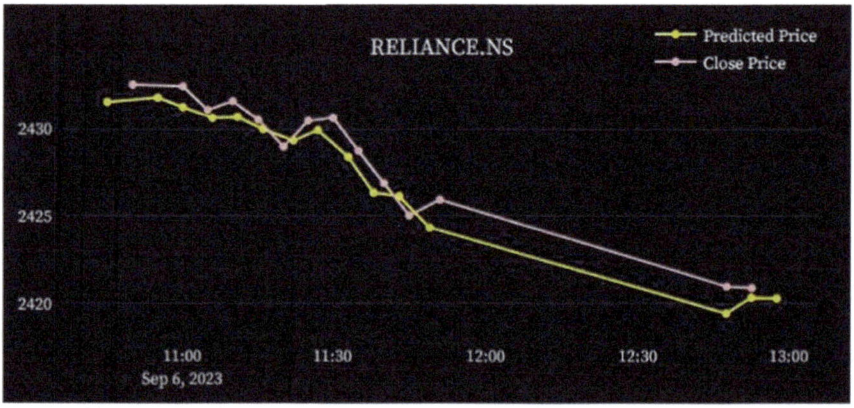

Fig. 9 Snippet of a live dashboard

7 Conclusion

With technological advancements, ease of access to information, awareness of financial trading, and the rise of online trading that offers low commissions, there is an increase in the interest and participation for trading in stocks. Therefore, it is important to understand the volatilities and have the potential to make rapid decisions on stock market.

In the proposed methodology, the simple linear regression model can capture the variations in the dataset and can predict the close price of stocks accurately in real time. Apache Kafka provided a high performance in live-streaming the prices without any latency and was able to handle huge amounts of data. The mean squared error of the models is very minimal compared to the scale of the data values, which proves that the model is better enough to learn the patterns in the data and predicting the stock prices. A stable economy is an indicator of less volatility in stock markets. During the tenure of the project, the stock markets did not have any major disruptions and the predicted results were reliable and accurate. Future works include working on liquid neural networks which are much suited for data collected over a period of time and have the ability to adopt even after training.

References

1. Prasad N (2023) 'Bright spot in global economy': India's GDP has touched $3.75 trillion-mark in 2023, says Nirmala Sitaraman. Retrieved from livemint.com: https://www.livemint.com/economy/bright-spot-in-global-economy-indias-gdp-has-touched-3-75-trillion-mark-in-2023-says-nirmala-sitharaman-11686564064530.html
2. Selvamuthu D, Kumar V, Mishra A (2019) Indian stock market prediction using artificial neural networks on tick data. Financ Innov 5(1):1–12. https://doi.org/10.1186/s40854-019-0131-7
3. Gururaj V, Shriya VR, Ashwini K (2019) Stock market prediction using linear regression and support vector machines. Int J Appl Eng Res. Retrieved from https://www.ripublication.com/ijaer19/ijaerv14n8_24.pdf
4. Wu S, Liu Y, Zou Z, Weng T-H (2021) S_I_LSTM: stock price prediction based on multiple data sources and sentiment analysis. Connection Sci. Retrieved from https://www.tandfonline.com/doi/full/10.1080/09540091.2021.1940101
5. Li Y, Pan Y (2021) A novel ensemble deep learning model for stock prediction based on stock prices and news. Int J Data Sci Anal. Retrieved from https://link.springer.com/article/10.1007/s41060-021-00279-9
6. Xiao D, Su J (2022) Research on stock price time series prediction based on deep learning and autoregressive integrated moving average. Sci Program 2022, Article ID 4758698, 12 p. https://doi.org/10.1155/2022/4758698
7. Aldhyani THH, Alzahrani A (2022) Framework for predicting and modeling stock market prices based on deep learning algorithms. Electronics 11:3149. https://doi.org/10.3390/electronics11193149
8. Mahajan V, Thakan S, Malik A (2022) Modeling and forecasting the volatility of NIFTY 50 using GARCH and RNN models. Economies 10(5):102. https://doi.org/10.3390/economies10050102
9. Liu H, Qi L, Sun M (2022) Short-term stock price prediction based on CAE LSTM method. Wirel Commun Mobile Comput. Retrieved from https://www.hindawi.com/journals/wcmc/2022/4809632/
10. Rishi T (2022) Stock market analysis using linear regression. In: Proceedings of the Jepson undergraduate conference on international economics, p 8. Retrieved from https://scholarworks.uni.edu/cgi/viewcontent.cgi?article=1016&context=jucie
11. Kanwal A, Lau MF, Ng SP, Sim KY, Chandrasekaran S (2022) BiCuDNNLSTM-1dCNN—a hybrid deep learning-based predictive model for stock price prediction. Expert Syst Appl 202:117123. https://doi.org/10.1016/j.eswa.2022.117123

12. Sahoo D (2023) ALGO TRADING DATA—Nifty 100 intraday data. Retrieved from Kaggle.com: https://www.kaggle.com/datasets/debashis74017/algo-tradingdata-nifty-100-data-with-indicators
13. Atwan T (2022) Time series analysis with Python Cookbook. packt. Retrieved from https://www.packtpub.com/product/time-series-analysis-with-pythoncookbook/9781801075541
14. Montgomery DC, Peck EA, Vining G (n.d.) Introduction to linear regression analysis. Wiley
15. James G, Witten D, Hastie T, Tibshirani R (n.d.) An introduction to statistical leaning with applications in R. Springer
16. Dua R, Ghotra MS, Pentreath N (2017) Machine learning with spark, 2nd edn. O'Reilly. Retrieved from https://www.oreilly.com/library/view/machine-learning-with/9781785889936/
17. Streamlit (n.d.) Retrieved from https://docs.streamlit.io/
18. Flatt C, Jacobs RL (2019) Principle assumptions of regression analysis: testing, techniques, and statistical reporting of imperfect data sets. Adv Dev Human Resour. Retrieved from https://www.researchgate.net/publication/335626297_Principle_Assumptions_of_Regresion_Analysis_Testing_Techniques_and_Statistical_Reporting_of_Imperfect_Data_Sets

RSS-Based Localization Using GRU and BiGRU Deep Learning Models in LoRaWAN-IoT Networks

R. Swathika and S. M. Dilip Kumar

Abstract Long Range Wide Area Network (LoRaWAN) has attracted much attention in research and business as a significant IoT enabler. It offers a desired choice for applications that utilize hundreds or thousands of actively linked devices to monitor a process or environment or aid in supervising a specific operation. Measurement of the Received Signal Strength Indicator (RSSI) of signals from numerous transmitters or reference sites is the foundation of Received Signal Strength (RSS)-based localization. The position of the devices or objects can be ascertained by contrasting the RSSI readings from these established locations. In this study, we investigated an indoor localization mechanism based on a Deep Learning (DL) algorithm called Gated Recurrent Unit (GRU) and by the strength of the received signal. To further improve the accuracy of the GRU, we have used the Bidirectional GRU (BiGRU), which gives the highest accuracy of 93%. The experimental results also analyzed the comparison of regression, classification, and softmax layers in the set of forward and backward layers in the BiGRU model.

Keywords LoRaWAN · DL · RSSI · GRU · BiGRU · Localization

1 Introduction

Internet of Things (IoT) systems and connected devices are being used increasingly, increasing the demand for efficient management of information and communication [1]. These systems frequently use low-power wide-area network technologies like LoRa and SigFox for longer-distance communication with low data rates. It opens up possibilities for cutting-edge technologies like indoor positioning systems and machine learning-based sensing devices, which can decrease power consumption and lower the price of battery-operated gadgets. These devices' data can have

R. Swathika (✉) · S. M. Dilip Kumar
University of Visvesvaraya College of Engineering, Bangalore University, Bengaluru, India
e-mail: swathikaramasamy@gmail.com

S. M. Thampi et al. (eds.), *Fifth International Conference on Computing and Network Communications*, Lecture Notes in Electrical Engineering 1219,
https://doi.org/10.1007/978-981-97-4540-1_11

a vast market influence because they can be used for localization, indoor tracking, extensive data research, and tailored services based on location or other user-specific information [2].

Industry research focuses primarily on enhancing the localization accuracy of IoT devices, especially in hazardous and demanding work environments. Due to its low cost, low energy consumption, and broad coverage, LoRa wireless technology is required to develop indoor applications further. Localization uses an RF signal to estimate the distance between a transmitter and a receiver. One of the following methods can be employed to do the ranging or localization, depending on the RF signal's measured property: RSSI, Time of Arrival (ToA), Time Difference of Arrival (TDoA), and Angle of Arrival (AoA) [3].

RSSI-based localization is an essential and efficient way of localization. Using RSS measurements to localize devices is known as RSS localization. These RSS measurements are gathered from the Gateways (GWs) dispersed throughout the LoRaWAN network, either using the trilateration approach, which is the intersection of three GWs, or the multilateration method, which employs more than three GWs. Additionally, various strategies, including fingerprinting methods, ML algorithms, DL algorithms, optimization algorithms, and many more, can be employed for RSS localization in LoRaWAN [3].

This paper examines the spectrum of localization errors produced by LoRaWAN IoT technology and its usability in RSSI-based localization applications. To improve the localization of the RSSI-based localization, logged RSSI values are averaged. This averaging can be done in linear, exponential, or powed RSS models and all these 3 models are taken into consideration during the simulation setup for calculating the localization accuracy. GRU and BiGRU models have been simulated to improve the accuracy of the RSSI-based localization and other parameters like error, R-squared, and evaluation time.

1.1 Paper Organization

The rest of the paper is formulated as follows: Sect. 2 provides the background details of this paper's related works with objectives, and limitations of the existing works. Section 3 briefs the research gap, problem statement and objectives of this paper. Section 4 demonstrates the GRU and BiGRU models algorithm for RSSI-based localization and associated metrics. Section 5 presents numerical results in tabular form related to GRU and BiGRU models, varying the layers like regression, classification, and softmax in the above models. Section 6 concludes with a conclusion and future work.

2 Background

This section explains the related works of this paper and summary of related works is updated in Table 1.

When evaluating accuracy and error in terms of distance and assessment time, Kwasme et al. [4] used ten different machine learning methods. The 340 m location estimation inaccuracy was accomplished utilizing the forest regression approach. Although its performance differs from the random forest regression point, even k-means generated the same accuracy. Anjum et al. [5] software-defined radios to determine the RSSI-based localization's performance characteristics. It is discovered that the vast variance in RSSI would impact the localization performance. Many methods can be developed to solve the aforementioned issue in the future.

To find the RSSI-based localization, Aarif et al. [6] used 2 different indoor and outdoor situations and specific ML techniques such as decision trees, support vector machines, and a few more methods. Using the trilateration approach to observe the outcomes experimentally, it contrasts LoRaWAN with ZigBee, WiFi, and Bluetooth. Environmental conditions substantially impact the signal quality. Therefore, the solution can be practically provided using DL models. Yoshitome et al. [7] attempted to

Table 1 Summary of related works

Paper	Objectives	Limitations
[4]	To minimize the error in location estimation	Still location estimation error can be improved
[5]	To assess the accuracy of LoRaWAN RSSI-based localization	High variance in RSSI
[6]	To conduct an empirical assessment of localization accuracy	Accuracy of LoRa is less when compared with other LPWAN technologies
[7]	To calculate RSSI in both NLOS and LOS scenarios	Power loss is high in NLOS scenarios
[8]	To use a GPS technique, wherever LoRa localization is not possible	GPS not performs well for indoor environment
[9]	To build a localization system that is economical and energy-efficient	Higher accuracy in LOS than in NLOS
[10]	To compare LoRaWAN, WiFi, ZigBee localization	WiFi proves to be more accurate than LoRaWAN
[11]	To use supervised machine learning techniques to train the RSSI localization	Decreasing the reference nodes will impact the accuracy
[12]	To calculate the average localization error in densely populated areas	Only applicable for few applications
[13]	To use machine learning techniques to attain localization accuracy	Accuracy can be improved further
[14]	To use DL models to localize the position in a dynamic environment	Only for smaller indoor space

measure an RSSI using LOS and NLOS in an indoor short-distance scenario. Compared to NLOS, the signal strength in LOS situations has suffered the most minor loss. Long-distance interior and outdoor environments can be used to test the future direction, and it can be compared to similar technologies like WiFi.

When the real localization strategy failed in some circumstances, Ingabire et al. [8] employed a GPS to localize the device placements. However, constantly operating a GPS technique will cost more money and energy. At that point, nodes without GPS can offer a localization estimate. The RNN technique was used by Sadowski et al. [9] to create a reduced power localization system for indoor applications. Comparing the LOS and NLOS scenarios to other relevant methodologies demonstrates that the localization error is minimized in the LOS and NLOS scenarios.

Maduranga et al. [10] examined wireless technologies like WiFi, BLE, ZigBee, and LoRaWAN for energy usage and localization accuracy. It localizes objects using RSSI data and the trilateration approach. To locate the localization and the precise RSSI values transmitted to the remote area's MQTT server, Piroddi et al. [11] employed an ML technique. Evaluation and discussion of the localization error between the expected and actual values. According to Perkovic et al.'s [12], localization is desired in very crowded metropolitan settings. It employs an ML technique with fingerprinting RSSI-based localization to estimate the average localization accuracy and energy usage.

Through the use of a machine learning (ML) strategy based on neural networks, Ali et al. [13]. Up to 98% accuracy is predicted by the device. In addition to using the ML approach in the microcontroller, it may be utilized with other ML approaches and compared to them. The neuron architecture is used by Hoang et al. [14] to determine the localization accuracy. To adapt to the network change, the accuracy is assessed between the actual and anticipated places. The suggested system adapts well to shifting environmental factors. The majority of the applications use supervised learning, which is predicated on the idea that the training and test sets of data are independent and identical [15]. This presumption is frequently broken in the actual world as a result of changes in distribution or domain between the source plus the desired information Therefore, it's crucial to make sure that the algorithms are capable of handling data that is not distributed evenly [16]. Domain generalization (DG) addresses the OOD-related problems in this context by learning models on various and unique sources [17].

3 Problem Statement and Objectives

The problem is to design an accurate indoor localization system in LoRaWAN, which is tedious. Traditional methods like GPS, triangulation, and multilateration techniques can only support a few devices, most suitable for outdoor localization only and less reliable when compared to the RSSI-based localization techniques. RSSI-based localization is ideal for both indoor and outdoor localization. However, using RSSI-based localization can be less accurate when the propagation of the signal is non-linear. Also, it is challenging during signal reflection and attenuation. DL models can be proposed along with the RSSI-based localization to maximize localization accuracy and support the devices' scalability. Here, we have considered a simple DL model called GRU, a neural network to solve the overcomes of the traditional localization methods. The objectives is to:

1. Compare GRU and BiGRU DL models to find the localization accuracy of the devices using RSS data.
2. Use the different variations in layers of the considered DL models to find the best model to predict the highest accuracy.
3. Predicting the highest accuracy, highest error, lowered R-squared and less evaluation time from the considered DL models.

4 Proposed Methodology

In this section, the algorithm of DL model GRU related to RSSI-based localization is discussed along with the required formulas and parameter variations.

4.1 Gated Recurrent Unit (GRU)

The GRU is a Recurrent Neural Network (RNN), and it can be employed for recording patterns and correlations in the RSSI data over time in the context of RSSI-based localization in LoRaWAN. It can be used as a variant of RNN to address some of RNN's drawbacks, including the vanishing gradient problem and its inability to capture long-term inter-dependencies. It is to locate temporal patterns among RSSI data acquired from LoRaWAN nodes [18]. The GRU model's output might be applied to fully connected layers of a future neural network to forecast latitude and longitude coordinates. The utilization of regression or classification techniques and the organization of the localization model will influence the formulation and overall performance. Additional post-processing procedures can be required to further transform the model's output into proper geographical coordinates. One of these steps involves converting projected values to geographic coordinates based on calibration or reference data [19].

The main idea of the GRU is the introduction of gating mechanisms, which enable the model to gather and disseminate pertinent information over lengthy periods while managing the information flow. The states in the GRU are hidden state (h_t), update gate (u_t), reset gate (r_t), candidate hidden state (c_t) and hidden state update (hu_t). It maintains a hidden state, designated as h_t, at each time step t that stores details about the sequence up to that moment. Every time step, this hidden state, which serves as the network's memory, is updated.

The update gate, abbreviated as (u_t), is a sigmoid activation function that specifies how much of the current hidden state should be added to the new candidate hidden state (h_t), and how much of the prior hidden state (h_{t-1}), is to be retained. The update gate evaluates both the most recent hidden state and the input at present is:

$$h_t = \sigma\left(lp_{u1} \times y_t + lp_{u2} \times h_{t-1} + lp_{u3}\right) \tag{1}$$

where y_t denotes the input at time t or RSSI input measurements. lp_{u1}, lp_{u2} and lp_{u3} represents the learning rate parameters and σ is the activation function. The reset gate, represented as r_t, is a sigmoid activation function that regulates how much of the candidate hidden state h_t should be updated and how much of the previous hidden state h_t should be forgotten and it is written as:

$$r_t = \tanh\left(lp_{r1} \times y_t + lp_{r2} \times h_{t-1} + lp_{r3}\right) \tag{2}$$

where y_t denotes the input at time t or RSSI input measurements, h_{t-1} is the previous hidden state, lp_{r1}, lp_{r2} and lp_{r3} represents the learning rate parameters and tanh is the activation function. The recommended update to the hidden state based on the current input and the reset gate is recognized as the candidate hidden state, denoted as (c_t):

$$c_t = \sigma\left(lp_{h1} \times y_t + lp_{h2} \times (r_t \times h_{t-1}) + lp_{h3}\right) \tag{3}$$

where y_t denotes the input at time t or RSSI input measurements, r_t is the reset gate, h_{t-1} is the previous hidden state, lp_{h1}, lp_{h2} and lp_{h3} represents the learning rate parameters and tanh is the activation function. The candidate hidden state c_t and the previously hidden state h_{t-1} are combined to create the new hidden state h_t or hu_t, which is managed by the update gate:

$$h_t = (1 - u_t) \times h_{t-1} + u_t \times c_t \tag{4}$$

Algorithm 1 outlines the GRU deep learning model for RSSI localization in LoRaWAN. It starts with the collection of data like RSSI values from LoRaWAN nodes. Then, preprocessing the data, set the GRU model with unit layers, input and output units. Adjust the weight and biases to train the GRU model. Evaluate the model's performance in terms of accuracy and other parameters.

The extended version of the BiGRU model is also used to improve the performance. Here, the forward GRU and backward GRU layers are used. Combining these two layers produces a Bidirectional layer. The input layer, output layer, and fully connected layer are used. The fourth layer used in this paper is the classification, regression, and softmax layers in different simulation scenarios. Combining all these layers gives the DL model.

Algorithm 1 GRU for RSSI localization in LoRaWAN

Load RSSI LoRaWAN data and locations.
Load X with actual data and Y with preprocessed code.
Train X (RSSI values).
Train Y (Device locations)
Normalize RSSI data.
Split into RSSI train and test data.
Define GRU model such as input units, hidden units and output size.
Create GRU model with hidden units, output mode and following layers:
    Create sequence input layer with input size.
    Create fully connected layer with output size.
    Create regression layer.
    Train the GRU model.
Predict the location using using trained model.
Denormalize the predicted locations.
Evaluate *accuracy, error, R-squared and time.*

5 Simulation Results

Setting up the environment, producing synthetic data, implementing localization algorithms, and assessing their effectiveness are all processes in creating a simulation environment for RSSI-based localization in LoRaWAN using MATLAB. The simulation environment used here is MATLAB R2020a with the inbuilt functions of all the 7 DL models considered in this paper. The dataset has 72 characteristics and 130,426 LoRaWAN message samples gathered in Antwerp, Belgium. Each row of a dataset represents a LoRaWAN message and includes information on the message's receiving base stations (BS), the time at which the message was received (RX Time), the LoRa spreading factor, and the transmitter's latitude and longitude at the time of transmission. The parameter setup for the simulation is tabulated in Table 2. GRU uses the regression layer as the last layer. BiGRU uses 2 sets of layers, namely forward layers and backward layers. The regression layer is used as the last layer in the initial setup in both the forward and backward layers.

Table 2 GRU and BiGRU parameter setup

Parameters	Values
Hidden units	64
Fully connected layer	2 classes
Max epochs	100
Mini batch size	27
Activation layer	Regression
Classification output	Crossentropyex
Output mode	Last
Sequence input	1 dimension

Table 3 Comparison of linear, exponential and powed RSS using GRU in LoRaWAN-IoT networks

Parameters	Linear RSS	Exponential RSS	Powed RSS
Accuracy	90.909	87.912	89.888
Error	9.0909	12.088	10.112
R-squared	0.91328	0.90325	0.90693
Time	8.7891	9.1933	8.7978

Table 4 Comparison of linear, exponential and powed RSS using BiGRU in LoRaWAN-IoT networks

Parameters	Linear RSS	Exponential RSS	Powed RSS
Accuracy	93.023	85.106	93.023
Error	6.9767	14.894	6.9767
R-squared	0.90615	0.91799	0.90615
Time	9.2207	8.46	9.2207

Table 3 compares linear, exponential, and powed RSS using the GRU deep learning model in LoRaWAN-IoT networks. From the comparison, it is observed that linear RSS has the highest accuracy of 90.909%, exponential RSS has the highest error rate of 12.088%, exponential RSS has the lowest *R*-squared and power RSS has the most down simulation time. Table 4 tabulates the comparison of linear, exponential, and powed RSS using BiGRU deep learning model in LoRaWAN-IoT networks. From the comparison, it is observed that linear RSS has the highest accuracy of 93.023%, exponential RSS has the highest error rate of 14.894%, exponential and power RSS has the lowest *R*-squared and exponential RSS has the less simulation time.

5.1 Layer Variations in GRU and BiGRU

Table 5 tabulates the linear, exponential, and powed RSS comparison using GRU layer variations deep learning model in LoRaWAN-IoT networks. From the comparison, it is observed that the regression layer has the highest accuracy of 93.023%, softmax has the highest error rate of 11.111%, regression has the lowest *R*-squared, and softmax has the less simulation time in linear RSS. Softmax layer has the highest accuracy of 87.912%, regression has the highest error rate of 14.894%, softmax has the lowest *R*-squared, and regression has the least simulation time in exponential RSS. The regression layer has the highest accuracy of 93.023%, classification has the highest error rate of 11.111%, lowest *R*-squared, and less simulation time in powed RSS. Figure 1 represents the comparison of performances using GRU layer variations.

Table 6 tabulates the comparison of linear, exponential, and powed RSS using BiGRU layer variations deep learning model in LoRaWAN-IoT networks. The for-

Table 5 Comparison of linear, exponential and powed RSS using GRU layer variations in LoRaWAN-IoT networks

RSS type	Parameters	Layers		
		Regression	Classification	Softmax
Linear	Accuracy	93.023	90.909	88.889
	Error	6.9767	9.0909	11.111
	R-squared	0.90615	0.90716	0.93664
	Time	9.2207	9.6192	8.9721
Exponential	Accuracy	85.106	86.022	87.912
	Error	14.894	13.978	12.088
	R-squared	0.91799	0.91945	0.90325
	Time	8.46	9.4773	9.1933
Powed	Accuracy	93.023	88.889	88.889
	Error	6.9767	11.111	11.111
	R-squared	0.90615	0.89784	0.91706
	Time	9.2207	8.9721	8.9933

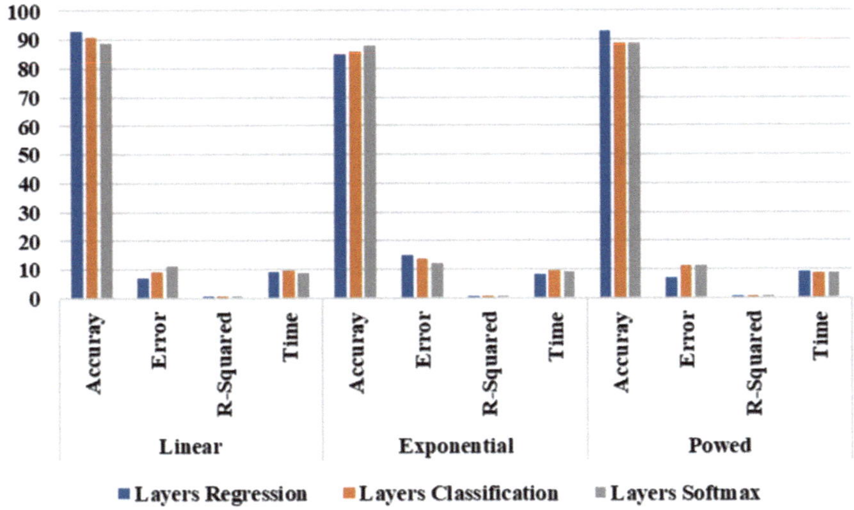

Fig. 1 Performance results comparison of linear, exponential and powed RSS using GRU layer variations

Table 6 Comparison of linear, exponential and powed RSS using BiGRU layer variations in LoRaWAN-IoT networks

RSS type	Parameters	Forward layers	Backward layers		
			Regression	Classification	Softmax
Linear	Accuracy	Regression	93.023	86.022	89.888
	Error		6.9767	13.978	10.112
	R-squared		0.90615	0.9139	0.89436
	Time		9.2207	9.1597	11
Linear	Accuracy	Classification	89.888	90.909	89.888
	Error		10.112	9.0909	10.112
	R-squared		0.9041	0.90716	0.91585
	Time		8.8956	9.6192	9.3483
Linear	Accuracy	Softmax	87.912	88.889	88.889
	Error		12.088	11.111	11.111
	R-squared		0.89733	0.89216	0.93664
	Time		8.6838	9.3551	8.9721
Exponential	Accuracy	Regression	85.106	90.909	88.889
	Error		14.894	9.0909	11.111
	R-squared		0.91799	0.92114	0.89421
	Time		8.46	9.4297	9.535
Exponential	Accuracy	Classification	89.888	86.022	88.889
	Error		10.112	13.978	11.111
	R-squared		0.8864	0.91945	0.91478
	Time		9.3065	9.4773	9.2305
Exponential	Accuracy	Softmax	87.912	90.909	87.912
	Error		12.088	9.0909	12.088
	R-squared		0.9184	0.91371	0.90325
	Time		9.6655	9.5118	9.1933
Powed	Accuracy	Regression	93.023	86.022	90.909
	Error		6.9767	13.978	9.0909
	R-squared		0.90615	0.89212	0.88549
	Time		9.2207	10.179	10.026
Powed	Accuracy	Classification	89.888	88.889	86.957
	Error		10.112	11.111	13.043
	R-squared		0.90853	0.89784	0.89701
	Time		9.3561	8.9721	8.8339
Powed	Accuracy	Softmax	87.912	86.957	88.889
	Error		12.088	13.043	11.111
	R-squared		0.91756	0.90356	0.91706
	Time		9.204	9.526	8.9933

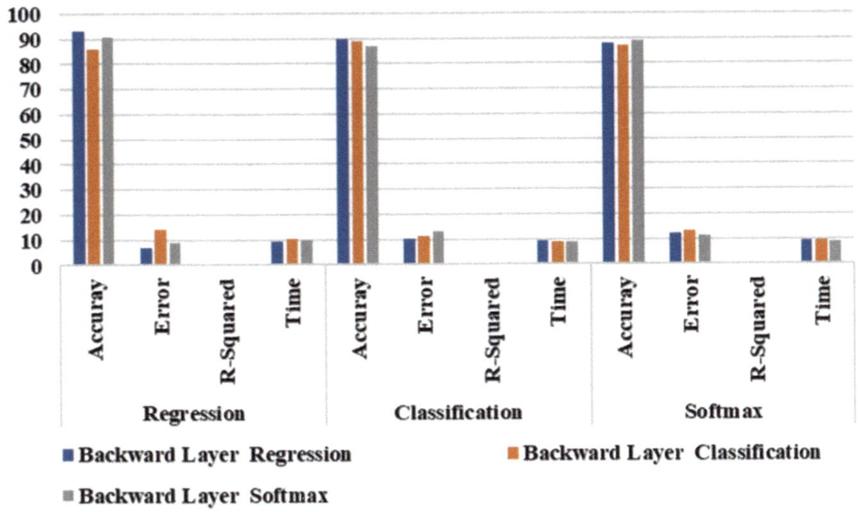

Fig. 2 Performance results comparison of linear RSS using BiGRU layer variations

ward and backward layers combine regression, classification, and softmax. From the comparison, it is observed that the forward layer as regression and the backward layer as regression have the highest accuracy of 93.023%, backward layer as a classification has the highest error rate of 13.978%, backward layer as softmax has the lowest *R*-squared and backward layer as a classification has the less simulation time in linear RSS. The forward layer as regression and backward layer as a classification has the highest accuracy of 90.909%, backward layer as regression has the highest error rate of 14.894%, backward layer as softmax has the lowest *R*-squared and backward layer as regression has the less simulation time in exponential RSS. The forward layer as regression and backward layer as regression has the highest accuracy of 93.023%, the backward layer as a classification has the highest error rate of 13.978%, the backward layer as softmax has the lowest *R*-squared and the backward layer as softmax has the less simulation time in powed RSS. Figures 2, 3, and 4 represents the comparison of performances using BiGRU layer variations with linear, exponential and powed RSS respectively.

The highest accuracy of the BiGRU model is achieved by using the regression as forward and the backward layer in linear RSS and powed RSS. Then, a forward layer is a regression, and the backward layer is a classification in exponential RSS. Regression, classification, or softmax output should be chosen depending on the particular requirements of your localization work. Regression is helpful if exact coordinate predictions are needed. If the location of the devices needs to be mapped into zones, classification is the best option. A softmax layer can be utilized if probability distribution over several possibilities.

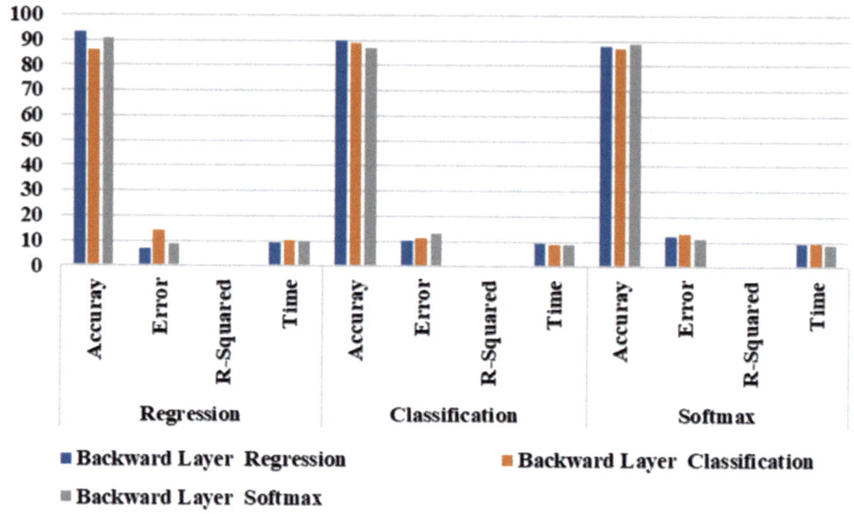

Fig. 3 Performance results comparison of exponential RSS using BiGRU layer variations

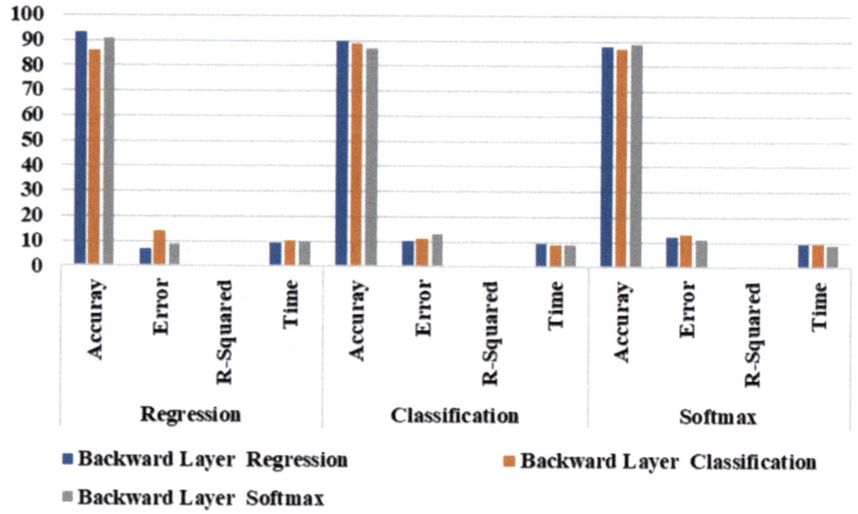

Fig. 4 Performance results comparison of powed RSS using BiGRU layer variations

6 Conclusion

Maximizing the RSS-based localization accuracy using GRU and BiGRU DL model is simulated and evaluated for the accuracy, error, R-squared, and evaluation time parameters. The layers in the model are varied concerning regression, classification, and softmax and also simulated for GRU and BiGRU models. The accuracy attained is 93% in the BiGRU model, which is higher when compared to the standard GRU model. The permutations and combinations of different layers are also analyzed and evaluated. The highest accuracy arrived for the BiGRU model is using regression as a forward and backward layer in linear RSS and powed RSS. Whereas regression is forward and classification is a backward layer in exponential RSS. But still the performance parameters can be greatly improved by considering the better approach of DL models and optimizer. The future work can include assessing the performance of some other DL models, hybrid models, and the use of optimizers like Adam, Adagrad, and RMSprop on top of DL models.

References

1. Akshatha PS, Dilip Kumar SM, Venugopal KR (2022) MQTT implementations, open issues, and challenges: a detailed comparison and survey. Int J Sens Wireless Commun Control 12(8):553–576
2. Swathika R, Dilip Kumar SM (2023) Optimizing throughput using effective contention aware adaptive data rate in LoRaWAN. In: Proceedings of international conference on frontiers in computing and systems. COMSYS 2022. Lecture notes in networks and systems, vol 690. Springer, Singapore
3. Swathika R, Dilip Kumar SM (2023) Analysis of BER performance over AWGN and Rayleigh channels using FSK and PSK modulation schemes in LoRa based IoT networks. In: 2023 international conference on intelligent and innovative technologies in computing, electrical and electronics (IITCEE), Bengaluru, India, pp 351–357
4. Kwasme H, Ekin S (2019) RSSI-based localization using LoRaWAN technology. IEEE Access 1
5. Anjum M, Khan MA, Hassan SA, Mahmood A, Qureshi HK, Gidlund M (2020) RSSI fingerprinting-based localization using machine learning in LoRa networks. IEEE Internet Things Mag 3(4):53–59
6. Aarif L, Tabaa M, Hachimi H (2022) Experimental test and performance of RSSI-based indoor localization in LoRa networks. Procedia Comput Sci 203:420–425
7. Yoshitome EH, da Cruz JVR, Monteiro MEP, Rebelatto JL (2021) LoRa-aided outdoor localization system: RSSI or TDoA? Internet Technol Lett 5(2):e319
8. Ingabire W, Larijani H, Gibson R, Qureshi A (2022) LoRaWAN based indoor localization using random neural networks. Information 13(6):303. https://doi.org/10.3390/info13060303
9. Sadowski S, Spachos P (2018) RSSI-based indoor localization with the internet of things. IEEE Access 6:30149
10. Maduranga P, Abeysekera R (2021) Supervised machine learning for RSSI based indoor localization in IoT applications. Int J Comput Appl 183(3):26–32
11. Piroddi A, Torregiani M (2023) Machine learning applied to LoRaWAN network for improving fingerprint localization accuracy in dense urban areas. Network 3(1):199–217
12. Perkovic T, Dujic Rodic L, Sabic J, Solic P (2023) Machine learning approach towards LoRaWAN indoor localization. Electronics 12(2):457

13. Ali IT, Muis A, Sari RF (2021) A deep learning model implementation based on RSSI fingerprinting for LoRa-based indoor localization. EUREKA Phys Eng 1:40–59
14. Hoang MT, Yuen B, Dong X, Lu T, Westendorp R, Reddy K (2019) Recurrent neural networks for accurate RSSI indoor localization. IEEE Internet Things J 6(6):10639–10651
15. Darbinyan R, Khachatrian H, Mkrtchyan R, Raptis TP (2023) ML-based approaches for wireless NLOS localization: input representations and uncertainty estimation. In: 2023 IEEE ninth international conference on big data computing service and applications (BigDataService), Athens, Greece, pp 87–94. https://doi.org/10.1109/BigDataService58306.2023.00019
16. Akrout M, Feriani A, Bellili F, Mezghani A, Hossain E (2023) Domain generalization in machine learning models for wireless communications: concepts, state-of-the-art, and open issues. arXiv e-prints. https://doi.org/10.48550/arXiv.2303.08106
17. Choi H, Oh J, Chung J, Alexandropoulos GC, Choi J (2023) WiThRay: a versatile ray-tracing simulator for smart wireless environments. IEEE Access 11:56822–56845. https://doi.org/10.1109/ACCESS.2023.3283610
18. Dey R, Salem FM (2017) Gate-variants of gated recurrent unit (GRU) neural networks. In: 2017 IEEE 60th international Midwest symposium on circuits and systems (MWSCAS), pp 1597–1600
19. Touzani Y, Douzi K (2021) An LSTM and GRU based trading strategy adapted to the Moroccan market. J Big Data 8(1):126

Locating and Routing for IoT-Based Fire Evacuation System

Vineeth Krishnan, Govind Santhosh Kumar, Prarthana Anand, and M. Rajesh

Abstract Fire in skyscrapers and high-rise building is inevitable due to huge volumes of thick gauged wires crawling all over the building, the rise in the number of electronic equipment inside and highly unpredictable acts of nature. These fires turn devastating when the latency in the response and evacuation system is not able to keep up with the enormity associated with modern towering buildings. As we move on to a more connected world, the amount of data generated by the equipment inside the building can now not only be used for their intended purpose. But together they provide a clearer and accurate view of the situation inside the building in real time. This Connected Fire Evacuation System takes in data from the various sensors as well as CCTV devices inside the building to present a real-time link for assisting trapped occupants by actively locating them and planning an escape route for them.

Keywords Dijakstra's algorithm · Fire evacuation system · Internet of Things · Rescue assistance · Video identification

1 Introduction

The acronym "OTT" refers to "Over-The-Top," portraying a contemporary digital content delivery method that skips traditional cable or satellite TV providers to distribute media content to consumers over the internet. According to a 195-nation analysis by Global Diseases Burden India recorded 27,027 deaths, every fifth fire-related death in the world took place in India out of the 9 million fire accidents and 1.2 lakh deaths across the globe.

V. Krishnan · G. Santhosh Kumar · P. Anand · M. Rajesh (✉)
Department of Computer Science and Engineering, Amrita School of Computing, Amrita Vishwa Vidyapeetham, Bengaluru, India
e-mail: m_rajesh@blr.amrita.edu

S. M. Thampi et al. (eds.), *Fifth International Conference on Computing and Network Communications*, Lecture Notes in Electrical Engineering 1219,
https://doi.org/10.1007/978-981-97-4540-1_12

A fire outbreak anywhere is a disaster and when inside a densely occupied turns into a traumatic event and the scale in which a disaster would happen is only going to increase as skyscrapers define the skyline of megacities with an ever-increasing population moving into urban areas and increase in urbanization in other areas as available land decreases. Another trend that can also be seen in past few years is huge increase in the number of connected devices in buildings. As more and more buildings turn smart, the Internet of Things (IoT) system inside such building also becomes more efficient and productive and opens new avenues of application and interest. Conventional fire alarm systems provide many small buildings with an adequate and cost-effective fire alarm system. In a more complex environment, a more 'intelligent' system is necessary to improve detection speed, accurate identification of incident spot and location of occupants.

In addition to the fire itself, a major cause of casualties is the panic that usually accompanies a poorly organized evacuation plan. The evacuation would require a large number of people to pass through narrow constricted hallways and corridors, which is a recipe for catastrophe if not organized systematically. By providing our users with a well-calculated and systematic evacuation plan that takes into consideration of their relative positioning to the fire and debris caused by the same, we ensure that there are no additional casualties credited to trampling. The system proposed in this paper ensures even distribution of foot traffic across all available exit nodes in order to ensure that these nodes do not get overly crowded and works in real time to find alternative routes to those blocked by fires and fallen debris.

On top of all these, additional and real-time information available to first responders would help perform in a more efficient manner. Lack of information about the building layout, blockages and alternative exit routes leads to an inefficient strategy for fire mitigation and safety services. Eyewitness information in such a chaotic situation can often be not very useful or will not help in coming with accurate strategies with pinpoint accuracy. Hence, there is a delay in getting the relevant information about the on-ground situation and this can lead to further delay in strategizing the recuse plan.

2 Related Works

In [1], the author talks about how we can use modern technology to notify authorities faster. The paper proposes to utilize an application that will notify the authorities and victims as soon as a fire is detected. This system fails to take advantage of modern technologies like IoT and Cloud Computing. IoT would give us a more targeted approach. It is planned to make use of edge computing and cloud-based connectivity for real-time performance and updates. This would give more detailed solution and would be an easier way to deal with the problem.

In [2], it discusses about the use of IoT devices to detect fires. This would utilize devices such as CCTVs and mics which are embedded into most modern-day structures to detect fires. It is planned to improve upon this system by also integrating a module that would enable to make use of graph-distance algorithms to guide the victims to safety. Dijkstra's algorithm is used to predict the most optimum path and help victims and firemen respond more efficiently.

In [3], it utilizes a flame and heat sensor to detect the fire. The system then notifies first responders of the emergency via GSM notification. The system then starts to direct the victims to the nearest exit from their current location. Although GSM would notify the authorities in a timely manner, the use of IoT device would help firemen to adopt a more targeted approach. Also, a system is proposed that works both ways, i.e., a system that not only provides the victims with an exit strategy but also the first responders with a route to nodes with trapped victims.

In [4], the author succeeds in realizing the potential of IoT in deciding paths for efficient evacuation. However, the author overlooks the advantage of using IoT devices to notify the authorities which would result in a faster response. Having an automated system to notify the first responders gives a tremendous advantage as it shaves off precious minutes from the rescue procedure and could be the determining factor between life and death.

In [5], it proposes an IoT-based evacuation system coupled with an exit strategy. However, the exit strategy is not dynamic and doesn't take into consideration paths blocked by fires, debris, etc. A fire emergency brings out any gaps in the structural integrity of the building and can make them more pronounced, often leading to a cave-in or collapse. Having a fixed exit path does not take these possibilities into consideration and can end up with the route blocked by such a cave-in. Our system accounts for debris and reroutes the user accordingly hence is more efficient and further reduces casualties.

3 Proposed System

The entire architecture can be divided into two parts:

3.1 The On-Premise Module

The on-premise module consists of the various sensors (Sound/Microphone, Smoke Sensor, Heat/Flame Sensor) and the edge computing device (A Raspberry Pi). These devices together formulate the components and act as the source of all information/data necessary for the system. This information is later on transferred to the off-premise module for processing. The on-premise system acts as a brain/smart component for all the sensors and the nodal point for connecting to the internet.

3.2 The Off-Site Module

The off-site module takes care of all the intensive computation task; the data is parsed through model on IBM cloud, and the dashboard is accessible through an Amazon S3 bucket. The IBM cloud is used to leverage the image recognition alongside OpenCV. The IBM cloud and AWS communicate through API calls among them. The user can also access the off-premise module through their personal device as well.

3.3 Architecture Diagram

Figure 1 portrays the architecture of the entire project. The data is fetched from the sensors and relayed to the cloud using the edge computing device. This information is processed using IBM Watson for the audio and visual classification. The dashboard and associated information can be accessed through a web app hosted on an AWS 3 bucket which interacts to the IBM cloud though multiple API calls.

Figure 2 describes the flow of data through the various interconnected modules that work in tandem to provide the necessary output for both the on-premise module as well as the off-site module.

The Raspberry Pi accumulates data from the CCTVs, smoke and sound sensors and regulates the data flow to the different modules in our system as well act as the nodal point for exchange of information from the in-house system to the cloud infrastructure that is located off-site. The data received is passed on to the respective

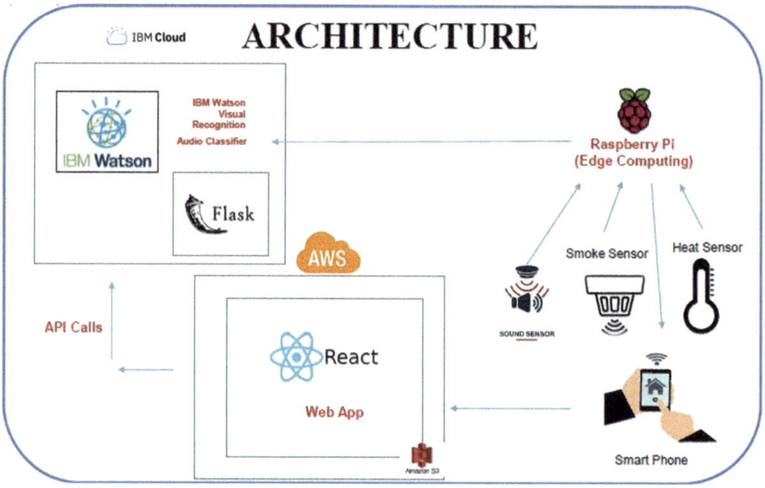

Fig. 1 Architecture diagram

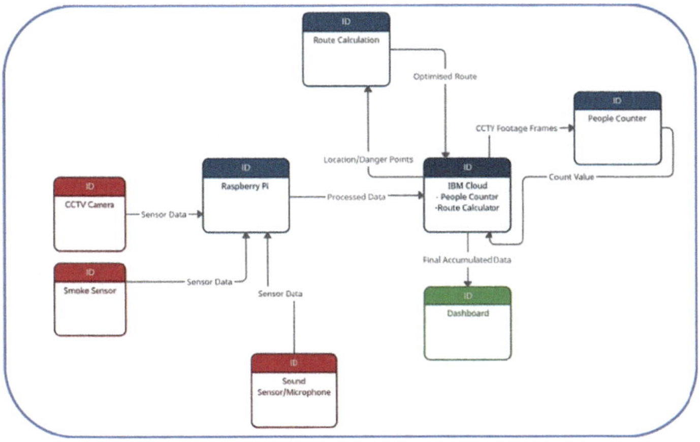

Fig. 2 Data flow diagram

modules after processing through the cloud platform. The route calculation module receives the location/danger points and give the optimized route as output. The people counter takes the CCTV footage frames and returns the count of occupants form each frame.

4 Implementation

When the fire alarm goes off, the first responders can access the dashboard to get the real-time updates of the happenings inside the building. The CCTV will be processed to automatically using computer vision to identify the zones that still contain a high number of occupants. The dashboard will also contain a route finder that computes the fastest and safest way to the exit from the current zone of the occupant [6, 7].

The fires will set off the sensors in the buildings which will in turn trigger the CCTVs and mics in the building to observe for trapped victims. This is done using the IBM Watson Video Recognition Software. In case of a very heavy fire, there are chances the CCTV device's line of sight could be obstructed by smoke [8]. This is where the IBM Watson AI audio classifier kicks in. It observes for any signs of human presence, screams, shouts, agitated voices, etc. These systems help in identifying whether there are people trapped in the general area. The positioning is made more accurate as we use Wi-Fi triangulation to triangulate onto the person's phone which we mostly carry on our person at all times.

Dijkstra's algorithm is used to help guide the victims to safety. The system selects the shortest path toward the nearest exit, thereby leading them to safety [9]. Raspberry Pi is used for edge computing, and the application is created using React. AWS is used to host the application [10, 11].

4.1 Module One: Route Finder

To help occupants during evacuation when a fire breaks out and to prevent them from getting trapped in a dead-end during evacuation, the route generator will compute the shortest as well as the safest path to exits taking into account the obstacles in the path as well as the location of the user [12].

To make the task of finding the shortest and safest path out of the building, we start by dividing the building to smaller areas/zones with each zone having a CCTV or smoke/fire sensor. The building blueprint/plan is then be used to generate the nodal graphs, showing clearly how the different area/zone is connected.

The module is constructed by using the NetworksX library based on Python. It takes in the current location and returns the safest path to the nearest exit by using the adjacency list and generating node pairs out of it.

4.2 Module Two: People Counter

In order to guarantee successful crowd control and estimation of the fastest route, real-time video identification can also be used to locate choke points on vital escape routes. It will also provide vital first-hand information to the first responders on the ground (i.e., security operations and neighborhood first responders to assist with smooth and secure evacuation attempts in times of panic. The first responders/fire and rescue department personnel who are rushing to the scene will also be fed this vital information.)

Live on-the-ground reports of the situation will allow the first responders/fire and rescue department personnel to better prepare and strategize its rescue operations on the way to the incident site, such as the number of people still in the building and their exact position or their last known location (if the cameras were damaged).

Trained model is used, which is able to detect multiple groups of objects, including individuals, to detect the total number of people present in each frame of a video. Other groups are excluded because we are only detecting individuals, so that real-time video recognition can only concentrate on human detection. The software will control and display the number of people in-frame. The storing of such video feeds in a database would also be a choice for post-event analysis at the same time [13].

Therefore, it is able to feed real-time occupant count data to dashboard for first responders/fire and rescue department staff as well as for real-time route optimization with this continuous update [14]. Each specific zone will be given its own threshold to determine if the region suffers from overcrowding. The dashboard that allows to make fast decisions will explain this information [15].

5 Results

After implementing the modules that calculate the routes, the count of occupants from CCTV footage and the dashboard to display this information alongside zone occupancy and equipment readiness, the following results were obtained. The obtained results were presented through a web app interface as well as through a formatted output generated out of a Python source code.

Figure 3 is that of the zone occupancy in the dashboard. This gives the user valuable insights on how many people are in each nodes allowing the first responders to get ample information to plan their strategy.

The above given figure (Fig. 4) displays the current health or status of the equipment used in our system. This helps us keep track of how efficiently the devices are working and lets us know when servicing and repair is due.

The above given figure (Fig. 5) demonstrates the output of our route selection program. The module can be executed in 2 different modes. The first mode involves finding the shortest escape route from the given node. Debris if noted/detected would result in removal of that node from the graph and this result in the second mode of operation where the shortest and safest path/exit strategy is for the victims by considering any debris that may have fallen. These modes of operation can be executed in a reverse execution to find an optimal/safest entry route for first responders to enter and reach a defined node from a point of exit. These values are calculated by the generation of the traversal graphs from the adjacency list which has been generated from the blueprint of the building.

The above given figure (Fig. 6) demonstrates the people counter module which uses the CCTV footage from the cameras embedded in the buildings to assess the frames and calculate how many people are trapped in which node of the building. This information is vital in the strategy planning phase that enables the first responders to target these nodes and plan their entry accordingly.

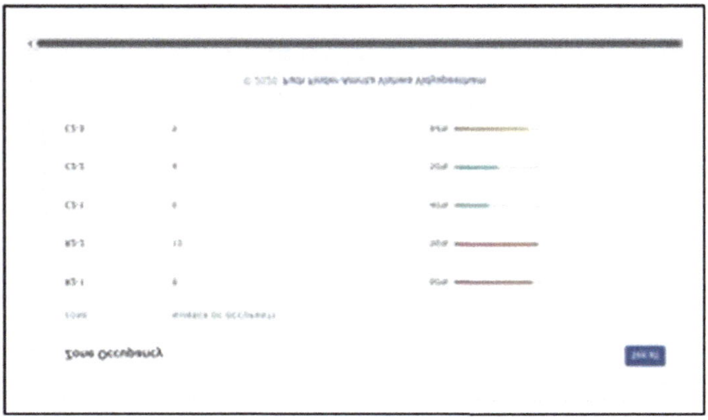

Fig.3 Zone occupancy

Fig. 4 Dashboard

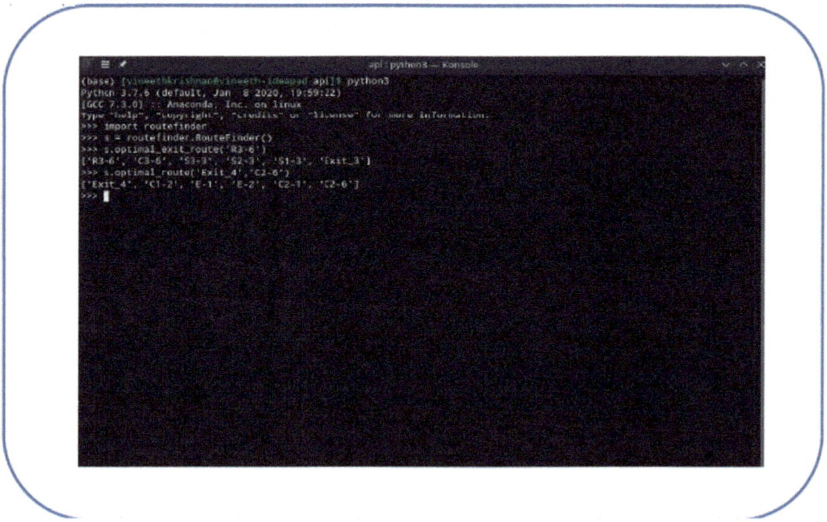

Fig. 5 Route selection

Fig. 6 People counter

6 Conclusion

It is already established how frequent fire accidents are in today's day and age. A system utilizing modern technology goes a long way in drastically reducing the number of casualties incurred in a fire accident. Proposed system can be further enhanced by adding a feature to automatically identify adjacent nodes and leverage this information to select a route for the user. This system can be put to good use in many real-world situations and can be further optimized for use by paramedics and other members of the first responders' team.

Wi-Fi triangulation also can be used to get a more accurate and a detailed idea of the path that the civilians trapped in the building and also the firemen could follow. The method is to set up the Wi-Fi inside the place where it need to navigate to any node, so the Wi-Fi signal is transmitted throughout the building so the signal from the Wi-Fi access point can be detected by smart phones or any other device which provides a similar purpose.

The area is divided into nodes based on the structural divisions of the same. Through Wi-Fi triangulation we can detect the position of the trapped victims by converging on the position of his cellular device which we usually carry on our self at all times. This information is then leveraged to identify which nodes potentially contain a victim and these nodes are tagged as those of interest and this information is relayed on to the first responders. This makes the area affected by the fire hazard more maneuvrable and this in turn increases the rescue efficiency, thereby working toward ultimate goal. The CCTV footage can be analyzed to automatically identify places (computer vision for areas where people with mobility impairments have been identified, e.g., wheelchair bound occupation).

References

1. Keerthi M, Aishwarya M, Atul V, Krupa P (2016) IoT based emergency evacuation system. Int J Eng Res Technol (IJERT) (Special Issue)
2. Moghaddam MT (2020) IoT based emergency evacuation systems, vol 5, issue 3. University of L'Aquila
3. Kumar N, Kumari P (2018) Intelligent fire detection and visual guided evacuation system using arduino and GSM. Int J Recent Trends Eng Res 4(12):2455–2457
4. Johnsaida N, Rahul LV, Shalini T (2018) IOT based smart fire emergency response system. Int J Adv Res Dev 3(2)
5. Xu Y, Wang Z, Zheng Q, Han Z (2012) The application of Dijkstra's algorithm in the intelligent fire evacuation system. In: 4th international conference on Intelligent Human-Machine Systems and Cybernetics (IHMSC), vol 1, pp 3–6
6. Rajesh M, Nagaraja SR (2019) An autonomous system of mobile robots for search in disaster affected sites. In: 2019 3rd International Conference on Electronics, Communication and Aerospace Technology (ICECA), pp 1190–1194
7. Wang T, Huang R, Li L Xu W, Nie J (2011) The application of the shortest path algorithm in the evacuation system. In: 2011 international conference on information technology computer engineering and management sciences (ICM), vol 2. IEEE, pp 250–253
8. Majumder S, O'Neil S, Kennedy R (2017) Smart apparatus for fire evacuation—an iot based fire emergency monitoring and evacuation system. Undergraduate Research Technology Conference (URTC). IEEE MIT, pp 1–4
9. Han Z, Weng W, Zhao Q, Ma X, Liu Q, Huang Q (2013) Investigation on an integrated evacuation route planning method based on real-time data acquisition for high-rise building fire. IEEE Trans Intell Transp Syst 14(2):782–795
10. Chalagulla D, Jayateertha J, Kudimi CR, Manitha PV, Sujith M (2018) Efficient energy utilization control with remote network access. In: 2018 International Conference on Communication, Computing and Internet of Things (IC3IoT), pp 262–266
11. Zualkernan IA, Aloul FA, Sakkia V, Noman HA, Sowdagar S, Hammadi OA (2019) An IoT-based emergency evacuation system. In: 2019 IEEE international conference on Internet of Things and Intelligence System (IoTaIS), Bali, Indonesia, pp 62–66
12. Thomas J (2019) IoT use case: optimal emergency escape route. Bosch Connected World Blog
13. Joshna V, Kashyap M, Ananya V, Manitha PV (2019) Fully autonomous robot to detect and degasify hazardous gas after flood disaster. In: 2019 2nd International Conference on Power and Embedded Drive Control (ICPEDC), pp 134–139
14. Roy C, Lakshmi D, Kumar G, Vishwas HN (2017) Smart environment using IoT. In: International conference on smart technologies for smart nation (SmartTechCon)
15. Reddy Mettu N, Sasikala T (2018) Intrusion detection in wireless sensor networks. In: 2nd International Conference on Inventive Communication and Computational Technologies (ICICCT 2018)
16. Sreevidya B, Rajesh M (2019) Design and performance evaluation of an efficient multiple access protocol for virtual cellular networks. In: International conference on computer networks and communication technologies. Lecture notes on data engineering and communications technologies, vol 15. Springer, Singapore

An Energy-Efficient Data Management Approach Using Sparse Compression in Future-Ready IoT Networks

Gemini George

Abstract IoT networks deployed for various industrial applications often consist of a large number of wireless sensors deployed over an area. Minimizing energy consumption and optimizing network longevity through efficient data traffic management are open research problems in such networks. Data transmission contributes to a major portion of sensor energy consumption. Hence, efficient data management schemes are to be deployed to improve the longevity and latency in such networks. Sparse compression is a technique by which the data can be sampled much below the Nyquist rate, enabling a compact data representation. In this work, we extend this approach to manage the data traffic in IoT networks and evaluate the performance of the sparse compression technique in data traffic management in future-ready IoT networks.

1 Introduction

IoT networks for many applications require the deployment of battery-powered sensors across a region. The sensors in the IoT networks are to be designed to consume low power to maximize network longevity. Once the batteries of the sensor nodes are drained out, replacing them may be often infeasible in many applications. Hence, reducing the energy consumption in the network is a major design issue [1]. The data sensed by sensor nodes are usually transmitted to a command office (CO). A major portion of the energy drain occurs during the transmission of the sensed data. If each sensor node directly transmits these data to the CO, a large amount of energy will be consumed in the transmission process. This demands the deployment of efficient data traffic management strategies in sensor networks [2].

G. George (✉)
Department of Computer Applications, St. George's College Aruvithura, Erattupetta, Kottayam, Kerala 686122, India
e-mail: gemini@sgcaruvithura.ac.in

S. M. Thampi et al. (eds.), *Fifth International Conference on Computing and Network Communications*, Lecture Notes in Electrical Engineering 1219,
https://doi.org/10.1007/978-981-97-4540-1_13

Sparse Compression (SC) [3–5] is an emerging theory of information acquisition that states that for sparse or compressible signals, the data can be sampled much below the Nyquist sampling frequency and then perfectly reconstructed by a non-linear reconstruction algorithm. SC can be simply defined as compressing at the time of sensing itself. An analog-to-digital converter based on the Shannon-Nyquist sampling theorem, which converts analog signals to digital, is used in a conventional communication system. The Shannon-Nyquist sampling theorem states that a signal must be uniformly sampled at least two times faster than its bandwidth to avoid losing information. In many applications, such as digital image and video cameras, the Nyquist rate may be so high that it results in too many samples, and these samples must be compressed to store or transmit them.

SC has emerged as a promising signal compaction strategy for highly correlated sensor signals compared to its peers, such as distributed source coding [6]. As per the theory of SC, a sensor data $\mathbf{u} = \{u_1, u_2 \dots u_N\}$can be effectively represented as $\mathbf{y} = \{y_1, y_2 \dots y_M\}$ via the transformation $\mathbf{y} = \mathbf{Au}$ where $\mathbf{A} = \mathbf{BF}$, where, $\mathbf{B} \in \mathbb{R}^{M \times N}$ is the measurement matrix, and $\mathbf{F} \in \mathbb{R}^{N \times N}$ is any sparsifying orthonormal basis $\mathbf{F}$ (for e.g., Fourier) for $\mathbf{u}$ such that $\boldsymbol{\theta} = \mathbf{Fu}$, where $\theta \in \mathbb{R}^N$ and $\mathbf{u} = \mathbf{F}^T \theta$. As $\mathbf{F}$ is a sparsifying basis, the energy of θ will be concentrated in a few coefficients. The signal recovery is governed by the ℓ_1 minimization, $\hat{\theta} = arg\ min\ \|\theta\|_1$ s.t. $\|\mathbf{y} - \mathbf{Au}\|_2 \leq \epsilon$, where $\|\|_1$ and $\|\|_2$ denote the ℓ_1 and ℓ_2 norms respectively, and ϵ is the acceptable limit of ℓ_2 norm error. The u is found from the relation $\mathbf{x} = \mathbf{u} = \mathbf{F}^T \theta$. If the measurement matrix obeys Restricted Isometric Property (RIP), then the ℓ_1 minimization recovery will be unique [7].

Efficient data traffic management is becoming an important design issue in modern IoT networks [1, 8]. Aiming for efficient data management, we extend the concept of SC for effective data traffic management in future-ready IoT networks in this work. Given a sensor distribution, we propose to design a data-efficient IoT network architecture. We simulate the proposed method with the SC incorporated and evaluate the performance.

The remaining sections are arranged as follows: Sect. 2 reviews the recent literature related to our work. The proposed approach of a traffic-efficient network architecture is presented in Sect. 3. Section 4 presents the performance evaluation of the proposed schemes, and the work is summarized in Sect. 5.

2 Literature Review

2.1 Principles of Sparse Compression

Suppose the given IoT network consists of N sensor nodes. So, the data x sensed by the sensor nodes is of length N and can be represented as $x = [x_1, x_2 \dots . x_N]^T$. The data of ith sensor node is represented as x_i. According to SC, a signal x, which is k-sparse in any of the transform domains can be recovered using any of the few linear combinations of x, provided certain conditions are satisfied [5]. The signal x is said to be k-sparse, if there exist only $k(k \ll N)$ non-zero entries in x. Since x is k-sparse, instead of sending N data to the CO, we send only a small number of projections of x to the CO which is given by:

$$y = A.x \tag{1}$$

where A is an $M \times N(M << N)$ random matrix called sensing matrix or measurement matrix and y is a vector of M projections and are called measurements. The sensing matrix can be any random matrix such as Gaussian or Bernoulli matrix. After y is collected at the CO, the original data x can be recovered by using l1-minimization or other heuristic algorithms, such as orthogonal matching pursuit [9]. The choice of M is also a crucial factor in the reconstruction phase. To perfectly recover a k-sparse signal vector x, M number of SC measurements are required which is given by [10]:

$$M \geq O(k.\log(N)) \tag{2}$$

Suppose the signal x is sparse in ψ domain. Then, x can be represented as $x = \psi s$, where ψ is an $N \times N$ transform matrix and s is a vector of coefficients. Therefore, y can be represented as [11]:

$$y = \phi\, \psi\, s = A\, s \tag{3}$$

where A is an $M \times N$ matrix. In order to fully recover a sparse signal x from y, the matrix A must satisfy either Restricted isometric property (RIP) or the Mutual incoherence property. A matrix A is said to satisfy RIP if there exists $\delta_k \in (0, 1)$ for all k-sparse signals s such that

$$(1 - \delta_k)\, \|s\|_2^2 \leq \|AS\|_2^2 \leq (1 + \delta_k)\, \|s\|_2^2 \tag{4}$$

2.2 *Data Management Schemes*

Data gathering in a network can be performed using different methodologies [12] such as centralized, tree-based approach, cluster-based approach, and in-network aggregation approach.

Centralized Approach In this approach, every node transmits its sensed data to the central node, which is usually the most powerful node Address Address-centric routing is used along with multi-hop algorithm by considering cost metric at every intermediate node. The central node's task is to aggregate the data from other nodes and report the same data to the CO. This approach suffers from the problem of high traffic due to a large number of messages being transmitted.

Tree-Based Approach In this approach, an aggregation tree is built initially, which is generally a minimum-spanning tree. In this tree, the root node acts as the CO, leaf nodes act as the source node and intermediate nodes act as parent nodes. The leaf node sends its sensed data to its parent node in a path discovered between the leaf node and the CO.

Cluster-Based Approach In this approach, the region of interest is divided into a number of clusters. A cluster head is elected within each cluster, whose main task is to aggregate the data. Each node senses the desired phenomena and reports the data to the cluster head of the same cluster rather than sending the data directly to the CO. Hence, it saves a lot of energy in a network.

In-Network Aggregation In this methodology, data is processed at intermediate nodes in order to reduce the consumption of critical resources such as energy and computation time. This approach also increases the network lifetime as it tries to reduce the energy consumption at every node. There are two further approaches for in-network aggregation: lossy aggregation with packet size reduction and lossless data aggregation without packet size reduction. In lossy aggregation, data is gathered from various source nodes and then some group function is applied to the gathered data. In this approach, size of the packet is reduced, as only calculated value of aggregate function is inserted into the packet after compression, rather than sending the whole packet of every node.

Energy efficiency and latency are important factors that should be taken care of at the time of design of wireless sensor network [13]. Energy-efficient routing protocols have been developed to reduce energy consumption since WSNs work on battery, which discharges quickly after each operation. Though many works available in literature optimize various network parameters, this report limits its discussion only to three important protocols, namely LEACH, SEP, and TEEN.

Micro-electromechanical systems and low-power design considerations Micro-electromechanical systems (MEMS) create tiny integrated devices and systems by combining electric and mechanical elements. MEMS of tiny sizes at low cost in massive volumes are produced using state-of-the-art microfabrication techniques. Smart IoT devices are designed by bundling processing units, storage devices, sensors, actuators, communication devices, and power sources.

While the IoT sensors capture the underlying physical phenomena, the IoT actuators perform tasks such as the rotation of motors. The processor analyzes the captured information from the sensors, issues control signals to actuators to drive required motors, and deploys different functionalities such as data compression. The communication devices receive control signals, organize into clusters, and transfer data packets. A smart IoT device relies on battery power; hence, its design is based on low-power consumption. Therefore, the power consumption of different micro-elements of the device is a critical design issue in determining its longevity. Theoretically, the overall power consumption is minimized to such an extent that an IoT device's battery should outlive the device itself. Energy harvesting from surroundings is an area that attracts research interest to prolong the life of smart devices further.

Though smart IoT devices vary widely in their applications, many macro trends are evolving in their features and characteristics. The major trends include reduced size, lower power consumption, and increased processing and communication capabilities. These trends lead to an ecosystem of sophisticated devices that perform complex tasks with better efficiency. However, the design of such smart devices includes complex objectives such as tiny operating system, low computational functionalities, limited storage, and low-power consumption at the micro-element level.

3 Proposed System

The basic idea of the proposed system is explained in Fig. 1. The entries in each row of the $M \times N$ dimension measurement matrix Φ are projected on N sensor nodes. The projection coefficient is multiplied by the locally collected data at each node and is passed on to the next node. The next node also performs this multiplication process and then adds the weighted data from the previous node. The weighted sum thus obtained is transferred to the next node. In this way, each node calculates and transmits a weighted sum along the route. Finally, when the weighted sum of all nodes reaches the CO, the CO gets a measured value. After each node transfers M times, the CO can get M measurements. Compressive data gathering converts the N samples needed by the CO into collecting M weighted sums of the local samples, which reduces the amount of data transmitted, decreases energy consumption, and extends the network lifetime.

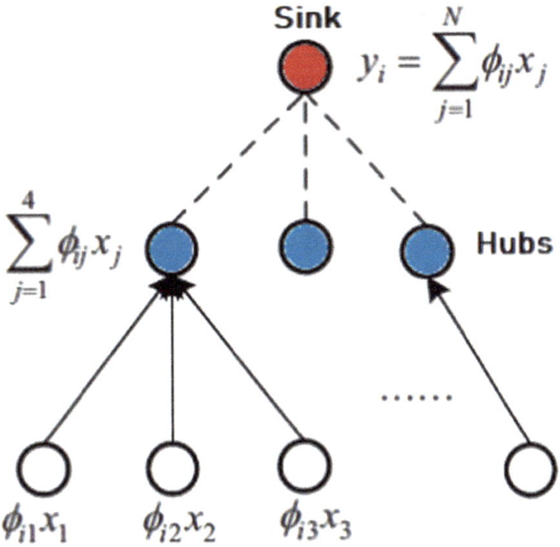

Fig. 1 Data collection model

The fundamental idea of SC allows to acquire only a certain number of their linear combinations of the original signal vector which can be expressed as $y_j = \sum_{k=0}^{N-1} \Phi_{j,k} x_k + \eta_j$ where η_j is a noise component, for $j = 0, ..., M - 1$ and with $M < N$, called measurements. Since $M < N$, the relationship $y = \Phi x + \eta$ does not, in general, allow a straightforward estimation of x. The method of solving linear equations requires that $M \geq N$ to find all the values of unknowns. But it is possible to reconstruct x from y, if x is known to be sparse with respect to the basis A. Then $y = \Phi A s + \eta$, where at most k components of s are known to be non-zero. Recent theoretical and algorithmic developments guarantees that exploiting this prior, depending on η and on specific properties of the matrix Φ, x can be effectively estimated with some known bounds on the estimation error. Beyond theoretical guarantees, we can say that most of the practical interest in the proposed method comes from a few key facts.

(1) The performance of estimation algorithms largely outperforms the theoretical constraints allowing an effective recovery of x from a small number of measurements, i.e., usually $M << N$.
(2) The mathematical conditions on Φ can be matched by using random matrices where the formal results depend on specific matrix distributions. But, in practice, a wide class of random matrices allows for effective signal recovery.
(3) It is possible to substantially improve acquisition performance by adapting the statistical distribution of A to that of the signal x, in particular to its second-order statistics in X. Once the distribution of A is chosen, a single instance can be drawn and used for sensing.
(4) The amount of processing required in SC is negligible as it amounts only to linear combinations. Thus, it perfectly suits the low resource budget of sensor nodes.

The network traffic could be reduced using SC as only M information is transmitted instead of N. But, computing the linear combination of the sensed data requires MN multiplications and $M(N - 1)$ additions for a Gaussian matrix. Thus, SC is slightly computationally complex.

The highest compaction and unique signal recovery have given CS a formidable advantage over its peers. The past few years witnessed efforts to extend the SC approach to distributed networks. Compressive Sparse Compression (CSC) is the first major step in that direction. In CSC, instead of transmitting the actual measurements, each sensor transmits M independent measurements obtained by a weighted combination of the actual measurements in each round, and the BO receives M measurements which are obtained by the weighted combination of measurements of all the sensors. This brings in load balancing across the network links as all sensors transmit the same amount of data leading to uniform energy consumption. Thus, for a network with N sensors, the total number of transmissions is in $O(MN)$ [14]. Hybrid Sparse Compression (HSC) schemes can be used to overcome the disadvantage of CSC whereby the farther nodes from the BO transmit more data than required to be in the Non-CS method. The data management in CSC and HSC are shown in Fig. 2

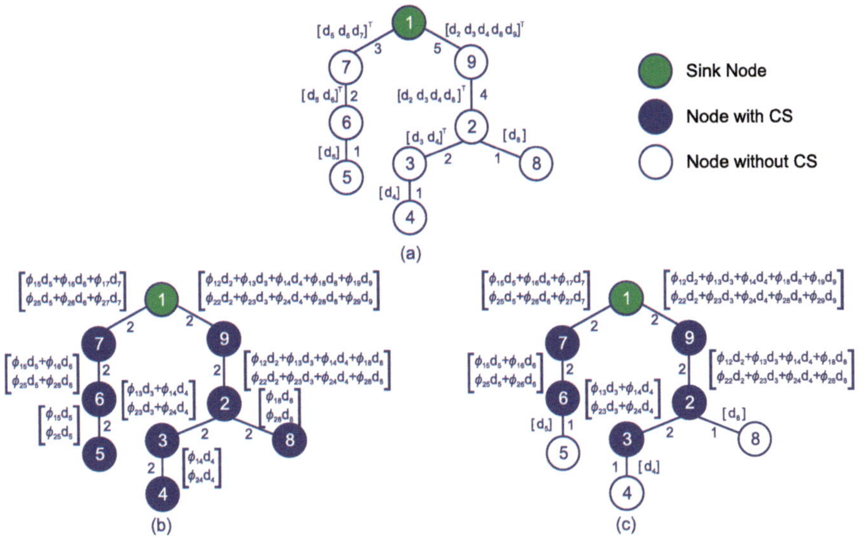

Fig. 2 Data management in SC variants **a** conventional **b** CSC **c** HSC

4 Simulation Results

The simulation setup is described in the following: We consider medium-range IoT applications consisting of devices that transmit using unlicensed radio technologies with a typical range of up to a few hundred meters. Therefore, we set a field of $800 \times 800\ m^2$ in which 250 sensors are distributed as the Poisson point process. The subregion size is taken as $1 \times 1\ m^2$, and a subregion is considered to be covered by a sensor if the center of the cell is in the coverage-range of the sensor. The CO is assumed to be located at the boundary corner at (800, 800). Each sensor is assumed to have started its operation with an energy of $1 J$. Operation proceeds in sequences, and in one sequence, every activated sensor generates 2000 symbols that are to be collected at the CO. Data transmission is the major source of energy consumption for the sensors. The energy consumption for data transfer is governed by the first-order radio model [15]. As the operation progresses, the remaining energy of the transmitting sensors starts depleting, eventually causing their death. Such sensors are excluded by the algorithm for the next rounds. Hence, as the operations progress, the number of live sensors decreases, which may affect the coverage of the subregions to be monitored. The network's energy is computed as the sum of the energies of the sensors alive. The network is considered to be inoperable when there are not enough sensors available to cover the subregions in the monitored area.

As already discussed, the proposed method operates in a cluster-based architecture where the CHs of each cluster gather the data from the cluster members and transmit the data to the CO via hopping through the heads of other clusters. We evaluate the performance of the proposed method with CSC and HSC modes. In CSC, each

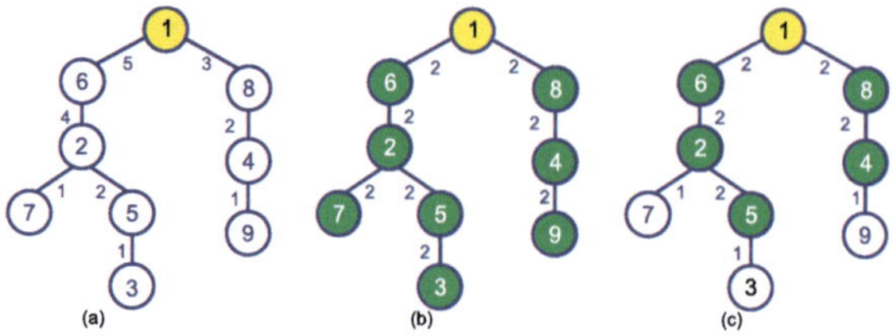

Fig. 3 Computation of data traffic in different schemes

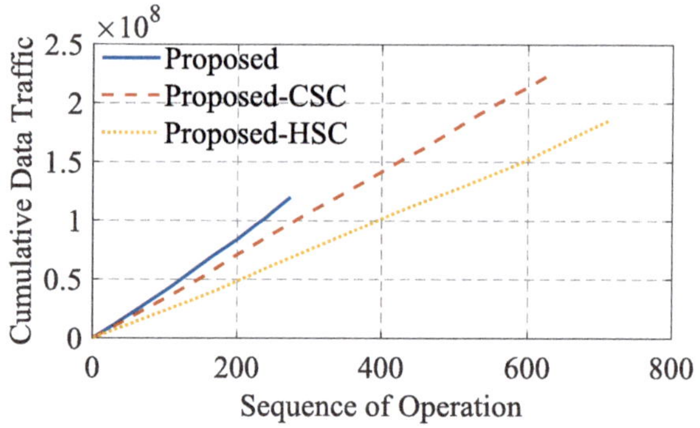

Fig. 4 Cumulative data traffic: proposed method with CSC and HSC

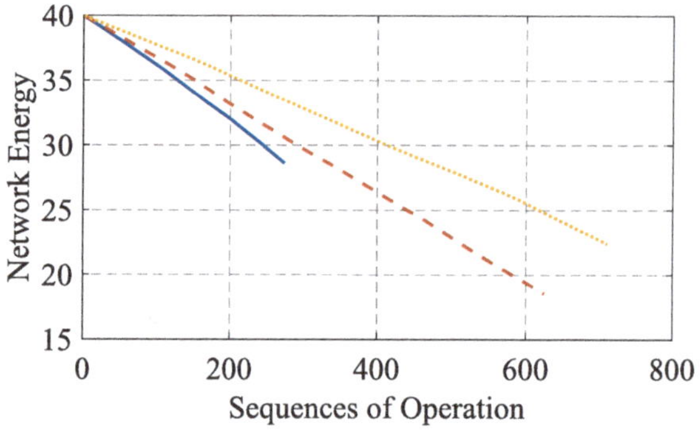

Fig. 5 Remaining network energy: proposed method with CSC and HSC

sensor transmits M independent measurements obtained by a weighted combination of the actual measurements in each sequence, and the BO receives M measurements which are obtained by the weighted combination of measurements of all the sensors. This brings load balancing across the network links as all sensors transmit the same amount of data, leading to uniform energy consumption. Thus, for a network with N sensors, the total data traffic is in $O(MN)$. Hybrid Sparse Compression (HSC) schemes can be used to overcome the disadvantage of CSC whereby the farther nodes from the BO transmit more data than required to be in the Non-SC method. The computation of data traffic in different schemes are shown in Fig. 3.

Simulation results show that the proposed method with HSC outperforms the other variants by reducing the data traffic because of the better traffic management adopted, by which the leaf nodes require to transmit only a lesser number of data.

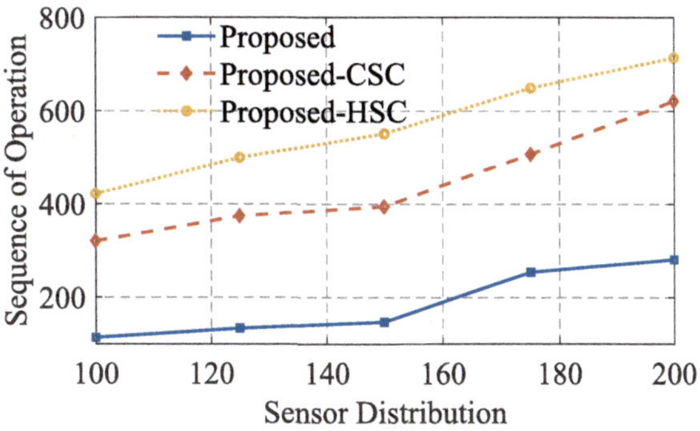

Fig. 6 Sequences of operation: proposed method with CSC and HSC

The data traffic of the methods are compared in Fig. 4. The reduction in the data traffic causes higher network energy for the proposed method with HSC followed by CSC and without SC as depicted in Fig. 5. The reduction in the data traffic and lesser energy spent in operation leads to an overall improvement in the network longevity for the proposed method with HSC followed by CSC and without SC as depicted in Fig. 6.

5 Conclusion and Future Works

Regardless of the applications the IoT networks serve, the data generated in the network eventually must be delivered to the CO. The process of data transmission and reception, especially over long distances, is energy-consuming. Data management strategies help to gather data in the IoT network efficiently. This work has demonstrated the importance of the sparse compression technique in reducing the data traffic and thereby causing a lesser energy consumption in the network, overall leading to network longevity. Further, the research in SC-based data gathering is only gaining momentum. The correlation among sensor readings is not completely exploited yet. Moreover, in many cases, such as surveillance networks, the sensor reading may follow some patterns. All such prior information can be used to reconstruct the sparse data.

References

1. Chandnani N, Khairnar CN (2022) An analysis of architecture, framework, security and challenging aspects for data aggregation and routing techniques in IoT WSNS. Theor Comput Sci
2. Manuel EM, Pankajakshan V, Mohan MT (2023) Linear programming models for the design of energy-efficient iot networks with transmission constraints. IEEE Trans Green Commun Network

3. Zhang X, He J, Zhou Y, Chi Y (2022) A subspace approach to sparse sampling based multi-attribute data aggregation in IoT. IEEE Internet Things J
4. Qin Z, Fan J, Liu Y, Gao Y, Li GY (2018) Sparse representation for wireless communications: a compressive sensing approach. IEEE Signal Proc Mag 35(3):40–58
5. Xu Y, Sun G, Geng T, Li Z (2019) Compressive multi-timeslots data gathering with total variation regularization for wireless sensor networks. IEEE Commun Lett 23(4):648–651
6. Sun P, Wu L, Wang Z, Xiao M, Wang Z (2018) Sparsest random sampling for cluster-based compressive data gathering in wireless sensor networks. IEEE Access 6:36,383–36,394
7. Candès EJ, et al (2006) Compressive sampling. In: Proceedings of the international congress of mathematicians vol 3. Madrid, Spain, pp 1433–1452
8. Das SK, Kapelko R (2021) On the range assignment in wireless sensor networks for minimizing the coverage-connectivity cost. ACM Trans Sens Netw (TOSN) 17(4):1–48
9. Mallat SG, Zhang Z (1993) Matching pursuits with time-frequency dictionaries. IEEE Trans Signal Proc 41(12):3397–3415
10. Manchanda R, Sharma K (2021) A novel framework for energy-efficient compressive data gathering in heterogeneous wireless sensor network. Int J Commun Syst 34(3):e4677
11. Prabha M, Darly SS, Rabi BJ (2021) A novel approach of hierarchical compressive sensing in wireless sensor network using block tri-diagonal matrix clustering. Comput Commun 168:54–64
12. Kavra R, Gupta A, Kansal S (2022) Systematic study of topology control methods and routing techniques in wireless sensor networks. Peer-to-Peer Network Appl 1–61
13. Donta PK, Amgoth T, Annavarapu CSR (2022) Delay-aware data fusion in duty-cycled wireless sensor networks: a q-learning approach. Sustain Comput: Inf Syst 33:100642
14. Cormen TH, Leiserson CE, Rivest RL, Stein C (2022) Introduction to algorithms. MIT press
15. Lee J-S, Kao T-Y (2016) An improved three-layer low-energy adaptive clustering hierarchy for wireless sensor networks. IEEE Internet Things J 3(6):951–958

A Framework for Developing IoT and Edge Computing-Based Smart Transportation Applications

Poojarini Mitra, Zeenat Rehena, and Nandini Mukherjee

Abstract Transportation is one of the main sectors responsible for negative environmental impacts. Integrating emerging technologies can help provide real-time, resilient, and intelligent solutions that are necessary for decarbonizing road transport. This paper presents a novel framework that combines the Internet of Things (IoT), Edge Computing, and Cloud Computing to deliver sustainable and efficient transportation solutions. The proposed framework facilitates the development of smart transportation applications that envision an efficient way of reducing emissions, optimizing traffic flow, and improving overall transportation efficiency by leveraging the huge storage and computational power of Cloud Computing and the real-time data processing capabilities of Edge Computing. In addition, the virtual sensor is employed by the framework in order to reduce the adaptability, interoperability, and heterogeneity constraints in the IoT environment. This paper examines the shortcomings of current systems, highlights the essential elements of the framework, and assesses the framework's effectiveness in providing effective and efficient smart transportation services.

Keywords Smart transportation system · Environmental sustainability · Internet of things · Edge computing · Cloud computing · Virtual sensors

---

P. Mitra (✉)
Techno Main Salt Lake, Kolkata, West Bengal, India
e-mail: poojarini.mitra@gmail.com

Z. Rehena
Aliah University, Kolkata, West Bengal, India
e-mail: zeenatrehena@aliah.ac.in

N. Mukherjee
Jadavpur University, Kolkata, West Bengal, India
e-mail: nandini.mukhopadhyay@jadavpuruniversity.in

S. M. Thampi et al. (eds.), *Fifth International Conference on Computing and Network Communications*, Lecture Notes in Electrical Engineering 1219,
https://doi.org/10.1007/978-981-97-4540-1_14

1 Introduction

From the beginning of civilization the art and science of moving from one place to another have been one of the most important necessities for humans. In the modern-day world, with the dramatic expansion of the urban dimension, one of the major concerning factors is the impact of transportation systems on the sustainability of the environment [1]. The transportation system is currently the highest contributor toward increased emission of greenhouse gases (GHG) and noise pollution resulting in global warming and rising health hazards [2]. Electric vehicles (EVs) can reduce the dependence on fossil fuels and decrease emissions and can be elected as an environment-friendly transport alternative. However, the promotion of new energy vehicles may not yield substantial results in the short-term owing to substandard technical proficiency, such as lack of charging infrastructure, the high upfront cost of EVs, and limited driving range [3]., Strategic transportation management planning with an integrated approach of organizing and execution, for example, smart route optimization, intelligent traffic light systems, agile parking systems, dynamic multi-modal selection, etc. can offer solutions for reduced carbon footprint. Smart Transportation Systems (STSs) provide a diverse array of solutions and are instrumental in advancing environmental sustainability within the transportation sector [4]. STS combines communication devices connected over the Internet, better known as the Internet of Things (IoT), and modern ubiquitous technologies, like, Edge and Cloud Computing to provide better mobility, reduce congestion, control pollution, and improve citizen's quality of life [5]. A recent report by the Global Enabling Sustainability Initiative (GeSI) [6] suggests that these technological innovations possess the capability to reduce up to 20% of emissions by 2030.

STS relies on IoT, Cloud Computing, and Edge Computing technologies to collect and analyze real-time data for traffic flow, vehicle movement, and road conditions. IoT devices generate large amounts of data, making traditional centralized cloud computing ineffective due to data generation and distance [7]. Edge Computing offers solutions by dispersing computing, control, storage, and networking capabilities near end-user devices, reducing dependence on cloud-based content processing and end-to-end mapping [8]. Virtualization, the process of abstracting hardware resources to virtual objects, offers benefits such as reduced hardware costs, improved scalability, flexibility, and easier maintenance. Virtual sensors can be customized to meet specific application needs, making them a versatile option for edge-cloud composite environments [9]. IoT, edge, and cloud computing systems can offer environmentally friendly transportation services by monitoring traffic, vehicle fleet, weather, and air quality in real-time. This allows for efficient route adjustments, reduced idle time, and improved efficiency. The system can also monitor parking spaces, predict traffic patterns, and adjust traffic lights to prioritize emergency vehicles. By utilizing IoT devices and advanced analytics, multi-modal transportation can be effectively tackled, reducing emissions and promoting sustainable options. This system provides instantaneous information on bus and train schedules and shared cars or bikes.

In this paper, we have reviewed the relevant literature based on two approaches, i.e., topical literature review and technical analysis. We aim to identify and deliberate upon the latest advancements in environmental sustainability applications within the realm of Smart transportation. The paper identifies some of the key challenges in the existing systems. In doing so, it proposes a framework based on a 4-Layer architecture, that integrates IoT devices with edge and cloud layers and correspondingly with the STS applications, in order to deliver environment-friendly, as well as, accurate information to users in the least service response time. The proposed framework has the potential to stimulate the demand for environmental sustainability solutions that can be provided by the IoT, edge, and cloud technologies.

The rest of the paper is organized as follows. Sect. 2 reviews the existing related works and discusses the various limitations of the existing systems. In Sect. 3 we present the proposed framework. The framework is evaluated in Sect. 4. Finally, we conclude the paper in Sect. 5.

2 Literature Survey

The transportation sector has a significant impact on society, the environment, and the economy. It is considered a crucial component for the efficient movement of people, goods, and services. To get the maximum utility, many research works have been carried out to address the various issues of a transportation system, like, efficient navigation, traffic, congestion, and pollution control. Table 1 summarizes the related works on the various transport-related issues addressed by the authors. In [10], the authors introduce the concept of vehicular networks, which are communication systems that allow vehicles to share data with infrastructure, and propose several ways in which vehicular networks can be used to reduce fuel consumption and emissions. In [11], the authors explore the concept of traffic signal optimization, which involves using advanced algorithms to adjust traffic signal timing to improve traffic flow and reduce congestion. Traffic signal optimization can reduce vehicle idling time and stop-and-go driving, which can lead to significant reductions in emissions and fuel consumption. The concept of the Internet of Vehicles (IoV) is utilized in [12], where the authors explore the various components of the IoV, including the intelligent grid, which involves the use of sensors and data analytics to manage traffic and improve safety, and the vehicular cloud, which enables vehicles to share data and resources such as processing power and storage. The authors in [13] provide a detailed overview of the various types of Intelligent Transportation System (ITS) technologies that can be used to reduce GHG emissions, including advanced traffic management systems, intelligent vehicles, and traveler information systems, and discuss the potential benefits of these technologies, such as reduced travel times, improved safety, and lower fuel consumption. In [14], the authors focus on the integrated approach to traffic control optimization, which involves using three different tools: VISSIM, a traffic simulation model; CMEM, a vehicle emissions model; and VISGAOST, a tool for optimization of signal timings. In [15], the authors explore

Table 1 Literature review on transport issues

References	Transport-related issues addressed		
	Pollution control	Traffic control	Efficient navigation
Alsabaan et al. [10]	✓		
Kwak et al. [11]	✓	✓	
Gerla et al. [12]	✓	✓	
Barth et al. [13]	✓		
Stevanovic et al. [14]	✓	✓	
Ji et al. [15]	✓	✓	
Peng et al. [16]	✓		✓
Zhao et al. [17]	✓	✓	

the signal coordination scheme based on traffic emission, which involves using traffic emission models to estimate emissions and optimize signal timing accordingly. The authors developed a new traffic emission model, the VT-Micro model, to estimate vehicular emissions at the micro-level and applied it to the signal coordination scheme. In [16], the authors describe a framework for an environmentally friendly low-carbon-emission ITS that is based on big data and machine learning methods. The framework includes several components, such as data collection and analysis, route prediction and control, and user feedback and decision-making. In [17], the authors describe the various components of ITS, such as intelligent vehicles, smart roads, and traffic management systems, and explain how these technologies can be used to reduce traffic congestion, improve safety, and lower carbon emissions.

In the context of STS, cloud and edge computing techniques have proven to be effective in facilitating the real-time transmission and processing of large volumes of data. Smart sensors networked with IoT devices provide these data, which are then transmitted to the cloud-based decision-making system. However, this can create a significant load on servers and cloud systems. To address this issue, edge computing has been developed to provide a more efficient infrastructure for the interaction between cloud computing and IoT devices, promoting their widespread use and optimizing resource utilization through data mining and analysis. Table 2 summarizes the existing works on the various technologies used for constructing a Smart Transportation System. In [18], the authors describe a cloud-based ITS architecture that includes a range of sensors and devices for collecting real-time data on traffic conditions, as well as a cloud computing platform for processing and analyzing this data. In [19], the authors discuss the cloud-based decision support system which involves using real-time data on bus locations and passenger demand to optimize route planning and bus scheduling. The system uses advanced analytics techniques, including machine learning algorithms and optimization models, to provide recommendations to bus operators and city officials. In [20], the authors describe an edge computing-based ITS architecture that includes a range of sensors and devices for collecting real-time data on traffic conditions, as well as edge computing devices

Table 2 Literature review on technologies used

References	Technologies used		
	IoT	Edge computing	Cloud computing
Ashokkumar et al. [18]	✓		✓
Mathirajan et al. [19]	✓		✓
Yan et al. [20]	✓	✓	
Zeng et al. [21]	✓		✓
Peyman et al. [22]	✓	✓	
Sood et al. [23]	✓	✓	✓
Ali et al. [24]			✓
Chavhan et al. [25]	✓	✓	

for processing and analyzing this data. The authors also discuss the use of machine learning and other artificial intelligence technologies for optimizing transportation systems and improving overall efficiency. In [21], the authors propose a framework that includes a range of sensors and devices for collecting real-time data on driving conditions, as well as a cloud-based platform for processing and analyzing this data. The authors also discuss the use of machine learning and other artificial intelligence technologies for optimizing driving performance and improving overall safety. In [22], the authors propose a system that includes a range of sensors and devices for collecting real-time data on traffic conditions, as well as an edge computing platform for processing and analyzing this data. The authors also discuss the use of machine learning and other artificial intelligence technologies for optimizing traffic flow and improving safety. In [23], the authors propose a solution that uses edge-cloud devices, which are located close to the users, to process and analyze the data generated by IoT sensors embedded in the vehicles. This approach reduces the response time and data transmission latency and also helps to preserve the privacy of the users. In [24], the authors discuss the integration of big data analysis and cloud computing into smart transportation systems and highlight the importance of data analytics for improving transportation system efficiency, safety, and sustainability. In [25], the authors propose an energy-efficient intelligent transportation system based on edge computing, artificial intelligence, and the Internet of Things that addresses challenges, such as traffic congestion, energy consumption, and emissions, to enable real-time data processing and analysis.

2.1 Limitation of Existing Systems

The transportation sector has been identified as a notable contributor to greenhouse gas emissions, air pollution, and energy consumption, which have detrimental effects on climate change, public health, and natural resources. However, the existing smart transportation system frameworks face several challenges that hinder their full effectiveness in their adaptation to the environment sustainability. The objective of an environment-friendly smart transportation system framework is to design and develop a comprehensive and sustainable transportation ecosystem that leverages advanced technologies to minimize its environmental impact.

Smart Transportation Systems (STSs) are being researched extensively, but their scope is limited to specific applications like pollution control or traffic control. Traditional transportation systems are often inefficient in terms of energy consumption and resource use, necessitating the implementation of energy-efficient technologies. IoT systems in transportation systems, including temperature, pressure, and air quality sensors, generate complex data streams. However, traditional sensors have limitations such as cost, noise, and interference, while IoT devices and sensors have limitations in computing power, storage, and battery capacity [7]. A successful approach involves a balanced combination of edge and cloud data analytics [26], leveraging their strengths for real-time responsiveness, scalability, complex analysis, and centralized management. Data quality is a challenge due to heterogeneity in IoT environments, and ensuring seamless interoperability among different technologies and devices is a challenge. Virtual sensors can help standardize data exchange, promote interoperability, and minimize the cost of implementation. Privacy and security in smart transportation systems also vary, and a well-designed user interface contributes to the overall success and effectiveness of a smart transportation system. Lastly, a well-designed and user-friendly user interface contributes to the overall success and effectiveness of a smart transportation system. User interface design in smart transportation systems plays a crucial role in enhancing user engagement, improve user satisfaction, and encouraging the adoption of eco-friendly transportation options.

3 Proposed Framework

The purpose of this research is to deliver explicit, environment-friendly traffic-related information with minimal service response delay, minimal redundancy, and minimum energy depletion. The proposed framework contains four layers. The layers are described as Sensing Layer (SL), Edge Layer (EL), Cloud Layer (CL), and Application Layer (AL). SL is responsible for the collection and transmission of traffic data. The IoT devices and sensors in the Sensing Layer perform all the basic functionalities of perceiving and gathering data as well as transmitting data to the Edge Layer. After receiving the data, EL activates the Virtual Sensors (VSs). The Virtual Sensor is a software-based sensor that operates independently to generate signals by amalgamating and consolidating signals received from physical or other virtual sensors, either synchronously or asynchronously [9]. Virtual Sensors are designed for sensors with reduced processing, memory, and battery capacities. Additionally, the virtual sensors provide an abstraction layer based on digital representations of sensor hardware, aggregating input from physical sensors to offer increased reliability, agility, and indirect measurement of non-measurable properties. They also facilitate collabo-

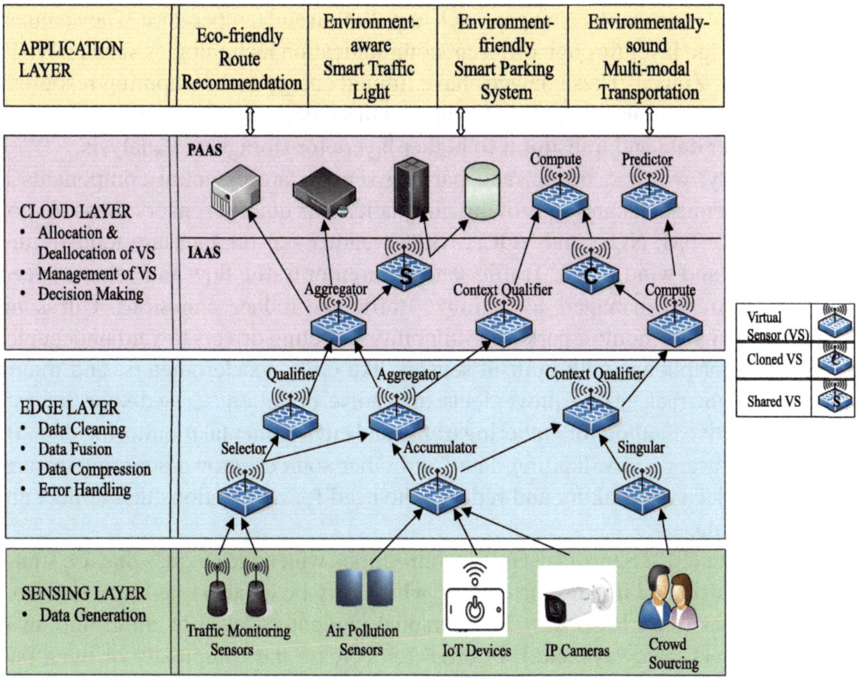

Fig. 1 Framework toward environmental sustainability in the context of smart transportation

ration on sensors, assets and organizational levels, improving accuracy and enabling service providers to offer services based on the same hardware [27, 28]. Moreover, this enormous volume of data presents the challenge of cost-effective storage and analysis. Effective data management, cleansing, and enrichment are necessary for generating valuable insights. Data Cleaning, Data Fusion, Data Compression, and Error Handling are performed in the Edge layer to mitigate these challenges. The data is then forwarded to the CL which contains the Infrastructure as a Service (IAAS) and the Platform as a Service (PAAS) layers. CL is responsible for Allocation and Deallocation of VS, Management of VS, and Decision Making. AL acts as an interface for the users. The functions of each of the layers are depicted in Fig. 1.

3.1 Description of the Four Layers of the Proposed Framework

1. Sensing Layer (SL): The sensing layer comprises heterogeneous, resource-constrained IoT devices (such as drones, smartphones, cameras) and strategically deployed sensors (such as temperature, humidity, air quality indicator) that gather and send real-time data regarding the number, speed, and type of vehicle as well

as detect environmental changes, such as pollution and temperature. They connect with the edge layer through different communication technologies such as Wi-Fi, Bluetooth, Zigbee. These devices have limited energy and computing resources. Heavy computations are not performed on these devices. Their primary function is to gather data and transmit it to higher layers for storage and analysis.
Air quality, weather, traffic, and parking sensors are essential components in reducing emissions and improving air quality. Air quality sensors measure pollutants like PM, NOx, and VOCs, while weather sensors measure temperature, humidity, and wind speed. Traffic sensors monitor traffic flow and vehicle speed, allowing traffic managers to optimize traffic and reduce emissions. Ultrasonic parking sensors monitor parking availability, directing drivers toward unoccupied spots. Smartphones with built-in sensors like GPS, accelerometers, and microphones can track traffic movements and noise pollution. Crowd-sourcing is a cost-effective method for gathering traffic and environmental monitoring data. By engaging users and validating data from other sources, crowd-sourcing ensures accurate decision-making and reduces the need for applications that collect and analyze data.
2. Edge Layer (EL): Sensors establish connections with edge devices that are strategically distributed in close proximity, which may be located one-hop away from IoT devices (e.g., Raspberry Pis, personal computers, etc.) or multi-hop away (e.g., routers, gateways, etc.). In order to decrease the complexity of interoperability and heterogeneity, our framework is employing VS as edge devices. The following is the list of virtual sensor types [29]:

    (a) Singular: This type facilitates one-to-one mapping of physical and virtual sensors, enabling end-users to configure sensor settings to obtain the data from a physical sensor.
    (b) Accumulator: This type accumulates sensing data within a specific area or time frame. For instance, multiple sensing units on a roadside can detect vehicle classification, speed, and direction. These parameters can be accessed as a unified data unit by an application.
    (c) Selector: This type allows for the selection of a specific physical sensor from a group of similar sensors. For example, a selector virtual sensor can choose one air quality analyzer sensor from multiple options at any given moment.
    (d) Aggregator: This type can perform statistical analysis on physical sensors and helps to overcome memory and computing limitations. For instance, an aggregator sensor can calculate the average, maximum, or minimum vehicle count in a specific region.
    (e) Qualifier: The singular type operates similarly, but the virtual sensor is only activated if the qualifier function is met. An example is a virtual sensor monitoring vehicle speed that generates an alert when the value exceeds a set threshold.
    (f) Context-qualifier: This type is similar to the qualifier but the function performs on a group of sensors. For example, a context qualifier sensor with object

recognition can provide detailed insights into the types of vehicles on the road and their movements.

(g) Predictor: This virtual sensor has the capability to make predictions on sensing values by analyzing past data. For example, a Traffic Flow predictor sensor can be used to predict traffic conditions for a short-term future period.

(h) Compute: This type employs a sophisticated algorithm to analyze data from multiple sensors and provide advanced insights. For instance, this virtual sensor can determine the car's condition by analyzing data from its sensors, such as engine temperature, oil level, and fuel level.

The Edge Layer performs the following functionalities:

(a) Data cleaning: Data Cleaning [30] involves identifying incomplete or inaccurate input and correcting errors in raw sensor data using rule- or model-based techniques. It is essential for sensor data as it can be noisy, incomplete, or contain outliers due to various factors, affecting data quality and analysis [31].

(b) Data fusion: Data Fusion [32] involves integrating data from diverse sources to produce a desired output variable or information. It may entail combining data from heterogeneous sources in terms of format, structure, and semantics. Data fusion enhances the precision and effectiveness of various applications. With the ever-increasing volume of data generated, data fusion is gaining significance.

(c) Data compression: Data Compression [33] is a technique that reduces storage space and data transfer by eliminating redundancy and using approximation methods. This involves dropping duplicated or irrelevant data and transferring compact models instead of raw data, without any loss.

3. Cloud Layer (CL): Cloud computing offers the necessary scalability, storage, and computing power to process and store large volumes of IoT data. Cloud platforms provide robust infrastructure, analytics tools, and services for managing and analyzing IoT data, including complex analytics tasks, machine learning models, data visualization, and long-term storage. Our framework enhances the computing and storage resources of IoT devices by supporting multiple cloud data centers in different geographic locations, providing location-independence for IoT applications. Additionally, our framework adopts the virtualization concept in the cloud, enabling the utilization of physical resources as virtual resources within cloud computing platforms. Cloud platforms handle the management and allocation of virtual sensors and other virtualized resources, ensuring their availability and performance. Virtual sensors can be incorporated into the cloud platform at both the Platform as a Service (PaaS) and Infrastructure as a Service (IaaS) layers, providing developers with seamless access to these resources.
4. Applications Layer (AL): The convergence of IoT, Edge, and Cloud technologies paves the way for the development of various intelligent and efficient transportation systems applications, like, Eco-friendly Route Recommendation,

Environment-aware Smart Traffic Light, Environment-friendly Smart Parking System, Environmentally-sound Multi-modal Transportation, etc.

(a) Eco-friendly Route Recommendation: The application employs a sophisticated blend of real-time data, intelligent algorithms, and user preferences to furnish individuals with optimal routes that effectively minimize environmental impact. By prioritizing eco-friendly transportation modes such as walking, cycling, or public transportation, this application serves to significantly reduce carbon emissions associated with individual vehicle usage.
(b) Environment-aware Smart Traffic Light: This application employs data analysis and real-time information to optimize traffic flow, reduce congestion, and minimize environmental impact. It prioritizes public vehicles, emergency services, and electric vehicles at traffic lights, enabling smooth movement through intersections. The application communicates with other smart city infrastructure components, such as smart parking systems, public transportation networks, and traffic management centers, to facilitate a coordinated and holistic approach to urban mobility.
(c) Environment-friendly Smart Parking System: This application, designed with an environmentally conscious approach, utilizes data analytics to furnish real-time information regarding parking availability. This feature empowers drivers to locate unoccupied parking spaces expeditiously, thereby curtailing the time spent in search of parking, as well as the associated fuel consumption and emissions.
(d) Environmentally-sound Multi-modal Transportation: This application endeavors to facilitate uninterrupted connectivity and foster the utilization of diverse modes of travel that cater to individual requirements and inclinations. The system seeks to promote sustainable transportation modes, including walking, cycling, and public transportation, with the aim of mitigating carbon emissions from private vehicles. By encouraging shorter trips to be undertaken on foot or by bicycle, and by offering efficient public transportation alternatives, the application can substantially reduce the volume of vehicles on the road and their corresponding emissions.

4 Evaluation of the Framework

As concerns about climate change and environmental sustainability continue to grow, the need for an environment-friendly smart transportation system becomes increasingly important. Our framework aims to reduce environmental impact and improve the efficiency of transportation. As such, we need to evaluate the impact of this framework by examining its various components and their contributions to effectiveness and sustainability. The discussion focuses on analyzing energy efficiency, network latency, affordability, the complexity of communication, data analytics, distributed computing, privacy and security, and user interaction. Table 3 describes these chal-

Table 3 Evaluation of the framework

Challenges	Contribution of the framework to overcome the challenges
Environment sustainability	The framework promotes and prioritizes environment-friendly transportation by integrating real-time data-driven decision
Energy efficiency	The deployment of the edge layer helps minimize energy consumption by bringing computing power closer to the data source and reducing the need for extensive data transmission to centralized servers or cloud platforms
Network latency	By processing data at the edge layer, the system can respond to real-time events and demands more quickly, reducing network latency
Affordability	Virtual sensors offer scalability and flexibility, allowing for cost-effective deployment across different domains and applications. They can be easily scaled up or down without significant additional investments making the framework affordable to the stakeholders
Heterogeneity	Virtual sensors at the edge and cloud layer can help address the heterogeneity of devices and protocols in IoT and cloud environments by providing a standardized interface and data abstraction layer
Data analytics	The framework harnesses the power of both edge and cloud computing. Edge devices perform initial data processing, filtering, and lightweight analytics, while more resource-intensive tasks or long-term analysis is offloaded to the cloud
Distributed computing	By separating the data pre-processing module at the edge layer and the decision-making module at the cloud layer, the framework enables workload distribution, ensuring that the energy load is balanced across the system
Privacy and security	Data minimization principles are followed at the edge layer to collect and process only necessary data reducing the amount of personally identifiable information (PII) being stored or transmitted
User interaction	The application layer provides user interfaces that allow end users to interact with the smart transportation system

lenges of smart transportation systems and it explains how our proposed framework overcomes the mentioned challenges. Through an extensive evaluation, we aim to provide insights into the potential areas of contribution of the framework and guide policymakers and stakeholders in creating more sustainable transportation systems.

5 Conclusion

The convergence of IoT, Edge Computing, and Cloud Computing technologies offers a promising framework for the development of Smart Transportation Applications for advancing environmental sustainability in the transportation sector. Through the implementation of this framework, we have the potential to optimize traffic flow, reduce carbon emissions, and enhance resource utilization in transportation networks.

By leveraging real-time data, advanced analytics, and intelligent decision-making, transportation systems can be optimized for energy efficiency, faster processing times, lower latency, improved system reliability, and enhanced user experiences. Our vision extends beyond theoretical possibilities to the tangible creation of innovative applications that will ensure seamless mobility for users. Our focus includes the development of smart transport management applications within the framework. These applications will employ cutting-edge algorithms and predictive analytics to address the critical need for environmental sustainability. We are dedicated to exploring the synergy between our framework and emerging technologies such as IoT, Edge Computing, and Cloud Computing. The applications we envision will harness the full capabilities of these technologies, ensuring rapid data transmission, real-time decision-making, and intelligent automation. As a future prospect, expect solutions that encourage eco-friendly travel choices, optimize routes for minimal environmental impact, and promote the use of sustainable transportation modes.

References

1. Klecha L, Gianni F (2018) Designing for sustainable urban mobility behaviour: a systematic review of the literature. In: Citizen, territory and technologies: smart learning contexts and practices: proceedings of the 2nd international conference on smart learning ecosystems and regional development-University of Aveiro, Portugal, 22–23, June 2017 2, Springer, pp 137–149
2. Khardi S, Bernoud-Hubac N (2022) Editorial for the special issue impacts of transport systems on air pollution and human health
3. Dua S (2022) 7 reasons not to buy e-car this year. https://economictimes.indiatimes.com/wealth/spend/7-reasons-not-to-buy-e-car-this-year/higher-cost/slideshow/92892326.cms. Online; Accessed 15 July 2022.
4. Shah KJ, Pan SY, Lee I, Kim H, You Z, Zheng JM, Chiang PC (2021) Green transportation for sustainability: review of current barriers, strategies, and innovative technologies. J Cleaner Prod 326:129392
5. Arthurs P, Gillam L, Krause P, Wang N, Halder K, Mouzakitis A (2022) A taxonomy and survey of edge cloud computing for intelligent transportation systems and connected vehicles. IEEE Trans Intell Transp Syst 23(7):6206–6221
6. Gesi homepage. https://gesi.org/
7. Rehena Z (2020) Internet of things: challenges and its applications. In: Interoperability in IoT for smart systems. CRC Press, pp 1–14
8. Hamdan S, Ayyash M, Almajali S (2020) Edge-computing architectures for internet of things applications: a survey. Sensors 20(22)
9. Martin D, Kühl N, Satzger G (2021) Virtual sensors. Bus Inf Syst Eng 63:315–323
10. Alsabaan M, Naik K, Khalifa T, Nayak A (2010) Vehicular networks for reduction of fuel consumption and co 2 emission. In: 2010 8th IEEE international conference on industrial informatics. IEEE, pp 671–676
11. Kwak J, Park B, Lee J (2012) Evaluating the impacts of urban corridor traffic signal optimization on vehicle emissions and fuel consumption. Transp Plann Technol 35(2):145–160
12. Gerla M, Lee EK, Pau G, Lee U (2014) Internet of vehicles: from intelligent grid to autonomous cars and vehicular clouds. In: 2014 IEEE world forum on internet of things (WF-IoT). IEEE, pp 241–246
13. Barth MJ, Wu G, Boriboonsomsin K (2015) Intelligent transportation systems and greenhouse gas reductions. Curr Sustain/Renew Energy Rep 2:90–97

14. Stevanovic A, Stevanovic J, Zhang K, Batterman S (2009) Optimizing traffic control to reduce fuel consumption and vehicular emissions: integrated approach with vissim, cmem, and visgaost. Transp Res Record 2128(1):105–113
15. Ji Y, Hu B, Hill G, Guo W, Blythe P, Gao L (2016) Signal coordination scheme based on traffic emission. IET Intell Transp Syst 10(2):89–96
16. Peng T, Yang X, Xu Z, Liang Y (2020) Constructing an environmental friendly low-carbon-emission intelligent transportation system based on big data and machine learning methods. Sustainability 12(19):8118
17. Zhao L, Jia Y (2021) Intelligent transportation system for sustainable environment in smart cities. Int J Electr Eng Educ 0020720920983503
18. Ashokkumar K, Sam B, Arshadprabhu R et al (2015) Cloud based intelligent transport system. Procedia Comput Sci 50:58–63
19. Mathirajan M, Devadas R, Ramanathan R (2021) Transport analytics in action: a cloud-based decision support system for efficient city bus transportation. J Inf Optim Sci 42(2):371–416
20. Yan G, Qin Q (2020) The application of edge computing technology in the collaborative optimization of intelligent transportation system based on information physical fusion. IEEE Access 8:153264–153272
21. Zeng Q, Duan Q, Shi M, He X, Hassan MM (2020) Design framework and intelligent in-vehicle information system for sensor-cloud platform and applications. IEEE Access 8:201675–201685
22. Peyman M, Copado PJ, Tordecilla RD, Martins LDC, Xhafa F, Juan AA (2021) Edge computing and IoT analytics for agile optimization in intelligent transportation systems. Energies 14(19):6309
23. Sood SK et al (2021) Smart vehicular traffic management: an edge cloud centric IoT based framework. Internet Things 14:100140
24. Ali MH, Jaber MM, Abd SK, Alkhayyat A, Albaghdadi MF (2022) Big data analysis and cloud computing for smart transportation system integration. Multimedia Tools Appl 1–18
25. Chavhan S, Gupta D, Gochhayat SP, Khanna A, Shankar K, Rodrigues JJ (2022) Edge computing AI-IoT integrated energy-efficient intelligent transportation system for smart cities. ACM Trans Internet Technol 22(4):1–18
26. Ciccarella G, Giuliano R, Mazzenga F, Vatalaro F, Vizzarri A (2019) Edge cloud computing in telecommunications: case studies on performance improvement and TCO saving. In: 2019 fourth international conference on fog and mobile edge computing (FMEC), pp 113–120
27. Gupta A, Mukherjee N (2017) Can the challenges of IoT be overcome by virtual sensors. In: 2017 IEEE international conference on internet of things (iThings) and IEEE green computing and communications (GreenCom) and IEEE cyber, physical and social computing (CPSCom) and IEEE smart data (SmartData). IEEE, pp 584–590
28. Bose S, Mukherjee N (2022) Senschedule: scheduling heterogeneous resources in sensor-cloud infrastructure. IEEE Trans Serv Comput 15(04):1825–1840
29. Mukherjee N, Bhunia SS, Bose S (2016) Virtual sensors in remote healthcare delivery: some case studies
30. Ridzuan F, Zainon WMNW (2019) A review on data cleansing methods for big data. Procedia Comput Sci 161:731–738
31. Wang X, Wang C (2019) Time series data cleaning: a survey. IEEE Access 8:1866–1881
32. Durrant-Whyte H (2016) Henderson. Multisensor data fusion. Springer handbook of robotics, T.C., pp 867–896
33. Sayood K (2017) Introduction to data compression. Morgan Kaufmann

A Mathematical Model for Data Traffic Management in Next-Generation IoT Networks

Gemini George

Abstract The rapid deployment of IoT networks in different industrial services has caused the emanation of a huge volume of data from sensors and monitors. The efficient analysis and compact representation of the big data generated is critical to managing its transfer, storage, and latency. In this work, we propose a novel graphical model that has the flexibility to adapt to different architectural features of distributed IoT systems with low-power sensors. Further, we consider the traffic load minimization problem in the network and map the problems of network coverage and clustering onto the graph to find an optimal distributed network structure that minimizes the traffic load. The proposed approaches are tested under different network conditions, and the simulation results emphasize the improved performance of the method.

Keywords Distributed IoT systems · Load minimization problem · Graphical model · Network coverage

1 Introduction

Being the backbone of IoT, wireless sensor networks (WSNs) consisting of battery-operated sensors are rapidly being integrated into applications such as environmental monitoring, healthcare, agriculture, infrastructure, logistics, smart cities, and many more. The sensors and monitors of such applications gather physical phenomena in their vicinity at immense speed, generating huge volumes of data. This poses a serious challenge not only to their storage and process but also to the network longevity as sensors consume energy majorly to transfer data [1, 2]. Thus, investigating efficient data traffic management strategies in WSNs attracts huge research interest.

Efficient data traffic management is an important design issue in next-generation IoT networks [3, 4]. Aiming for this, we propose a mathematical framework that

G. George (✉)
Department of Computer Applications, St. George's College Aruvithura,
Erattupetta, Kottayam, Kerala 686122, India
e-mail: gemini@sgcaruvithura.ac.in

S. M. Thampi et al. (eds.), *Fifth International Conference on Computing and Network Communications*, Lecture Notes in Electrical Engineering 1219,
https://doi.org/10.1007/978-981-97-4540-1_15

is capable of capturing various architectural characteristics of IoT networks of different structures and topologies. Such a framework helps the researchers map their network design issues to a common platform and find a solution using the framework's features. In addition, we address the problem of efficient data gathering in an IoT network and design an approach to find a network structure for efficient data management. We consider a cluster-based three-layer network where each sensor transfers its data to its respective cluster-head (CH) within a cluster, and the inter-cluster data transfer to the base office (BO) happens via hoping the intermediate CHs.

The paper is organized as follows: Sect. 2 presents the fundamental data management approaches in IoT systems. The proposed mathematical model of an IoT network is presented in Sect. 3. An approach for efficient data collection is reported in Sect. 4. Section 5 presents the simulation results, and we conclude our work in Sect. 6.

2 Literature Review

Sensors in the IoT networks are organized into clusters to improve the network longevity [5, 6]. In a cluster-based network structure, the sensors in the cluster cover the cluster area. One of the sensors (on a rotation basis) is chosen as a cluster head (CH). All non-CH sensors transfer the generated data to the CH of their clusters. The CHs transmit the data collected to the BO. Such transmissions to the BO can be done either by direct transmission or in multiple transmissions via other CHs [7]. The second approach is better in terms of energy efficiency as the direct transmission from CH to the BO causes faster energy drainage from CH. However, this approach requires all CH nodes to be in a connected network referred to as overlay network [8]. Clustered networks are generally considered more energy-efficient [9] than unclustered networks for the following reasons.

- Energy saving: Only the sensors chosen as the CHs are involved in routing, scheduling, and relaying the data to the BO.
- Reduced collisions: The sensors chosen as the non-CH nodes are not involved in channel access. Further, clustering ensures that no collisions are happening within a cluster.
- Better bandwidth utilization and system capacity: Clusters provide frequency reuse.
- Exploiting correlations: The data collected by the CHs are often highly correlated. Hence, using a compression scheme at the CHs can reduce data traffic to the BO.

However, clustering results in more energy drain at CHs. Therefore, efficient CH election and rotation schemes [10] policies are to be designed. The number of clusters and selection of the CHs are to be carefully designed to minimize the overall energy spend and to optimize network longevity. Further, an activity that triggered in the monitoring region can be detected by a more than one sensors which can belong to different clusters, resulting in redundant data traffic. Efficient switching of sensor states between 'on' and 'off' modes alleviates such issues.

The *hybrid energy-efficient and distributed* (HEED) protocol [11] improves the network longevity by balanced selection of CHs, reducing the message overhead, and finalizing the clustering process within a limited number of iterations. The HEED protocol has transformed through the years with various improvements and is now one of the dominant IoT protocols. The major variants of HEED are:

1. *Distributed weight-based energy-efficient hierarchical clustering protocol* (DWEHC): DWEHC [12] is a HEED variant that yields a better cluster balancing and intra-cluster topology. DWEHC deploys a multilevel grading of sensors that helps terminate the clustering process in very limited iterations compared to HEED.
2. *Unequal clustering-HEED* (U-HEED): U-HEED solves the hot-spot problem by generating clusters with size proportional to the distance between the BO and the cluster CH. Thus, the clusters close to the BO posses lesser members than those that lie farther away. The cluster radius is computed using the competition radius formula [13]. The performance metrics of the U-HEED is better than the HEED, LEACH, and unequal LEACH.
3. *Rotated hybrid energy-efficient and distributed* (R-HEED): R-HEED [14] is a variant of HEED with rotations of the CHs. The clustering is divided into rounds, where each round consists of setup and the steady-state phases. After the setup phase, the CH of a cluster broadcasts a schedule of the CH rotation to its cluster members which eliminates the requirement of re-clustering.
4. *Heterogeneous HEED* (H-HEED): In H-HEED [15], some selected sensors is equipped with higher energy than others. The sensors are categorized into normal energy nodes and advanced energy nodes. CHs are chosen from the advanced energy nodes in the clustering phase.
5. *Rotated unequal clustering-HEED* (RUHEED): RUHEED [16] is a better version of the U-HEED with CH rotation within a cluster. The RUHEED is designed in three stages: CH election, cluster formation, and CH rotation. Once the clusters are initialized, the sensor with the maximum residual energy is selected as the CH within the cluster. The cluster re-configuration takes place when a node failure occurs in any of the clusters. The CHs are rotated in every round within a cluster based on their energies.
6. *Dynamic weight clustering-based algorithm* (DWCA): DWCA [17] enables balanced network traffic and network scalability. In the setup phase, each sensor calculates its residual energy, number of neighbors, and distance from the BO, computes a weight value based on the aforementioned parameters, and broadcasts the information to its neighbors. The sensors with the maximum weight values are chosen as CHs, and the non-CH sensors associate with their nearest CH.

3 Mathematical Modeling of General Wireless Sensor Network

Let $N = \{1, 2 \dots n\}$ represent the set of network devices, and b represent the BO. Let A and B denote the set of activated sensors and cluster heads, respectively. Let $\mathcal{R} = \{1, 2, \cdots, r\}$ represent the set of non-overlapping sub-regions of the area to be monitored. With respect to the above notations, we define our mathematical model as follows.

Definition 1 (*Sensor Range Function*) The Sensor Range Function is the function $\Omega : (N - B) \times (N - B) \rightarrow \{T, F\}$ defined between sensor devices to check whether a sensor device is in the range of another. Let α and β be any two devices in $(N - B)$, then the Sensor Range Function $\Omega(\alpha, \beta)$ gives a $True$ value if and only if the sensor α can reach sensor β in single hop.

Definition 2 (*Sensor Cluster-head Range Function*) The Sensor Cluster-head Range Function is the function $\Psi : (N - B) \times B \rightarrow \{T, F\}$ defined between non-cluster head sensor devices to a cluster-head to check whether a sensor device is in the range of the cluster-head. Let α and β be any two devices in $(N - B)$ and B respectively, then the Sensor Cluster-head Range Function $\Phi(\alpha, \beta)$ gives a $True$ value if and only if the sensor α can reach the cluster-head β in a single hop.

Definition 3 (*Cluster-head Range Function*) The Cluster-head Function is the function $\Gamma : B \times B \cup \{b\} \rightarrow \{T, F\}$ defined either between two cluster-heads or between a cluster-head and a sink to check whether the latter is in the range of the former. Let α and β be any two devices in B and $B \cup \{b\}$ respectively, then the Cluster-head Range Function $\Gamma(\alpha, \beta)$ gives a $True$ value if and only if the cluster-head α can reach the cluster-head (or sink) represented by β in a single hop.

Definition 4 (*BO Range Function*) The Sink Range Function is the function Ψ : $\{b\} \times B \rightarrow \{T, F\}$ defined between the sink and a cluster-head to check whether the BO is in the range of the cluster-head. Let b denote the BO, and α be any cluster-head in B, then the Sink Range Function $\Psi(b, \beta)$ gives a $True$ value if and only if the cluster-head α can reach the BO b in a single hop.

Definition 5 (*IoT Network*) An IoT Network is a 7-tuple $(N, A, b, C, B, \triangle, \triangledown)$, where

- N is the set of network devices in the IoT network
- A is the set of activated devices in the network
- b is the sink node in the WSN
- C is the set of clusters in the WSN
- B is the set of cluster-heads in the WSN
- $\triangle$ is the sensor cluster-head range characteristic set defined by $\triangle = \{(\alpha.\beta), \Phi(\alpha, \beta) = T) | (\alpha, \beta) \in (N - B) \times (N - B)\}$
- $\triangledown$ is the cluster-head range characteristic set defined by $\triangledown = \{(\alpha.\beta), \Gamma(\alpha, \beta) = T) | (\alpha, \beta) \in B \times B \cup \{b\}\}$

Definition 6 (*Cluster*) A cluster is defined as a pair (C,h) where $C \subset N$ represents the set of sensor nodes within the cluster and $\alpha \in B$ is a special sensor node that acts as the head of the cluster.

Now, we analyze a typical IoT network architecture and its implication on data traffic regulation. Concerning this architecture, we note the following:

- Instead of restricting the Sensor Range Function Ω to have the domain $N - BXN - B$, we may allow Ω to be defined for any pair of nodes i and j $\in N \bigcup\{b\}$, to indicate whether either one falls into the range of the other.
- In common wireless sensor networks, the CHs are capable of transmitting into farther distances to enable transmission to the CC in lesser hops. Therefore, we may think of the Cluster-head Range Function Γ to be an extended restriction of Ω in that $\Gamma(i, j) \geq \Omega(i, j)\ \forall i \in BXB \bigcup\{b\}$.
- Φ and Ψ can be computed from Ω and Γ.

These observations lead us to have a more compact formal representation for cluster-based IoT architecture where the underlying sensor network can be represented by a 7 tuple $(N, A, b, C, B, \triangle, \triangledown)$ and hence the corresponding IoT network graph has edges corresponding to only two sets $\triangle$ and $\triangledown$. In the forthcoming section, we use this mathematical framework to find an efficient network configuration for integrated energy and data harvesting for a clustered network.

4 Finding an Efficient IoT Network Structure

In this section, we are going to find an efficient network architecture, the steps of which follow:

4.1 Node Scheduling Phase

In this phase, we select the set of active nodes. The activated sensors should cover the monitoring region spanned by $\mathcal{R}$. We first unmark all the given regions in $\mathcal{R}$. Let U denote the corresponding set of unmarked sets. Further, we initialize the active nodes A set as set ϕ. Now, let i be any node, $R(i)$ denotes the region coverage of the node i. Now, we sort the set of sensors in U in the non-decreasing order of their $|R(i)|$ values, resulting in a list L_a. Let i^* be the sensor in L_a such that $|R(i^*)| = max\{|R(i)| \mid i \in U\}$. We designate i^* as being selected and include it in A. Further, we mark all the cells in $|R(i^*)|$, and update the set U by eliminating all the cells in $|R(i^*)|$ from $\mathcal{R}$. We repeat the above procedure until U becomes ϕ. Then, we have a set A consisting of the active sensors ensuring the coverage of the non-overlapping sub-regions in $\mathcal{R}$.

Algorithm 1 Node Scheduling Function

Input: Set of nodes N and the region $mathcal R$
Output: Set of scheduled nodes A

function SCHEDULE(N, Γ)
  Set $A := \phi$;
  Unmark all nodes in N
  Set U:=ϕ
  while $N \neq 0$ **do**
    for each node i in N **do**
      $R(i)$={$r \mid r \in \mathcal{R}$ and i covers the sub-region r}
    max:= -∞ and i^*:= NULL;
    for each node i in N **do**
      if $max \leq R(i)$ **then**
        Set $i^* := i$ and $max := R(i)$;
    Set $A := A \bigcup \{i^*\}$
    Mark all cells in $R(i^*)$
    Set $U := U \setminus R(i^*)$;
  Return A

4.2 *Cluster Formation*

In this step, we find the set of sensors in A to be designated CHs. We start with unmarking all the nodes in A and initialize the set V with these unmarked sets of active nodes. Further, we initialize the CH set B to be ϕ. Let i be any node and let $L_c(i) = \{j \mid \Omega(i, j) = 1\}$. Now, we find i^* such that $|L_c(i^*)| = max\{|(L_c(i))|\}$. Then, i^* is chosen as a CH and $L_c(i^*)$ as members, forming a cluster. The node i^* is included in the list B. The iteration completes until each sensor is either designated as CH or becomes a part of a cluster.

4.3 *Finding the Routes*

We form a graph out of the active sensors chosen by the above steps, and we add edges in the graph so that two nodes i and i' are connected if they are in the Γ range. Then, we run the shortest path algorithm from each active node, keeping the single destination as the CCC. This establishes the shortest paths from each active node to the sink.

Algorithm 2 Cluster Form Function

Input: Set of active nodes A, the set of sensors N, and the sensor range function Ω
Output: Set of cluster heads B

function FORMCLUSTER(A, N, Ω)
  Set $B := \phi$;
  Unmark all sensors in A and set $V := A$;
  while a node remains not a part of any cluster **do**
    for each node i in $N \bigcap V$ **do**
      $L_c(i) := \{j \mid j \in (S \bigcap V)), \text{ and } \Omega(i, j) = 1\}$
    $max := -\infty$ and $i^* := NULL$;
    for each node $i \in N \bigcap V$ **do**
      if $max \leq L_c(i)$ **then**
        Set $i^* := i$ and $max := L_c(i)$;
    Set $B := B \bigcup \{i^*\}$;
    For a new cluster with CH as i^* and all the nodes in $L_c(i^*)$ connected to i^* as non-CHs
    Set $V := V \setminus (Ł_c(i^*) \bigcup \{i^*\})$;
  Return B

Algorithm 3 Algorithm establishing the routes

Input: Set of active nodes A, the set of sensors N, and the range functions Ω and Γ
Output: A graph representing the network structure

Form the graph $G = (A, E)$ where $E := \{(i, j) | i, j \in H, \Gamma(i, j) = 1\}$;
$dest := z$;
$dist[dest] := 0$;
$done[dest] := TRUE$;
for each node $i \in N$ **do**
  $dist[i] := \infty$;
  $done[dest] := FALSE$;
EnquePQ(Q,0,dest);
while isnotempty(Q) **do**
  current:=dequeuePQ(Q);
  for $i \in N$ **do**
    if $(i, current) \in E$ **then**
      if (done[i]=FALSE & dist[i]$\geq$ dist[current]+$W_{AN}((i, current))$) **then**
        $dist[i] := dist[current] + W_{AN}((i, current))$;
        $EnquePQ(Q, dist[i], i)$
        next[i]:=current;
  $done[current] := TRUE$;
  for $i \in N$ **do**
    $route := NIL$;
    $GETROUTE(next, i, route, dest)$
    for each edge $(k, l) \in route$ **do**
      $E := E \bigcup \{(k, l)\}$
return G

Algorithm 4 Algorithm to find route

function GETROUTE(next, i, route, d)
  function APPEN(route,i)
    if $(i \neq \delta)$ **then**
      function GETROUTE(next, next[i], route, d)

5 Simulation Results

We evaluate the performance of the proposed method in a monitoring region of sizes $D \times D\ m^2$ wherein sensors are deployed according to uniform distribution. The sensors are assumed to start operating with an initial energy of $E'J$. The BO is assumed to be located at a corner point (D, D) of the monitoring region. Each sensor generates 5000 symbols in every sequence of operation. The maximum inter-cluster and intra-cluster distances are taken as 30% and 20% of D. As the operations proceed, the energies of the activated sensor nodes start decreasing according to the standard radio communication model [18]. The sum of the energies of the active sensors is taken as the network energy. The system ceases its operation the active sensors fails to cover the monitoring region in $\mathcal{R}$.

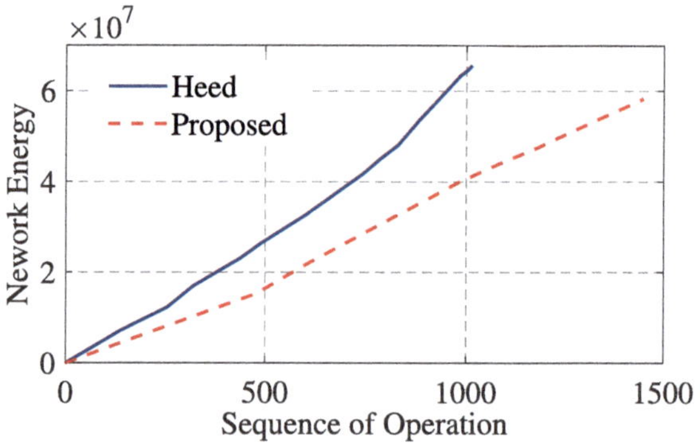

Fig. 1 Cumulative data traffic: modified leach and proposed method

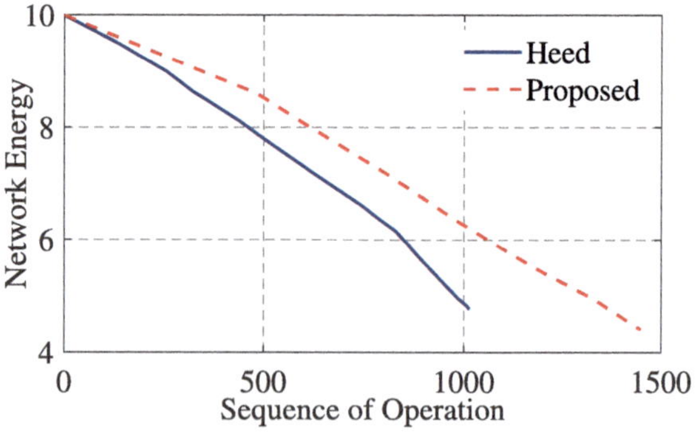

Fig. 2 Remaining network energy: modified leach and proposed method

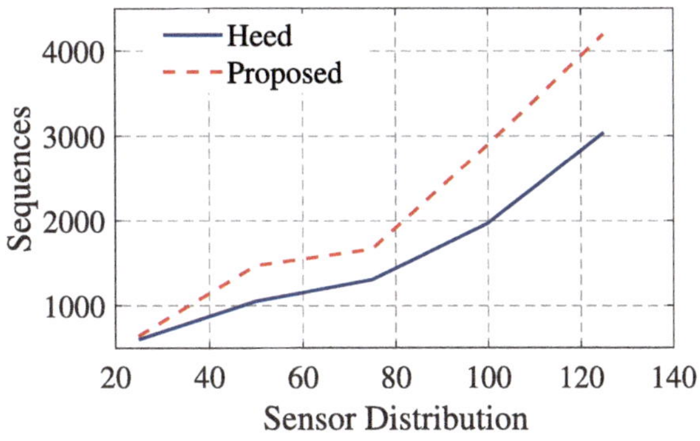

Fig. 3 Sequences of operation: modified leach and proposed method

HEED is a major protocol for energy-efficient IoT systems. The characteristics of the HEED protocol are:

1. A sensor is open to join any cluster based on its local data,
2. The CHs are distributed well over the monitoring region,
3. Intra-cluster communications involve single hop and inter-cluster communications involve multihop data traffic, respectively,
4. Selection of CHs is based on the energy remaining in the sensors and intra-cluster communication cost.

Therefore, we choose the latest version of HEED [19] for comparing the performance of the proposed method. Simulation results show that the proposed method outperforms the HEED by reducing the data traffic because the better clustering approach adopted. The data traffic of the methods are compared in Fig. 1. The reduction in the data traffic and lesser energy spent in operation, leads to an overall improvement in the network longevity. Further, the proposed method yields a lower energy spend as shown in Fig. 2, and provides improved life for the network as depicted in Fig. 3.

6 Conclusion

The proposed mathematical framework captures the salient features of the next generation IoT networks. Hence, researchers can map the underlying network onto the mathematical framework and use established algorithms to find solutions to different networking problems. The proposed framework can be extended to map application specific features.

References

1. Thomson C, Wadhaj I, Tan Z, Al-Dubai A (2021) Towards an energy balancing solution for wireless sensor network with mobile sink node. Comput Commun 170:50–64
2. Du R, Gkatzikis L, Fischione C, Xiao M (2017) On maximizing sensor network lifetime by energy balancing. IEEE Trans Control Netw Syst 5(3):1206–1218
3. Chandnani N, Khairnar CN (2022) An analysis of architecture, framework, security and challenging aspects for data aggregation and routing techniques in IoT WSNs. In: Theoretical computer science
4. Das SK, Kapelko R (2021) On the range assignment in wireless sensor networks for minimizing the coverage-connectivity cost. ACM Trans Sensor Netw (TOSN) 17(4):1–48
5. Daneshvar SMH, Mohajer PAA, Mazinani SM (2019) Energy-efficient routing in wsn: a centralized cluster-based approach via grey wolf optimizer. IEEE Access 7:170019–170031
6. Behera TM, Mohapatra SK, Samal UC, Khan MS, Daneshmand M, Gandomi AH (2019) Residual energy-based cluster-head selection in WSNs for IoT application. IEEE Internet Things J 6(3):5132–5139
7. Manuel EM, Pankajakshan V, Mohan MT (2023) Linear programming models for the design of energy-efficient IoT networks with transmission constraints. IEEE Trans Green Commun Netw
8. Chamam A, Pierre S (2009) On the planning of wireless sensor networks: energy-efficient clustering under the joint routing and coverage constraint. IEEE Trans Mobile Comput 8(8):1077–1086
9. Salman HM (2014) Survey of routing protocols in wireless sensor networks. Int J Sensors Sensor Netw 2(1):1–6
10. Rawat P, Chauhan S (2021) Particle swarm optimization-based energy efficient clustering protocol in wireless sensor network. Neural Comput Appl 33:14147–14165
11. Younis O, Fahmy S (2004) Heed: a hybrid, energy-efficient, distributed clustering approach for ad hoc sensor networks. IEEE Trans Mobile Comput 3(4):366–379
12. Ding P, Holliday J, Celik A (2005) Distributed energy-efficient hierarchical clustering for wireless sensor networks. In: International conference on distributed computing in sensor systems. Springer, Heidelberg, pp 322–339
13. Li C, Ye M, Chen G, Wu J (2005) An energy-efficient unequal clustering mechanism for wireless sensor networks. In: IEEE international conference on mobile adhoc and sensor systems conference. IEEE, p 8
14. Mardini M, Yassein MB, Khamayseh Y, Ghaleb BA (2014) Rotated hybrid, energy-efficient and distributed (r-heed) clustering protocol in WSN. WSEAS Trans Commun 13:275–290
15. Kour H, Sharma AK (2010) Hybrid energy efficient distributed protocol for heterogeneous wireless sensor network. Int J Comput Appl 4(6):1–5
16. Aierken N, Gagliardi R, Mostarda L, Ullah Z (2015) Ruheed-rotated unequal clustering algorithm for wireless sensor networks. In: 2015 IEEE 29th international conference on advanced information networking and applications workshops. IEEE, pp 170–174
17. Essa A, Al-Dubai AY, Romdhani I, Eshaftri MA (2017) A new dynamic weight-based energy efficient algorithm for sensor networks. In: Smart grid inspired future technologies. Springer, Heidelberg, pp 195–203
18. Lee J-S, Kao T-Y (2016) An improved three-layer low-energy adaptive clustering hierarchy for wireless sensor networks. IEEE Internet Things J 3(6):951–958
19. Ullah Z (2020) A survey on hybrid, energy efficient and distributed (heed) based energy efficient clustering protocols for wireless sensor networks. Wireless Personal Commun 112(4):2685–2713

Crowdsourced Mobile Applications for Disaster Relief: A Case Study of Landslide Tracker and Amrita Kripa App Deployments in Sikkim Schools

Ramesh Guntha, Raji Pushpalatha, Sudesh Kumar Wadhawan, and Balmukund Singh

Abstract Effective disaster relief efforts hinge on proactive community involvement empowered by scientific understanding and cutting-edge technological tools, throughout the disaster lifecycle, spanning pre-disaster preparations, during-disaster rescue and evacuations, and post-disaster relief and rehabilitation. Crowdsourced mobile applications have emerged as powerful tools to facilitate communication among stakeholders, including community members, non-governmental organizations (NGOs), local administrators, and government agencies. This paper showcases Amrita's two crowdsourced apps: the Landslide Tracker app, a multilingual app designed to enable community members to document landslides with geo-tagging, images, and landslide characteristics, and the Amrita Kripa app, facilitating real-time communication between disaster victims and relief volunteers and agencies. The Amrita Kripa app enables victims to request rescue, supplies, medical assistance, and services, with requests geo-tagged and instantly communicated to all relevant stakeholders and relief providers. The present research involved the enhancement and deployment of these apps in a case study area of Sikkim, a Northeastern state of India, which is particularly vulnerable to multiple geohazards. This focused group research could examine the receptivity of these two apps among higher-secondary and undergraduate school students and faculty through a two-day workshop that has been revealed to be beneficial in strengthening their communication and cooperation

R. Guntha (✉) · S. K. Wadhawan · B. Singh
Center for Wireless Networks and Applications (WNA), Amrita Vishwa Vidyapeetham, Amritapuri, India
e-mail: rameshg@am.amrita.edu

S. K. Wadhawan
e-mail: wadhawansk.leo@gmail.com

B. Singh
e-mail: balmukundsingh@am.amrita.edu

R. Pushpalatha
Amrita School for Sustainable Futures, Amrita Vishwa Vidyapeetham, Amritapuri, India
e-mail: rajip@am.amrita.edu

S. M. Thampi et al. (eds.), *Fifth International Conference on Computing and Network Communications*, Lecture Notes in Electrical Engineering 1219,
https://doi.org/10.1007/978-981-97-4540-1_16

to mitigate a possible calamity. This case study aims to illustrate a path forward for enhancing community resilience and effective engagement in disaster relief efforts.

Keywords Disaster relief · Crowdsourced mobile applications · Community engagement

1 Introduction

India is exposed to a wide range of multi-hazards, falling into four main categories: (1) water-related hazards such as landslides, floods, and cyclones; (2) geophysical hazards like earthquakes and tsunamis; (3) environmental changes leading to sea-level rise and heat waves; and (4) threats from nuclear, biological, and chemical disasters and terrorism. These hazards result in an Average Annual Loss (AAL) of about $9.8 billion. A significant part of India's territory is prone to natural disasters, including 68% of its cultivable area being at risk of drought. Sikkim, in the northeastern Himalayas, is particularly susceptible to landslides and earthquakes due to its geography and climate. Factors like heavy rainfall, tectonic movements, and human activities such as unplanned urbanization intensify geohazard risks in the area. This leads to frequent landslides in Sikkim, causing substantial human and economic losses and underscoring the urgent need for effective disaster management and mitigation measures [1].

The National Disaster Management Act 2005 [2] emphasizes prioritizing preparedness and mitigation in disaster management over ad hoc rescue and relief efforts. In India, the disaster management phase model includes Mitigation, Preparedness, Response, and Recovery. Mitigation focuses on pre-disaster measures like infrastructure enhancement, while Preparedness involves community training for disaster scenarios. The Response phase addresses immediate humanitarian needs during disasters, and recovery focuses on post-disaster community and economic stabilization. These phases are institutionalized across various government ministries and departments [3].

Five key pillars underpin community resilience: Self-awareness, Mindfulness, Self-care, Positive Relationships, and Purpose [4]. Self-awareness involves understanding personal traits, strengths, and weaknesses. Mindfulness focuses on present awareness and non-judgmental observation of thoughts and feelings. Self-care is about maintaining effective functioning and confidence in daily challenges. Positive Relationships are about nurturing supportive connections, while Purpose involves serving beyond oneself, found in faith, family, or affiliations. These pillars, combined with disaster management phases, create a holistic strategy for enhancing community resilience, helping communities better manage and recover from crises. Additionally, there's growing interest in integrating traditional indigenous knowledge and new technologies for improved resilience and disaster risk reduction [5].

The Landslide Tracker application, essential during monsoon seasons, enables users to record landslides by logging GPS coordinates, occurrence times, images, and geological data, contributing to a comprehensive landslide database and supporting early warning systems [6]. With over 500 global events documented, including in Kerala, Tamil Nadu, and the Himalayas, it's accessible on various platforms like Google Play Store and iOS App Store. The "Amrita Kripa" app, developed under the Indo-German 'FloodEvac' project, is a multilingual, crowdsourced disaster relief platform that allows victims to request aid and volunteers to offer help, with information geo-tagged on an interactive map [7]. Used extensively during the 2018 and 2019 Kerala floods, it facilitated communication and assistance among victims, volunteers, and relief agencies. Amrita University's 24-h call center, supporting the app, assisted 100,000 victims and handled over 25,000 calls. Despite its effectiveness, the app faced challenges like unreasonable requests and misuse [7].

The "Landslide Tracker" and "Amrita Kripa" apps are crucial for disaster management, allowing communities to report landslides and seek aid during emergencies. Promoting these apps, particularly through educational institutions and among youth, is essential for wider adoption and effective community engagement. A 2-day Disaster Resilience workshop in Sikkim, organized by entities including the UNESCO Chair on Experiential Learning and the Sikkim State Disaster Management Authority, trained 85 participants from local institutions. The workshop focused on enhancing disaster awareness and preparedness, teaching about hazards, vulnerabilities, and the use of these apps. Feedback showed increased disaster awareness and resource identification within the community. Plans involve tailoring this training to various Indian regions, aiming to boost nationwide community resilience and preparedness.

This paper comprehensively examines disaster management and technology, beginning with an overview of related works that focus on the disaster lifecycle and the use of crowdsourced apps in recent global disasters. It then provides an in-depth analysis of the "Amrita Kripa" and Landslide Tracker apps, detailing their roles throughout the disaster relief lifecycle. The paper also discusses a two-day workshop in Sikkim schools, presenting feedback from students and teachers. It moves on to outline future plans aimed at enhancing community awareness of these apps, concluding with final reflections and acknowledgments.

2 Related Works

Community-centric approaches in disaster management emphasize collaboration between the public, scientists, and experts, enhancing the collective creation of scientific knowledge [8, 9]. The growth of global citizen science initiatives is closely linked to advancements in AI and specialized tools [10], fostering democratic participation and raising awareness about societal challenges [11]. These approaches are increasingly reliant on multimedia resources from various platforms, including text, photos, and videos [12], with locals in landslide-prone areas serving as crucial human sensors. Volunteer organizations support this by facilitating cost-effective geo-data collection [13, 14].

Despite the benefits, challenges such as data accuracy, inconsistent terminology, language barriers, and gaps in data remain [15]. Addressing these requires specialized techniques for extracting meaningful insights and rigorous data quality assessments [16]. Tools like NASA's landslide reporter and the Landslip Landslide tracker aid volunteers in logging data, stored on cloud servers and visualized on GIS platforms [4]. The literature highlights the importance of citizen science tools in various disaster management phases [12, 17].

The efficacy of these tools varies based on geographic context, technological proficiency, and participant motivation [18]. For instance, the disaster relief apps used during the 2023 Turkey earthquake [19] and other global disasters from 2010 to 2019 [20] demonstrate their varied applications. These include social media platforms like Twitter, Facebook, and WhatsApp, mapping tools like Google Maps and OpenStreetMap, and mobile applications like Amrita Kripa and Ushahidi [21–24]. These tools serve various purposes, from emergency information dissemination to crisis mapping and managing requests for rescue and supplies.

Despite the low download rates of many disaster management apps [25], their integrated features like geolocation and communication sensors are highly beneficial [26]. Familiarizing stakeholders with these apps during the preparedness phase is crucial for their effective use in subsequent disaster stages [27]. The deployment of the Amrita Kripa app during the Kerala floods in 2018 and 2019 is a case in point [6]. Although the app was advertised on TV, many people preferred to submit requests through the call center. However, the call center volunteers effectively used the app to manage requests, demonstrating the importance of training and awareness in maximizing the utility of these tools [28].

Research in automated AI-based computer vision algorithms has made significant strides, particularly in applications like detecting flood depth [29] and tracking objects [30]. These advancements present promising opportunities for applications like Landslide Tracker, where such algorithms could be employed to automatically detect and analyze landslide characteristics, including the height, severity, material composition, and extent of damages caused by landslides. Similarly, in the context of Amrita Kripa, AI-driven image analysis could be utilized to assess the needs of disaster victims more efficiently. By analyzing the uploaded images, the system could estimate the requirements for rescue operations, medical assistance, supplies, and shelter, thereby streamlining the response process and enhancing the effectiveness of disaster management efforts.

3 Disaster Lifecycle and Crowdsourced Applications

The urgent impacts of climate change demand unified, fair, and inclusive action to protect people and the planet. This discussion focuses on enhancing public resilience and participation throughout the disaster lifecycle in Sikkim, covering pre-disaster preparations, active disaster response, and post-disaster relief and rehabilitation, despite the inherent challenges posed by various geological, geographical, and socio-cultural contexts. While complete disaster prevention is unattainable, dedicated efforts can minimize risks and manage crises, preventing hazards from becoming disasters. Disaster Resilience involves stakeholders' ability to adapt to and recover from hazards without compromising long-term development and cultural prospects. Crowdsourced applications like Landslide Tracker and Amrita Kripa can assist stakeholders during all disaster relief phases.

3.1 Landslide Tracker

Maintaining comprehensive landslide inventories is crucial for accurately modeling a region's susceptibility to landslides and predicting future events. The swift recording of landslides is vital due to their traces being rapidly erased by either human intervention or natural processes, such as erosion and vegetation growth. The Landslide Tracker app leverages crowdsourcing, enabling users ranging from ordinary citizens to experts to contribute to real-time, comprehensive landslide inventories. The information recorded is globally accessible, and the app is a pivotal tool in global landslide tracking and management.

The "Landslide Tracker" app employs a structured database, amalgamating diverse data types including text, numbers, dates, multiple choices, locations, and images to facilitate meticulous, real-time landslide tracking. It captures detailed landslide information, including type, material, size, impact, area, geology, and geomorphology. The app introduces multi-level user accounts—Tracker, Investigator, and Expert—each providing varying levels of access to tracking forms and ensuring enhanced data reliability with increasing user expertise. The user interface is divided into multiple screens for detailed event capture and tracking. Additional details are collected using a methodology adapted from the 40 questions used by the Geological Survey of India for landslide event collection through the collaborative effort of LANDSLIP partners. Once submitted, landslide data becomes immediately available for all users to view and can be accessed through the landslide image listing, Google Maps interface, or tabular listing (Fig. 1). Moreover, the app, designed for intuitive and rapid use, aims to elevate landslide risk awareness through integrated short documents and videos, offering guidelines for safety, image capturing, mapping, and data selection [31]. This methodology uniquely enhances community participation and seamlessly integrates event data collection, organization, spatial and temporal summarization, and validation of landslide events and impacts, thereby aiding in initiating appropriate disaster management activities to reduce risk levels.

The architecture of Landslide Tracker is centered around three technical themes: security, performance, and network resilience (Fig. 2). Security is addressed through robust authentication and authorization measures, ensuring that only registered users can record and view landslides, while safeguarding data from unauthorized access.

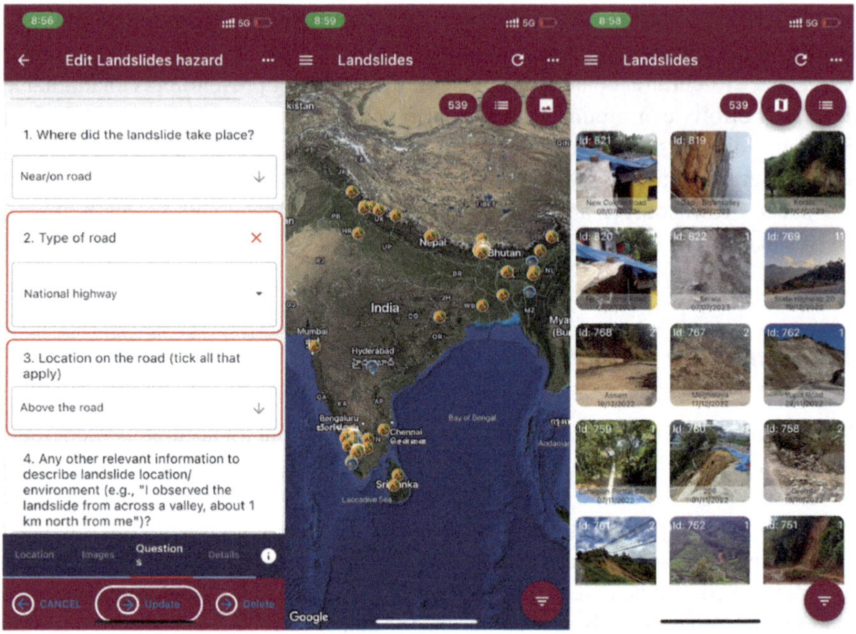

Fig. 1 Landslide tracker for capturing and viewing Landslides

This is enforced both on the client (mobile and web apps) and server sides. Performance enhancements are achieved through the design of high-performance data structures and databases, along with client and server-side caching. This caching mechanism, coupled with cache synchronization, ensures low latency data retrieval and prevents data in the cache from becoming stale. The app's data persistence in the mobile database allows for faster loading upon reopening. Network resilience is bolstered by local storage on the app, enabling landslide recording even in the absence of an internet connection, with automatic updates pushed to the server once connectivity is restored. Our performance testing showed a significant reduction in loading times, with over a 300% decrease in time taken to load 2000 landslides when using cache mode compared to no-cache mode.

3.2 Amrita Kripa

The "Amrita Kripa" app emerges as a crucial tool in disaster relief operations, ensuring the provision of accurate, timely, and detailed information, which is pivotal during crises. The app meticulously addresses the need for dynamic information updates during relief efforts, requiring victims to provide specific, detailed information, such as location, urgency, and medical condition [20].

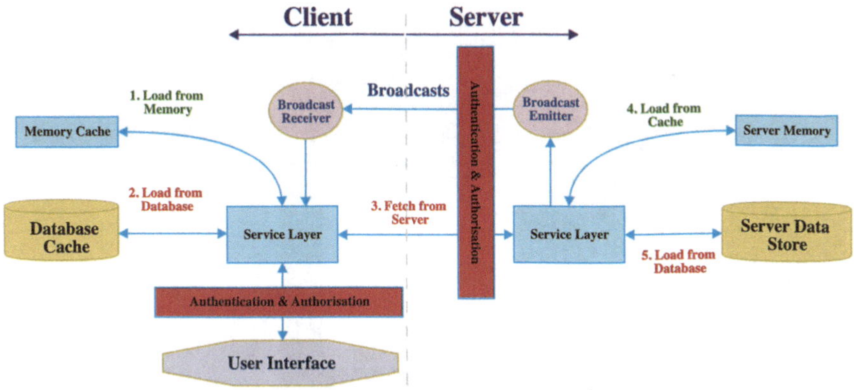

Fig. 2 Architecture diagram of landslide tracker

The humanitarian components of the app cater to the diverse needs of victims and volunteers. It enables victims to request various relief needs, which are visible on a map interface to relief agencies and government bodies. It allows victims to specify details about their rescue needs, medical conditions, and essential item requirements while also facilitating volunteers and agencies to offer help. The app ensures each request or offer contains GPS geolocation for accessibility and manages the status of each through various stages of their lifecycle. It also provides a multilingual interface, a catalog of precautions for various situations, and requires user registration, providing a scoring system to assist in decision-making during relief operations.

Organizational components focus on systematic processes for planning and executing relief efforts at a regional level, catering to agencies like Governments and NGOs. This includes systematic classification and summarization of user requests and offers, enabling victims to choose a predefined medical problem type, allowing victims to specify their service and supply needs in detail, and providing flexibility to revise type, subtype, or attribute choices as per evolving needs. It supports summarization of requests and responses along geographical hierarchies and provides summaries to aid agencies and users in planning effective relief, enabling support for multiple simultaneous disasters across the globe (Fig. 3).

Technical components ensure the system's functionality amid unstable power and network conditions during disasters, including offline mode, which allows users to log in and submit requests/offers without network connectivity, synchronizing data when the network is available, and a local database on the mobile/web app, enabling local caching of relevant user-level data to minimize unnecessary server calls, preserving network and battery resources. This comprehensive approach ensures the app effectively supports all stakeholders throughout the disaster relief lifecycle. The Amrita Kripa architecture, as depicted in Fig. 4, features a high-performance, flexible database on both server and client sides, complemented by hierarchical caching and two-way, automatic, event-driven cache synchronization. This design ensures data integrity, manageability, and rapid development adaptability, with separate tables for

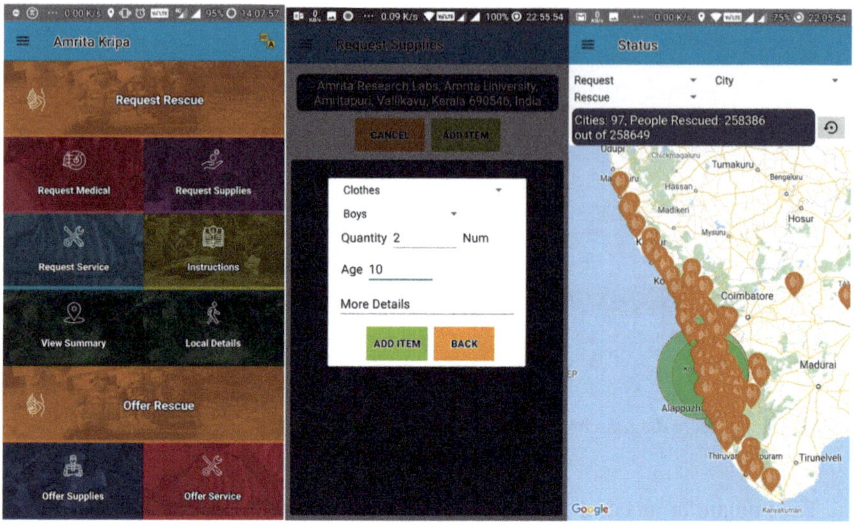

Fig. 3 Amrita Kripa screens showing the requests, offers, supplies request, and the summary view showing the city-level rescue requests

each functionality and indexed columns for enhanced performance. The server-side caching mechanism, tailored for heavy-read entities and geographical summaries, facilitates efficient data management by auto-removing cached child entities when a parent is deleted. The mobile database supports offline mode and seamless synchronization with the server, even during network or power disruptions. Additionally, push notifications are used to efficiently disseminate updates, further optimizing performance.

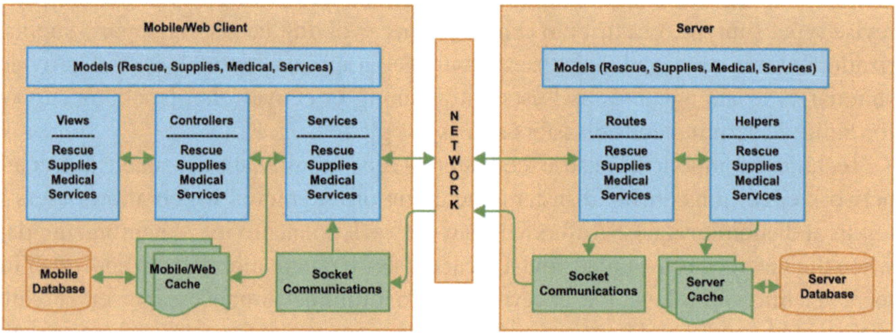

Fig. 4 Amrita Kripa component architecture for server and client

3.3 Disaster LifeCycle

The disaster management lifecycle encompasses four phases: Mitigation, Preparedness, Response, and Recovery. These phases represent a chronological sequence of actions undertaken by stakeholders to effectively minimize the impact of a disaster.

Mitigation: Mitigation strategically aims to either diminish the likelihood of disaster occurrences or mitigate their impacts, employing measures such as clearing drainages, reinforcing river embankments to prevent urban flooding, or fortifying vulnerable slopes and evacuating inhabitants from potential landslide zones to minimize landslide impacts.

Preparedness: This phase is dedicated to empowering the community to adeptly handle impending disasters by implementing the early warning systems for landslides [32], familiarizing them with disaster relief apps, conducting mock drills, preparing shelters, and stockpiling medical supplies, food, and rations.

Response: The Response stage encompasses activities executed by stakeholders during the disaster, including rescue and sheltering operations, provision of medical assistance, distribution of food and clothing, and offering psychological and emotional counseling.

Recovery: During the Recovery phase, stakeholders endeavor to restore normalcy in the aftermath of the disaster, involving cleanup of debris, restoration of infrastructure, and resumption of daily activities in homes, schools, and workplaces.

3.4 Usages of Landslide Tracker and Amrita Kripa During Disaster LifeCycle Phases

The table lists the possible ways the Landslide Tracker and Amrita Kripa can help during the various phases of the disaster lifecycle (Table 1).

4 Deployment Case Study

Sikkim, a multiethnic and multilingual state in India, borders Tibet, Bhutan, Nepal, and West Bengal and is the least populous and second smallest Indian state, spanning 7096 km^2. Although agriculture is the economic backbone for 80% of its population, the state, divided into East, North, South, and West Sikkim, frequently experiences landslides due to changing climatic conditions, complex geology, soil cover, deforestation, infrastructure expansion, and active tectonics [33]. The prevalent climatic changes, alterations in land use, and recurrent floods and landslides underscored Sikkim as a pertinent location for landslide training.

Table 1 Crowdsourced applications and disaster phases

Disaster phase	Applications	Usages
Mitigation	Landslide tracker	Historical landslide data allows for strategic mitigation, such as fortifying vulnerable slopes with nets and structures, clearing water passages, or enacting temporary local evacuations based on landslide frequency and severity. This data-driven approach aims to minimize future landslide impacts
Preparedness	Landslide tracker & Amrita Kripa	Leveraging data on landslide frequency and intensity, shelters can be established, and communities informed through workshops and mock drills. Utilizing previous disaster data from Amrita Kripa, collection centers can be preemptively stocked, medical staff and facilities pre-arranged, and government authorities prepared to manage typical service requests, such as debris clearance, enhancing overall disaster preparedness and response
Response	Amrita Kripa	As detailed in Sect. 3.2, Amrita Kripa's features, including request, offer, real-time communication, and summarization, facilitate stakeholders in executing timely and efficient disaster relief efforts
Recovery	Landslide tracker & Amrita Kripa	Landslide Tracker enables capturing details of regional landslides, including location, images, and characteristics like slope and soil type. Amrita Kripa allows citizens to report necessary repair works in their communities, while its summarization feature assists local officials in swiftly estimating and allocating necessary recovery resources and personnel

A 2-day Disaster Resilience training workshop was held in Sikkim, orchestrated by the UNESCO Chair on Experiential Learning, SSDMA, ICL, and IPL, aiming to enhance community resilience. Engaging 85 participants from Nar Bahadur Bhandari Government College, Deorali Govt Girls' Senior Secondary School, and Tadong Senior Secondary School in Gangtok, along with four teachers aged 28 to 40, the workshop focused on building civilian preparedness and response capabilities (Table 2).

Table 2 Participants demographics

Participants	Number	Age group	Institution
Students group 1	15 + 15	16–18	Tadong Senior Secondary School in Gangtok, Sikkim (TSSS); Deorali Govt Girls' Senior Secondary School (GSSS)
Students group 2	55	18–22	Nar Bahadur Bhandari Government College (NBBGC)
Teachers	4	28–40	TSSS, GSSS, and NBBGC

4.1 Overview of Workshop Schedule and Topics Covered—Day 1

The topics and activities are based on the below objectives.

To increase awareness of the disasters: Introduction to disaster resilience encompasses understanding various hazards and vulnerabilities, including landslides, earthquakes, and floods, along with strategies for their mitigation. The discourse extends to the three pivotal components of disaster resilience, the comprehensive disaster lifecycle, and the crucial role of stakeholder participation in disaster preparedness. Furthermore, it explores the intertwining relationship between disaster resilience and the establishment of robust community resilience.

To enable the usage of participatory approaches and tools for the disaster lifecycle: Participatory engagement focuses on comprehending community challenges, resources, and resilience regarding disaster preparedness. This involves understanding dos and don'ts for various disasters, identifying where to find additional information (websites, government announcements), observing environmental cues, and utilizing simple methods to measure crucial triggers. The approach encourages logging data into social media apps, correlating observations with potential disasters, fostering community inclusivity, and leveraging collective efforts for widespread community benefits. Additionally, tutorials on rainfall thresholds for landslides and flood events enhance preparedness.

Activity: Mapping the community's resources and vulnerabilities to various hazards as mind mapping diagrams to bring awareness of local communities (Fig. 5).

4.2 Learning the Tools for Participating in the Disaster LifeCycle—Day 2

Training on Landslide Tracker and Amrita Kripa: The training provides a thorough walkthrough of the registration processes and functionalities of the apps,

Fig. 5 The authors conducted workshop on Day 1. Dr. Wadhawan and Dr. Raji (top row) and Mr. Bal Mukund and Mr. Ramesh (bottom row)

complemented by a live demonstration. After the demonstration, participants are guided through downloading, installing, and registering for the Landslide Tracker and Amrita Kripa. They are then prompted to submit dummy landslide reports and engage in role-play as victims, submitting test requests for rescue, supplies, and medical emergencies. Additionally, participants are encouraged to submit supply offerings and are navigated through the entire disaster relief lifecycle.

Certificate distribution: The participants were awarded certificates on behalf of the UNESCO Chair on Experiential Learning for Sustainable Innovation & Development Sikkim State Disaster Management Authority (SSDMA), Government of Sikkim International Consortium on Landslides (ICL), and International Program on Landslides (IPL).

4.3 Observations and Feedback

This training enhanced the students' understanding of risk response and developed skills for effectively connecting with diverse audiences to elevate community engagement and motivate risk-reducing actions. Participants learned to navigate varied

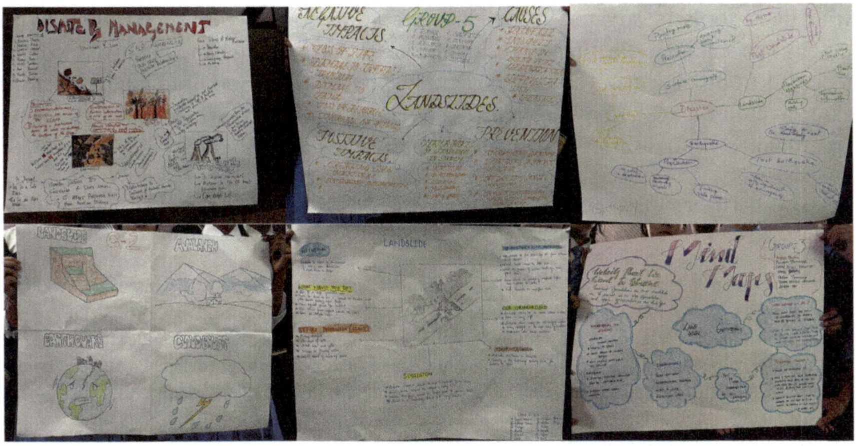

Fig. 6 The students drew mind maps of the disasters, community resilience, and disaster relief on Day 2

values and risk perceptions, apply social science and risk communication principles in challenging scenarios, respond confidently to difficult questions, and craft effective risk communication strategies. The training emphasized disaster awareness, participatory approaches and tools for preparedness, operational training on the Landslide Tracker and Amrita Kripa apps, and risk awareness, and underscored the significance of collaboration, coordination, and community engagement. Post-training, participants became community trainers, furthering disaster resilience education within their communities.

Feedback from the students and teachers: Students and teachers exhibited enthusiasm in learning about the disaster relief lifecycle, easily relating to the topic due to personal experiences with landslides, either directly or through relatives. Expressing gratitude for our outreach efforts, they particularly enjoyed the mind-mapping activities. Their profound knowledge about disasters and local community awareness was surprising and impressive. Some students even advocated for a more action-oriented approach, emphasizing the importance of moving beyond discussion (Fig. 6).

5 Conclusions

This paper explores the disaster lifecycle, emphasizing the critical role of community awareness and participation across all disaster phases. The significance of crowdsourced applications in facilitating timely and accurate information flow throughout the disaster cycle is highlighted, with a focus on two mobile applications: Landslide Tracker and Amrita Kripa. These applications' utility across various disaster lifecycle phases is discussed, along with the vital need to acquaint communities with these tools in advance. This also details a two-day workshop conducted in the Sikkim school, reflecting on the enthusiastic participation and positive feedback received from students. The enthusiastic response is motivating to contemplate an extensive tour across Kerala schools, conducting numerous workshops to acquaint communities with disaster management approaches.

Acknowledgements Deep gratitude is extended to Sri Sadguru Mata Amritanandamayi Devi (Amma), a globally renowned humanitarian and spiritual leader, and Chancellor of Amrita University, for her ongoing guidance in community service. Appreciation is also conveyed to all contributors to this research paper and the "Amma C20 Sikkim Event," which seamlessly integrated Sikkim's cultural ethos with Amma's teachings, made possible by the unwavering support of schools and the Sikkim Government. Special acknowledgment is given to the dedicated educators, students, and staff from participating schools, whose enthusiasm and commitment enriched the experience for all. The Sikkim Government's crucial support in organizing the event is gratefully recognized. Finally, sincere thanks are expressed to Dr. Maneesha Vinodini Ramesh for her steadfast support and exemplary leadership throughout the project.

References

1. Agrawal SK, Kumar D (2018) Recent trends in disaster mitigation and management: vulnerability atlas of India. Adv Indian Earthq Eng Seismol: Contrib Honor Jai Krishna 383–410
2. Sarkar S, Sarma A (2006) Disaster management act, 2005: a disaster in waiting?. Econ Polit Wkly 3760–3763
3. National Disaster Management Authority (2016) National policy on disaster management. Ministry of home affairs, Government of India
4. de Graaf-van Dinther R, Ovink H (2021) The five pillars of climate resilience. Climate resilient urban areas: governance, design and development in coastal delta cities pp 1–19
5. Wadhawan SK, Singh B, Ramesh MV (2023) Understanding multi-hazards and enhancing capacity for disaster resilience—catchment based case studies in parts of Himalaya and Western Ghats, India. Jour. Geological society of India, (In Press)
6. Ramesh VM, Guntha R, Arnhardt C, Singh G, Kr V, Rao P, Halan G, Malamud B (2021) Spatial temporal tracking of landslide events: a crowdsourced mobile app. In: EGU general assembly conference abstracts, pp EGU21–16104
7. Guntha R, Rao SN, Ramesh MV (2020) Deployment experiences with Amrita Kripa: a user-friendly feature rich crowdsourced humanitarian application. Procedia Comput Sci 171:302–311
8. Haklay M (2013) Neogeography and the delusion of democratisation. Environ Plan A 45(1):55–69
9. Haklay MM, Dörler D, Heigl F, Manzoni M, Hecker S, Vohland K (2021) What is citizen science? The challenges of definition. Sci Citizen Sci 13
10. Can R, Kocaman S, Gokceoglu C (2019) A convolutional neural network architecture for auto-detection of landslide photographs to assess citizen science and volunteered geographic information data quality. ISPRS Int J Geo Inf 8(7):300
11. Cieslik K, Shakya P, Uprety M, Dewulf A, Russell C, Clark J, Raj Dhital M, Dhakal A (2019) Building resilience to chronic landslide hazard through citizen science. Front Earth Sci 7:278
12. Kocaman S, Anbaroglu B, Gokceoglu C, Altan O (2018) A review on citizen science (CitSci) applications for disaster management. Int Arch Photogramm Remote Sens Spat Inf Sci 42:301–306

13. Parajuli BP, Shakya P (2019) Empowering citizens about emerging technologies for resilience building to landslide risk reduction in mountainous rural parts of Nepal. In: Geophysical research abstracts, vol 21
14. Rohan T, Wondolowski NA, Shelef E (2020) Landslide susceptibility analysis based on citizen reports to a 311 system. In: AGU fall meeting abstracts, vol 2020, pp NH031–07
15. Goodchild MF (2007) Citizens as sensors: the world of volunteered geography. GeoJournal 69:211–221
16. Can R, Kocaman S, Gokceoglu C (2020) Development of a CitSci and artificial intelligence supported GIS platform for landslide data collection. Int Arch Photogramm Remote Sens Spat Inf Sci 43:43–50
17. Kocaman S, Candan G (2019) CitSci as a new approach for landslide researches. Intell Syst Crisis Manag: Gi4DM 2018 11:161–183
18. Franceschinis C, Thiene M, Mattea S, Scarpa R (2020) Do information and Citizens characteristics affect public acceptability of landslide protection measures? A latent class approach. Handb Clim Serv 503–513
19. https://interestingengineering.com/innovation/turkish-developed-apps-helping-earthquake-victims
20. Guntha R, Rao SN, Ramesh MV (2020) Architectural considerations for building a robust crowdsourced disaster relief application. In: 2020 international conference on communication systems & networks (COMSNETS). IEEE, pp 638–641
21. Pánek J, Marek L, Pászto V, Valůch J (2017) The crisis map of the Czech Republic: the nationwide deployment of an Ushahidi application for disasters. Disasters 41(4):649–671
22. Morrow N, Mock N, Papendieck A, Kocmich N (2011) Independent evaluation of the Ushahidi Haiti project. Dev Inf Syst Int 8(2011):111
23. https://apps.apple.com/lt/app/earthquake-turkey/id1539771064
24. McAtee KJ, Robert B, Lakis DC, Bachmann DJ, Kman NE (2022) Emergency preparedness and disaster response: there's an app for that 2.0. Prehospital Disaster Med 37(1):117–123
25. Hoehle H, Venkatesh V (2015) Mobile application usability. MIS Q 39(2):435–472
26. Tan ML, Prasanna R, Stock K, Hudson-Doyle E, Leonard G, Johnston D (2017) Mobile applications in crisis informatics literature: a systematic review. Int J Disaster Risk Reduction 24:297–311
27. Nilsson EG, Stølen K (2011) Generic functionality in user interfaces for emergency response. In: Proceedings of the 23rd Australian computer-human interaction conference, pp 233–242
28. Antoniou N, Ciaramicoli M (2013) Social media in the disaster cycle–useful tools or mass distraction?. In: International astronautical congress
29. Narayanan RK, Lekshmy VM, Rao S, Sasidhar K (2014) A novel approach to urban flood monitoring using computer vision. In: Fifth international conference on computing, communications and networking technologies (ICCCNT). IEEE, pp 1–7
30. Jurado F, Palacios G, Flores F, Becerra HM (2014) Vision-based trajectory tracking system for an emulated quadrotor UAV. Asian J Control 16(3):729–741
31. Guntha R, Vinodini Ramesh M (2021) Technical considerations for building a landslide tracker mobile app. In: EGU general assembly conference abstracts, pp EGU21–6019
32. Hemalatha T, Ramesh MV, Rangan VP (2019) Effective and accelerated forewarning of landslides using wireless sensor networks and machine learning. IEEE Sens J 19(21):9964–9975
33. Gupta V, Chauhan N, Penna I, Hermanns R, Dehls J, Sengupta A, Bhasin RK (2022) Geomorphic evaluation of landslides along the Teesta river valley, Sikkim Himalaya, India. Geol J 57(2):611–621

Design and Modeling of Modular Wheelchair for Therapeutic Exercises

Baishnavi Mahanta, Jyotindra Narayan, Mohamed Abbas, and Santosha K. Dwivedy

Abstract Conventional wheelchairs, the well-known technology transfer devices, always have certain limitations in providing proper therapeutic exercises to the upper and lower limbs of the human body. Therefore, in this work, the concept of a modular wheelchair is proposed to allow shoulder, elbow, and knee joint movements in addition to the wheeler mobility requirements. The proposed device is designed in FUSION 360 software for three therapy modules: shoulder abduction/adduction, elbow flexion/extension, and knee flexion/extension. The former two modules are designed with spring-based passive actuation methods, while the latter one is provisioned with linear actuators. The leg rest is designed with a telescopic arrangement to make the device eligible for people of different leg widths. Taking the critical design parts, the loading conditions are computed and specified with boundary constraints. Thereafter, dynamic modeling of the rear-arm rest, forearm rest, and leg rest is formulated mathematically using Newtonian approach. Finally, finite element analyses (FEAs) for the critical components are carried out based on the loading and boundary conditions. Moreover, the dynamic modeling results are plotted with different design parameters, demonstrating the motion behavior of different therapy modules. Such modular wheelchair devices could be a potential solution to reduce the burden of physiotherapists by automatically performing the tiresome and repetitive rehabilitation process.

B. Mahanta · J. Narayan (✉) · M. Abbas · S. K. Dwivedy
Department of Mechanical Engineering, Indian Institute of Technology Guwahati, Guwahati, Assam 781039, India
e-mail: n.jyotindra@gmail.com

M. Abbas
e-mail: abbas@iitg.ac.in

S. K. Dwivedy
e-mail: dwivedy@iitg.ac.in

J. Narayan
Department of Mechanical Engineering, Indian Institute of Technology Patna, Patna, Bihar 801106, India

M. Abbas
Department of Design and Production, Al-Baath University, Homs, Syria

S. M. Thampi et al. (eds.), *Fifth International Conference on Computing and Network Communications*, Lecture Notes in Electrical Engineering 1219,
https://doi.org/10.1007/978-981-97-4540-1_17

Keywords Therapeutic exercises · Modular wheelchair · Therapy modules · Finite element analyses · Dynamic modeling

1 Introduction

Mobility is a fundamental element of an individual's daily life, allowing them to perform the basic activities of daily living (ADLs). However, due to neurological disorders, degenerative diseases, and aging, some people experience certain mobility constraints which prevent them from performing their basic daily needs [1]. Over the past two decades, a substantial increase in neurological diseases such as stroke, epilepsy, spinal cord injury, multiple sclerosis, and Parkinson's disease has been reported for different age ranges. A study by Global Burden of Disease (GBD) [2] affirmed the 'stroke' as the second principal cause of disability in 2019 for elderly people (> 50 years). On the other hand, the stroke incidence for the pediatric population varies from 2.5–13 per 100,000 per year [3]. Assistive technology promotes independence, reduces reliance on formal support services, and enhances participation in education and civic life, minimizing the impact of diseases on individuals and society. Despite the use of wheelchairs for mobility, achieving functional independence and limb rehabilitation remains challenging. Conventional wheelchairs limit daily activities, leading to the degradation of limb performance. Manual therapy by physiotherapists is cumbersome and costly. Researchers explore innovative mobility assistive devices like exoskeletons, modular wheelchairs, and reconfigurable canes to address these challenges [4–6].

Wheelchairs are a medical device which is used to improve the functional mobility of people suffering from neurological disorders, impaired nerves, and weak muscles. They come in various types, such as manual and electric, including designs like pediatric, positioning, sports, all-terrain, and standing power wheelchairs [7]. Massey University engineers created the Ezy Wheels [8], employing a push-and-pull movement for optimal stroke efficiency, reducing chronic shoulder and wrist issues. In run mode, a three-geared mechanism facilitates smooth travel on long journeys and uneven terrain. Notably, a handle-based propulsion (HBP) mechanism [9] in manual wheelchairs improves ergonomics, utilizing hand gears on armrests for natural upper body movement. The manual standing wheelchair Leeding E.D.G.E. [10] focuses on a sit-to-stand module, enhancing accessibility and addressing shoulder injury concerns. Krishnan and Pugazhenthi [11] developed a robot-assisted wheelchair with a turntable rotary mechanism for independent transfers from the wheelchair to the commode seat.

Researchers have explored wheelchair designs integrating lower and upper-limb rehabilitation modules for ergonomic comfort during push-and-pull operations [12–17]. Notable examples include the TEK Robotic Mobilization Device (TEK RMD), the world's smallest motorized device for sit-to-stand motion in lower limb disability [12], and the ReChair, combining mobility, gait rehabilitation, and voice-controlled posture monitoring [13]. Aljohani et al. [14] introduced portable manual exercise equipment for full-body training in paraplegia, focusing on range-of-motion improvement and injury prevention. Hwang and Jeon [15] developed a wheelchair-integrated lower limb rehabilitation system with accessible automated training programs. A low-cost motorized wheelchair exoskeleton was proposed for mobility assistance and sit-to-stand rehabilitation in paraplegic subjects [16]. Additionally, research by Shaikh-Mohammed et al. [17] involved the design and prototyping of a wheelchair with various stages of development for sit-to-stand rehabilitation.

Performing therapeutic exercises in conventional wheelchairs presents certain limitations that need consideration. The fixed design of standard wheelchairs may restrict the range of motion, make it challenging to engage specific muscle groups during exercises, and impede tailoring therapeutic interventions to individual needs. While several assistive wheelchairs focus on efficient locomotion and multi-posture drive, a therapeutic wheelchair with exercise modules for upper and lower limb rehabilitation remains absent. Therefore, this work introduces a passive-active powered therapeutic wheelchair designed for mobility assistance and rehabilitation of an average young male subject (23.8 years, 73 kg, 173.1 cm). Key contributions include:

1. A low-cost wheelchair is designed with passive spring-based arrangements for abduction/adduction of the shoulder joint and flexion/extension of the elbow joint.
2. A telescoping leg rest system is proposed to adjust varying calf widths and enable knee joint flexion/extension with two independent linear actuators.
3. The static analysis of the critical load-bearing components is carried out using the finite element modeling (FEM) approach, and the safety factor is identified.
4. The coupled dynamic formulation for the human arm and armrest and human leg and leg rest models is presented using the Newtonian approach.

The rest of the work includes therapeutic wheelchair design; identifying boundary conditions for critical parts; conducting dynamic modeling for therapy modules; and demonstrating FEM and dynamic modeling to visualize structural strength and motion behavior. Finally, concluding remarks are briefly presented.

2 Design of the Therapeutic Wheelchair

2.1 Design Requirements

This section outlines the detailed design of a therapeutic wheelchair, adhering to WHO guidelines for wheel-based mobility and ergonomic seating [18]. Key considerations include robust seat support, effective wheel type, ergonomic arm/footrests, and optimal push rim placement. Design must align with user-specific functional and posture support requirements, considering environmental constraints [19]. For physiological therapy, arm and leg rests should align with human limbs, accommodating varying widths. The design prioritizes non-compromising therapy modules, cost-effectiveness, and minimal weight. Proposed features are documented and

Table 1 Weight of parts based on the percentage of total human body weight [20]

Extremity	Segments	Percentage of body weight (%)
Base extremity	Head	8.26
	Whole trunk	55.1
	Thorax	20.1
	Abdomen	13.06
	Pelvis	13.66
Upper extremity	Total arm	5.7
	Upper arm	3.25
	Forearm	1.87
	Hand	0.65
	Forearm and hand	2.52
Lower extremity	Total leg	16.68
	Thigh	10.5
	Leg	4.75
	Foot	1.43
	Leg and foot	6.18

implemented using design software like FUSION 360 and SolidWorks. CAD models undergo finite element and dynamic modeling for strength and actuator validation.

2.2 Design Specifications

The circumferential dimensions of the spoked wheels and attached push rim are specified according to the seating height given in [20, 21]. A 270° castor-type wheels are considered for the front support. The distance between two spoked wheels is decided based on the width of the seating frame. The upper and lower-body dimensions are specified following the average human male, as shown in Table 1 [20]. To realize the effectiveness of the proposed design, a dummy of the average human male is designed with body height and weight values specified in Tables 1 [20] and 2 [21]. To better understand body parameters, different body segments of the dummy are marked in Fig. 1a, b and defined as per the serial number in Table 2. However, an outer bodycover is designed to visualize the upper- and lower-extremity muscles in the present work. Three joint movements are assessed: shoulder abduction/adduction, elbow flexion/extension, and knee flexion/extension. Flexion bends a limb toward the body, while extension straightens it. Abduction and adduction move the limb away and toward the body midline. Figure 1c illustrates the ranges of motion (ROMs) for elbow, shoulder, and knee joints in this study.

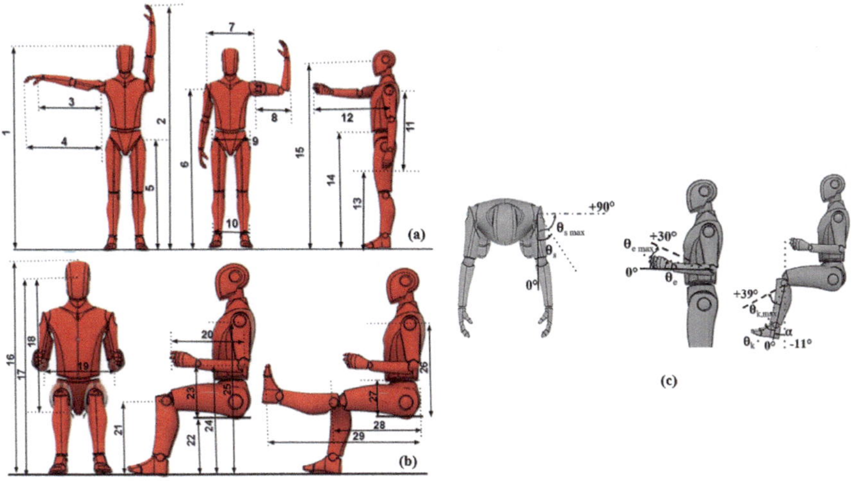

Fig. 1 Human body segments in **a** standing pose, and **b** sitting pose used to design the therapy modules (marked numbers are annotated as serial numbers in Table 2), and **c** range of motions (ROM) for shoulder (a/d), elbow (f/e), and knee joint (f/e) (from left to right)

2.3 CAD of the Proposed Therapeutic Wheelchair

The wheelchair comprises a frame, rear and front wheels, leg and forearm rests, and actuation mechanisms. The drawings with appropriately selected dimensions are shown in Fig. 2a. Two therapy modules for shoulder and elbow joints, along with knee joint movement, are depicted in Fig. 2b, c. Using FUSION 360, CAD is conducted for different parts (Fig. 3). The rear wheel shown in Fig. 3a, designed for comfort, has a push rim diameter of 58 cm. The shaft diameter, connecting the push rim, rear wheels, and chair frame with each other, is selected as 6 cm. The chair frame shown in Fig. 3b, based on human dimensions, measures 53.5×44.5 cm^2. Back handles are placed 50.5 cm apart, 48.45 cm above the seat. The castor-type front wheel (Fig. 3c) has an 8.7 cm center diameter and 270° movement. Rear and forearm rests (30 cm long, 40 cm apart) allow separate therapeutic exercises (Fig. 3d, e). Lightweight and cost-effective, helical springs enable passive actuation for arm exercises (Fig. 3f). A leg rest with telescopic links accommodates different calf widths (Fig. 3g). The length of each leg rest is 43.5 cm and is separated by an outer distance of 32.2 cm. Inclined at 11°, the leg rests use ball-screw linear actuators for flexion/extension (Fig. 3h). The CAD of these elements aligns with human joint movements (Fig. 3c).

Table 2 Length of major human body part dimensions in an average male [21]

No.	Quantity under measure	Male body dimensions, cm (50 percentile)
1	Total height	174.4
2	Vertical arm-reach	215.6
3	Side arm-reach	62.2
4	Outstraightened side arm length	72.4
5	Standing leg length (distance from edge bone to floor)	85.1
6	Shoulder height	146.7
7	Shoulder width	48.7
8	Shoulder-elbow length	32.8
9	Hip width	38.3
10	Leg position width	83
11	Total arm length	75.5
12	Forward stretched arm length	74.5
13	Dactylion height	62.3
14	Waistline height	108.3
15	Eye-level height	165.9
16	Sitting height	141.7
17	Sitting eye-level height	133.2
18	Sitting vertex eye-level height	81.8
19	Elbow-elbow breadth	41.9
20	Forearm length	46.8
21	Kneecap height	54.4
22	Popliteal space height	43.9
23	Elbow to seat height	25.4
24	Sitting elbow to floor height	69.5
25	Sitting shoulder to floor length	109.3
26	Shoulder to seat length	65.1
27	Sitting hip height	14.5
28	Hip length (buttock to kneecap length)	59.2
29	Sitting leg length (buttock to heel distance of an outstretched forward leg)	109.6

3 FEM: Meshing and Boundary Conditions

Finite element modeling or static structural analysis is performed to find the strength of various critical components under specific loading conditions. The critical components selected are sitting chair frame, rear-arm rest, forearm rest, and leg rest. Based on the market availability, AISI 304 Stainless Steel has selected which properties

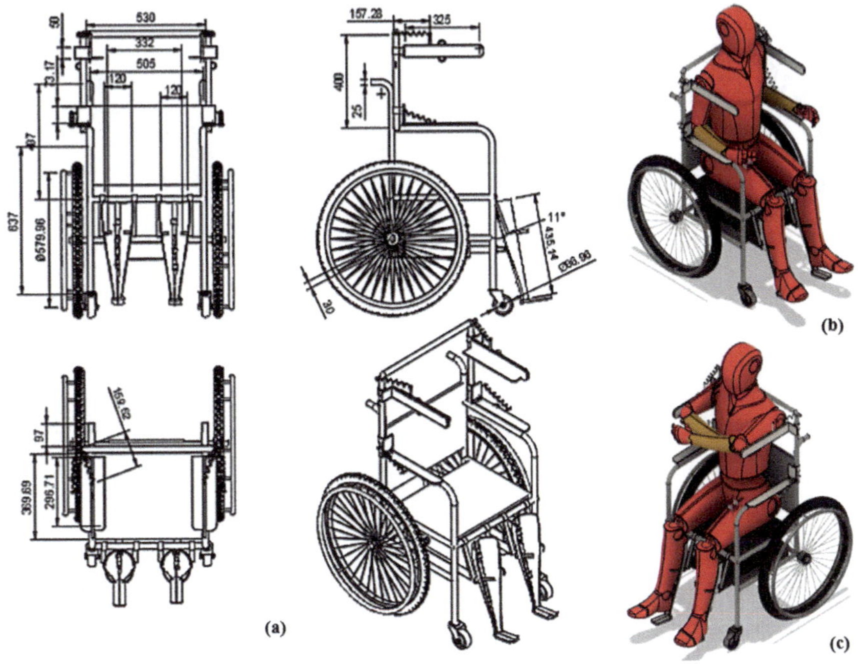

Fig. 2 Wheelchair's **a** dimension/drawings, **b** CAD in forearm therapy mode, and **c** rear-arm therapy mode along with leg therapy mode

are as follows [22]: melting point: 1400–1455 °C, specific heat capacity: 0.5 J/g °C, thermal conductivity: 16.2 W/m K, ultimate tensile strength: 505 MPa, yield tensile strength: 215 MPa, and modulus of elasticity: 193–200 GPa. After assigning material properties, meshing is carried out with an appropriate number of nodes and elements. Thereafter, the maximum applied loads within different parts of the wheelchair system are evaluated based on the location of fixed connections.

3.1 Meshing

A solid 3D tetrahedral mesh is utilized for critical components in the wheelchair frame analysis. The frame mesh comprises 75,921 nodes and 40,569 elements (maximum element size: 20 mm). The rear-arm rest mesh includes 4326 nodes and 2217 elements (maximum element size: 15 mm). The forearm rest mesh is formed with 19,758 nodes and 11,795 elements (element size: 15 mm). The leg rest analysis uses a mesh with 19,874 nodes and 9764 elements (maximum element size: 20 mm).

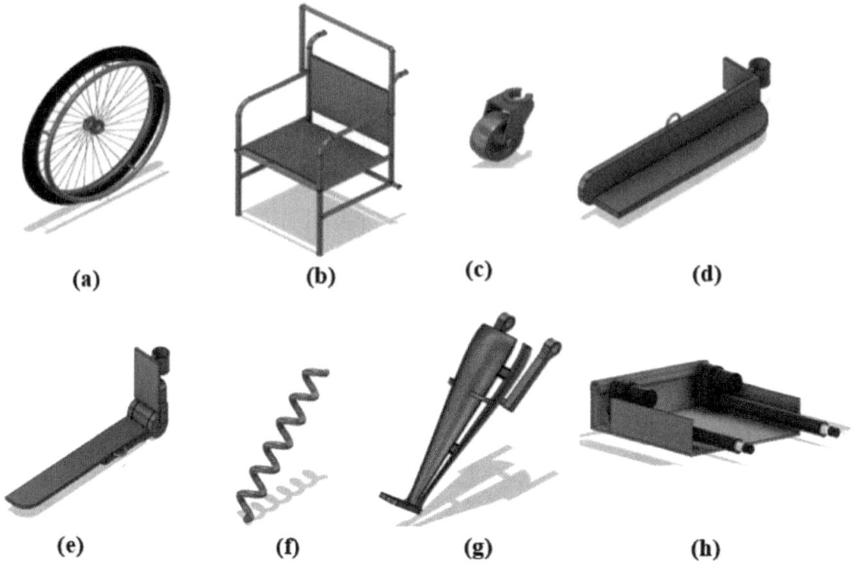

Fig. 3 CAD of different parts of the therapeutic wheelchair **a** rear wheel, **b** chair frame, **c** from the wheel, **d** rear-arm rest, **e** forearm rest, **f** helical spring, **g** leg rest, and **h** linear actuator

3.2 Boundary Conditions

After forming a solid mesh with specified nodes and elements, boundary conditions of fixing at one location and applying load at another location are implemented for each critical component. The applied forces for all critical components are as follows:

Chair Frame: The net downward force applied in the COM of the frame,

$$F_{\mathrm{ch}} = (m_{\mathrm{h}} + M_{\mathrm{f}})\, g = (20 + 70) \times 9.81 = 882.9\ \mathrm{N} \tag{1}$$

where m_{h} denotes the average human male mass, M_{f} represents the total wheelchair mass, and g is the acceleration due to gravity.

Rear-Arm Rest: The net downward force applied in the COM of the rear-arm rest,

$$\begin{aligned} F_{\mathrm{r}} = (m_{\mathrm{r}} + M_{\mathrm{r}})\, g &= ((5.7\% \text{ of } 70) + 2.353) \times 9.81 \\ &= 62.224\ \mathrm{N} \end{aligned} \tag{2}$$

Maximum torque applied at the hinge joint assuming rear-arm along with armrest moves $\theta_{\mathrm{s}} = 90°$ movement in $(t = 3\ \mathrm{s})$,

$$\tau_{\mathrm{r}} = (m_{\mathrm{r}} + M_{\mathrm{r}}) \times \left(\frac{2 \times r \times \theta_{\mathrm{s}}}{t^2}\right) \times r \quad (3)$$
$$= ((5.7\% \text{ of } 70)\ + 2.353) \times \left(\frac{2 \times 0.15 \times \frac{\pi}{2}}{3^2}\right) \times 0.15$$
$$= 0.049 \text{ Nm}$$

where m_{r} denotes the mass of human rear-arm, M_{r} represents the mass of rear-arm rest, and r signifies the forearm rest COM distance from a hinge point.

Forearm Rest: The net downward force applied in the COM of the rear-arm rest,

$$F_{\mathrm{f}} = (m_{\mathrm{f}} + M_{\mathrm{f}})\, g = ((2.52\% \text{ of } 70) + 2.251) \times 9.81 \quad (4)$$
$$= 39.387 \text{ N}$$

Maximum torque applied at the hinge joint assuming forearm along with armrest moves $\theta_{\mathrm{e}} = 30°$ movement in ($t = 1$ s),

$$\tau_{\mathrm{f}} = (m_{\mathrm{f}} + M_{\mathrm{f}}) \times \left(\frac{2 \times f \times \theta_{\mathrm{e}}}{t^2}\right) \times r \quad (5)$$
$$= ((2.5\% \text{ of } 70) + 2.251) \times \left(\frac{2 \times 0.15 \times \frac{\pi}{6}}{1^2}\right) \times 0.15$$
$$= 0.0946 \text{ Nm}$$

where m_{f} denotes the mass of the human forearm, M_{f} represents the mass of forearm rest, and f signifies the rear-arm rest COM distance from a hinge point.

Leg Rest: The net downward force applied in the COM of the leg rest,

$$F_{\mathrm{ka}} = (m_{\mathrm{k}} + M_{\mathrm{k}})\, g = ((m_{\mathrm{ka}} + m_{\mathrm{af}}) + M_{\mathrm{k}})\, g \quad (6)$$
$$= ((4.75\% \text{ of } 70) + (1.43\% \text{ of } 70) + 3.431) \times 9.81$$
$$= 76.10 \text{ N}$$

Maximum force due to actuator placement,

$$F_{\mathrm{ka}} = (m_{\mathrm{k}} + M_{\mathrm{k}})\, g \times \tan\left(\alpha + \theta_{\mathrm{k,max}}\right) \quad (7)$$
$$= 76.10 \times \tan(11° + 39°) = 90.69 \text{ N}$$

where m_{k} denotes the mass of the human leg (knee-ankle, m_{ka} and ankle-foot, m_{af}), M_{k} represents the mass of leg rest, α signifies fixed inclination angle, and $\theta_{\mathrm{k,max}}$ is the maximum knee joint ankle covered (refer to Fig. 1c).

The chair frame assumes fixed connections with wheels, experiencing a downward force (882.9 N) at its center of mass (COM) (Fig. 4a). Rear-arm rests are hinge-fixed

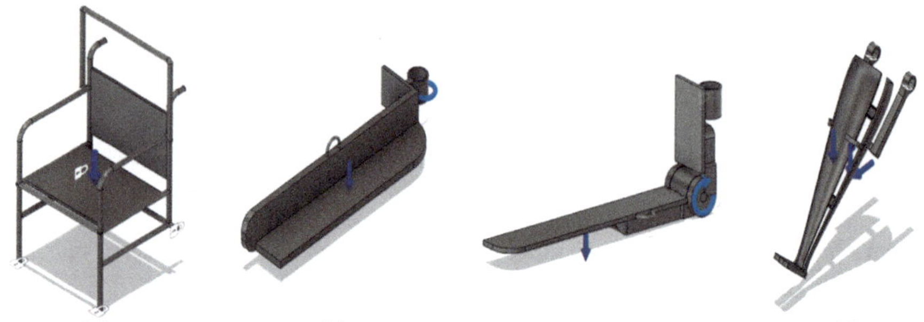

Fig. 4 Loading locations and directions applied for **a** chair frame, **b** rear-arm rest, **c** forearm rest, and **d** leg rest

to vertical edges, subjected to a load of 62.224 N and 0.049 Nm torque at the COM (Fig. 4b). Forearm rests, hinge-fixed to horizontal edges, undergo a downward force of 39.387 N and 0.095 Nm torque at the hinge connection (Fig. 4c). Leg rests, hinge-fixed to the chair frame, receive a downward force of 76.10 and 90.69 N directed backward (Fig. 4d).

4 Dynamic Modeling: Equations of Motions

Now, the dynamic model of the therapy modules is formulated using the Newtonian principle [23] to examine the motion and evaluate the required force during the motion. The equations of motions for the rear-arm rest, forearm rest, and leg rest are as follows.

4.1 Rear-Arm Rest

Assuming the COM of coupled rear-arm rest follows the straight-line motion, one can obtain and balance the net force ($F_{\mathrm{net,r}}$) from Fig. 5a, b,

$$F_{\mathrm{net}} = -\left(m_{\mathrm{r}} + M_{\mathrm{r}}\right) a_{\mathrm{r}} = k_{\mathrm{sr}} x_{\mathrm{r}} \sin\left(\theta_{\mathrm{sr}}\right) + b_{\mathrm{sr}} v_{\mathrm{r}} \tag{8}$$

where a_{r}: resultant acceleration for the coupled system (human rear-arm); k_{sr} and b_{sr}: spring and damping constants of the rear-arm rest spring; x_{r}: displacement of rear-arm rest (spring elongation); v_{r}: armrest velocity; θ_{sr}: angle formed by spring with rear-arm rest on the xy plane. Rewriting Eq. (8) as

Fig. 5 Load applied in rear-arm rest **a** CAD model, **b** free-body diagram (FBD)

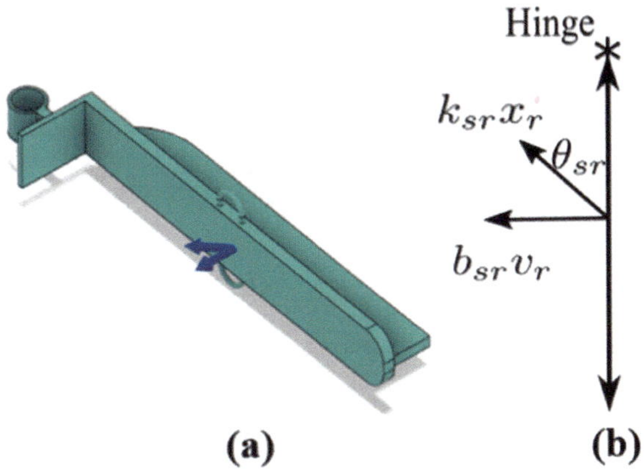

$$\frac{d^2x_r}{dt^2} + \frac{k_{sr}\sin(\theta_{sr})}{m_r + M_r} + \frac{b_{sr}}{m_r + M_r}\frac{dx_r}{dt} = 0 \tag{9}$$

Considering an exponential function as follows:

$$x_r = A_r e^{\lambda t} \tag{10}$$

where A_r is amplitude constant in the case of rear-arm rest, λ is the time constant, and t is the time period.

Differentiating $x_r = A_r e^{\lambda t}$ for the first and second time and substituting into Eq. (9), one may obtain

$$\lambda^2 + \frac{b_{sr}}{m_r + M_r}\lambda + \frac{k_{sr}\sin(\theta_{sr})}{m_r + M_r} = 0 \tag{11}$$

Solving Eq. (11) for the time constant λ, one can get

$$\lambda = -\frac{b_{sr}}{2(m_r + M_r)} \pm \left(\frac{{b_{sr}}^2}{4(m_r + M_r)^2} - \frac{k_{sr}\sin(\theta_{sr})}{m_r + M_r}\right)^{1/2} \tag{12}$$

Further, the natural angular frequency can be expressed as

$$\omega_{o_{sr}} = \left(\frac{k_{sr}\sin(\theta_{sr})}{m_r + M_r}\right)^{1/2} \tag{13}$$

The angular frequency of damped harmonic oscillations can be obtained as

$$\omega_{sr} = \left(\omega_{o\,sr}^2 - \left(\frac{-b_{sr}}{2(m_r + M_r)}\right)^2\right)^{1/2} \tag{14}$$

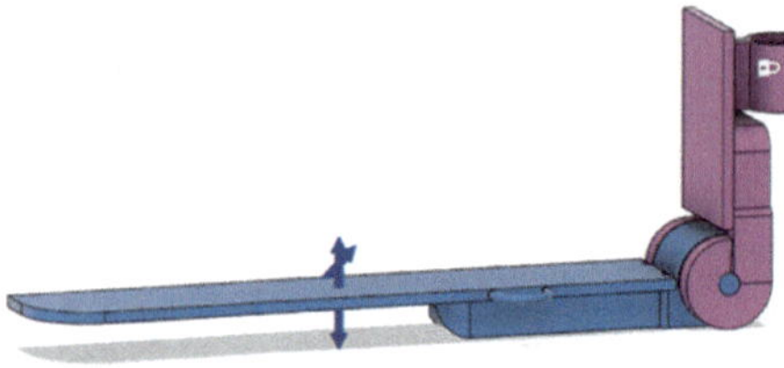

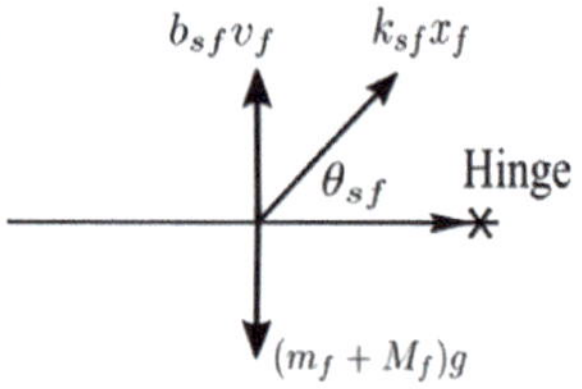

(a) (b)

Fig. 6 Load applied in forearm rest **a** CAD model, **b** free-body diagram (FBD)

Substituting ω_{sr} from Eq. (14) to Eq. (10) and expanding after neglecting the imaginary part,

$$x_{\text{r}} = A_{\text{r}} \text{e}^{-\frac{b_{\text{sr}}}{2(m_{\text{r}}+M_{\text{r}})}t} \cos(\omega_{\text{sr}} t) \tag{15}$$

Therefore, the differential equation (9) has similitude to a cosine curve oscillating in the envelope of an exponential function, i.e., $E_{\text{r}} = A_{\text{r}} \text{e}^{-\frac{b_{\text{sr}}}{2(m_{\text{r}}+M_{\text{r}})}t}$.

4.2 *Forearm Rest*

Assuming the forearm rest and the human forearm move in a straight line, Fig. 6a, b help determine and balance the net force ($F_{\text{net,f}}$).

$$F_{\text{net,f}} = (m_{\text{f}} + M_{\text{f}})\, a_{\text{f}} = (m_{\text{f}} + M_{\text{f}})\, g - k_{\text{sf}} x_{\text{f}} \sin(\theta_{\text{sf}}) + b_{\text{sf}} v_{\text{f}} \tag{16}$$

where a_{f} is the coupled system-human rear-arm acceleration, with k_{sf} and b_{sf} as the spring and damping constants, x_{f} as rear-arm rest displacement, v_{f} as armrest velocity, and θ_{sf} as the angle between the spring and rear-arm rest on the xy plane.

At equilibrium (from Fig. 9),

$$(m_{\text{f}} + M_{\text{f}})\, g = k_{\text{sf}} A_{\text{o}} \sin(\theta_{\text{sf}}) \tag{17}$$

where A_{o} is the elongation amplitude of the spring in the equilibrium state.

Substituting Eq. (17) to Eq. (16), one may obtain

$$(m_{\text{f}} + M_{\text{f}}) \frac{\text{d}^2 \acute{x}_{\text{f}}}{\text{d}t^2} + k_{\text{sf}} \acute{x}_{\text{f}} \sin(\theta_{\text{sf}}) - b_{\text{sf}} \frac{\text{d}\acute{x}_{\text{f}}}{\text{d}t} = 0 \tag{18}$$

where $\acute{x}_{\text{f}} = (x_{\text{f}} - A_{\text{o}})$.

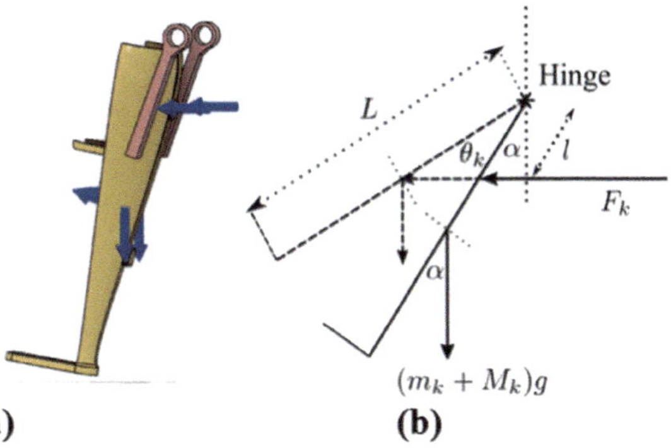

Fig. 7 Load applied in leg rest **a** CAD model, **b** free-body diagram (FBD)

Now, invoking the steps of finding time constant, one can find the equations as follows:

$$x_{\mathrm{f}} = A_{\mathrm{o}} + A_{\mathrm{f}} \mathrm{e}^{-\frac{b_{\mathrm{sf}}}{2(m_{\mathrm{f}}+M_{\mathrm{f}})}t} \cos(\omega_{\mathrm{sf}} t) \tag{19}$$

Therefore, the differential equation (18) has similitude to a cosine curve oscillating in the envelope of an exponential function, i.e., $E_{\mathrm{f}} = A_{\mathrm{f}} \mathrm{e}^{-\frac{b_{\mathrm{sf}}}{2(m_{\mathrm{f}}+M_{\mathrm{f}})}t}$ with A_{f} as amplitude constant in case of forearm rest.

4.3 *Leg Rest*

Considering the actuator force F_{k} at the leg rest as shown in Fig. 7a, b,

$$F_{\mathrm{k}} = (m_{\mathrm{k}} + M_{\mathrm{k}})\, g \times \tan(\alpha + \theta_{\mathrm{k}}) \tag{20}$$

where α signifies fixed inclination angle, and θ_{k} is the knee joint angle covered.

Evaluating the net torque applied on the leg rest,

$$\tau_{\mathrm{net,1}} = F_{\mathrm{k}} \cos(\alpha + \theta_{\mathrm{k}}) l - (m_{\mathrm{k}} + M_{\mathrm{k}})\, g \times \sin(\alpha + \theta_{\mathrm{k}})(L/2) \tag{21}$$

where l is the distance at which actuator force is applied, and L is the total length.

Exploiting Eq. (20) into Eq. (21), one may get

$$\tau_{\mathrm{net,1}} = (m_{\mathrm{k}} + M_{\mathrm{k}})\, g \times \sin(\alpha + \theta_{\mathrm{k}})\,(l - (L/2)) \tag{22}$$

Therefore, the above equation represents a sinusoidal curve with a variation of θ_{k}.

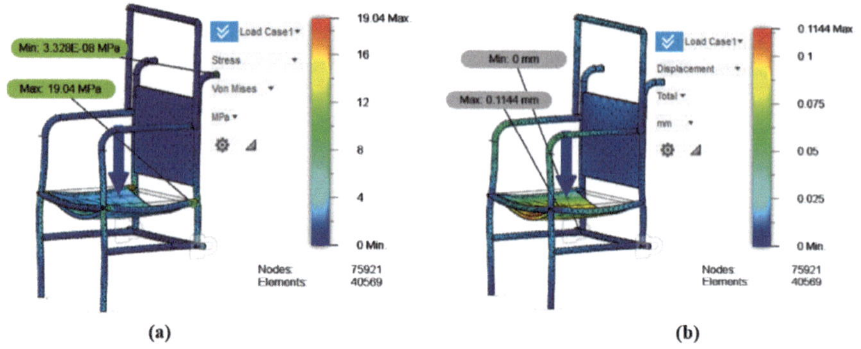

Fig. 8 **a** Von Mises stress distribution, **b** total deformation of chair frame

5 Results and Analysis

This section presents and discusses results from static structural and dynamic modeling. Finite element analyses determine strength and deformation under applied boundary conditions. Additionally, dynamic modeling plots show spring and actuator specifications for rear-arm rest, forearm rest, and leg rest.

5.1 Results I: Finite Element Modeling

After applying load and boundary conditions, simulations reveal Von Mises stress and total deformation for the chair frame (Fig. 8a, b). Maximum stress is 19.04 MPa at corner joints, causing 0.114 mm deformation at the front edge. Rear-arm rest analysis (Fig. 9a, b) indicates 13.44 MPa stress at the vertical hinge joint, with 0.1403 mm deformation at the distal end. Forearm rest results (Fig. 9c, d) show 21.54 MPa stress and 0.567 mm displacement at the chair frame's vertical plate and forearm rest's distal end. Leg rest simulations in Fig. 10a, b show a max stress of 5435 MPa and a 0.085 mm deformation. The designs are considered safe as the maximum stress for each component is within the yield limit of steel material (215 MPa).

5.2 Results II: Dynamic Modeling

Equations (15) and (19) depict the time-dependent displacement of springs, as illustrated in Fig. 11a, b. Design parameters for rear-arm and forearm rests, including mass, stiffness, and damping constants, are $m_{\mathrm{r}} = 3.99$ kg, $M_{\mathrm{r}} = 2.353$ kg, $k_{\mathrm{sr}} = 9.12$ Nm, $m_{\mathrm{f}} = 1.764$ kg, $M_{\mathrm{f}} = 2.251$ kg, $k_{\mathrm{sf}} = 5.80$ Nm, $b_{\mathrm{sr}} = b_{\mathrm{sf}} = 0.5$, and $\theta_{\mathrm{sr}} = \theta_{\mathrm{sf}} = 20°$. Amplitude constants for armrests are $A_{\mathrm{r}} = A_{\mathrm{f}} = 0.05$ m, and

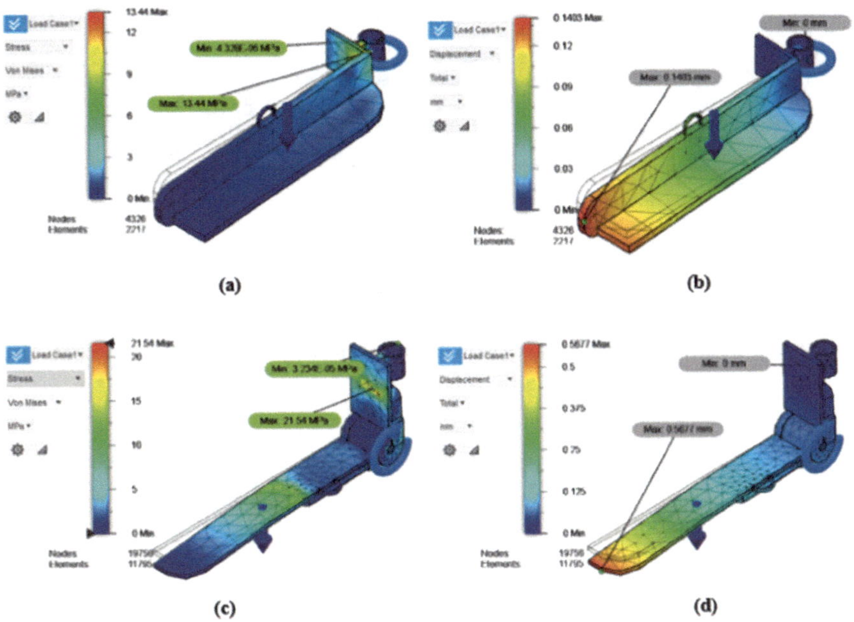

Fig. 9 **a** Von Mises stress distribution of rear-arm rest, **b** total deformation of rear-arm rest, **c** Von Mises stress distribution of forearm rest, **d** total deformation of forearm rest

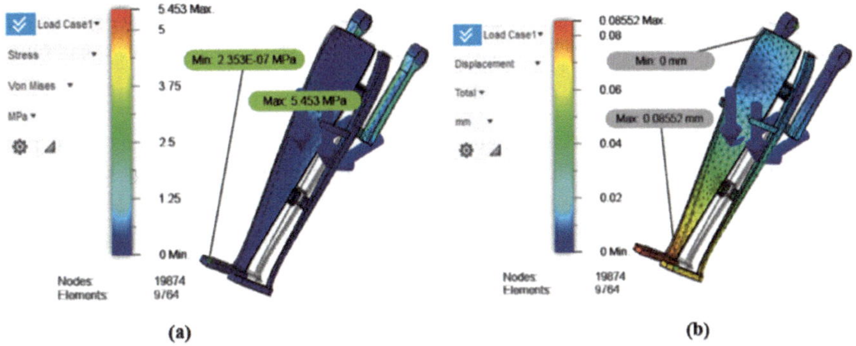

Fig. 10 **a** Von Mises stress distribution of leg rest, **b** total deformation of leg rest

$A_o = 0.025$ m. These displacements represent the precise motion of rear-arm and forearm during rehabilitation training, resembling cosine curves within exponential function envelopes. Figure 11c displays the envelope plots (E : E_r, E_f) over time, emphasizing nearly straight-line motion for a brief duration ($t \leq 3$ s).

In the case of leg rest, following Eq. (20), the actuator force (F_k) is plotted with a variation of human lower leg mass (m_k) in Fig. 12a and with a variation of knee joint angle (θ_k) in Fig. 12b. In telescopic leg rests with different widths,

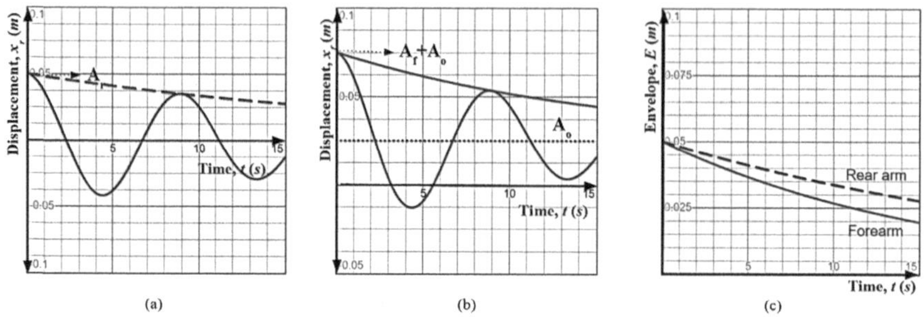

Fig. 11 Motion of armrests during therapeutic exercises **a** displacement of rear-arm rest versus time, **b** displacement of forearm rest versus time, **c** exponential envelope by armrest motions versus time

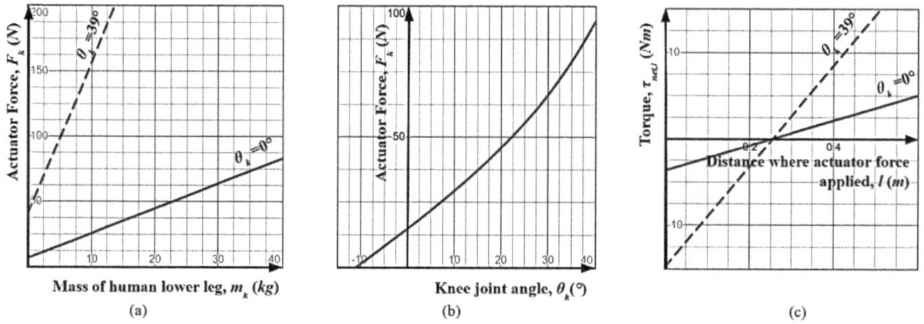

Fig. 12 **a** Leg rest actuator force required versus the mass of human lower leg, **b** leg rest actuator force needed versus knee joint angle, **c** leg rest actuator torque versus distance of force application

actuator force linearly increases with lower leg mass and steepens as knee joint angle varies from 0 to 39°. Conversely, the force tangentially increases with knee joint angle for fixed lower leg and leg rest mass ($m_{\mathrm{k}} = 4.326\ \mathrm{kg},\ M_{\mathrm{k}} = 3.431\ \mathrm{kg}$). Figure 12b shows residual forces (< 15 Nm) representing actuator holding force at an initial inclination of 11°. The net torque on the leg rest $\left(\tau_{\mathrm{net},1}\right)$ is estimated, considering the length where the actuator force is applied, with additional parameters: $L = 0.51\ \mathrm{m},\ g = 9.81\ \mathrm{m/s^2},\ \alpha = 11°$. The torque curve steepens at maximum knee joint angle, as depicted in Fig. 12c.

6 Conclusion

This paper introduces a therapeutic wheelchair featuring three modules for rear-arm, forearm, and lower leg therapy. The design allows users to move these body parts through specified movements. Upper-limb modules employ passive actuation, while lower limb modules use linear actuators. Finite element modeling assesses critical components, and dynamic modeling illustrates module motion during rehabilitation.

The design aims to ease physiotherapists' workload by enabling subjects to perform repetitive exercises manually. A prototype is currently in production, with details to be presented in future work.

Acknowledgements The authors would like to thank Mechatronics and Robotics Laboratory, IIT Guwahati, for supporting this research work which is internally funded.

References

1. Abbas M, Narayan J, Dwivedy SK (2021) Event-triggered adaptive control for upper-limb robot-assisted passive rehabilitation exercises with input delay. Proc Inst Mech Eng Part I J Syst Control Eng 09596518211047008
2. Feigin VL, Stark BA, Johnson CO, Roth GA, Bisignano C, Abady GG, Abbasifard M, Abbasi-Kangevari M, Abd-Allah F, Abedi V (2021) Global, regional, and national burden of stroke and its risk factors, 1990–2019: a systematic analysis for the global burden of disease study 2019. Lancet Neurol 20(10):795–820
3. Kupferman JC, Zafeiriou DI, Lande MB, Kirkham FJ, Pavlakis SG (2017) Stroke and hypertension in children and adolescents. J Child Neurol 32(4):408–417
4. Kalita B, Narayan J, Dwivedy SK (2021) Development of active lower limb robotic-based orthosis and exoskeleton devices: a systematic review. Int J Soc Robot 13(4):775–793
5. Choukou M-A, Best KL, Potvin-Gilbert M, Routhier F, Lettre J, Gamache S, Borisoff JF, Gagnon D (2019) Scoping review of propelling aids for manual wheelchairs. Assist Technol. https://doi.org/10.1080/10400435.2019.1595789
6. Arefin P, Habib MS, Arefin A, Arefin MS (2020) A review on current mechanical and electronic design aspects and future prospects of smart canes for individuals with lower limb difficulties. Mater Sci Res India 17(1):25–33
7. Inun HB, Siagian CD, Kadir SB, Zamlan BABM, Abd Halim S (2020) Electric wheelchair controlled by joystick and Android/IoS smartphone. J Sains Sos Pendidik Tekn I J Soc Sci Techn Educ (JoSSTEd) 1(1):53–59
8. Wheelchair revolution. Accessed 2022-01-06. https://www.massey.ac.nz/about/news/wheelchair-revolution/
9. Kurup NBR, Puchinger M, Gfoehler M (2019) A preliminary muscle activity analysis: handle based and push-rim wheelchair propulsion. J Biomech 89:119–122
10. Standup wheelchair. Accessed 2022-01-06. https://www.yankodesign.com/2010/08/30/stand-up-wheelchair/
11. Krishnan RH, Pugazhenthi S (2017) Design and development of a robotic self-transfer device for wheelchair users. J Enabl Technol. https://doi.org/10.1108/JET-12-2016-0025
12. Tek robotic mobilization device helps paraplegics stand, huge improvement over the wheelchair. Accessed 2022-01-06. https://www.theverge.com/2012/3/19/2885741/tek-rmd-paraplegic-motorized-standing-movement-device
13. Cao W, Yu H, Wu X, Li S, Meng Q, Chen C (2021) Development and evaluation of a rehabilitation wheelchair with multiposture transformation and smart control. Complexity 2021
14. Aljohani K, Enderli T, Guan S, Hillman D (2015) Wheelchair-mounted exercise equipment for paraplegics
15. Hwang B, Jeon D (2013) Development and preliminary testing of a novel wheelchair integrated exercise/rehabilitation system. In: 2013 IEEE 13th international conference on rehabilitation robotics (ICORR). IEEE, pp 1–6
16. Gupta M, Narayan J, Dwivedy S (2020) Modeling of a novel lower limb exoskeleton system for paraplegic patients. In: Advances in fluid mechanics and solid mechanics: proceedings of the 63rd congress of ISTAM 2018. Springer, pp 199–210

17. Shaikh-Mohammed J, Dash SS, Sarda V, Sujatha S (2021) Design journey of an affordable manual standing wheelchair. Disabil Rehabil Assist Technol 1–11
18. Guidelines on the provision of manual wheelchairs in less resourced settings. Accessed 2022-01-06. https://www.who.int/publications/i/item/9789241547482
19. Visagie S, Duffield S, Unger M (2015) Exploring the impact of wheelchair design on user function in a rural South African setting. Afr J Disabil 4(1):1–8
20. Plagenhoef S, Evans FG, Abdelnour T (1983) Anatomical data for analyzing human motion. Res Q Exerc Sport 54(2):169–178
21. Main anthropometrical human body physical dimensions. Accessed 2022-01-06. http://macoshdesign.com/en/theory/articles/anthropometrical
22. Hudok D (1990) Properties and selection: irons, steels, and high-performance alloys. In: Metals handbook, vol 1, pp 200–211
23. Marion JB (2013) Classical dynamics of particles and systems. Academic Press, New York

Analyzing Data Traffic in IoT Networks: A Novel Graph-Based Architecture

Gemini George

Abstract The rapid deployment of IoT networks in different industrial services has caused the emanation of a huge volume of data from sensors and monitors. The efficient analysis and compact representation of the big data generated are critical to managing its transfer, storage, and latency. In this work, we propose a novel graphical model that has the flexibility to adapt to different architectural features of distributed IoT systems with low-power sensors. Further, we consider the traffic load minimization problem in the network and map the problems of network coverage and clustering onto the graph to find an optimal distributed network structure that minimizes the traffic load. The proposed approaches are tested under different network conditions, and the simulation results emphasize the improved performance of the method.

Keywords IoT · Sensor network · Graph based architecture · Network construction

1 Introduction

The Internet of Things (IoT) is being rapidly deployed in a variety of services and applications. Such deployments have resulted in the generation of large volume and variety of data from sensors and actuators. The analytics of data can derive important insights that can be used for the precise and efficient control of the underlying processes. Cloud computing is the naturally preferred technology to store, manage, and analyze the data gathered by such systems. However, the presumption of a stable IoT ecosystem with unlimited connectivity to the cloud is often not realistic. This is because the variety, volume, and velocity of the data generated by the IoT devices are so large that the transfer of such traffic to the cloud storage often outstrips the network bandwidth, and the associated latency makes it unsuitable for real-time and

G. George (✉)
Department of Computer Applications, St. George's College Aruvithura, Erattupetta, Kottayam, Kerala 686122, India
e-mail: gemini@sgcaruvithura.ac.in

S. M. Thampi et al. (eds.), *Fifth International Conference on Computing and Network Communications*, Lecture Notes in Electrical Engineering 1219,
https://doi.org/10.1007/978-981-97-4540-1_18

remote applications. This necessitates the need for efficient load management and synthesis strategies to minimize the volume of the data to be transferred.

In low-power IoT networks, the sensors are designed to consume low power [1], as replacing them can often be infeasible in many applications once the sensors' batteries are drained out. Therefore, reducing energy consumption is a major design issue in WSN to maximize the network life. A major portion of the energy consumption occurs during the processing and transmission of the sensed data [2]. The data these nodes sense must finally reach the central command center (CCC). If each sensor directly transmits the data to the CCC, much energy will be consumed in the transmission process [3]. This further demands the deployment of efficient network configurations and hop-based transmission strategies.

The issues related to an efficient load regulation network are to be addressed in the following seven aspects:

1. Efficient routing and clustering that reduce the traffic load.
2. Cognitive scheduling (activation and deactivation) of sensor states that guarantee proper coverage.
3. Balancing of energy consumption of the sensors nodes.
4. Compact data representation and compression schemes that reduce the transmission of redundant data.
5. Addressing the flow and capacity constraints of the links.
6. Precise transmission power control for low energy consumption.
7. Efficient data link layer protocols that address access control and re-transmission strategies.

The scope of this work is limited to the first four aspects. Efficient load balancing schemes contribute to future-ready WSNs in three major different ways. First, such schemes improve data storage, management, and analytics. Further, the transmission of compressed data improves network latency and bandwidth utilization. Finally, such schemes improve the lifetime of low-power wireless networks, wherein the sensors are battery-operated, and the energy spending is proportional to the number of transmissions.

Since every network has an underlying inherent graphical structure, the proper graphical model helps to represent the network problems as graph-theoretic problems and to find solutions. The overall aim of this work is to design an efficient graphical framework for managing the data traffic in the IoT networks. The novel features of the proposed graphical model, its benefits, and major contributions of our work are as follows:

1. We develop a general graphical model that covers different architectural features of the IoT networks operating in low-power conditions. The model can be used for cluster/non-cluster-based and hierarchical/non-hierarchical-based architectures. Thus, this graphical model can be utilized by researchers in several flexible ways to formulate their problems and to find solutions.
2. In the proposed graphical model, the graph nodes represent the sensor devices, and the edges represent the possible connectivities among the devices. The edges

are to be cognitively selected according to the architecture under consideration to obtain the final graph, representing the optimal network configuration that leads to minimal data traffic under the particular selected architecture.
3. Modeling of a given IoT network as a graph helps to apply the well-known combinatorial optimization techniques and thus obtain sophisticated approaches that can be effectively used to analyze and solve the network problems.

The rest of the paper is organized as detailed. In Sect. 2, we discuss the existing works in the domain. In Sect. 3, we formally define our graphical model representing the IoT network. We present the approach to find an optimal network structure in Sect. 4, and the results are discussed in Sect. 5.

2 Literature Review

We describe the relevant literature related to our proposed work in the following. To improve the life of a network, factors such as routing, network latency, clustering, fault tolerance, and energy balancing among sensors need to be considered. The earlier research efforts to improve the performance of IoT networks started with the optimization of single or partial factors separately. Energy-efficient routing process [4], fault tolerance [5], network latency [6] clustering process [7], coverage and connectivity [8], energy consumption [9], and guaranteeing area coverage [10] are some examples of such works. The paper [11] and its references provide a detailed survey on multi-objective optimization techniques in WSNs. The problem of maximizing sensor network lifetime under all such integrated constraints is rarely attempted.

However, optimizing a few factors without considering the effect of other factors does not lead to an overall efficient solution. For example, switching sensor states should ensure factors such as the coverage of the field, connectivity of every sensor, and connected graph formation of cluster-head (CH) nodes. If such factors are optimized separately, other parameters may violate requirements. For example, a solution that guarantees coverage and connectivity need not necessarily guarantee CH connectivity and can lead to routing failure. This demands the need for global planning admitting all stake-holding parameters.

Even two decades after its inception in 2002, the *low energy adaptive clustering hierarchy protocol* (LEACH) still counts as one of the dominant generic energy-efficient WSN protocols. Several variants of the LEACH continue to emerge, improving the efficiency in different performance parameters. This section presents a brief of different LEACH protocols.

1. *Standard LEACH protocol*: The main objective of the LEACH protocol [12] is to improve the life of the network by energy efficiency derived from the combination of the probability-based CH selection and rotation schemes. The LEACH works in rounds with two phases: the setup phase and the steady-state phase. The setup phase includes processes such as clustering, CH selection, and

TDMA slot scheduling. Any sensor denoted by n is designated with a random probability, and it is chosen as a CH if its value is less than a threshold value given by,

$$T(n) = \begin{cases} \frac{P}{1-P\cdot\left(r \mod \frac{1}{P}\right)}, & \text{if } n \in G \\ 0, & \text{otherwise} \end{cases}, \tag{1}$$

where P and r denote the desired proportion of the CHs and the round number, respectively, and G denotes the set of sensors not involved in past $\frac{1}{P}$ rounds of CH selections. Thus, once a sensor is selected as a CH, it doesn't participate in further CH selection process for $\frac{1}{P}$ rounds. The selected CH sensors broadcast an advertisement packet to the non-CH sensors. Based on the received signal strength, the non-CH sensors associate with the nearest CH to form clusters. Each round prompts the CH rotation, ensuring the proper energy balancing among the sensors. The transmission of the data occurs in the steady-state phase. Single hop transmissions are used for data transfer within the cluster, while the inter-cluster transmissions involve data hopping via direct transmission to the CCC in another TDMA slot schedule. The main advantages of LEACH are:

- The distributed nature of the LEACH alleviates the requirement for global information.
- The non-CH members transmit their data to their CH within a cluster, reducing energy consumption. The CHs transmitting longer distances to the control center are liable for rotation in every round and, thus, contribute to energy balancing.

2. *LEACH-centralized* (LEACH-C): LEACH-C [13] is a centralized version of the standard LEACH, where the control station coordinates the networking processes such as CH selection, cluster formation, and scheduling. The sensors are equipped with a global positioning system (GPS) receiver, and each sensor transmits its geographical position and residual energy value to the control center in every round. Based on these, the control center forms optimum clusters using a simulated annealing clustering scheme. The sensors with energy below an energy threshold E_{av} are removed from CH selection process, where

$$E_{\text{av}} = \begin{cases} \frac{\sum_{n=1}^{N} E_n}{N}, & \text{if } n \in G \\ 0, & \text{otherwise} \end{cases}, \tag{2}$$

where N is the cardinality of the set of sensors, G is the set of sensors that have not been involved in the previous round of CH selection, and E_n is the residual energy of nth sensor. The cluster formation based on the simulated annealing technique reduces the energy consumption of the non-CH sensors within each cluster. The control center broadcasts the network configuration to every sensor. The selected CH sensors broadcast the TDMA schedule to their cluster members based on which the intra-cluster transmissions take place. The LEACH-C suffers from the disadvantage of the requirement of GPS at every sensor.

3. *Security-based LEACH* (SLEACH): SLEACH [14] is the first attempt to introduce security to LEACH using the *sensor protocol for information via negotiation* (SPIN) [15]. Equipping the sensors with cryptographic security features is challenging because of the tiny operating system and low functional capabilities of the battery-operated sensors. The SLEACH provides security only from external attacks and assumes that the control center is trusted. Security measures based on data authentication and data freshness are incorporated in SLEACH. The former authenticates the origin of the message, while the latter secures the system from old messages. However, the SLEACH does not provide security during sensor data acquisition and cluster formation. Further, there is no provision to update the security key. The disadvantages of the SLEACH are resolved in [16].
4. *Security-LEACH* (Sec-LEACH): Sec-LEACH [17] secures the LEACH from sinkholes and selective forwarding attacks. A set of keys is assigned to the sensors, with the pair of keys shared with the control center.
5. *Quadrant cluster-based LEACH* (Q-LEACH): Q-LEACH [18] combines the quadrant-based directional routing protocol (Q-DIR) scheme [19] with the LEACH protocol. Q-LEACH integrates location-based routing and limited flooding. The monitoring field is divided into four quadrants, and clusters are formed quadrant-wise. Route request packets (RREQ) are used for inter-CH communication and to determine the shortest transmission paths. Though Q-LEACH improves the network lifetime, it increases the delay and congestion. These limitations are overcome in the improved version, namely *enhanced Q-LEACH* [20]. *Quadrature LEACH* [21] is another improvement to enhance the performance of the Q-LEACH.
6. *Armor-LEACH* (A-LEACH): A-LEACH [22] improves the SLEACH by incorporating a *time-controlled clustering algorithm* (TCCA) [23]. *The time to live* (TTL) part of the TCCA provides better clustering with lesser energy consumption. A-LEACH secures the system from jamming, spoofing, selective forwarding, sinkhole, and replay attacks.
7. *LEACH-hybrid* (LEACH-H): LEACH-H [24] derives synergy from the combination of LEACH and LEACH-C, where the uncertain number of CHs in LEACH is replaced with a fixed number of CH selection technique. During the setup phase, the control center computes the optimal number of clusters using the simulated annealing technique, and the number of clusters remains the same in every round. The LEACH-H ensures an even distribution of CH compared to LEACH and LEACH-C. LEACH-H is the first attempt to find a reasonable solution to networks of large sizes. However, the protocol suffers from large message overheads.
8. *Unequal clustering-LEACH* (U-LEACH): The hotspot problem in basic LEACH causes the early death of the CHs far from the control center because of their

large energy consumption for transmitting their aggregated data over a large distance to the control center. The U-LEACH [25] alleviates this issue by reducing the cluster size with an increase in the distance between the cluster center and the control center. The approach results in a lesser volume of aggregated data at CHs farther from the control center compared to the CHs near.

9. *Modified-LEACH* (MOD-LEACH): MOD-LEACH [26] uses two different signal boosting for intra-cluster and inter-cluster transmissions in comparison with the same signal boosting in LEACH for both types of transmissions. This approach is based on the fact that the intra-cluster transmissions are regionally confined compared to the direct transmission from CHs to the control center. The scheme provides considerable savings in energy consumption.
10. *Intra-balanced LEACH* (IB-LEACH): IB-LEACH [27] introduces one more level in signal aggregation by designating some non-CH members in a cluster as signal aggregators. Thus, the signal aggregation at the CH is divided and a part of the job is done by the aggregator nodes. The aggregators and CH within a cluster directly transmit their aggregated data to the control center, while the rest of the cluster management is handled by the CH itself. Thus, IB-LEACH alleviates the high energy consumption and faster death of the CHs.
11. *Multihop LEACH* (MH-LEACH): As per the first order radio model [28], the energy consumption for transmitting the data is proportional to the square of the Euclidean distance between the transmitter and the receiver for distances below a threshold distance, and proportional to the fourth power of the Euclidean distance if the distance is above the threshold. According to this model, the LEACH and its aforementioned variants are inefficient in terms of energy consumption. MH-LEACH aims to resolve this problem by introducing multihop transmission between the CHs and the control center. The setup phase of MH-LEACH is the same as that of LEACH, but in the steady-state phase, the CHs away from the control center transmit their data to the control center via intermediate CHs, while the adjacent CHs transmit directly. Therefore, the distant CHs have to choose a route for hopping their data to the control center. Though the MH-LEACH is highly scalable, the multihop transmissions via intermediate nodes increase the complexity and message overhead.
12. *Two level-LEACH* (TL-LEACH): TL-LEACH [29] evenly distributes the traffic load by adopting a two-level hierarchy of clusters, and the data from the sensors are transmitted to the control center via this hierarchy. The CHs are classified into two levels: upper-level and lower-level. The upper-level CHs are referred to as the primary CHs, while the lower-level CHs are referred to as the secondary CHs. The secondary CHs perform the partial local computation of data, while the primary level CHs perform the complete local computation. TL-LEACH extends the lifetime of a sensor network by even distribution of energy among sensor nodes. For large-area networks, TL-LEACH outperforms

other LEACH variants presented so far. The disadvantage of the TL-LEACH is that the primary CHs near the control center suffer from the hotspot problem.

13. *LEACH-Mobile* (LEACH-M): The LEACH version for the mobile sensors is the LEACH-M [30], which supports networks with stationary CHs and mobile non-CHs. The setup phase of LEACH-M is the same as that of the basic LEACH, while in the steady state, each CH verifies whether the data sent is received in the allotted TDMA time slot. The sensors which are unable to receive are removed from a table of CHs, and the time slot is allotted to a newly joined sensor if any. The CHs presume that the sensors not responding to the messages have shifted from its cluster to another, and the CH in the latter cluster accommodates the mobile sensor via an advertisement message. The LEACH-M improves the reliability of data transfer in mobile networks at the cost of message overheads. *LEACH-Mobile extended* [31] supports mobility issues of both CHs and non-CH sensors.
14. *Multihop and single hop routing LEACH* (MS-LEACH): The analysis of the size of clusters based on the single hop and multihop transmissions is presented in [32]. The MS-LEACH proposes a combination of single hop and multihop data transmission within a cluster based on the size of the cluster. The difference with the basic LEACH is that in the steady-state phase, the position of the sensors and the cluster area are determined, based on which the nature of transmission (single hop or multihop) is decided from every non-CH sensor to the CH. The multihop transmission path is determined by the execution of the Dijkstra algorithm, and the CH broadcasts the path information within the cluster.
15. *LEACH with routing based on ant colony optimization*: The MH-LEACH is improved by introducing a routing scheme based on *ant colony optimization* in [33]. In this work, the optimal number of CHs is computed based on the energy consumption in one round, and *Voronoi diagrams* [34] are constructed with CHs as their cores. The sensors in each Voronoi diagram form a cluster. The approach reduces the energy consumed for intra-cluster transmissions. Further, the ant colony algorithm helps to reduce the energy consumption in inter-cluster transmissions.
16. *Field observation interest-LEACH* (FOI-LEACH): FOI-LEACH [35] is an improvement on the LEACH protocol for field observation instruments. The advantage of FOI-LEACH is that it selects CHs such that the premature death of CHs is reduced, thereby increasing the network lifetime. The method does not suffer from hotspot problems. However, FOI-LEACH is not appropriate for networks of large sizes.

3 Modeling of Sensor Network Graph

Defining a graphical model for a general WSN is motivated by the fact that every network has an underlying graphical structure, modeling of which helps deeper investigation of the network. Any such graphical model must map the parameters of the concerned network to appropriate entities of the graphical model, thereby enabling further research on the network using graph-theoretic algorithms.

We consider an IoT network consisting of a set of N stationary sensors $S = 1, 2, \ldots, N$ of dense deployment, and the CCC is denoted by f. Let A denotes the subset of the sensors which are active at any given instance. Let the monitoring area consist of subregion denoted by the set $\mathcal{R} = \{1, 2 \ldots R\}$, which are to be covered by the set A of active nodes. Each of the active sensors is assumed to be capable of capturing the events that occur in their neighborhood of radius δ, termed as the coverage range. Let the range ω_1 denote the maximum intra-cluster transmission range, and the range ω_2 represent the maximum inter-cluster transmission range [10]. Inter-cluster transmission is designed to occur via hopping through the CHs. Let H be the set of CHs. Further, it is assumed that $\delta < \omega_1 < \omega_2$, and the ratio $\gamma = \frac{\omega_1}{\delta}$ is termed as the *connectivity-coverage* ratio. We present formal definitions associated with the proposed multihop graph referred to as *IoT Network Graph* in this section.

Definition 1 [*Sensor Device*]: A sensor device is a device which captures the physical measurement in its vicinity. The set of sensor devices is denoted by S.

Definition 2 [*Sink*]: The device that consumes the aggregated data in an IoT network is called the sink. It is normally located at the CCC. The data generated by the sensors in an IoT network is to be gathered efficiently at the sink. The consumed data is used for deriving significant parameters of the underlying process sensed by the IoT network. We denote the sink by z.

Definition 3 [*Network Device*]: Any sensor device or sink is referred to as a network device.

Definition 4 [*Active Devices*]: At any given point in time, a subset of the set of sensor devices captures the physical parameters from the surroundings. Those set of sensor devices that capture the data from the surroundings together with the sink form the set of active devices. The set of active devices is denoted by the notation A.

Definition 5 [*Coverage Matrix*]: The coverage matrix ρ of size $|S| \times |R|$ defines the coverage of the sensors over the monitoring region $\mathcal{R}$. The jth element in the ith row is $\top$ if sensor i covers the subregion j, and $\perp$, otherwise.

Definition 6 [*Cluster-Heads*]: A CH is a network device that collects data within a cluster in an IoT network. The CH can be a sensor device with its own sensed data. Clearly, the set of CHs is a subset of the network devices, and all network devices except the CH and sink together form the set of non-cluster-head devices. The set of CHs is denoted by the notation H.

Let

- S is the set of network devices in the IoT network.
- A be the set of active network devices, $A \subseteq S$.
- z is the sink in the IoT network.
- H is the set of CHs. Clearly, $H \subset S$.
- C is the set of clusters in the WSN.

Definition 7 [*Sensor Range Function*]: The Sensor Range Function is the function $\Omega : (S - H) \times (S - H) \rightarrow \{T, F\}$ defined between sensor devices to check whether a sensor device is in the range of another. Let α and β be any two devices in $S - H$, then the Sensor Range Function $\Omega(\alpha, \beta)$ gives a True value if and only if the sensor α can reach sensor β in single hop.

Definition 8 [*Sensor Cluster-Head Range Function*]: The Sensor Cluster-head Range Function is the function $\Phi : (S - H) \times H \rightarrow \{T, F\}$ defined between non-cluster-head sensor devices to a cluster-head to check whether a sensor device is in the range of the cluster-head. Let α and β be any two devices in $S - H$ and H respectively, then the Sensor Cluster-head Range Function $\Phi(\alpha, \beta)$ gives a True value if and only if the sensor α can reach the cluster-head β in single hop.

Definition 9 [*Cluster-Head Range Function*]: The Cluster-head Function is the function $\Gamma : H \times H \cup \{z\} \rightarrow \{T, F\}$ defined either between two cluster-heads or between a cluster-head and a sink to check whether the latter is in the range of the former. Let α and β be any two devices in H and $H \cup \{z\}$ respectively, then the Cluster-head Range Function $\Gamma(\alpha, \beta)$ gives a True value if and only if the cluster-head α can reach the cluster-head (or sink) represented by β in single hop.

Definition 10 [*Sink Range Function*]: The Sink Range Function is the function $\Psi : H \times \{z\} \rightarrow \{T, F\}$ defined between the sink and a cluster-head to check whether the sink is in the range of the cluster-head. Let z denote the sink and α be any cluster-head in H, then the Sink Range Function $\Psi(\beta, z)$ gives a True value if and only if the cluster-head α can reach the sink z in single hop $\mathcal{C}\ \mathcal{C}$.

Definition 11 [*Cluster*]: A cluster is defined as a pair (C, h) where $C \subset S$ represents the set of sensor nodes within the cluster and $h \in H$ is a special sensor node which acts as the head of the cluster. The special node h and the sensor nodes in C satisfy the following rules.

- Coverage: $\phi(\alpha, h) = T | \forall\, \alpha \in C$.

Algorithm 1 Complete algorithm

Input: Set of sensors S, coverage matrix ρ, the sensor CH function Ψ, the cluster-head range function Γ

Output: The graph G_{OPT}

```
Set A := ACTIVEGRAPH(S, ρ);
Set H := CLASSIFY(A, < S^1, S^2 ... S^{t-1} >, Ψ);
Unmark all sensors in A and set V_A := A;
Set E_H := {(i, j)|i, j ∈ H and Γ(i, j) = 1}
Set W_H(i, j) := 1 ∀ (i, j) ∈ E_H
Form the graph G_H = (H, E_H, E_H)
G_A := (A, E_A) where E_A = φ
dest := z;
dist[dest] := 0;
done[dest] := TRUE;
for each node i ∈ S do
    dist[i] := ∞;
    done[dest] := FALSE;
    EnquePQ(Q, 0, dest);
    while isnotempty(Q) do
        current := dequeuePQ(Q);
        for i ∈ S do
            if (i, current) ∈ E then
                if (done[i] = FALSE and dist[i] ≥ dist[current] + W_AN((i, current))) then
                    dist[i] := dist[current] + W_AN((i, current));
                    EnquePQ(Q, dist[i], i)
                    next[i] := current;
        done[current] := TRUE;
        for i ∈ S do
            route := NIL;
            GETROUTE(next, i, route, dest)
            for each edge (k, l) ∈ route do
                E_A := E_A ⋃{(k, l)}
return G_A
```

Definition 12 [*Active Network Graph* G_{A}]: An active network graph is a weighted, directed, and vertex-colored graph $G_{\mathrm{A}} = (V_{\mathrm{A}}, E_{\mathrm{A}}, W_{\mathrm{A}}, C_{\mathrm{A}})$, where

- $V_{\mathrm{A}} = A$
- $E_{\mathrm{A}} = \{(i, j) | (i, j) \in A \times A\}$ and $\Omega_k(i, j) = \top$ for $k = 1$ or $2\}$
- $W_{\mathrm{A}}(i, j) = 1$ if $(i, j) \in E_{\mathrm{A}}$ and $\Omega_2(i, j) = 1$.
- $C_{\mathrm{A}}(i) = \begin{cases} G, & \text{if the vertex } i \in V_{\mathrm{A}} \text{ is the CCC,} \\ R, & \text{if the vertex } i \in V_{\mathrm{A}} \text{ is a CH,} \\ B, & \text{otherwise} \end{cases}$.

4 Finding Efficient Network Configuration for a Given Sensor Deployment

In this section, we design a polynomial time algorithm to find an efficient network configuration for a cluster-based multihop architecture from a given sensor network deployment. We propose a polynomial time algorithm to compute an efficient network configuration for data and energy harvesting. Taking sensor deployment graph G_S as input, we perform the following steps to get an overlay network graph G_O that corresponds to the desired network configuration.

Step 1: Active Sensor Graph Construction: In this step, we take the sensor graph G_A as input and select a minimal set of vertices (except the sink) that cover the field of deployment denoted by the set of sub-regions $\mathcal{R} = \{1, 2 \ldots R\}$. Using the coverage matrix ρ defined in Sect. 3, we select vertices one by one in the non-increasing order of the number of sub-regions they cover. We repeat this procedure until all the sub-regions in $\mathcal{R}$ are covered. We now construct the active sensor graph G_A with the chosen vertices and the sink with no edges.

Step 2: Vertex Classification: In this step, we identify the CH and non-CH nodes in the active sensor graph G_A. We take each node of G_A iteratively and find the number of active nodes that are in the sensor CH range of the node. In each layer, the assignment of a sensor i as a cluster-head is done in the non-increasing order of the number of sensors that can be connected in its sensor CH range. Once the sensor i is assigned as a CH, a new cluster is formed, and the associated sensors that are in the sensor CH range of the sensor i are assigned as non-CH nodes of the newly formed cluster with sensor i as the CH.

Step 3: Path Selection in the Graph: After the above three steps, there can be multiple paths from every node to the sink. The least-hop path can be found via running the shortest path algorithm from each node to the sink.

The complete algorithm is given in 1.

5 Results and Discussions

We consider a monitoring region of dimensions $L \times L$ m^2 and a uniform distribution of the sensors. The sensors are initially equipped with a starting energy of E_0 J. The CCC is considered to be located at a corner point (L, L) of the monitoring area. Sensor readings are assumed to be converted to 4000 symbols in each sequence of operation. The maximum inter-cluster and intra-cluster distances are taken as 40 and 30% of L. As the sequences of operation proceed, the energies of the transmitting sensor nodes start decreasing according to the standard radio model [28]. The remaining energy of the network is taken as the sum of the energies of the sensors alive. The network fails when there are not enough sensors alive for the full coverage of the monitoring region in $\mathcal{R}$.

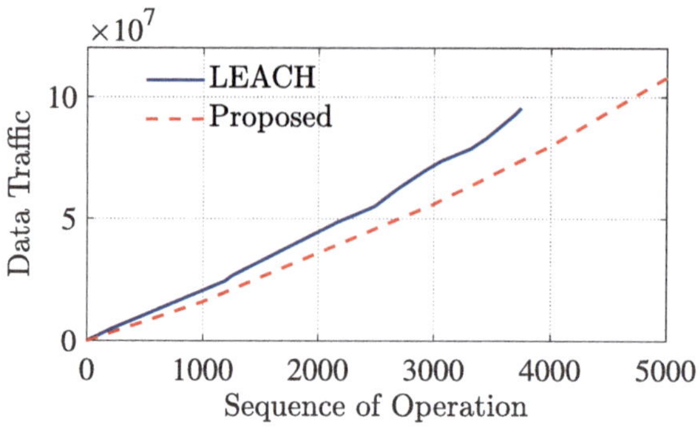

Fig. 1 Cumulative data traffic: LEACH and proposed method

LEACH is the most popular protocol for energy-efficient WSNs, but it is not governed by the intra-cluster and inter-cluster transmission ranges. We consider the latest version of the LEACH [35]. LEACH aims at prolonging the network lifetime by including the influence of the residual energy in the CH selection process. The threshold expression is given by

$$T(n) = \begin{cases} \frac{P}{1-P\cdot(r \mod \frac{1}{P})}\frac{E_n}{E_I}, & \text{if } n \in G \\ 0, & \text{otherwise} \end{cases}, \tag{3}$$

where E_n is the residual energy of nth sensor, and E_I is the initial energy of the sensor. The modification improves the network lifetime by 30% compared to standard LEACH, as the residual energy of the sensors is considered in CH selection compared to the purely random selection in standard LEACH.

For a fair comparison with the LEACH protocol, we modify the proposed method by relaxing such constraints. The proposed method results in lesser data traffic because the non-CH members transmit directly to the CCC, compared to the hopping via CHs in the peer protocol. This results in reduced data traffic and lesser energy spent in operation, as shown in Fig. 1, leading to an overall increased number of sequences of network operation. Further, the superior clustering method in the Proposed method compared to the probabilistic selection in the LEACH results in selecting a lower number of clusters and CHs. This results in a lesser direct transmission to CCC, amounting to a lower energy spend as in Fig. 2, and provides improved life for the networks to operate as depicted in Fig. 3.

6 Conclusion

The proposed graphical framework covers the features of generic IoT networks. Hence, researchers can use the framework to apply powerful combinatorial algorithms to find solutions to various network design problems. The proposed graphical

Fig. 2 Remaining network energy: LEACH and proposed method

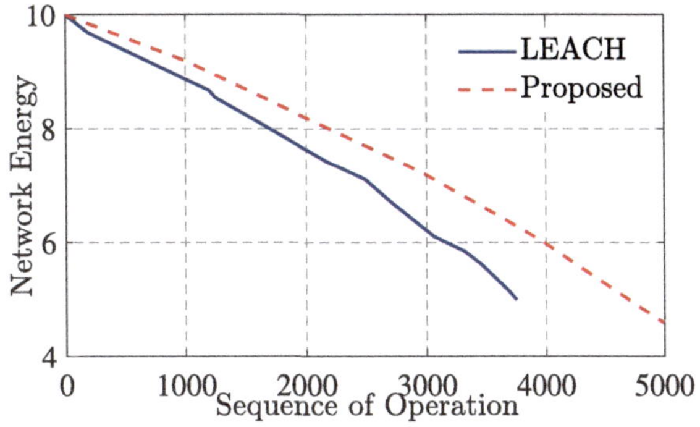

Fig. 3 Sequences of operation: LEACH and proposed method

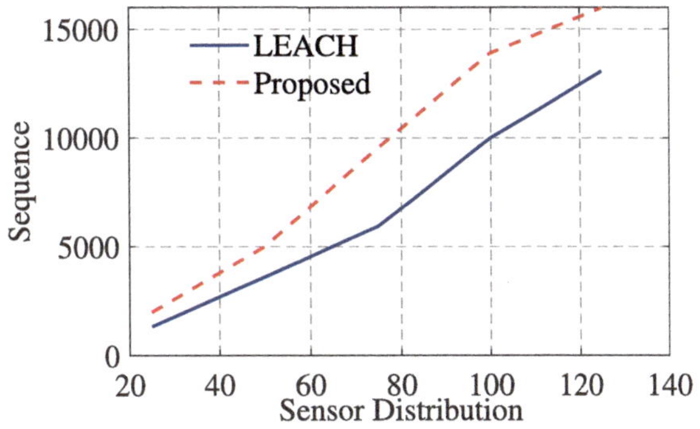

structure can be extended to capture the features of other applications, such as target tracking, surveillance, and environmental monitoring.

References

1. Ullah Z (2020) A survey on hybrid, energy efficient and distributed (HEED) based energy efficient clustering protocols for wireless sensor networks. Wireless Pers Commun 112(4):2685–2713
2. Sheriba S, Rajesh DH (2021) Energy-efficient clustering protocol for WSN based on improved black widow optimization and fuzzy logic. Telecommun Syst 77(1):213–230
3. Du X, Zhou Z, Zhang Y, Rahman T (2020) Energy-efficient sensory data gathering based on compressed sensing in IoT networks. J Cloud Comput 9(1):1–16
4. Panchal A, Singh RK (2021) EHCR-FCM: energy efficient hierarchical clustering and routing using fuzzy C-means for wireless sensor networks. Telecommun Syst 76(2):251–263
5. Bhuiyan MZA, Wang G, Cao J, Wu J (2013) Deploying wireless sensor networks with fault-tolerance for structural health monitoring. IEEE Trans Comput 64(2):382–395
6. Haenggi M, Puccinelli D (2005) Routing in ad hoc networks: a case for long hops. IEEE Commun Mag 43(10):93–101
7. Xiuwu Y, Qin L, Yong L, Mufang H, Ke Z, Renrong X (2019) Uneven clustering routing algorithm based on glowworm swarm optimization. Ad Hoc Netw 93:101923

8. Das SK, Kapelko R (2021) On the range assignment in wireless sensor networks for minimizing the coverage-connectivity cost. ACM Trans Sens Netw (TOSN) 17(4):1–48
9. Doostali S, Babamir SM (2020) An energy efficient cluster head selection approach for performance improvement in network-coding-based wireless sensor networks with multiple sinks. Comput Commun 164:188–200
10. Chamam A, Pierre S (2009) On the planning of wireless sensor networks: energy-efficient clustering under the joint routing and coverage constraint. IEEE Trans Mob Comput 8(8):1077–1086
11. Fei Z, Li B, Yang S, Xing C, Chen H, Hanzo L (2016) A survey of multi-objective optimization in wireless sensor networks: metrics, algorithms, and open problems. IEEE Commun Surv Tutor 19(1):550–586
12. Heinzelman WR, Chandrakasan A, Balakrishnan H (2000) Energy-efficient communication protocol for wireless microsensor networks. In: Proceedings of the 33rd annual Hawaii international conference on system sciences. IEEE, p 10
13. Heinzelman WB, Chandrakasan AP, Balakrishnan H (2002) An application-specific protocol architecture for wireless microsensor networks. IEEE Trans Wireless Commun 1(4):660–670
14. Ferreira AC, Vilaça MA, Oliveira LB, Habib E, Wong HC, Loureiro AA (2005) On the security of cluster-based communication protocols for wireless sensor networks. In: International conference on networking. Springer, pp 449–458
15. Perrig A, Szewczyk R, Tygar JD, Wen V, Culler DE (2002) Spins: security protocols for sensor networks. Wireless Netw 8(5):521–534
16. El_Saadawy M, Shaaban E (2012) Enhancing S-LEACH security for wireless sensor networks. In: 2012 IEEE international conference on electro/information technology. IEEE, pp 1–6
17. Oliveira LB, Wong HC, Bern M, Dahab R, Loureiro AAF (2006) SecLEACH—a random key distribution solution for securing clustered sensor networks. In: Fifth IEEE international symposium on network computing and applications (NCA'06). IEEE, pp 145–154
18. Gnanambigai J, Rengarajan N, Anbukkarasi K (2013) Q-LEACH: an energy efficient cluster based routing protocol for wireless sensor networks. In: 2013 7th international conference on intelligent systems and control (ISCO). IEEE, pp 359–362
19. Latiff L, Fisal N, Arifin S, Ahmed AA (2010) Directional routing protocol in wireless mobile ad hoc network. IntechOpen
20. Deepa S, Marimuthu C, Dhanvanthri V (2015) Enhanced Q-LEACH routing protocol for wireless sensor networks. ARPN J Eng Appl Sci 10(9):4036–4041
21. Manzoor B, Javaid N, Rehman O, Akbar M, Nadeem Q, Iqbal A, Ishfaq M (2013) Q-LEACH: a new routing protocol for WSNs. Procedia Comput Sci 19:926–931
22. Abuhelaleh MA, Mismar TM, Abuzneid A (2008) Armor-LEACH-energy efficient, secure wireless networks communication. In: 2008 proceedings of 17th international conference on computer communications and networks. IEEE, pp 1–7
23. Serna Oliver R (2010) An approach to timeliness in wireless sensor network communications. Ph.D. dissertation, Technische Universität Kaiserslautern
24. Wang W, Wang Q, Luo W, Sheng M, Wu W, Hao L (2009) LEACH-H: an improved routing protocol for collaborative sensing networks. In: 2009 international conference on wireless communications & signal processing. IEEE, pp 1–5
25. Ren P, Qian J, Li L, Zhao Z, Li X (2010) Unequal clustering scheme based LEACH for wireless sensor networks. In: 2010 fourth international conference on genetic and evolutionary computing. IEEE, pp 90–93
26. Mahmood D, Javaid N, Mahmood S, Qureshi S, Memon AM, Zaman T (2013) MODLEACH: a variant of LEACH for WSNs. In: 2013 eighth international conference on broadband and wireless computing, communication and applications. IEEE, pp 158–163
27. Salim A, Osamy W, Khedr AM (2014) IBLEACH: intra-balanced LEACH protocol for wireless sensor networks. Wireless Netw 20(6):1515–1525
28. Lee J-S, Kao T-Y (2016) An improved three-layer low-energy adaptive clustering hierarchy for wireless sensor networks. IEEE Internet Things J 3(6):951–958

29. Loscri V, Morabito G, Marano S (2005) A two-levels hierarchy for low-energy adaptive clustering hierarchy (TL-LEACH). In: IEEE vehicular technology conference, vol 62, no 3. IEEE, p 1809
30. Kim D-S, Chung Y-J (2006) Self-organization routing protocol supporting mobile nodes for wireless sensor network. In: First international multi-symposiums on computer and computational sciences (IMSCCS'06), vol 2. IEEE, pp 622–626
31. Kumar GS, Vinu PM, Jacob KP (2008) Mobility metric based LEACH-mobile protocol. In: 2008 16th international conference on advanced computing and communications. IEEE, pp 248–253
32. Qiang T, Bingwen W, Zhicheng D (2009) MS-LEACH: a routing protocol combining multi-hop transmissions and single-hop transmissions. In: 2009 Pacific-Asia conference on circuits, communications and systems. IEEE, pp 107–110
33. Sun Y, Dong W, Chen Y (2017) An improved routing algorithm based on ant colony optimization in wireless sensor networks. IEEE Commun Lett 21(6):1317–1320
34. Aurenhammer F, Klein R (2000) Voronoi diagrams. Handb Comput Geom 5(10):201–290
35. Huo J, Deng X, Al-Neshmi HMM (2020) Design and improvement of routing protocol for field observation instrument networking based on LEACH protocol. J Electr Comput Eng 2020

Privacy, Security, Forensics

A Lightweight Privacy-Preserving and Authentication Model (LiPAM-FSG) for Fog-Based Smart Grids

Neha Kaliya, Digambar Pawar, and Adesh Raut

Abstract The utilization of fog-based technology within the realm of smart grids stands out as one of the most promising advancements. Smart grids, being repositories of highly sensitive data, carry the inherent risk of potentially exposing user's private information. Traditional security models often rely on complex cryptographic operations, resulting in high computational complexity and latency. In response to this challenge, we propose the adoption of a lightweight encryption technique known as the PICO cipher. This cipher not only consumes fewer resources, including flash and RAM, when compared to other cryptographic algorithms but also exhibits a greater avalanche effect. The overall computation time for various functionalities (data aggregation, dynamic pricing, and load forecasting) should be less. To maintain data integrity, hash-based message authentication code (HMAC) authentication technique is used. We introduced the concept of the message queue telemetry transport (MQTT) publish-subscribe model for data aggregation. The model is evaluated in iFogSim simulator to understand real-time parameters. Also, we implemented PICO cipher and other similar cipher algorithms to study time and complexity. Upon thorough examination, our proposed smart grid design model outperforms a conventional fog-based smart grid architecture, demonstrating its enhanced capabilities in preserving data privacy and minimizing computational overhead.

Keywords Smart grid · PICO Cipher · Fog privacy · MQTT protocol · HMAC authentication · Fog authentication

N. Kaliya (✉) · D. Pawar · A. Raut
University of Hyderabad, Prof. CR Rao Road, Hyderabad 500046, India
e-mail: 20mcpc01@uohyd.ac.in

D. Pawar
e-mail: dpr@uohyd.ac.in

A. Raut
e-mail: 21mcmt25@uohyd.ac.in

S. M. Thampi et al. (eds.), *Fifth International Conference on Computing and Network Communications*, Lecture Notes in Electrical Engineering 1219,
https://doi.org/10.1007/978-981-97-4540-1_19

1 Introduction

A smart grid is an electrical grid network which incorporates digital advanced technologies to manage and monitor electricity transport from distribution center to end users. Nowadays, with the initiative of smart city implementation, it is also a major research area [1]. It is an extension of traditional grid architecture with various advanced features such as limiting the electricity usage, generating electricity bills with privacy preservation, and setting threshold for electricity consumption and monitoring it . Accumulating smart meters data from all over the country and storing it for further processing give rise to a significant challenge of storing big data [2].

General fog-based smart grid architecture is shown in Fig. 1. Tier 1 consists of smart meters, which are responsible to collect total electricity consumption data at ground level. Further this collected data is sent to fog nodes for quick processing. Here, each smart meter is connected independently to fog nodes. Fog nodes lie in tier 2. Data is sent at regular intervals to fog nodes from smart meters. Finally these fog nodes are connected to the cloud layer which lies in in Tier 3, where large amounts of data are stored. The control center of the smart grid collaborates with the cloud layer for storing its customer data.

Introducing fog layer between cloud and smart meters helps in generating quick responses. The average latency, computation cost, storage cost, and network usage of cloud-based smart grids is more as compared to fog-based smart grids [3], and the architecture of smart grid includes connection of smart meters (IoT devices) to fog layers for fast processing of data. Most of the architectures are using data aggregation technique to reduce data traffic at fog layer [4–6]. There is a need of

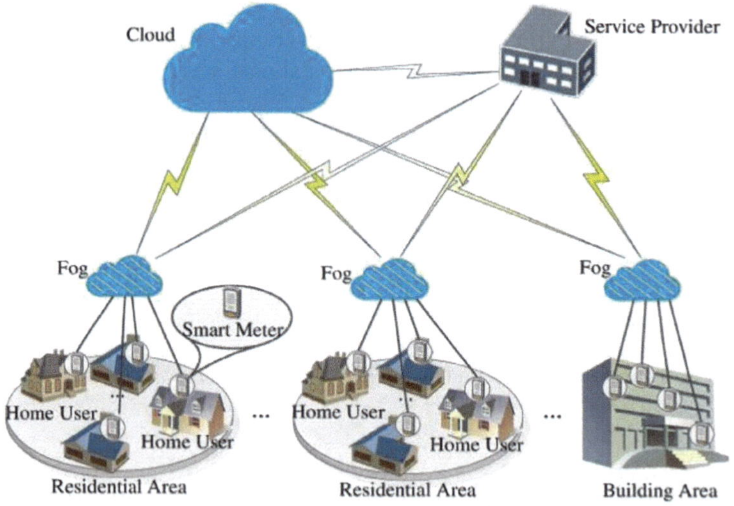

Fig. 1 Smart grid architecture

privacy preservation of user data stored at fog layer to ensure confidentiality and integrity of user's energy usage information and lifestyle habits.

1.1 Motivation and Contribution

As data transitions from smart meters to fog nodes for the purposes of data aggregation and subsequent processing, significant security concerns emerge. This presents an opportunity for malicious actors to compromise data integrity, potentially violating user privacy. In this context, the information and communication technology (ICT) framework of the new smart grid will evolve into a comprehensive information system, significantly expediting the processing of metering information on a large scale. However, significant concerns regarding smart grids revolve around potential security challenges stemming from the transmission of end-user metering data over the network and the collection of extensive micro-data, which could introduce new threats to customer privacy. Furthermore, security issues arise due to the crucial necessity of safeguarding the power grid from denial-of-service attacks, as such attacks could substantially disrupt modern living infrastructure.

To establish a resilient and secure communication framework, it is imperative to implement robust data encryption and authentication mechanism. Current proposed solutions are ensuring secure communication but latency and computation overheads increased. Due to the resource-constraint nature of smart meters, there is a need of lightweight security mechanism. This will ensure low computation overhead and low latency. As communicating entities will not be busy in computing parameters to pass security check for long time, lightweight mechanism can ensure high packet delivery ratio and low energy consumption.

In this context, we proposed a mechanism of privacy preservation in fog-based smart grids using lightweight PICO cipher, HMAC authentication, and MQTT pubsub model. Fog-based smart grid model is comprised of multiple components such as data aggregation, generating bills, data collection, maintaining big data, securing the communicating entities, and maintaining data and user's privacy. We are focusing on security mechanism to provide data privacy and confidentiality (using PICO cipher algorithm) and to provide user authentication (using HMAC code).

2 Related Work

Intelligent power supply is a mandatory initial step to build and address the challenges of smart infrastructures such as smart home, smart city, or smart buildings. In order to provide optimal scheduling, smart meters are deployed to record real-time electricity consumption data usage. If this data is disclosed, user's privacy may leak as application usage patterns can be tracked by consumption profile. Hence, it is very important to preserve privacy of data and user. Recently various techniques have been proposed to solve the problem of privacy preservation in smart grids. We have listed below few major related techniques and their privacy preservation mechanisms.

Meng et al. [7] proposed an improved ECC-AES hybrid encryption algorithm. Here initial key transfer and communication phase is set up by elliptic curve cryptography (ECC) algorithm. This is done to enhance the speed of encryption and decryption. Once the communication is established using same parameters, the key is transferred to communicating entities for further AES encryption. ECC is used to encrypt the AES key and to improve the security of the algorithm. Experimental evaluation proved that using ECC algorithm, computing time, memory usage, and CPU occupancy is reduced. Similarly, [8] also proposed triple DES algorithm for data encryption. But 3-DES is also not much secure and time taken for processing is high.

Abood et al. [9] presented a survey related to all significant algorithms used for cryptography in smart grids. They made a comparison between AES, DES, TDES, EDES, RSA, and BLOWFISH to give an idea about smart grid security system. The performance evaluation of these algorithms is done using MATLAB.

A complete end to end system was proposed by Li et al. [10] called SecGrid. They have given both hardware and software details. They have used an Intel SGX-enabled CPU in it. The minimal requirement specification needed here is that the CPU should support AES algorithm for encryption and decryption. Encrypted data by AES was used to protect the privacy of the user in smart grid. The results show that SecGrid is much faster than the already available smart grid systems. This work does not provide any research work based on other modules of smart grid such as load forecasting and dynamic pricing.

In [11], authors presented a secure power usage data aggregation scheme in smart grid. They explained how internal attackers can take advantage of the information if privacy is not in place. So, the aggregation scheme is designed in such a way that suppliers get to know about the electricity usage in a total neighborhood for arranging the electricity supply instead of knowing individual power consumption. They addressed the issue of internal attackers and gave a solution using batch verification.

Guan et al. [12] proposed a privacy preservation and efficient data aggregation scheme. They used the concept of block chain and bloom filter for privacy, integrity, and authentication of users. Users were divided in different groups, and each group has a private block chain to record its member's data. In addition to this, bloom filter is adopted to authenticate the blocks in block chain.

Gong et al. [13] proposed a privacy-preserving scheme for incentive-based demand response in the smart grid. The scheme creates pseudonym using discrete logarithm and achieves anonymity through a ring signature during the registration process of a pseudonym.

In [14], authors presented a novel approach for ensuring data privacy through differential privacy. This method involved introducing noise derived from a virtual rechargeable battery, all the while preserving the accuracy of billing information. The core of this model leverages fog computing data aggregation, employing lightweight cryptographic techniques to safeguard the integrity and confidentiality of data generated by resource-constrained devices.

Khan et al. [15] proposed a fog-enabled data aggregation scheme with privacy preservation and fault tolerance. To safeguard metering data privacy, author's used the Boneh-Goh-Nissam (BGN) cryptosystem, while source authentication is ensured through the use of the elliptic curve digital signature algorithm (ECDSA). ECDSA's swift signature generation, facilitated by its smaller key sizes, makes it well-suited for resource-constrained devices like those in advanced metering infrastructure (AMI).

In [16], authors proposed a data aggregation scheme for a smart grid with user anonymity ensuring privacy. They used homomorphic re-encryption for smart meters. Here smart meters transmit the cipher texts anonymous, and later proxies are distributed by re-encrypting the aggregated data so that only the designated recipients can access it. This leads to more network usage, more computations, and more noise in the network. Studies also commented on privacy and differential privacy mechanisms on smart grid parameters [17].

Wang et al. [18] have created a detailed four-tier data aggregation system. In this model, fog nodes (FNs) are used for the collection and aggregation of electricity consumption data. They have used the El-Gamal cryptosystem. A distributed decryption process is proposed that allows for fine-grained access control and real-time price-based billing.

All the above-mentioned schemes were designed to achieve privacy in one or other way at smart meters and fog layer. But we need to ensure that the designed system should have least computation and energy overhead. Using symmetric encryption cannot ensure less computation and good security [19]. The computational capabilities of a smart meter are fairly low, and hence, to encrypt the data at this device, lightweight cipher algorithm is required. We have done an extensive survey on lightweight cipher algorithms, and based on the experimental results, we proposed a lightweight privacy-preserving mechanism for fog-based smart grid using PICO cipher, HMAC authentication, and MQTT publish–subscribe model.

3 Proposed Model

In the proposed architecture, smart meters are deployed at ground level to record and store the electricity usage readings. The data collected by smart meters is timely stored at the database at server level. The smart meters will consist of an Arduino Duo microcontroller. The data at smart meters will be stored in encrypted form using PICO cipher, and later, hash-based message authentication code (HMAC) will be generated at ground layer itself to ensure data integrity and authentication. We have introduced MQTT publish–subscribe model to reduce traffic and latency at fog layer. Data aggregation in our model is done at ground layer itself using MQTT model. Here the smart grid system is divided into various areas. Each area has some smart meters in its range as shown in Fig. 2. In each area, every smart meter is connected

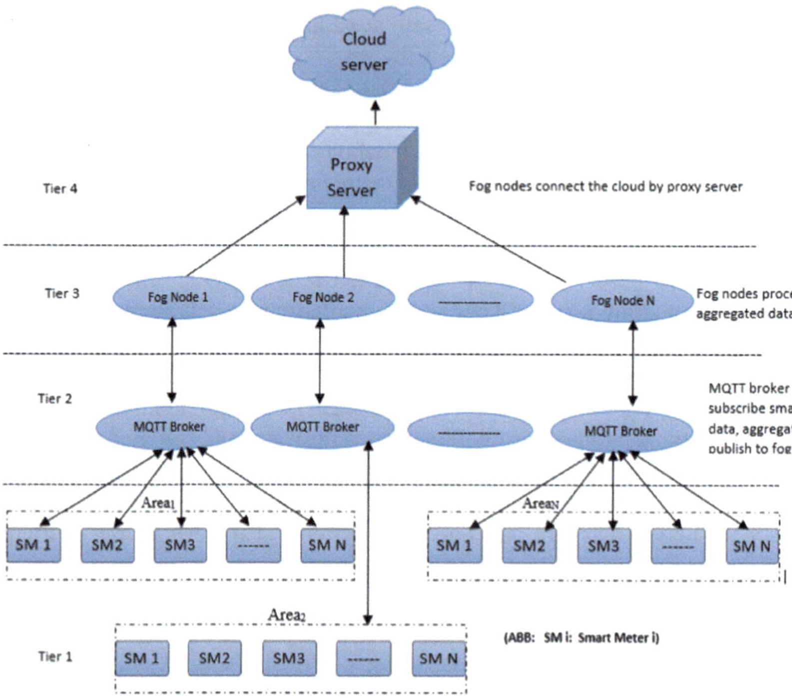

Fig. 2 Proposed system model

by MQTT broker and sends the data. MQTT broker will aggregate the received data and send it to fog node for further processing.

3.1 Hardware Specification of the Model

For the experimental purpose, we used 32-bit ARM **Cortex-M3 processor** which is known for high-performance, real-time processing in cost-constrained applications and can handle complex tasks. Any Arm® Cortex®-M3 microcontroller offers high scalability combined with an optimal trade-off between performance and cost. The small footprint of the core allows it to be used either as a single core in small devices or as an additional embedded companion core when specific hardware isolation or task partitioning is required.

The Arduino Due is a microcontroller board based on the Atmel SAM3X8E ARM Cortex-M3 CPU. It is the first Arduino board based on a 32-bit ARM core microcontroller. It has 54 digital input/output pins (of which 12 can be used as PWM outputs), 12 analog inputs, 4 UARTs (hardware serial ports), a 84 MHz clock, an

USB OTG capable connection, 2 DAC (digital to analog), 2 TWI, a power jack, an SPI header, a JTAG header, a reset button, and an erase button.

3.2 *Software Specification of the Model*

Cipher Algorithm

Arduino microcontroller is capable of performing encryptions and suitable for the criteria designed for cost and energy consumption. We have done a comparative study of traditional cryptographic algorithms used till now in different models to study their performance detailed in Table 1. From this study, we found that there is a strong need to use lightweight cipher algorithm in smart grids to reduce computations over a smart meter. In this direction, to select an optimal cipher (best fit) for our model, we have done a detailed study of lightweight ciphers (cryptographic algorithms) as detailed in Table 2.

Table 1 Performance evaluation of traditional cryptographic algorithms

Algorithm	Key size (bits)	Block Size (bits)	No. of rounds	Structure	Encryption time (sec)	Decryption time (sec)
DES	64	46	16	Festial	0.7137	0.1263
E-DES	1024	128	16	Festial	0.6358	0.4325
T-DES	112/168	64	48	Festial	0.4472	0.2203
RSA	1024/4096	128	1	PKA	0.70385	0.64986
Blowfish	32–448	64	16	Festial	0.2371	0.1125
AES	128/192/ 256	128	10/12/14	Substitution/ permutation	0.1754	0.1862

Table 2 Comparative study of lightweight cryptographic algorithms

Cryptographic algorithms	Block size	Key size	Energy req (Enc/Dec)	RAM	ROM (KB)	Processor
Speck Cipher [20, 21]	64 bit	64/96/ 128/192 / 256 bits	78 μJ/byte	1.5 KB	4	ARM Cortex-M4
Twine Cipher [22]	64 bit	80/128 bits	3.3 μJ/byte	N/A	8.5	ARM Cortex-M0
PICO cipher [23]	64 bit	128 bits	1.17 μJ/byte	3.3 KB	6.5	ARM Cortex-M3
PRESENT cipher [24]	64 bit	80/128 bits	N/A	272 bytes	2.5	N/A

From the above study, we considered PICO lightweight cipher most suitable as per our requirement. It can transmit 64 bit block size data, with key size of 128 bits. It just requires 1877 gate equivalents. PICO consumes only 2504 bytes of flash memory which is less than the ultra-lightweight cipher PRESENT. It has a very strong substitution layer (S-box) which makes the design robust and introduces a great avalanche effect. PICO has a strong and compact key scheduling. PICO consumes less dynamic power (28 mW) than PRESENT cipher (38 mW). Along with its lightweight properties, it is resistant against various security attacks such as differential and linear cryptanalysis, algebraic and biclique attack. Cipher with good avalanche effect has least probability to be vulnerable against any attack. Even the speck cipher can be considered for encryption/decryption of data, but its energy consumption for cryptographic operations is far more than that of PICO cipher. The memory requirement of PICO cipher can be easily supported by the microcontroller Arduino Due which we have used in smart meters.

Hash-Based Message Authentication Code (HMAC) Authentication

HMAC algorithm is used to check the authenticity and integrity of messages. The main aim of the HMAC is to work in one direction, i.e., it easily generates output hash value from input message, but it is not the same other way round. It uses a shared secret key and hash function together.

Initially a message is padded with a shared secret key, and its combined output is given as input to the HMAC. The output (hash value) of HMAC is sent to the central server along with the data. On the server side the server receives the message and applies HMAC on the message. If the generated hash value matches with the hash value which was sent from client side, then we can say that the message was not tampered. Only thing to be taken care of here is that the shared secret key and hash function should be kept secret, and these two things should be known only to client and server.

In our model, an HMAC is generated for the encrypted data (using PICO cipher) of smart meter. This unique HMAC value is sent to the MQTT broker along with the actual encrypted data. MQTT broker generates a HMAC from an already shared secret key and hash function and compare it with received HMAC value from a particular smart meter. If the calculated hash matches with the received HMAC, it proves data integrity and authenticity.

Figure 3 shows complete flowchart of data encryption and hash mechanism carried out in proposed model.

MQTT Model

MQTT stands for message queuing telemetry transport. It is a lightweight protocol. It is used in IoT applications, low bandwidth and unreliable networks, resource-constrained devices, real-time data streaming, decentralized communication, and

Fig. 3 Flowchart of proposed model

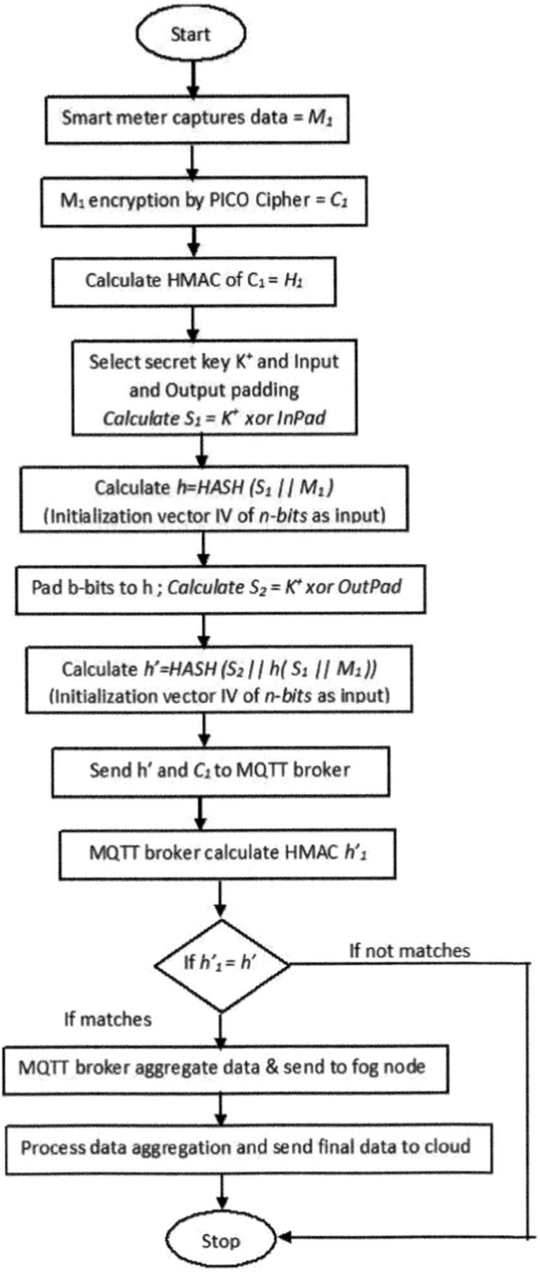

reliable message delivery. Generally it is used for efficient, lightweight, and reliable communication. MQTT supports both real-time messaging and asynchronous messaging.

In asynchronous messaging, messages are stored at the broker until the subscriber becomes available. Later the subscriber can retrieve the data from the broker whenever it becomes available. By this, the applications can function at their own speed without human interference. Few of the use case examples of asynchronous messaging are sensor data aggregation, remote command execution, data logging and batch processing, event-driven systems, etc.

In our proposed smart grid model, MQTT is used for the task of data aggregation, and the structure of the same consists of the following components:

1. Subscribing to smart meter data
2. Data aggregation
3. Publishing aggregated data to fog node.

Each smart meter publishes its stored encrypted value with HMAC to the specific MQTT broker. This can be done periodically or continuously, depending on the model requirement. An MQTT broker lies at an intermediate state between smart meters (publishers) and fog nodes (subscribers). The fog nodes will receive the data from MQTT brokers in a single go. The fog nodes will subscribe to the MQTT broker to retrieve smart meter data. They will receive data from smart meters through MQTT broker at regular intervals. Hence, smart meters will send data to MQTT broker instead of fog nodes, which will significantly decrease traffic at fog nodes.

MQTT's lightweight nature minimizes network overhead, making it suitable for resource-constrained smart meters (IoT devices). The asynchronous nature of MQTT allows the central system to process data at its own pace, accommodating varying processing capabilities and intermittent connectivity. We have implemented and tested the above scenario using Eclipse Paho MQTT client library. The output results are shown below in Fig. 4a and b.

Since there are a large number of smart meters (users) installed, we need to store the data on a large scale. Introducing the fog layer for storing these large amounts of data will reduce latency and increase throughput.

(a)

(b)

Fig. 4 **a** Subscription to home smart meters **b** Publishing data to fog nodes

We have successfully implemented the designed scenario on iFogSim simulator and observed that the response time using the fog layer is quite less as that of cloud layer. Cloud layer collaborates with the control center for implementation of smart grid.

Also, we observe that our proposed model performs better than the already available model.

4 Experimental Results

4.1 Experimental Algorithm Evaluation

We implemented various lightweight ciphers and recorded the time and memory consumed by them as detailed in Table 3. Along with time taken and memory factors, we need to consider other factors like hardware requirements, avalanche effect, and resistance against various types of attacks to select the most suitable cipher algorithm for the model we designed. We programmatically tested the performance of every cipher algorithm, and output of these algorithms is shown in Figs. 5, 6, and 7.

Table 3 Time and memory consumed by cipher algorithms

Cipher algorithm	Time consumed (Encryption/Decryption) (s)	Sbox memory consumed (bits)
Twine cipher	0.000171	8
PICO cipher	0.00020	4
PRESENT cipher	0.000135	4
SPECK cipher	0.00022	8

Fig. 5 Twine cipher output

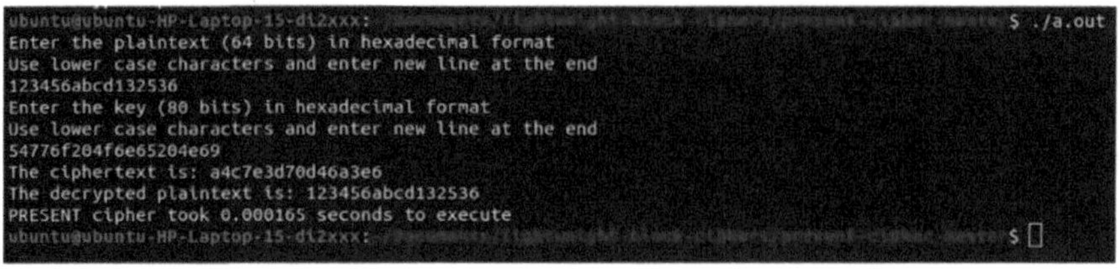

Fig. 6 Present cipher output

Fig. 7 PICO cipher output

From the above experimental observations, we confirmed that the PICO cipher is the most suitable lightweight algorithm for our smart meter design. Major description of PICO cipher is listed below:

1. Sbox memory consumed for PICO cipher is 4 bits, whereas for Twine/Speck cipher it is 8 bits.
2. PICO cipher has got very good avalanche effect.
3. PRESENT cipher is proved to be useful for hardware constrained devices, but PICO cipher provides optimum performance on both hardware as well as software platforms.
4. PICO consumes 28 mW of dynamic power which is less than the PRESENT cipher (38 mW).

4.2 System Model Implementation on iFogSim Simulator

We implemented our system model using iFogSim to check network parameters. In our proposed design, smart meters are connected individually to MQTT broker and MQTT broker is connected to fog nodes. We implemented two more scenarios in iFogSim to compare the performance of system designed. One scenario was smart grid design where smart meters are directly connected to the cloud without fog nodes for aggregation. Second one was smart grid design by introducing fog layer without using MQTT broker for data aggregation. To make a comparative study, we named the systems as: cloud-based smart grid (CBSG), fo-based smart grid (FBSG), and fog-based smart grid with MQTT model (FBSGMM). The comparative results are shown in Table 4, and graphical results are shown in Figs. 8 and 9. The experimental results clearly conclude that FBSGMM takes least execution time and loop delays. This ensures that the secure data is delivered to higher ends with highest throughput.

Also, FBSGMM has lowest cloud cost execution, which motivates to incorporate fog layer to reduce packet delivery ratio. From the above experiments, it is fairly concluded that smart meters with MQTT concept and lightweight cipher algorithm will definitely a better option to improve system efficiency.

Table 4 Comparative study of all models in iFogSim

System model	Execution time (ms)	Total network usage	Application loop delays (ms)	Cloud cost execution
CBSG	635	47,151.8	248.23	95,088.8
FBSG	433	760	49.04	0
FBSGMM	313	380	48.69	0

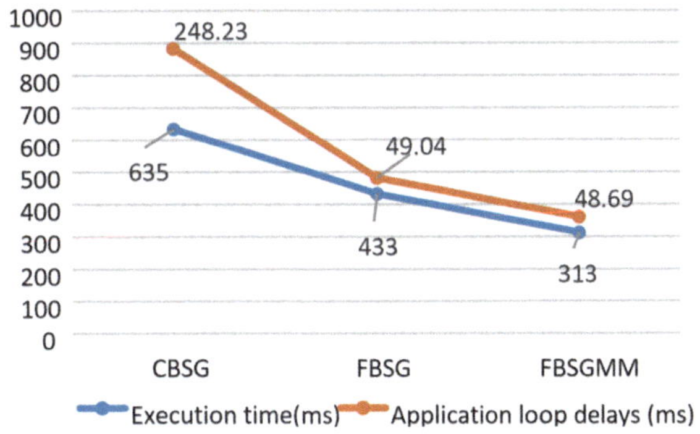

Fig. 8 Graphical analysis of execution time and application loop

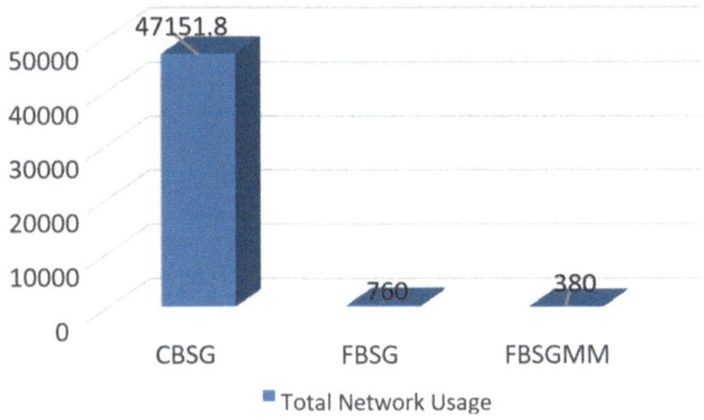

Fig. 9 Graphical analysis of network usage and cloud cost execution

5 Conclusion and Future Work

For maintaining low latency communication in time-sensitive application areas including smart grids, smart homes, IoV, health care, etc., fog computing has recently emerged as a demonstrably effective option. The demand for a prompt response has increased along with the amount of data being collected by smart meters. To that end, we created an energy-efficient privacy-preserving fog-based smart grid architecture.

We proposed a lightweight privacy-preserving and authentication model (LiPAM-FSG) in this research. To protect data confidentiality on the smart meter device, we employed the PICO cipher, and to ensure data integrity, we used the hash-based message authentication code (HMAC). Smart meters (IoT devices) and fog nodes are connected via the MQTT publish–subscribe model. Multiple smart meters data is published to a fog node by a MQTT broker that subscribes to the data that smart meters provide. The experimental results proved that this novel technique takes less execution time, low latency, provides better avalanche effect, and has better time and space complexity as compared to existing models.

Investigating which smart meter can be made as a broker would be the future work on this. Optimization of a variety of hardware and software platforms employed to implement the functionality of the smart grid is also required. Several additional lightweight cryptography techniques can be investigated for safeguarding smart meter (customer) data.

References

1. Wang W, Xu Y, Khanna M (2011) A survey on the communication architectures in smart grid. Comput Netw 55(15):3604–3629
2. Khan S, Parkinson S, Qin Y (2017) Fog computing security: a review of current applications and security solutions. J Cloud Comput 6(1):1–22
3. Wang Z, Liu Y, Ma Z, Liu X, Ma J (2020) LiPSG: Lightweight privacy-preserving Q-learning-based energy management for the IoT-enabled smart grid. IEEE Internet Things J 7(5):3935–3947
4. Zhao S, Li F, Li H, Lu R, Ren S, Bao H, Han S (2020) Smart and practical privacy preserving data aggregation for fog-based smart grids. IEEE Trans Inf Forensics Secur 16:521–536
5. Bao H, Lu R (2015) Comment on privacy-enhanced data aggregation scheme against internal attackers in smart grid. IEEE Trans Industr Inf 12(1):2–5
6. Liu JN, Weng J, Yang A, Chen Y, Lin X (2019) Enabling efficient and privacy-preserving aggregation communication and function query for fog computing-based smart grid. IEEE Trans Smart Grid 11(1):247257
7. Meng Y, Liangliang Z, Yao R, Yongxian Y, Jiaji L (2021) Research on fast encryption method for smart energy management system in smart gird. In: 2021 international conference on communications, information system and computer engineering (CISCE). IEEE, pp 76–80
8. Manoharan AM, Sumithra MG, Nagarajan B, Venkatesan C, Basuvaiyan S (2021) Realization of full parallelism triple AES algorithm for data encryption. In: 2021 5th international conference on electronics, communication and aerospace technology (ICECA). IEEE, pp 663–670

9. Abood OG, Elsadd MA, Guirguis SK (2017) Investigation of cryptography algorithms used for security and privacy protection in smart grid. In: 2017 nineteenth international Middle East power systems conference (MEPCON). IEEE, pp 644–649
10. Li S, Xue K, Wei DS, Yue H, Yu N, Hong P (2019) SecGrid: a secure and efficient SGXenabled smart grid system with rich functionalities. IEEE Trans Inf Forensics Secur 15:1318–1330
11. Shibutani K, Isobe T, Hiwatari H, Mitsuda A, Akishita T, Shirai T (2011) Piccolo: an ultra lightweight blockcipher. In: CHES vol 6917, pp 342–357
12. Guan Z, Si G, Zhang X, Wu L, Guizani N, Du X, Ma Y (2018) Privacy-preserving and efficient aggregation based on blockchain for power grid communications in smart communities. IEEE Commun Mag 56(7):8288
13. Gong Y, Cai Y, Guo Y, Fang Y (2015) A privacy-preserving scheme for incentive-based demand response in the smart grid. IEEE Trans Smart Grid 7(3):1304–1313
14. Kserawi F, Al-Marri S, Malluhi Q (2022) Privacy-preserving fog aggregation of smart grid data using dynamic differentially-private data perturbation. IEEE Access 10:43159–43174. https://doi.org/10.1109/ACCESS.2022.3167015
15. Khan HM, Khan A, Jabeen F, Rahman AU (2021) Privacy preserving data aggregation with fault tolerance in fog-enabled smart grids. Sustain Cities Soc 64:102522
16. Wu L, Zhang W, Zhao W (2022) Privacy preserving data aggregation for smart grid with user anonymity and designated recipients. Symmetry 14(5):847
17. Almaleh A, Lahiq S, Al-Shehri F (2023) Smart grid privacy via differential privacy of smart metering data. In: 2023 3rd international conference on computing and information technology (ICCIT), Tabuk, Saudi Arabia, pp 556–561. https://doi.org/10.1109/ICCIT58132.2023.10273939
18. Wang H, Gong Y, Ding Y, Tang S, Wang Y (2023) Privacy-preserving data aggregation with dynamic billing in fog-based smart grid. Appl Sci 13(2):748
19. Yang C, Wu J, Wang L, Zhang X, Li L, Liu S (2020) Smart grid monitoring systems based on advanced encryption standard and wireless local area network. IOP Conf Ser: Mater Sci Eng 719(1):012056. IOP Publishing
20. Beaulieu R, Shors D, Smith J, Treatman-Clark S, Weeks B, Wingers L (2015) The SIMON and SPECK lightweight block ciphers. In: Proceedings of the 52nd annual design automation conference, pp 1–6
21. Beaulieu R, Shors D, Smith J, Treatman-Clark S, Weeks B, Wingers L (2013) The SIMON and SPECK families of lightweight block ciphers. Cryptology e-print archive
22. Suzaki T, Minematsu K, Morioka S, Kobayashi E (2011) Twine: a lightweight, versatile block cipher. In: ECRYPT workshop on lightweight cryptography vol 2011
23. Bansod G, Pisharoty N, Patil A (2016) PICO: an ultra lightweight and low power encryption design for ubiquitous computing. Defence Sci J 66(3)
24. Bogdanov A, Knudsen LR, Leander G, Paar C, Poschmann A, Robshaw MJ et al (2007) PRESENT: an ultra-lightweight block cipher. In: Cryptographic hardware and embedded systems-CHES 2007: 9th international workshop, Vienna, Austria, September 10–13, 2007. Proceedings vol 9. Springer Berlin Heidelberg, pp 450–466

Time Series Analysis and Rule Mining for Detecting Industrial Control System Data Injection Attacks

Merwa Mehmood, Zubair Baig, and Naeem Syed

Abstract Cyber-physical systems (CPSs) are complex interconnections of control systems that operate in tandem to carry out industrial processes and activities. A research problem of interest is the lack of cyber-attack models for a CPS, which hinders the design of robust defense mechanisms. We aim to generate variant attack patterns through the application of association rule mining to a commonly available dataset, namely SWaT. Our methodology comprises the generation of synthetic attack variants to foster the development of robust intrusion detection models. Synthetic samples created by this process can be used to create adversarial samples, which have not yet been addressed in the field of CPS. In this work, we have also generated more than six hundred classification rules and applied time series analysis to validate potential data injection attack variants for the SWaT dataset.

Keywords Attack variants · Rule mining technique · Time-series analysis · CPS

1 Introduction

Cyber-physical systems (CPSs) are complex interconnected physical systems that are logically operating to undertake critical industrial tasks. Some examples of CPS are water treatment and distribution systems, power generation plants, oil refineries, and smart grid [1]. With the common adoption of the Internet of Things (IoT) paradigm in a CPS, many systems and devices are automated and communicate without human intervention, improving operational efficiency.

M. Mehmood (✉) · Z. Baig · N. Syed
Center for Cyber Resilience and Trust (CREST), Deakin University , Geelong, Australia
e-mail: mehmoodme@deakin.edu.au

Z. Baig
e-mail: zubair.baig@deakin.edu.au

N. Syed
e-mail: naeem.syed@deakin.edu.au

S. M. Thampi et al. (eds.), *Fifth International Conference on Computing and Network Communications*, Lecture Notes in Electrical Engineering 1219,
https://doi.org/10.1007/978-981-97-4540-1_20

CPS data is collected in the form of equipment logs and data streams generated by IoT sensors. A significant amount of real-valued data is time series-based in nature [2], i.e., they are categorized and processed based on time stamps. A CPS generates a large volume of data which makes it vulnerable to data-centric cyber-attacks. Traditionally, statistical techniques such as Cumulative Sum (CUSUM) and Exponentially Weighted Moving Average (EWMA) control charts have been adopted to detect anomalies [3]. However, with the integration of IoT devices in CPS systems, i.e., Information Technology (IT)/Operational Technology (OT) systems, it is tedious to detect anomalies through the deployment of conventional cyber security countermeasures. The most efficient method used for analyzing anomalies in communication messages is through machine learning-based Intrusion Detection Systems (IDS). These intelligent methods detect real-time anomalies more efficiently [4] when compared to statistical methods. ML systems learn normal and anomalous patterns from training data to help identify intrusion attempts by an adversary. Recent studies [5, 6] have focused on the potential of cyber-physical attacks in the context of ML-based Intrusion Detection Systems (IDS). These systems are vulnerable to adversarial ML attacks, which are defined to bypass the IDS by introducing maliciously designed input vectors. This can lead to incorrect predictions or classifications by the ML engine. In Industrial Control Systems (ICS), adversarial attacks involve manipulating input, sensor readings, or actuator states to deceive an anomaly detection systems.

Recent literature [7] shows that ML-based IDS have an advantage over conventional IDS as they adopt autonomous extraction of patterns from data through the adoption of techniques such as unsupervised rule mining. Generating real-world attacks for an ICS requires a complex understanding of underlying processes, and it can also cause physical damage to test environments. The use of association rule mining-based techniques fosters the extraction of rules that uncover complex correlations between features, which can then be applied to generate synthetic attack samples. This does not require a deep understanding of the underlying system and therefore is less intrusive during training time. In the context of generating synthetic samples, the work reported in [8, 9] addresses anomaly detection in CPS, by using generative models for understanding the temporal patterns of a dataset. This shows the importance of temporal patterns in detecting attacks in a system. In our contribution, we generate then validate attack rules through temporal pattern analysis, i.e., a time series analysis of individual CPS process features.

This paper is organized into the following sections: Sect. 2 briefly discusses the SWaT system and a detailed discussion on communication networks that underlie a CPS. Section 3 comprehensively explains the SWaT data injection attacks. Section 4 explains the generation of attack variants, and Sect. 5 presents the time series analysis of attack data. Section 6 elaborates on the validation of these generated rules through the temporal consistency of features. Section 7 concludes our work.

2 Background

In [10], the authors used a rule mining technique to generate variants of known attacks. The issue of limited attack patterns in the SWaT dataset was reported. The technique overcomes the limitations of the manual generation of attack rules. It also addresses an attack scenario of tank overflow and underflow of a typical OT facility water treatment plant tank, which is represented through the SWaT dataset. Another work presented in [11] comprises an a priori algorithm of rule mining that automatically identifies the attack state from unlabeled data. They have proposed an approach to automatically detect the fault state of a system through rule mining and labeling of anomalous data samples as *attack*.

SWaT has several stages of operation translated into a dataset. Stage-2 is associated with the chemical properties of the water. Previous literature [12] addressed the issue of the generation of attack variants for tank overflow and underflow attack scenarios across the various stages of the SWaT system. The authors focus on the physical properties of the system as the impact of an attack can be easily represented as anomalous readings in the physical properties of the system. There is less reported work related to the generation of attack variants of chemical processes due to the complexity of these processes in a typical water treatment plant ICS, and the authenticity of these unknown or zero-day attack variants was thus not validated.

In [13, 14], the authors present a specific type of attack, namely a classifier evasion attack. Temporal consistency is an important factor in crafting a synthetic attack sample as it explains any change in sensor reading according to process physics concepts and system control actions. We aim to generate attack variants using an unsupervised mining technique and validate them with the temporal aspect (time stamps) of system data. These validated generated rules are then presented as variant system attack samples.

Overview of SWaT: A Secure Water Treatment Plant The Secure Water Treatment Plant (SWaT) is a fully operational scaled-down version of an industrial water treatment system located at the Singapore University of Technology and Design (SUTD). This testbed has a capacity of producing 18.93 L of water per hour. The SWaT data collection process took 11 days with continuous system operation of 24 h per day. In the first 7 days, routine OT operations generating normal data were collected from the water treatment plant comprising all six steps of the filtration process, and data capture was committed without any disruption. For the next four days, cyber-attacks were launched against the OT facilities with intermittent attack traffic, ranging between a few minutes to an hour, and comprising both single and multiple attack points (sensors or actuators). The total samples of the SWaT dataset are 449,919 across 51 features and one target column representing the sample label as either attack or normal.

There are six stages in a SWaT system shown in Fig. 1, Stage-1 represents raw water features, and Stage-2 comprises the pretreatment process through chemical dosing depending on the quality of the water. Stage-3 is the ultrafiltration process and Stage-4 comprises dechlorination before it is passed to Stage-5 for reverse osmosis.

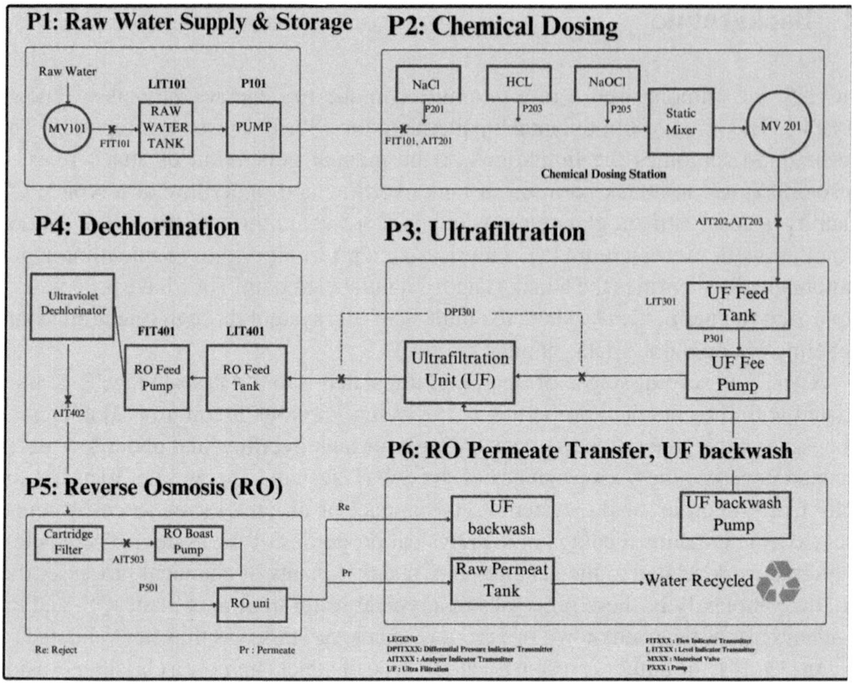

Fig. 1 Six stages of the water treatment plant

The last stage (Stage-6) comprises the distribution of the treated water and back washing of the same to clean up the ultrafiltration unit. The main purpose of this experimental testbed is to enable researchers to validate research for the design of a secure CPS.

3 Control and Communication Network

In Fig. 1, we illustrate the SWaT dataset stages of operation. The SWaT system has two levels of communication networks. Level 1 network includes a Supervisory Control and Data Acquisition (SCADA) server and a historian as part of a star topology that connects Operation Technology (OT) Programmable Logic Controllers (PLCs) across each of the six stages of the process. At each process stage, the PLCs are connected to local sensors and actuators of Level 0. A total of 68 sensors and actuators [15] are included, with some of the actuators working as a standby component, such as the Stage-1 actuator (P102), which is on standby and is activated only when the actuator (P101) fails. There is a separate layer of communication operating between the different PLCs, which functions in tandem with the OT system. The Level 0 communication network is either wired or wireless and comprises sensors, actuators, and PLCs. Manual switches are deployed for control of these devices.

The SWaT control system is configured with dual PLCs, one serving as a primary PLC and the other as a secondary and takes over if the primary PLC fails. Each PLC collects data from sensors and actuators of a given stage (one of six stages, as described previously) of the SWaT system. For instance, water level sensors attached to each water tank measure the level of water once every second. Several other sensors measure the physical and chemical properties of water flowing through each of these six stages of the SWaT system. The SWaT testbed also models false data injection attacks, which are then translated into data samples for inclusion in the training dataset. The wireless communication between the PLCs and sensors, actuators, and SCADA servers is compromised when false data is injected into any of the stages; i.e., a false data injection attack is carried out on the testbed.

4 False Data Injection Attacks

In the context of the SWaT system, false data injection attacks manipulate the communication message between sensors/actuators and control units. The false data introduces anomalies in the system resulting in a disruption of temporal patterns of the features. An experimental investigation of false data injection attacks was carried out on the SWaT dataset [16]. These attacks comprise the injection of fabricated data to disrupt the system's operation. In the case of the SWaT dataset, they include communication messages exchanged between the SCADA servers and PLCs. A total of 36 cyber-attacks were launched at different stages of the SWaT system to understand the system's response to cyber threats, assess the vulnerabilities of the system, and evaluate the robustness of detection systems against these attacks. These false data injection attacks are further categorized into four types of attacks based on the entry points to the SWaT system, to foster exploration of the different attack points exploited by attackers to compromise the integrity of a network. The Single-Stage Single-Point (SSSP) has one attack point in a CPS and includes 26 attack samples of the original dataset. The Single-Stage Multiple-Point (SSMP) has two or more attack entry points but only one stage and includes four attacks. The Multistage-Single-Point (MSSP) has similar attack entry points to SSSP, but these attacks are conducted at multiple stages with two attack types. The Multistage Multi-Point attack is like SSMP, but it is performed at two or more stages with four attacks. The attacks were implemented on the testbed with different attack duration from a few minutes to an hour. The system was allowed to stabilize itself before consecutive attacks or the attacks were launched one after another.

In a real-world scenario, the attacker could apply several strategies to inject false data in PLCs, disrupting the control system or carrying out a reconnaissance for a subsequent attack. Many traditional IDS are signature-based and rely only upon signature databases. Hence, the IDS trained on the current SWaT dataset would be incapable of identifying new attack variants, unknown or zero-day attacks. These attacks could cause undesirable responses or result in component damage. Therefore, relying on fewer attack variants in the original dataset would make the IDS system less robust in detecting variants of attacks. Hence, it is important to understand existing attacks launched against the SWaT testbed and to study the impact of existing or known attacks on all interconnected sensors/actuators in the SWaT testbed. This will enable an understanding of the deviation of variants from known attacks.

One common observation about cyber-attacks launched against the SWaT dataset is that sensors or actuators that are performing the same function at different stages respond to an attack in a similar manner. This suggests a correlation among features of different stages performing the same functions. The underlying correlations between different sensors and their generated data can be exploited to detect anomalies in the system. The main focus of our work is to validate newly generated attack rules based on the rule mining technique that adhere to the temporal consistency of correlated features. This will foster a generation of synthetic samples of existing attacks for the SWaT system, inherently addressing the issue of zero-day attack detection. These attack rules can further be applied in generating adversarial samples in the SWaT dataset. The scope of this work is limited to the validation of attack rules, specifically the features of Stage-2.

Table 1 shows existing false data injection attacks of Stage-2 (water pretreatment phase) which are included in the SWaT system to depict the disruption of water quality. The first attack targeted the entry point (AIT202 pH analyzer) which is a single-stage single-point attack that relies upon a single feature of Stage-2 of the water treatment system. This pH analyzer (AIT202) measures and monitors the concentration of chemicals (HCL) in water. For instance, in this specific interference, the attack point is on a sensor that manipulates the measurement and sets the sensor (AIT202 = 6). As a consequence, the water becomes more acidic. Figure 2a, b shows the original time series of the pH analyzer and P203, the point at which the pH analyzer (AIT202 = 6) causes the actuator state to (P203) to be open, this manipulation in the actuator state will result in a spike in pH after some time, and the values of AIT02 reaches above the high threshold.

The second false data injection attack in Stage-2 targeted the actuators (P203 and P205) which are turned off thus stopping the dosing of chemicals (HCl and NaOCl).

Table 1 Cyber-attacks presented for the SWaT system

P2 chemical dosing station attack				
Attack point	Attack category	Description	attack	Impact
AIT202	SSSP	pH analyzer	Set AIT202 = 6	Altered pH
P203, P205	SSMP	Dosing pumps	Turn off P203, P205	Change in water quality
P201,P203,P205	SSMP	Dosing pumps	Turn on P201, P203, P205	Change in water quality

SSSP-Single-Stage Single-Point Attack
SSMP-Single-Stage Multi-Point Attack

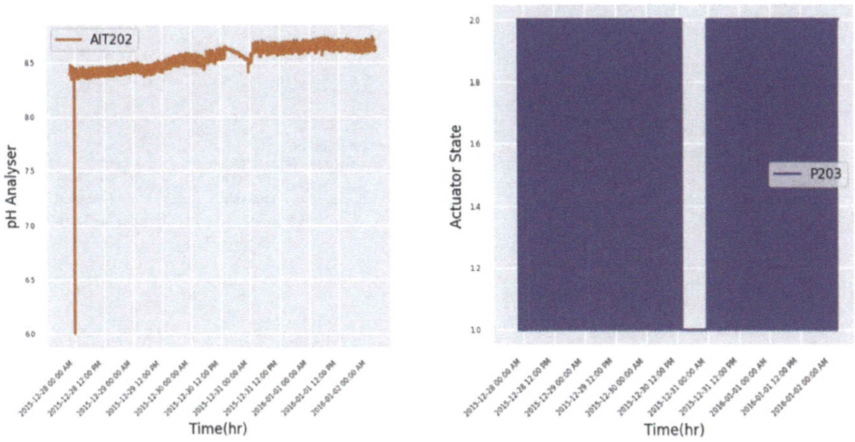

(a) Time series of pH analyzer (AIT202)

(b) Time series of Pump (P203)

Fig. 2 Original time series of pH analyzer and pump

The false information reported to PLC about high concentrations of pH analyzer (AIT202) and ORP analyzer (AIT203) thus indicating above the threshold resulted in the turning of P203 and P205 to OFF state. These disturbed levels of chemicals affect the water treatment process. The last attack on the actuators (P201, P203, P205) is a single-stage multi-point attack targeting two or more attack points/features (sensors or actuators) of the same stage. These attacks manipulate the actuator's state (P201, P203, P205) by turning on these dosing pumps. The control center receives misleading information about low concentrations of chemicals (HCl and NaOCl) below the control design threshold resulting in turning the pumps to an ON state. To enhance the security of the SWaT system by addressing these data false injection we propose a framework elaborated in the next section.

5 Temporal Correlation and Rule-Mining Framework

We propose a method that analyzes the SWaT data injection attack by combining time series analysis with the attack rules generated by the rule mining technique. One of the major concerns in anomaly detectors in CPS is the insufficient number of attack samples. Both researchers and skilled domain experts simulated 36 known attack patterns [17], and these patterns were collected in the SWaT testbed over the period of 4 days by continuously running the OT plant. However, the manual generation of attack patterns produces a limited number of attack samples.

To address this issue, we have used a rule mining technique that was able to generate, i.e., synthesize novel attack rules. These attack rules are based on frequently occurring itemsets in the dataset. These itemsets were then validated with the time

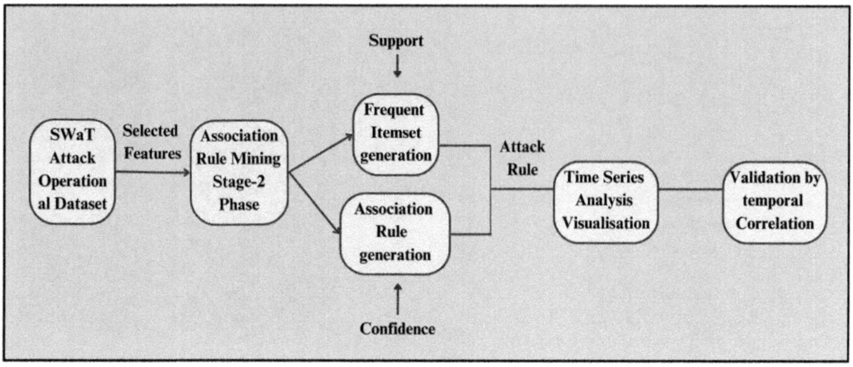

Fig. 3 Temporal correlation and rule-mining framework

series analysis of features of Stage-2. This will help to illustrate the authenticity of the generated rules and will aid in generating realistic synthetic attack samples.

We also aim to investigate whether these resultant attack samples that are not present in the original dataset will be a good indicator of real-world cyber-attacks. To address the issue of the generation of synthetic samples in ICS, our proposed attack model generates an exhaustive list of attack rules and checks the robustness of the anomaly detectors based on the [18] attack generator framework.

The initial step of the proposed framework includes the collection of attack operational samples shown in Fig. 3. There are 54,621'Attack' samples in the SWaT dataset. We aim to investigate Stage-2 (pretreatment phase) of the SWaT system for the generation of synthetic samples. To reduce the feature space and make the problem traceable, the correlation matrix presented highly correlated features of Stage-2 of the SWaT testbed. We reduced the feature space and selected only the highly correlated features for analysis.

5.1 Generation of Attack Rules

For a generation of attack rules, we have selected correlated features. Stage-2 is a pretreatment process that starts with checking the quality of the water before processing it. The raw water (Stage-1) passes through the dosing station, and it has three dosing pumps (P201) for NaCl, (P203) for HCl, and (P205) for NaOCl; these three chemicals are analyzed by the sensors conductivity analyzer (AIT201), pH analyzer (AIT202), and ORP analyzer (AIT203), to maintain water quality within established bounds. There are eleven features (FIT201, MV201, AIT201, P201, P202, AIT202, P203, P205, AIT203, P205, P206) in Stage-2. In our current analysis, the feature MV201 is not significant as we are considering the attack on the chemical analyzer which affects the respective pumps used for specific MV201 chemical dosing, and there were three backup pumps P202, P204, P206 that were also excluded as there was no change in their state (values) across the dataset.

The second step involves the generation of data variants of attack samples by association rule mining technique. The frequent itemsets and rules are generated based on our chosen min-support = 0.5 and min-confidence = 0.5. A total of 67 frequent items and 612 rules are generated by using the FP growth algorithm, which is an efficient algorithm for pattern identification in large datasets. We discretized these features to facilitate the generation of attack rules. The features were discretized based on the design thresholds of the system to facilitate their presentation as inputs to the association rule engine.

6 Time Series Analysis

We perform a time series analysis on six selected features of Stage-2 to understand anomalous behavior and its impact on the sensors and actuators in the SWaT system over time as well as to identify the disturbances in the normal periodic pattern that serve as a means for detecting potential threats to the SWaT system. It also captures the temporal correlation between sensors and actuators helping to understand the system behavior deeply. Combining the time series analysis and rule mining for SWaT data injection attacks effectively identifies the interdependencies among features. To understand the interdependencies among actuators (P201, P202, P203) a multi-point single-state attack shown in Fig. 4 is an attack against the actuators, which is an entry point in Stage-2 of the SWaT system; disrupting the chemical concentrations in water measured by three different analyzers.

The effect of the above attack can be visualized in Fig. 5 that shows a time series plot with highlighted attack area indicating the attack period (three days) of sensors AIT201 (conductivity analyzer), AIT202 (pH analyzer), and AIT203 (ORP analyzer) and their corresponding actuators P201, P203, and P205. In the AIT201 and P201

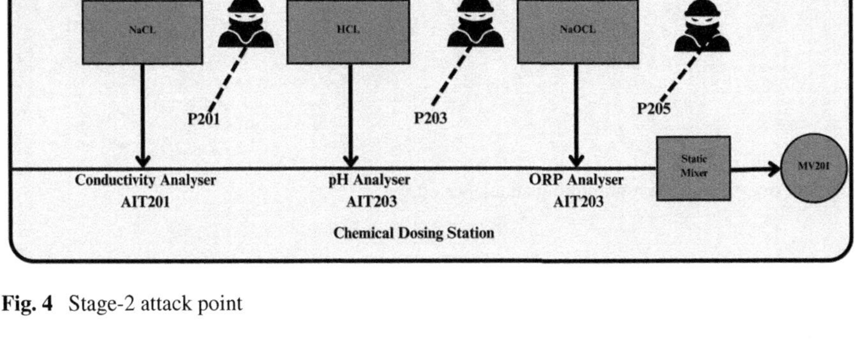

Fig. 4 Stage-2 attack point

plots, a known attack in the SWaT system shows manipulation of the state of P201 by sending a false sensor reading of (AIT201 > = 260) which is a high concentration of the chemical (NaCl). In actuality, the AIT201 was less than a high threshold level which is the highlighted part in the graph as sensor values fluctuate below the design threshold (AIT201 < = 250). At the start of the process, the AIT201 was within high (AIT201 < = 260) and low threshold (AIT201 < = 250) limits, but an attack on the actuator state distorts the chemical dosing.

The AIT202 and P203 plots show a periodic pattern with fluctuations in values above the high pH threshold (AIT202 > = 7.05). The decreased concentration recorded by AIT201 shows that there was a decrease in the concentration of chemical (NaCl) which also affected the pH of the next sensor. This resulted in the pH of water becoming too high. At this specific time, P203 is turned off blocking the chemical dosing of HCl.

Similar behavior can be observed in the AIT203 and P205 plots, with a spike in value above the high threshold (AIT203 > = 380). All these processes run in a cyclic manner, so manipulation in one actuator state affects the chemical analyzer. This temporal consistency shows interdependence, trends, and cyclic patterns of different features of a single stage.

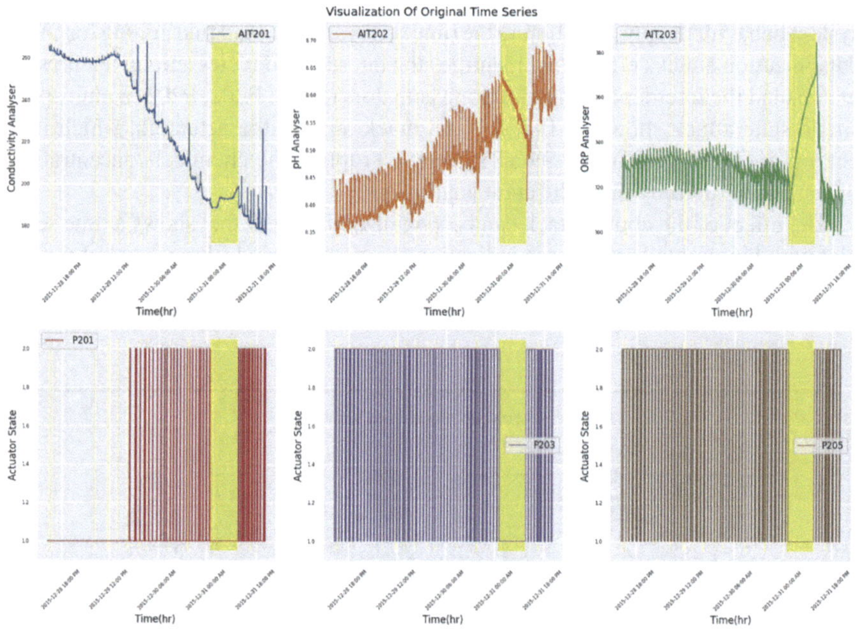

Fig. 5 Time-series plot highlighting attack period

6.1 *Validation of Generated Rules*

The last step of our proposed framework includes filtering attack rules based on the highest confidence values and corresponding support values. We have visualized the plot for confidence, support values of generated rules, and frequency distribution frequent itemsets with their discretized states in Fig. 6a–c. For this defined threshold, the support value is between 0.7 and 0.8 and the majority of rules have confidence values of 1. We can infer from these values the relevance of the generated synthetic rules.

Table 2 shows the generated attack rules for Stage-2 (water pretreatment phase). We aim to analyze the behavior of the pH analyzer in correspondence with the

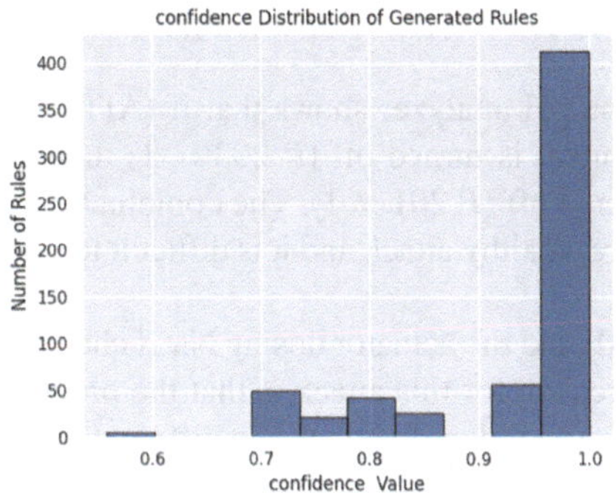

(a) Generated Attack Rules

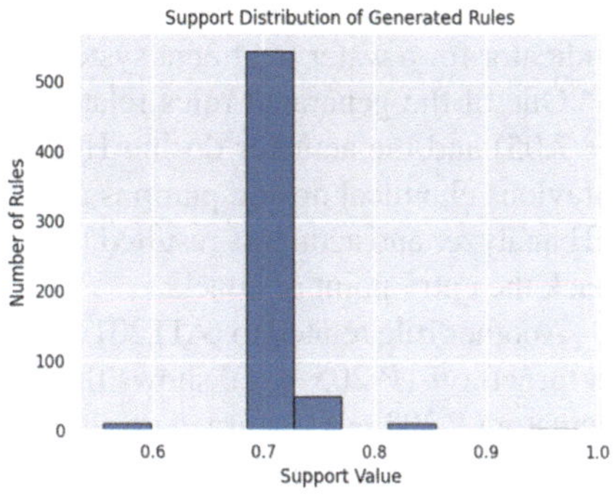

(b) Frequently Occurring Itemset

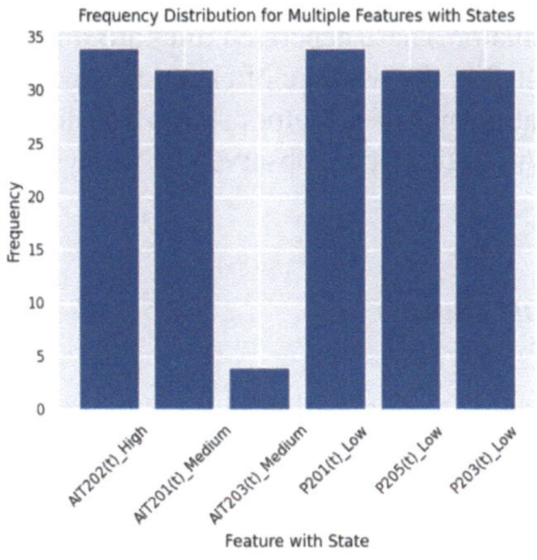

(c) Frequency Distribution of Frequent Itemset

Fig. 6 Generated attack rules and frequent itemset support and distribution

Table 2 Stage-2 generated attack variants

Antecedent	Consequent	Support	confidence
P203(t) = Close	P201(t) = Close	0.732850	1.0
$AIT\ 202(t) >= 7.05$,' P203(t) = Close	'P201(t) = Close	0.7296	1.0
P205(t) = Close	P203(t) = Close	0.729591	1.0
P201(t) = Close,'P205(t) = Close	P203(t) = Close'	0.72959	1.0
'P205(t) = Close', $'AIT\ 201(t) >= 260'$	P203(t) = Close	0.7295	1.0

actuator P203 to show that attack rules generated by the association rule can also lead to a real-world attack. The chemical analyzer (AIT202) measures the pH of the water which is an important factor that makes the water suitable for use. Above and below the high threshold indicates the change in pH of water which is not a suitable indicator for a water treatment system.

One of the generated rules related to the pH analyzer shows that if (AIT202 $>= 7.05$) and the actuator dosing HCl chemical is turned off ($P\ 203 = 1$) then the previous chemical dosing pump is also turned off ($P\ 201 = 1$). The combination of pH analyzer and actuators resulted in a more stealthy attack and it is difficult to trace back the entry point of attack.

Another rule related to (AIT201 $>= 260$) and the actuator dosing NaCl chemical is turned off ($P\ 205 = 1$), shows that the reason for this attack is that the previous actuator ($P\ 203 = 1$) is turned off. It is difficult to identify attacks on different points of one stage. To recover from a multi-point attack is difficult, as it becomes difficult to trace back the attack point entry. Generated rules related to actuators (P201, P203, and P205) show that the state for actuators is turned off which means that if the attacker intends to change the state of P203 then our first actuator is also closed and not dosing chemicals in the water. Similarly, the closed state of P205 results in the closing of the P203 actuator. These generated rules are similar to our existing attacks on actuators that resulted in the wastage of chemicals by not adding any chemicals to the water. This relationship of actuators shows that it is easy to manipulate the state of the actuator with little impact observed by the system.

6.2 *Temporal Correlation*

Figure 6c shows a frequency distribution plot for the frequent itemsets with their frequently occurring states. These states are discretized based on design thresholds. For example, ($P\ 201 - low$) implies a closed state of P201. It is evident that the closed state of the pump resulted in an attack. High values of AIT202 and the values between low and high thresholds resulted in an attack in AIT203 and AIT201. It is validated by the time series analysis of these selected features shown in Fig. 5. By comparing both Figs. 6c and 5 it is evident that the frequently occurring itemsets having high frequency in data are related to the attack points identified in the time series analysis of these features. (AIT202-High) has the highest frequency which shows that a pH value above the design threshold has an attack with the corresponding actuator (P201-low) in the closed state. (AIT203-medium) is a feature that occurs less frequently among all features. The frequency of AIT201 and AIT202 is almost similar, showing that these features are correlated with each other. The interruption in the first chemical analyzer (AIT201) directly affected the second chemical analyzer (AIT202), and this effect was reduced in the case of the third chemical analyzer (AIT203). The generated attack rules are variants of existing cyber-attacks that show temporal consistency.

7 Conclusion

This paper presents an effective approach to understanding SWat data injection attacks by combining temporal analysis and association rule mining techniques. It further validates the newly generated attack variants for Cyber-Physical systems based on a rule mining technique. Through this analysis, we have observed the temporal consistency in our generated attack rules. We were able to detect potential synthetic samples that could lead to previously unknown (zero-day) attacks. We also identified those samples that appeared frequently in the attack sample set that showed temporal consistency. These samples that appear often will be used to create adversarial attacks as part of our future work.

References

1. Wang Z et al (2021) A survey on recent advanced research of CPS security. Appl Sci 11(9):3751
2. Ahmed CM, Palleti VR, Mathur AP (2017) WADI: a water distribution testbed for research in the design of secure cyber physical systems. Proceedings of the 3rd international workshop on cyber-physical systems for smart water networks
3. Sun B et al (2013) Building energy doctors: an SPC and Kalman filter-based method for system-level fault detection in HVAC systems. IEEE Trans Autom Sci Eng 11(1):215–229
4. Kwon D et al (2019) A survey of deep learning-based network anomaly detection. Cluster Comput 22:949–961
5. Chen J et al (2020) Generating adversarial examples against machine learning-based intrusion detector in industrial control systems. IEEE Trans Dependable Secure Comput 19(3):1810–1825
6. Gomez ALP et al (2021) Crafting adversarial samples for anomaly detectors in industrial control systems. Procedia Comput Sci 184:573–580
7. Al-Garadi MA et al (2020) A survey of machine and deep learning methods for internet of things (IoT) security. IEEE Commun Surv Tutorials 22(3):1646–1685
8. Li D et al (2019) MAD-GAN: multivariate anomaly detection for time series data with generative adversarial networks. International conference on artificial neural networks. Springer International Publishing, Cham
9. García Gonzalez G et al (2021) On the usage of generative models for network anomaly detection in multivariate time-series. ACM SIGMETRICS Perform Eval Rev 48(4):49–52

10. Umer MA et al (2021) Attack rules: an adversarial approach to generate attacks for industrial control systems using machine learning. Proceedings of the 2th workshop on CPSIoT security and privacy
11. Khalili A, Sami A (2015) SysDetect: a systematic approach to critical state determination for industrial intrusion detection systems using Apriori algorithm. J Process Control 32:154–160
12. Adepu S, Aditya M (2016) An investigation into the response of a water treatment system to cyber attacks. 2016 IEEE 17th international symposium on high assurance systems engineering (HASE). IEEE
13. Aoudi W, Mikel I, Magnus A (2018) Truth will out: departure-based process-level detection of stealthy attacks on control systems. Proceedings of the 2018 ACM SIGSAC conference on computer and communications security
14. Erba A, Tippenhauer NO (2022) Assessing model-free anomaly detection in industrial control systems against generic concealment attacks. Proceedings of the 38th annual computer security applications conference
15. Umer MA et al (2020) Generating invariants using design and datacentric approaches for distributed attack detection. Int J Crit Infrastruct Prot 28:100341
16. Goh J et al (2017) A dataset to support research in the design of secure water treatment systems. Critical information infrastructures security: 11th international conference, CRITIS 2016, Paris, France, October 10–12, 2016, Revised selected papers 11. Springer International Publishing
17. Adepu S, Aditya M (2016) Distributed detection of single-stage multipoint cyber attacks in a water treatment plant. Proceedings of the 11th ACM on Asia conference on computer and communications security
18. Mehmood M, Baig Z, Syed N. Securing industrial control systems (ICS) through attack modelling and rule-based learning'. 16th international conference on communication systems and networks (COMSNETS)

Radius-Based Authentication: UFrames' Novel Approach to Prevent Account Sharing

D. Sai Vardhan, Doddapaneni Tanmayi, Kakaraparthi Rakesh, and B. Sreevidya

Abstract OTT platforms such as Netflix and Amazon Prime have been in the market for more than a decade. They have come to full-fledged usage especially during the lockdown. The main reason for their magnanimous success is ease of use and feasibility to access unlimited content anywhere and anytime with basic Internet connection. The features in OTT have made accessing the content and viewing the content easy and delightful for the users. However, the "sharing account" option enabled for a single user to access the content from multiple devices is instead being used to share one account among multiple users, negatively affecting the business of OTT creators. In this paper, the focus is to build an OTT platform that proves advantageous to both users and providers. Besides the prevailing features that are found across platforms, including sign-in and sign-up functionalities, categorized content selection, and payment options, the proposed system enables automatic sign-out when the non-primary device strays beyond range from the primary device, content uploading, and more. Furthermore, there is a proposal to implement a GPS technology-based mechanism aimed at preventing users from sharing their accounts with others. The proposed approach conceptually and experimentally proves to be beneficial to the OTT platforms from a business perspective. The scheme is able to prevent users from sharing the account beyond a geographical limit, thus improving the profit. At the same time, these stringent requirements are developed in such a way that a balance is made between user experience, ease of use experience along with improved security measures.

Keywords Data mining · Prediction · Naïve Bayes · ID3 · WEKA · R studio · Confusion matrix

---

D. S. Vardhan · D. Tanmayi · K. Rakesh · B. Sreevidya (✉)
Department of Computer Science and Engineering, Amrita School of Computing, Amrita Vishwa Vidyapeetham, Bengaluru, India
e-mail: b_sreevidya@blr.amrita.edu

S. M. Thampi et al. (eds.), *Fifth International Conference on Computing and Network Communications*, Lecture Notes in Electrical Engineering 1219,
https://doi.org/10.1007/978-981-97-4540-1_21

1 Introduction

The acronym "OTT" refers to "Over-The-Top," portraying a contemporary digital content delivery method that skips traditional cable or satellite TV providers to distribute media content to consumers over the Internet. OTT platforms provides a diverse range of content including movies, TV series, live sports, and exclusive programming; they are mostly subscription-based. As long as the viewers/users have an Internet connection, viewers may use OTT platforms to watch content at their convenience. These platforms also facilitate provision to access the content on the platform of viewers' choices as they are available across variety of devices including smartphones, tablets, smart TVs, and streaming devices.

The media landscape has been significantly altered by OTT platforms, which have changed how content is created, shared, and consumed. For instance, several OTT platforms have now ventured into making their own content; this has upended the established studio structure, thus giving new opportunities to the performers and directors. The availability of OTT platforms on a variety of devices, such as smart TVs, smartphones, tablets, and streaming gadgets like Roku and Apple TV, is another benefit. Whether they are at home or on the road, viewers can easily access their favorite material on their preferred device thanks to this.

OTT platforms have made significant investments in creating unique content that is only available on their platform. This tactic has worked well because viewers are ready to pay various OTT subscriptions to obtain original programmes. Also had global reach that they are able to provide material to audiences in many nations and locations. Because many OTT platforms cater their content to local consumers, this has helped to create a broad and global audience.

Due to its adaptability, affordability, original content, device compatibility, global reach, and data-driven insights, OTT platforms have found success. These elements have upended the conventional TV business and facilitated consumers' access to top-notch entertainment on their terms. It is obvious that OTT platforms have emerged as a major player in the entertainment sector and will continue to influence how we consume media content as their popularity soars.

There may be many a number of motivations for developing an OTT platform. Since OTT platforms are becoming more and more popular, creating one can represent a valuable business opportunity. Successful OTT platforms might generate income via partnerships, advertising, and subscriptions, making them a desirable alternative for business owners trying to get into the digital media market. The initiative can be motivated by a love of media material and a want to provide consumers a better watching experience. The development of an OTT platform allowed the project's authors to promote their preferred material while providing viewers with a more personalized and flexible watching experience. The ambition to innovate and develop new technologies that enhance the way we consume media content may be the driving force behind the development of an OTT platform.

The idea could be driven by a desire to encourage social impact by giving marginalized voices and stories a platform. Project creators might support social change by providing a platform that encourages varied and inclusive content, which would assist to subvert the media industry's established power structures. In general, the reason for developing an OTT platform as a project may be fueled by a desire to launch a profitable business, advance cutting-edge technology, create a love of media content, or enhance social effect.

There are numerous OTT platforms available today in the market such as Netflix, Amazon Prime, and Disney Plus Hotstar. They have most of the features in common except for a few such as display and subscription plans. So, coming up with an OTT platform, which not only includes the features that are most common and appreciated but also additional features that never existed before, was a difficult task. The implementation of user location tracking and updating feature was a little challenging as accessing the location by running the app in background and also updating the location in the app while it is not in use was critical due to permissions restrictions.

The payment gateway integration procedure has also got complicated due to limited resources since it is still a small-scale project.

2 Related Works

The paper [1] discusses the design and implementation of a video streaming platform using a distributed environment in a cloud platform. The suggested platform is intended to handle the resource constraints and heavy network traffic of conventional video streaming platforms using the AWS services such as Amazon S3 and LightSail. The platform is made up of a load balancer that distributes the workload among the servers and a number of servers positioned in various places. In order to prevent overload or underutilization, the load balancer is intended to make sure that the workload is distributed equally among the servers.

A content distribution network (CDN) is also used by the suggested platform to guarantee high-quality video streaming. In order to decrease network latency and enhance the overall video streaming experience, the CDN caches video content on edge servers that are closest to the end users. The proposed work also uses an adaptive bitrate streaming algorithm to adjust the streaming quality of video content based on the network signals. The experiment findings used to gauge the performance of the platform under consideration are presented in the paper. The outcomes demonstrate that the platform can efficiently and quickly manage numerous requests for simultaneous video streaming. This paper gave a firm understanding of the power of the cloud-based services and the usage of open sources such as AWS and FireBase to come up with an efficient solution to store the datasets that are part of the OTT platform.

The study in this paper [2] clearly demonstrates that the current generation of gadgets includes an integrated GPS system, which we can use for authentication for more secure considerations as well as for the approach. Verifying a user's or system's identity through authentication includes checking a set of credentials, like a username and password. To ensure that only authorized individuals or systems can access sensitive information, security is crucial in authentication.

We use two factors of authentication, OTP and location, to provide additional secure authentication. Many factors make one-time password (OTP) a more secure authentication approach than conventional login and password-based authentication. Because OTP is a one-time use password, it can only be used once and for a short amount of time. Traditional usernames and passwords, on the other hand, can be misused or stolen.

To further increase the security of your system or application, location-based authentication can be a helpful complement to OTP authentication. These are a few major advantages of location-based authentication. By forcing users to be at a certain area in order to access the system or application, it adds an extra degree of protection to your authentication process. Even with the right OTP, this makes it harder for attackers to acquire illegal access. Location is used here for security as well, and the strategy we are taking for the project is to limit password sharing. To do this, we access the location of the main device and look for any login attempts. The user is logged in if the new device is within the threshold radius.

The historical development and mathematical methods employed in astronomy and navigation are examined in the article "From Navigation to Star Hopping: Forgotten Formulae" [3]. The author contends that conventional methods and formulas that were once employed by ancient mariners and astronomers have been neglected in contemporary navigational and astronomical practices. The introduction of the article discusses the evolution of navigation and the use of celestial objects for positioning, including navigation using the North Star and other stars. The author then goes on to talk about how astronomy has evolved, including how star charts are used and how the motion of the stars is measured.

Despite their value in astronomy and navigation, the author contends that many of these mathematical methods have been lost or forgotten over time. This article helped to narrow down the algorithm to calculate distance to implement the restriction on sharing accounts option feature in the OTT platform.

The study "Application of Firebase in Android App Development" [4] examines the use of Firebase, a platform for building cloud-based mobile and web apps, in Android app development. The authors talk about Firebase's numerous benefits, including real-time database, cloud storage, authentication, and cloud messaging. The authors also go over how these elements can be used when creating Android apps and give instances of when they have been applied in actual applications.

The writers concluded by providing a case study of an Android application created with Firebase. They mentioned about how Firebase was used to execute the app's features, such as real-time synchronization of user data and the use of Firebase Cloud Messaging for push notifications. This paper helped to understand and successfully implement the FireBase in the backend of the OTT application as the database to store movie-related information, users data, and devices data.

The paper "Building Interactive Web Applications with Streamlit" [5] gives a basic overview of Streamlit, a Python framework for creating web applications. The authors highlight the salient characteristics of Streamlit, such as its straightforward syntax, capacity for rapid app prototyping and deployment, and interactive widgets for data visualization. The paper offers a step-by-step tutorial on how to create a straightforward Streamlit programme, including data import, data visualization, and interactive widget addition. The writers also go into more complicated elements like cloud deployment and service integration for the app.

By showing a number of sample apps, such as a data exploration app, a machine learning app, and a real-time collaborative app, the authors demonstrate the possibilities of Streamlit. The exploration of Streamlit's potential applications in business and academics, as well as a request for community involvement and continued development, round out the study. Overall, the paper provides a useful manual for creating interactive web applications using Python while offering a thorough introduction to Streamlit and its features.

T. Jiang and Q. Wang's paper "Analysis of User's Willingness to Pay for Video Content on Over-the-Top Platforms" [6] looks into users' willingness to pay for video content on OTT platforms. The writers begin by outlining the OTT market and its expansion before going over the various pricing strategies employed by OTT providers, such as subscription-based, advertising-based, and pay-per-view models. The findings of a survey done among Chinese OTT users to determine their readiness to pay for video con-tent are then presented by the writers. According to the poll, people are typically willing to pay for video content on OTT platforms, with the vast majority of participants saying they would be happy to pay less than 50 yuan (about $7) per month. Users who are happier with the OTT platforms' variety of content and level of service are more likely to be willing to pay.

In the realm of user authentication, there is a considerable body of research investigating various techniques. Currently, textual password-based schemes are the most widely adopted method for authentication, and they have been extensively studied. However, these textual passwords have been found to suffer from several drawbacks, as reported in prior works. Specifically, issues related to usability [10], memorability [11], and vulnerability [12] have been identified. To overcome the limitations of textual passwords, researchers have turned their attention to graphical passwords as a potential alternative. Graphical passwords have gained significant attention due to their ease of use and improved memorability compared to textual passwords. They rely on users selecting images, patterns, or symbols as their authentication credentials. Despite their advantages, graphical passwords are not without their vulnerabilities.

Notably, they are susceptible to shoulder surfing and observation attacks [7, 8], where malicious actors may gain unauthorized access by spying on users' actions during the authentication process. Overall, while graphical passwords show promise as a user-friendly alternative to textual passwords, it is essential to consider their security implications and explore ways to mitigate the risks they present.

The authors employ statistical research to pinpoint important variables like service quality, content diversity, and convenience that affect users' willingness to pay. Additionally, they talk about how user willingness to pay is affected by piracy and offer recommendations for OTT providers on how to combat it, such as by lowering prices and enhancing the caliber of their legal content. In their last section, the authors explored the significance of their findings for OTT providers and provided advice on how to boost user engagement and income on OTT platforms by, among other things, personalizing recommendations and enhancing user interface. In general, the study offered perceptions into user willingness to pay for video content on OTT platforms and emphasized on the factors which determine the user's spending pattern. This information is so valuable that it helped to make decisions on how to create pricing plans and raise user engagement on the services on the UFrames OTT platform.

3 System Description

Fig. 1 describes the low-level architecture design of the UFrames application (Table 1).

It covers all the modules in the application such as log in, watch movies, subscription plans, and managing devices. The users can log into the app using phone number and OTP sent to the number. The application first checks in the background if the user is new or already existing based on the records stored in Firestore database in the "users" collection. If the user is new, a new account is created, and the credentials are stored in the Firebase as the primary user. Otherwise, the account sharing restriction feature is implemented. It first checks if the user trying to log in is the primary user or not. If yes, they are allowed to log in without any restriction. If not, it checks if the device is in within 20 kms radius of the primary device. If not, they are not allowed to log in into the app. If yes, they can use the features of the app. If there are no primary users for the account, the device trying to login is made the primary user.

On successful login, the user is navigated to the main page where movies are segregated based on popularity, release date, genre, and language. The user can watch a movie by simply clicking on it. Once the user clicks on a movie, its name and release year are sent as keys to the Firestore database. The movie-related information such as language, release year, rating, and overview along with a play button are displayed on the screen. When the user clicks on the play button, the code first checks if he/she is authorized to access the video content. This is decided based on the subscription plan chosen and the movie's type. If the user is eligible to view the content, the movie is retrieved from the AWS S3 bucket and starts playing on the screen which the user can control using media controls such as play, pause, slider, and full screen mode. If the user is not eligible to access the content, he/she is redirected to the subscription plans page.

When a user subscribes for a plan, the code first checks the data in the Firebase to verify if he/she is already on a plan. If a plan already exists but time has expired or no plan exists, the user needs to play the full amount. If the user tries to upgrade a plan before the current plan expires, only a partial amount to fulfill the payment requirement of the upgraded plans should be paid. When a user logs out, the app first checks if it is a primary user trying to perform the action or not. If yes, then the

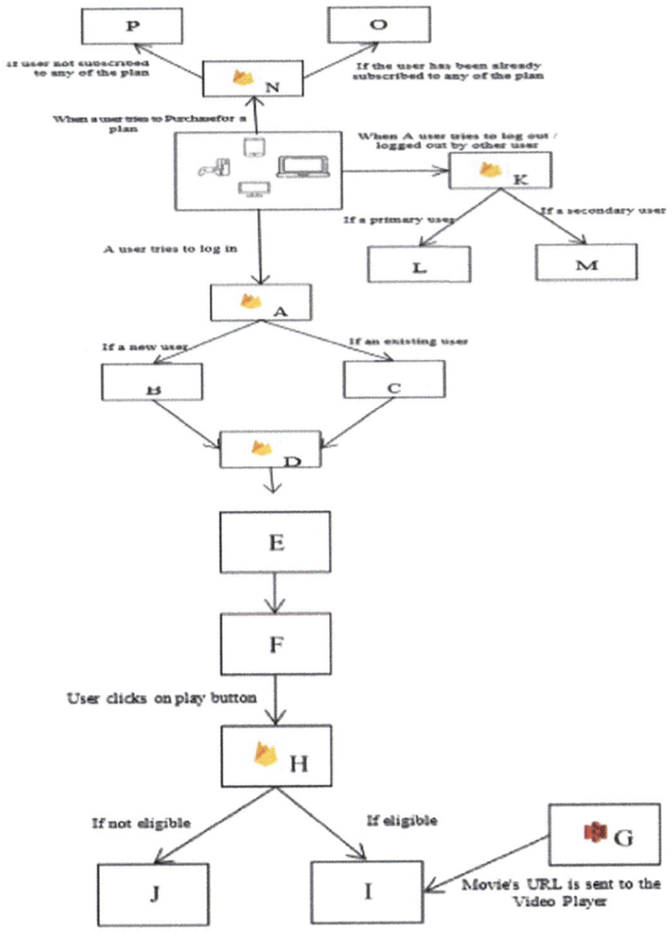

Fig. 1 Low-level architecture design

user's credentials are completely removed from the database, and a secondary user takes his/her place. Otherwise, the user's credentials are simply removed from the database.

The admin can handle the operations in the app via the Web site. The admin can upload a movie in the "Upchatload" section in the site by filling basic details such as name, language, release year, video file URL, and movie's type and date from which the movie should be made available in the app. Once the form is submitted, the code checks if the movie with the same name, language, and year already exists in the database. If it does, then the movie will not be uploaded, and an error message "Movie already ex-ists" will be displayed on the screen.

If the movie does not exist, the code first fetches the additional information not provided in the form from the The Movie Database (TMDb) API and stores it in the

Table 1 Description of symbols in Fig. 1

Symbol	Description
A	Determine if an existing account or new account based on database (Firebase)
B	Create an account, log in, and make the device a primary user in the FireBase
C	If there are no existing devices logged in that account (or) if the device carries a registered mobile number then that device will be updated as a primary device and logs in In other cases, the device is treated as secondary device and logs in, only if the account's primary device is within its radius and also based on the account subscription model
D	The app's main page is displayed on the screen, including movies segregated in terms of genre and language. The movies in each section are ordered based on popularity
E	User requests for a movie
F	Entire movie-related information is displayed on the screen with a play button
G	Movie files are stored in Amazon S3 can be accessed through a URL
H	Checks if the user is eligible to watch the movie based on his/her subscription plan
I	Video content retrieved from the S3 bucket starts playing on the screen
J	The user is redirected to purchase the subscription plans page
K	Checks if the user is a primary user in FireBase
L	Sign-out the primary device and update the first secondary user as the new primary user
M	Sign-out from UFrames
N	Checks if the user already has a plan or not
O	The plan is upgraded. Only a partial amount is requested, based on the amount paid for the existing plan and the expiry date
P	The user can subscribe to the selected plan (basic/super) by paying the respective amount

Firestore database. The movie is not stored in the Firebase, instead uploaded into the AWS S3 bucket, and only the movie video's URL is stored in the database. The admin can also handle the subscription plans and their prices in the "plans" section of the Web site.

4 System Implementation

The Haversine formula is a key component of our UFrames application. It allows us to accurately calculate the distance between two points on the Earth's surface, considering its curvature. In UFrames, we leverage the Haversine formula to enforce proximity restrictions for secondary device access. By using latitude and longitude coordinates, we ensure that secondary devices remain within the designated radius of the primary device. This enhances account security and prevents unauthorized

Haversine formula:

$$a = \sin^2(\Delta\varphi/2) + \cos\varphi_1 \cdot \cos\varphi_2 \cdot \sin^2(\Delta\lambda/2)$$

$$c = 2 \cdot \mathrm{atan2}(\sqrt{a}, \sqrt{(1-a)})$$

$$d = R \cdot c$$

where φ is latitude, λ is longitude, R is earth's radius (mean radius = 6,371km); note that angles need to be in radians to pass to trig functions!

JavaScript:

```
const R = 6371e3; // metres
const φ1 = lat1 * Math.PI/180; // φ, λ in radians
const φ2 = lat2 * Math.PI/180;
const Δφ = (lat2-lat1) * Math.PI/180;
const Δλ = (lon2-lon1) * Math.PI/180;

const a = Math.sin(Δφ/2) * Math.sin(Δφ/2) +
          Math.cos(φ1) * Math.cos(φ2) *
          Math.sin(Δλ/2) * Math.sin(Δλ/2);
const c = 2 * Math.atan2(Math.sqrt(a), Math.sqrt(1-a));

const d = R * c; // in metres
```

Fig. 2 Implementation

access. The Haversine formula plays a vital role in providing a seamless and secure streaming experience for our users (Fig. 2).

Example scenario: There is a customer named "A" who lives in Hyderabad. He has a friend named "B" who lives in Bangalore. In regular OTT platforms, they both can use the same account. On the other hand, in UFrames, user "B" will be able to access user "A"'s account only if he/she is in 20KM radius of user "A". Hence, user "B" will be automatically signed out from the account.

The scenario discussed in Fig. 3 is implemented using the code shown in the Fig. 4 every time the user tries to use the application as can be shown in Fig. 4 an alert message saying "Due to UFrames Account Violations, Your Session has been Expired" will be popped up on the screen. On pressing "OK", the user gets automatically signed out.

5 Results and Discussion

The user can log in into the app via phone number as shown in the left of Fig. 5. If the phone number is valid, an OTP is sent to the user's mobile. The user is automatically redirected to the OTP page as shown in center of the Fig. 5. The login button remains disabled as long as a valid OTP is entered. Once the OTP is verified, the user is logged in, and the home page shown in the right of Fig. 5 is displayed on the screen where movies are segregated based on popularity, genre, language, and type.

When clicked on a particular language, all the movies under that language would be displayed on the screen as shown in the left of Fig. 6. Similarly, when clicked on

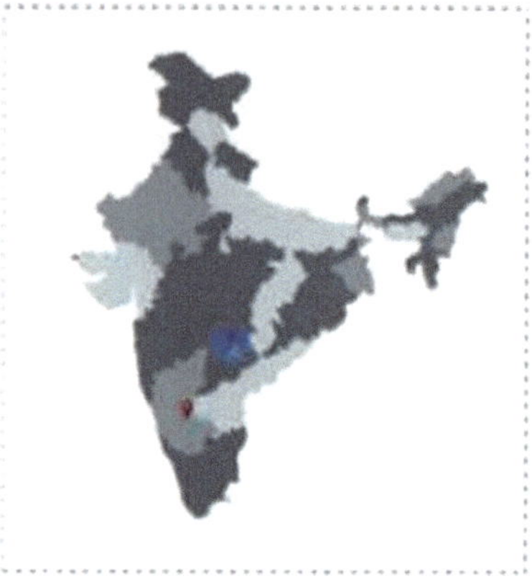

Fig. 3 Example of account sharing restriction mechanism

Fig. 4 Implementation of account sharing restriction mechanism

a particular genre, all the movies under that genre would be displayed on the screen as shown in center of Fig. 6. The user can also search for a particular movie of their choice by searching for it in the search page as shown in the right of Fig. 6. The page by default would display all the movies in a three-column format.

The movies with the most similar name as the one searched for would be displayed on the screen. On the left of Fig. 7, the user gives the input as "P", and the movies whose name starts with "P" are displayed on the screen. Once the user decides which movie to watch, he can simply click on it. The information related to the movie such as title, rating, language, release year, and overview along with a play button would be displayed on the screen as shown in the center of Fig. 7. The user can click on the play button to start watching the movie. If the user is eligible to watch the movie, the movie starts playing in the landscape mode by default as shown in the right of Fig. 7. Otherwise, they are directed to the subscription plans page.

The video can be played in fullscreen mode as well by clicking on the fullscreen button as shown in Fig. 8. The user can also pause the film and can also control the point from which video should be played using the slider as shown in the right of Fig. 8.

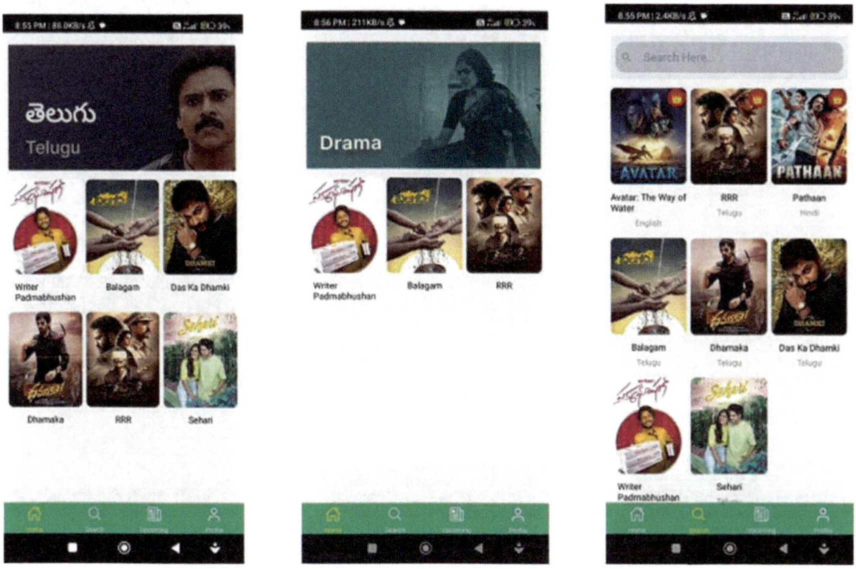

Fig. 5 Login, OTP, and homepage

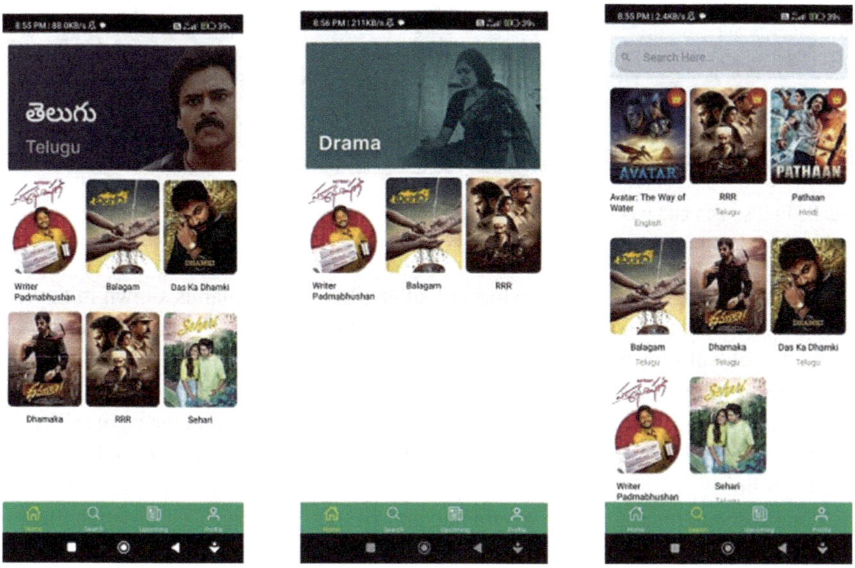

Fig. 6 Language, genre, and search page

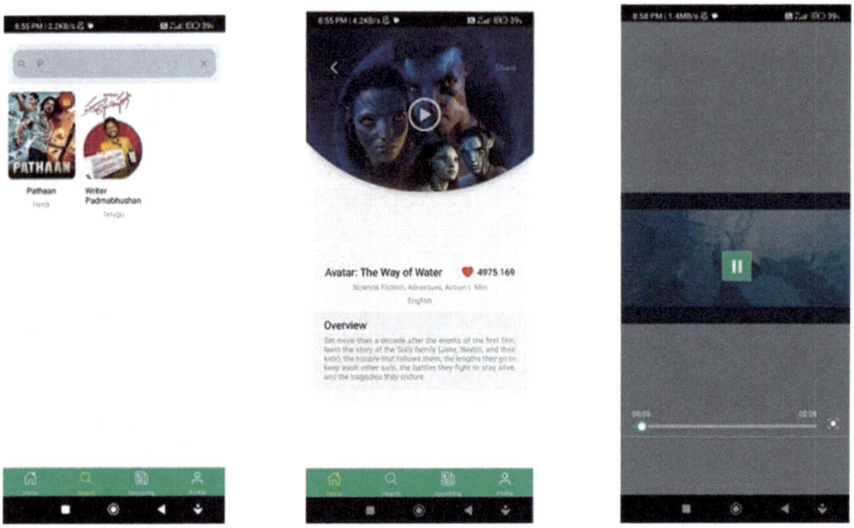

Fig. 7 Search result, movie display page, and landscape mode in video player

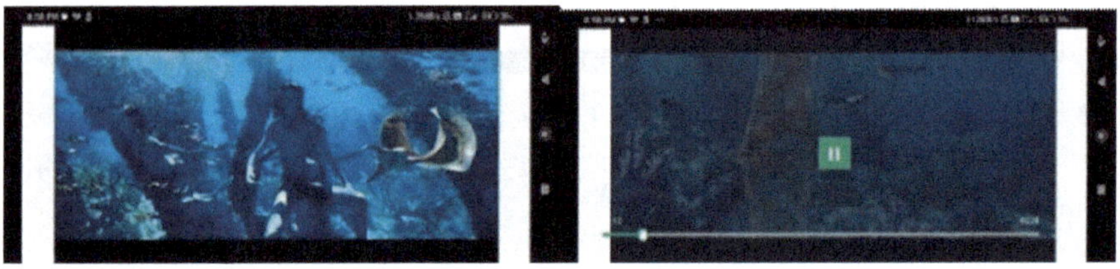

Fig. 8 Full screen and pause mode in video player

The profile page of a user who has not subscribed to any plan is shown in the left of Fig. 9. The default user subscription plan would be "free". The profile page of an account with subscription plan is shown in the center Fig. 9. The subscription plan type along with the number of days left for it to expire are displayed on the screen. The manage devices pages are displayed on the screen as shown in the right of Fig. 9. The primary user is marked with a green symbol while other users are marked with a red symbol.

The subscription plans page would be displayed on the screen as shown in the left of Fig. 10. The user can pay for a plan by simply clicking on the Pay Now button. The user is then redirected to the payments page as shown in the right of Fig. 10. The desired plan can be activated by providing a few bank details.

Fig. 11 showcases a prominently displayed alert message within the UFrames application interface. The alert is titled "Session Expired" and serves as a clear indication that the user has contravened UFrames' policy. It provides a concise and informative message notifying the user of the violation. The alert prompts the user to

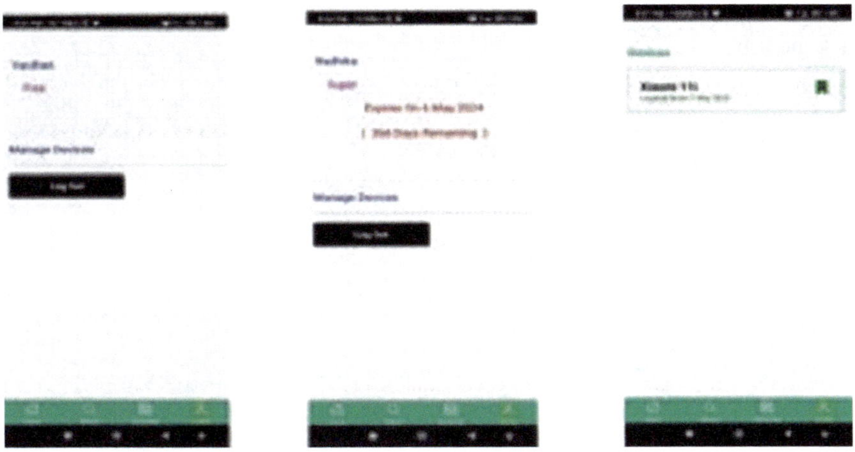

Fig. 9 Profile page with subscription plans and manage devices page

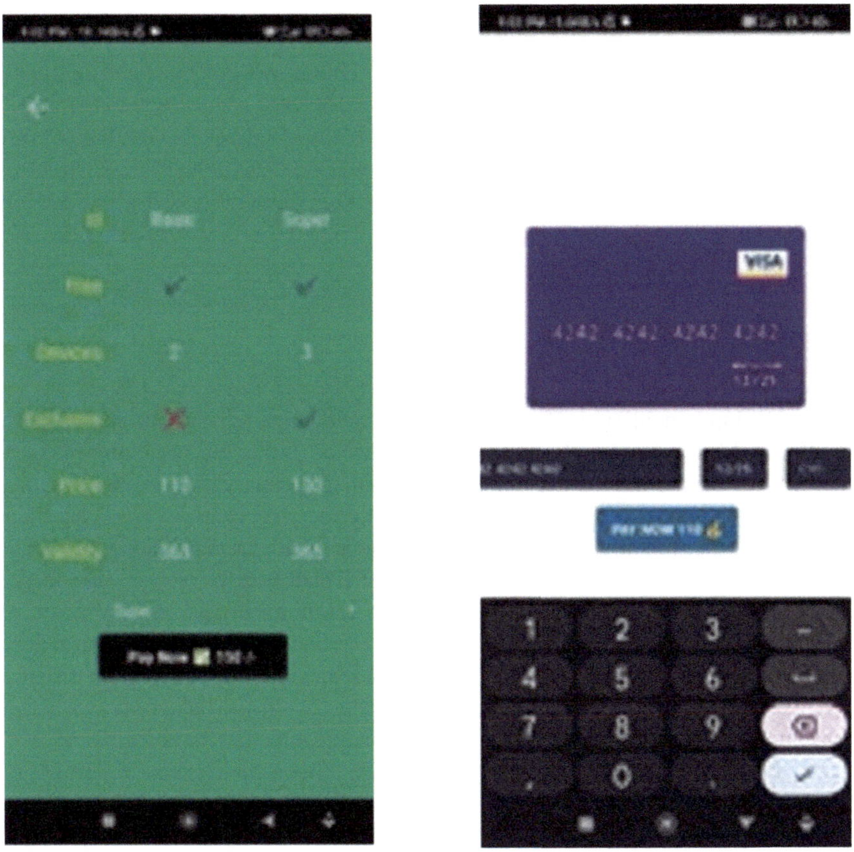

Fig. 10 Subscription and payments page

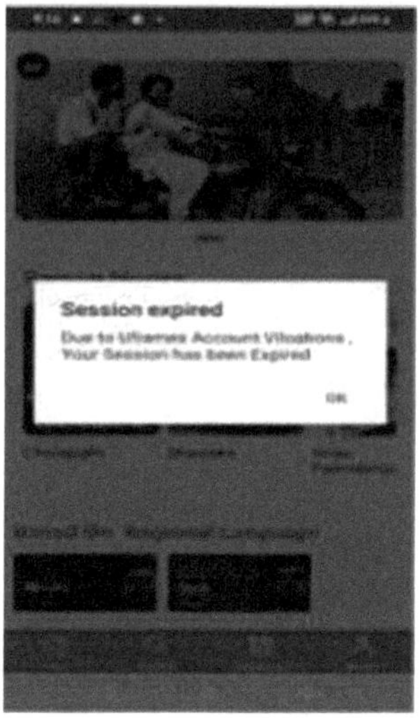

Fig. 11 UFrames policy violation alert

acknowledge the violation by clicking the "Ok" button, which triggers an automatic logout from their UFrames account. This alert is triggered when a secondary user is either intentionally removed from the account by any of the associated devices or if they have exceeded the permissible distance from the primary device. By enforcing this policy, UFrames ensures strict adherence to their guidelines on account sharing and maintains the integrity of their services.

6 Conclusion

UFrames' implementation of a radius-based authentication system represents a significant development in deterring account sharing within the streaming industry. By verifying the proximity of secondary devices to the primary account owner's location, UFrames aims to offer a unique and controlled account sharing experience. This comparative analysis highlights the contrasting account sharing policies of regular OTT platforms and UFrames; while regular OTT platforms prioritize convenience and flexibility, UFrames adopts a stricter approach to ensure account security. As the streaming industry continues to evolve, innovations in account sharing practices, like UFrames' radius-based authentication, will shape the future of content distribution while maintaining a sustainable business model, ensuring a balance between user convenience, personalized experiences, and revenue protection for content providers.

In UFrames, while the current implementation utilizes the Haversine formula for distance calculations, there are alternative methods that can be explored to enhance accuracy and efficiency. For example, integrating advanced geolocation algorithms or incorporating machine learning algorithms could potentially improve the precision of distance measurements between primary and secondary devices.

To optimize the performance and minimize the load on the application, the frequency of device location updates can be dynamically adjusted based on factors such as device activity, network conditions, and user preferences. This adaptive approach would help strike a balance between real-time updates and resource utilization, ensuring a smooth user experience.

These potential future enhancements aim to improve the accuracy, efficiency, user experience, and account management capabilities of UFrames. By leveraging innovative technologies and user-centric features, UFrames can continue to evolve and provide an enhanced streaming experience for its users.

References

1. Korupolu M, Jannabhatla S, Kommineni VS, Kalyanam H, Vasantham V (2021) Video streaming platform using distributed environment in cloud platform. 7th international conference on advanced computing and communication systems (ICACCS), vol 1. IEEE, pp 1414–1417
2. Nisar F (2019) Location based authentication service using 4G/5G devices. International conference on communication technologies (ComTech). IEEE, pp. 120–126
3. Shylaja BS (2015) From navigation to star hopping: forgotten formulae. Resonance 20(4):352–359
4. Khawas C, Shah P (2018) Application of firebase in android app development—a study. Int J Comput Appl 179(46):49–53
5. Khandelwal U, Kevin C, Dan J, Lukasz K (2019) Sample efficient text summarization using a single pre-trained transformer. arXiv preprint arXiv:1905.08836
6. Jiang T, Wang Q (2019) Analysis of user's willingness to pay for video content on over-the-top platforms. IEEE Trans Broadcast 65(2):380–390
7. Suo X, Zhu Y, Owen GS (2005) Graphical passwords: a survey. Computer security applications conference, 21st annual IEEE
8. Tari F, Ozok A, Holden SH (2006) A comparison of perceived and real shoulder-surfing risks between alphanumeric and graphical passwords. Proceedings of the second symposium on Usable privacy and security, ACM, pp 56–66
9. Rathgeb C, Uhl A (2011) A survey on biometric cryptosystems and cancelable biometrics. EURASIP J Inf Secur 2011(1):1–25
10. Adams A, Sasse MA (1999) Users are not the enemy. Commun ACM 42(12):40–46
11. Florencio D, Herley C (2007) A large-scale study of web password habits. Proceedings of the 16th international conference on world wide web, ACM, pp 657–666
12. Bonneau J (2012) The science of guessing: analyzing an anonymized corpus of 70 million passwords. 2012 IEEE symposium on security and privacy (SP), IEEE, pp 538–552
13. Sai Nikhil MMV, Sarrin A, Nair GS, Supriya M (2022) Design and implementation of e-commerce website using automata theory 2022 6th international conference on trends in electronics and informatics (ICOEI)
14. Usha SI, Choudary KR, Kavitha CR, Sasikala T (2018) Data mining techniques used in the recommendation of e-commerce services, 2018 second international conference on electronics, communication and aerospace technology (ICECA)
15. Shrivastava A, Rajesh M (2018) Automatic irrigation system with data log creation, 2018 second international conference on inventive communication and computational technologies (ICICCT)
16. Praveen A, Nanda K, Rajith N, Giriraj N, Radhika R, Mahesh N, Vishnu K, Anjali T, Sarath S (2020) Conference room booking application using flutter, 2020 International conference on communication and signal processing (ICCSP), Chennai, India

17. Thangam S, Ramasamy G, Sathya P, Ezra K (2016) A web based intelligent orchestration tool for smart city. Int J Control Theory Appl 9:109–113
18. Dhanush GA, Divya R, Harini G, Teja TPVK, Priyanka K (2021) Develop a scalable and serverless client-based application using agile methodology. ICCET
19. Tannu K, Jayanthi Kannan MK (2022) A study on impact of privacy and password/account sharing of OTT platforms. J Emerg Technol Innovative Res 9(6):422–427

Deep Neural Network-Based Secure Energy-Efficient Power Allocation in an Interference Network

Shahid Mehraj Shah, Faraz Nassar, Azam Iftikhar, Majed Haddad, and Mohammed Zafar Ali Khan

Abstract A general communication network can be modeled as an interference channel, wherein multiple transmitters communicate with multiple receivers. With the inclusion of the Internet of Things (IoT) in a cellular network, e.g., in a 6G communication system, there are two major challenges: one is to allocate optimal resources (power, etc.) to each IoT node and the second is to ensure that resource allocation is energy efficient, as IoT nodes have limited energy resources. Conventional techniques, like the water-filling-based solution for power allocation work, are computationally tractable for a few users. In a typical IoT interference network, the number of users is large, and conventional approaches do not scale up as the number of users increases. In this paper, we use ensemble deep neural network (DNN) and genetic algorithm (GA)-based power control methods to address the non-convex optimization problem of maximizing the sum rate and energy efficiency for the interference channel. We also consider an interference channel with an eavesdropper and compute the secrecy sum rate and energy efficiency using the same AI methods. For the deep neural network (DNN), we have considered an unsupervised learning strategy as the ground truth is missing in the power control problem, and we directly maximize the objective function by minimizing the loss in the training

---

Faraz Nassar and Azam Iftikhar were with the Department of ECE, National Institute of Technology Srinagar when part of the work was done.

S. M. Shah
Department of ECE, National Institute of Technology Srinagar, Srinagar, India
e-mail: shahidshah@nitsri.ac.in

F. Nassar
Department of ECE, Indian Institute of Technology Delhi, Delhi, India

A. Iftikhar
Data Scientist, Privacy Identity, Baltimore, MD, USA

M. Haddad
Department of EE, Avignon University, Avignon, France
e-mail: majed.haddad@univ-avignon.fr

M. Z. A. Khan (✉)
Department of EE, Indian Institute of Technology, Hyderabad, Kandi, India
e-mail: zafar@ee.iith.ac.in

S. M. Thampi et al. (eds.), *Fifth International Conference on Computing and Network Communications*, Lecture Notes in Electrical Engineering 1219,
https://doi.org/10.1007/978-981-97-4540-1_22

phase. To improve the performance of DNN, we have used the ensemble method for deep neural networks, as it takes the least amount of time. We observe that GA and DNN outperform conventional optimization techniques like sequential quadratic programming (SQP).

Keywords Deep learning · Energy efficiency · Interference network · Genetic algorithm · Physical layer security

1 Introduction

With the emergence of 5G and beyond 5G communication system, more and more devices will need to be connected, for example unmanned air vehicles (UAVs) and Internet of Things (IoT) [1]. As the number of devices, which communicate over the cellular and IoT network, increases exponentially, two important aspects become very important (1) security and (2) utilizing energy efficiently to drive these devices [2]. For simple channel models like Multiple Access Channel (MAC, which models the up-link scenario of cellular communication system), the problem of optimal power allocation among different transmitters can be solved using conventional techniques called water-filling-based approach [3]. An iterative water-filling approach for MAC has been reported very recently in [4, 5]. Since then, there have been many extensions to this technique to several other multi-user channel models like fading multiple access wiretap channel [6–8]. Multiple Access Wiretap channel with finite constellation has been studied in [9]. Some more extensions for MAC-WT have also been reported [10–12]. Similar results have been reported for the problem of optimal power allocation in broadcast channel [13]. However, the most general model for a typical communication system, wherein several transmitters are communicating with several receivers, is modeled as an Interference channel in network information theory [14]. The optimal power allocation in the fading interference channel (which models the wireless interference channel) is hard to solve, owing to the non-convex nature of the achievable rate region, in general. Most of the papers take 2 user case to study the power control of fading interference channel and provide a structure of the power allocation policies, like in [15]. However, in the context of IoT network, the number of transmitters and receivers will be very huge, as IoT sensors are being deployed at large scale. Hence, the problem of power allocation becomes intractable using conventional techniques.

1.1 Related Work

In the last decade, there have been various efforts to use techniques from artificial intelligence (like deep neural networks) for optimal power allocation in complex communication networks [16]. In this direction, the authors of [17] propose an ensemble deep neural network-based power control for fading interference channel. In this paper, the authors propose a DNN-based power control algorithm, which they call $PCNet$, and show that a variant of $PCNet$ outperforms the conventional existing solutions with low complexity. There have been several other directions for optimal power allocation using diverse techniques from the artificial intelligence paradigm: reinforcement learning-based techniques [18] and deep Q-learning-based power control [18]. For a detailed survey refer to [19].

1.2 Our Contribution

In this paper, we use the $PCNet$ algorithms of [17] to study two important aspects of large-scale IoT network: (1) the energy efficiency of the interference network, (2) achievable secrecy sum rate and energy efficiency in presence of an eavesdropper. We further use a genetic algorithm to compute the sum rate, and energy efficiency of the interference channel, both with and without an eavesdropper. We also use a conventional optimization technique called sequential quadratic programming (SQP) to compute the optimal power allocation. We have used the functions defined in [20].

The rest of the paper is organized as follows: We introduce the channel model in Sect. 2. In Sect. 3, we study the energy efficiency of the interference network. A deep neural network and genetic algorithm are introduced in Sect. 4. In Sect. 5, we discuss the simulations and numerical results and we conclude the paper in Sect. 6.

2 Channel Model

A single antenna interference channel with K users is considered. We assume that all receiver and transmitter pairs are synchronized. The signal that the ith receiver receives is given by

$$y_i = h_{ii}x_i + \sum_{j\epsilon K/i} h_{ji}x_j + n_i \tag{1}$$

where $h_{i,i}$ denotes the direct-link channel for the ith user, $h_{j,i}$ denotes the cross-link and the channel between the jth transmitter and the ith receiver and $n_i \sim \mathcal{CN}(\mu, \sigma^2)$ denotes the receiver noise (with circularly symmetric normal distribution) which is independent across both time and users.

Receiver i only intends to decode x_i. For simplicity, we assume that all receivers have the same noise power σ^2. In the case of the channel with the eavesdroppers, we assume that K eavesdroppers attempt to intercept the packets transmitted from users to all receivers.

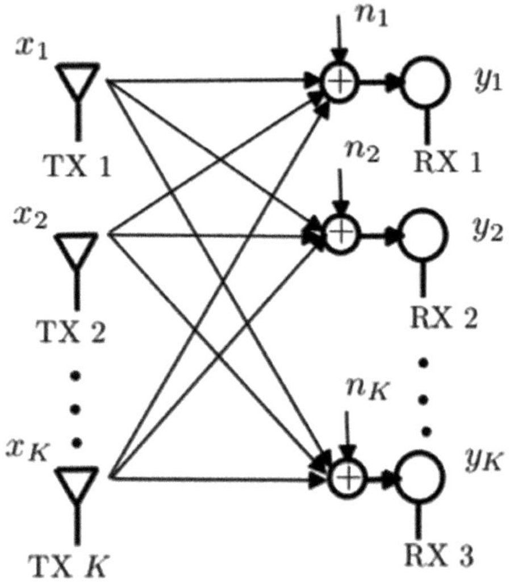

Fig. 1 K-user interference channel

Let g_{ii} denotes the direct-link channel gain for the ith eavesdropper, g_{ji} denotes the cross-link channel gain between the jth eavesdropper to the ith user. We can similarly model the noise received at the jth eavesdropper as denoted by n_{ej}, as $n_{ej} \sim \mathcal{CN}(\mu, \sigma_e^2)$. The signal received is then given by

$$z_i = g_{ii}x_i + \sum_{j\epsilon K/i} g_{ji}x_j + n_{ei} \tag{2}$$

The channel coefficients are assumed to be fixed in one time slot but vary independently from one time slot to the next, implying a block fading channel model. We do not limit the distribution of h_{ii} since our technique is adaptable enough to work with a wide range of channel models. It is assumed that the code-books are random Gaussian. There is no interference cancelation at each receiver, and encoding is done independently for each user. As a result, the transmitter-receiver pairs do not work together, and multi-user interference is classified as additive noise (Fig. 1).

3 Secure Energy Efficiency for Interference Network

3.1 Without Security

Let $P = \{P_1, P_2, P_3, \ldots, P_k\}$ be the power profile of all the K users, and all receivers are assumed to have the same noise power σ^2. Then, the achievable rate of the ith receiver is

$$R_i(P) = \log\left(1 + \frac{P_i \|h_{ii}\|^2}{\sigma^2 + \sum_{j\epsilon K/i} P_i \left\|h_{ji}\right\|^2}\right). \quad (3)$$

The channel coefficients are preset for each sample, and the power control algorithm generates power profiles for each sample. The achievable power efficiency is defined as:

$$\frac{\sum_{i=1}^{K} R_i(P)}{P_c + \sum_{i=1}^{K} P_i}$$

where P_c is the total circuit power. The channel coefficients are preset for each slot, and the power control algorithm generates a power profile.

The goal of power control is to identify the near-optimal power control solutions, i.e., power profiles P for all users to maximize the system performance while keeping the power limits in mind. The power control problem can be formulated in a variety of ways depending on the performance measurements and restrictions used. This work focuses on the power efficiency maximization problem.

The power efficiency maximization problem is formally given as

$$\max_{P} \frac{\sum_{i=1}^{K} R_i(P)}{P_c + \sum_{i=1}^{K} P_i} \quad (4)$$

$$\text{s.t.} \mid 0 \le P_i \le P_{\max} \quad (5)$$

3.2 *Secure Energy Efficiency*

Let us now look at the energy efficiency of the interference network in the presence of an eavesdropper. Let $P = \{P_1, P_2, P_3, \ldots, P_k\}$ be the power profile of all the K users. We further assume that all receivers have the same noise power, σ^2, and also all the eavesdroppers have the same noise of σ_e^2. Then, the achievable rate of the ith receiver is

$$R_i(P) = \log\left(1 + \frac{P_i \|h_{ii}\|^2}{\sigma^2 + \sum_{j\epsilon K/i} P_i \left\|h_{ji}\right\|^2}\right) \quad (6)$$

and the achievable power efficiency can be defined as

$$\frac{\sum_{i=1}^{K} R_i(P)}{P_c + \sum_{i=1}^{K} P_i}$$

where P_c is the power efficiency coefficient. The maximum rate that is individually achievable for the ith eavesdropper is

$$R_i^{\rm e}(P) = \log\left(1 + \frac{P_i \left\|g_{ii}\right\|^2}{\sigma_{\rm e}^2 + \sum_{j\epsilon K/i} P_i \left\|g_{ji}\right\|^2}\right). \tag{7}$$

Now we first define secrecy sum rate for interference channel with an eavesdropper [21],

$$R_{\rm s} = \left[R_i(p) - R_i^{\rm e}\right]^+ \tag{8}$$

where $[x]^+ \equiv \max(0, x)$. Hence, the secrecy sum rate maximization problem can be formulated as:

$$\max_P \sum_{i=1}^{K} \{R_i(P) - R_i^{\rm e}(P)\} \quad \text{s.t} \quad 0 \le P_i \le P_{\max}. \tag{9}$$

Consequently, the secure energy efficiency can be computed as (Fig. 2):

$$\max_P \frac{\sum_{i=1}^{K} R_i(P) - R_i^{\rm e}(P)}{P_{\rm c} + \sum_{i=1}^{K} P_i} \tag{10}$$

$$\text{s.t.} \mid 0 \le P_i \le P_{\max}. \tag{11}$$

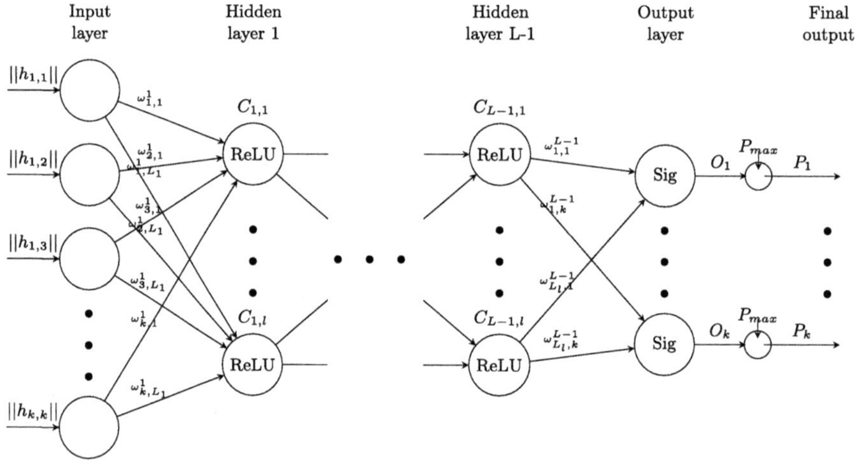

Fig. 2 Deep neural network architecture

4 DNN Network Design and Genetic Algorithm Architecture

4.1 Network Design

One input layer of $K \times K$ nodes, one output layer of K nodes, and $L - 1$ completely linked hidden layers make up the network. The network's input layer is created by aligning all $||h_{i,j}||$ into a column vector indicated by h. L fully connected layers, including $L - 1$ hidden layers and one output layer, process the input vector.

Output of kth hidden layers is calculated as follows:

$$c_k = \text{ReLU}\left(\text{BN}(W_k c_{k-1} + b_k)\right),$$

where c_{k-1} and c_k are the previous and current layers' output vectors. W_k is the weight matrix and b_k is the $l_{k\times 1}$ bias vector, and the batch normalization is denoted by BN(·).

The transmit power of all transmitters is determined by the output layer and is calculated as follows:

$$c_L = \text{Sig}(W_L C_{L-1} + b_L).$$

The transmission power should be between $[0, P_{\max}]$. To regulate the output, sigmoid function is used as the output of sigmoid function is between 0 and 1. For user i, transmit power is calculated as follows:

$$P_i = P_{\max} c_{L,i}.$$

For the simulations, i.i.d. Rayleigh fading is assumed for all channels. Therefore, we have $h_i \sim \mathcal{CN}(0, 1) \forall i \in K$. Each receiver's noise power is set to the same value of σ^2.

4.2 Ensembling

For a particular channel realization h, the deep neural network cannot ensure that the outputs are globally optimal power profiles $P(h)$. As a result, the trained model may reach a local maximum. The concept of ensemble learning is used to improve performance and get closer to the global optimum (Fig. 3).

The ensemble method is a machine learning approach that mixes multiple different base models to improve performance. We have used N models trained for a given interference channel model [22]. All the networks have the same structure, but each of the models is trained on different data sets. These steps are being done to increase the variety of the N local learners, which is a key tenet in ensemble learning. As

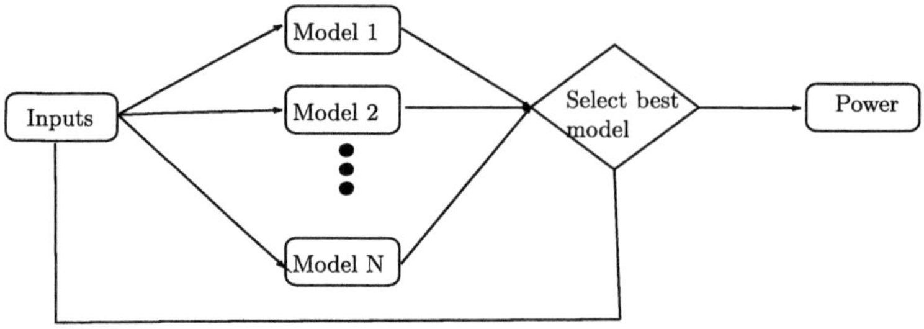

Fig. 3 Ensemble with N deep neural networks

shown in Fig. 2, the network input is first fed into all models in the ensemble, each model outputs a power profile, and the loss function is computed for the models from N outputs and the input vector. The selector compares all the power profiles and then outputs the power profile which corresponds to the lowest loss value.

Algorithm 1: Ensemble learning algorithm

Require: Input: $\{||h_{i,j}||\}_{1 \leq i,j \leq K}$
Result: Best model
Divide data set into N sub-sets; **for** *each data set i* **do**
  Obtain optimal power using DNN ; ▷ `By minimizing cost function`
end
return The best model

4.3 Genetic Algorithm-Based Power Allocation

Genetic algorithm (GA) is a search-based optimization technique based on the principles of genetics and natural selection. We have implemented a genetic algorithm in Python. During the implementation, we have used a mutation rate of 0.4, an elite size of 40, and the algorithm is run for 10 generations. We have used a population of size 200, and the initial population is populated with completely random solutions.

Chromosome Representation: Each chromosome contains a solution point which contains K number of power and can be represented as $1 \times K$ array where K is the number of users.

Crossover: Multi-point crossover is used which is used to get new off springs.

Algorithm 2: Genetic algorithm

Require: Input data: $\{||h_{i,j}||\}_{1\leq i,j\leq K}$
Result: Best solution found for maximizing energy efficiency: The power allocation vector $\mathbf{P} = (P_1, \ldots, P_K)$

Population initialization: Assign random values to initial values of power; **for** *population size* **do**
  execute phases ; ▷ Mutation and crossover
  if *elitism* **then**
    population[0] = fittest ; ▷ Next evolution
  end
end
return fittest

5 Simulations and Numerical Results

In this section, we study the results obtained by simulation of various methods of power control. The results are presented for $K = \{5, 10, 15, 20, 25, 30, 35\}$, $P_{\max} = 1$ and EsN0 as 10 dB, where EsN0 is defined as

$$\text{EsN0} = 10 * \log\left(\frac{P_{\max}}{\sigma^2}\right)$$

The number of nodes in each layer, which is denoted succinctly as $\{l_0, l_1, l_2, \ldots l_L\}$. For each value of K, the network structures are different. The number of nodes in each layer is directly proportional to the number of users which is denoted as K. The network structure for K number of users is given as: $\{k, 1.6 * k, 3.2 * k, 6.4 * k, 3.2 * k, 1.6 * k, k\}$. The performance of the ensemble method improves with the increase in the ensemble size as it approaches the global optimum. Figure 4 shows the comparison of the Ensemble method with time and performance for power efficiency, where $K = \{5, 10, 15, 20, 25, 30, 35\}$, EsN0 = 10 dB, $P_c = 0.1$ and ensemble size of $\{1, 2, 3, 4, 5\}$. The trend in (a) can be understood from the fact that as we increase the ensemble size, we can get closer to the minima of the loss function and therefore the maximum sum rate. From (b), we can see that as the ensemble size increases, the computational time increases as the computational complexity increases. We can see from Fig. 5 that the sum rate achieved using DNN method is higher than that achieved using other techniques like SQP, GA, WMMSE for an increasing number of users. From Fig. 6, we can see that GA outperforms the other techniques throughout. It can be explained as the objective function keeps on changing with the increase in the number of users. We can conclude that GA performs better overall in the presence of an evesdropper. We can see from the results shown in Fig. 7 that as the number of users increases, the secrecy capacity decreases. It can be easily understood from the general expression of the capacity. Results shown in Fig. 8 represent the effect of depth, i.e., number of hidden layers in a deep neural network on the performance

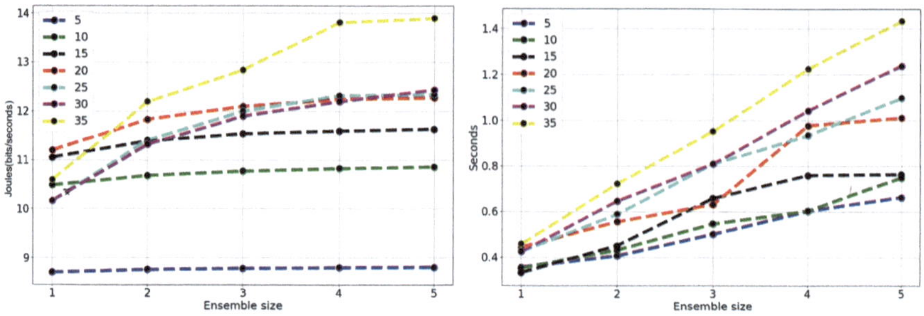

(a) Ensemble size vs power efficiency (b) Ensemble size vs time

Fig. 4 Comparison of ensemble method with time and performance for power efficiency

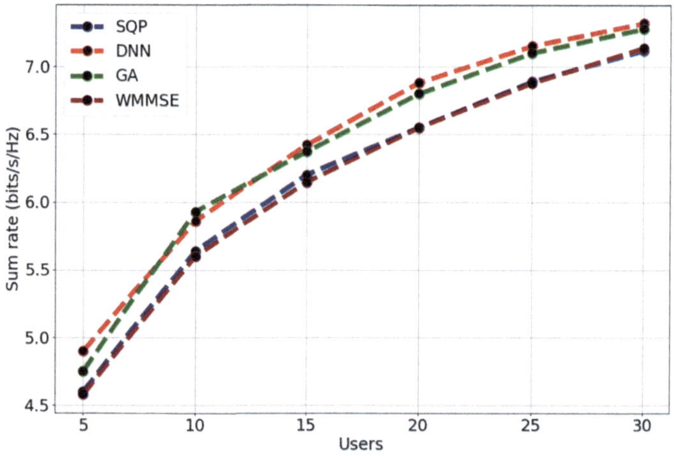

Fig. 5 Comparison of performance of SQP, DNN, WMMMSE, and GA for sum rate maximization problem

of the existing neural network. We can see as the number of layers increases, the performance of the code increases for a small number of users but degrades for a large number of users. This result cannot be generalized as it just plots the effect of several layers ignoring other parameters like activation function and number of nodes. The result given in Fig. 9 compares the energy efficiency achieved for Rician with that of previously generated Rayleigh. We can see from the plot that for a large number of users, the energy efficiency achieved using Rician reduces as compared to that of Rayleigh.

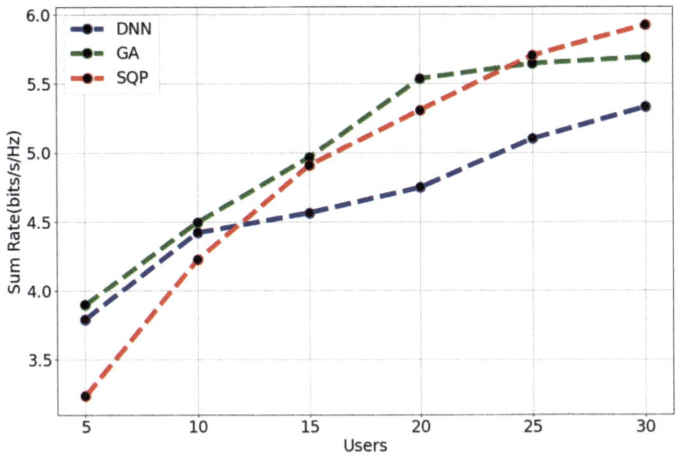

Fig. 6 Comparison of performance of DNN, SQP, and genetic algorithm for SRM in the presence of eavesdropping constraint

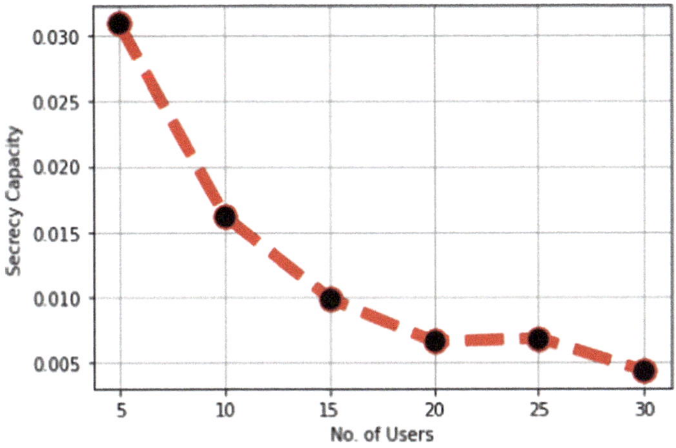

Fig. 7 Effect of increasing the number of users on the secrecy capacity

5.1 Parameters Used for Simulation

Deep neural network was implemented in Keras which is a Python library for DNN. The parameter was set as in [1], and they utilize Xavier initialization to initialize the network parameters and ADAM optimizer to train the network, since it is standard practice while training a neural network. The gradient is calculated using stochastic gradient descent with a batch size of 10^3 and a run time of 50 epochs. The training data is created at random based on the channel model, with a total of 10^8 channel samples generated for the network's training. A validation data set of 2×10^3 samples is utilized to test the network performance during training to track it.

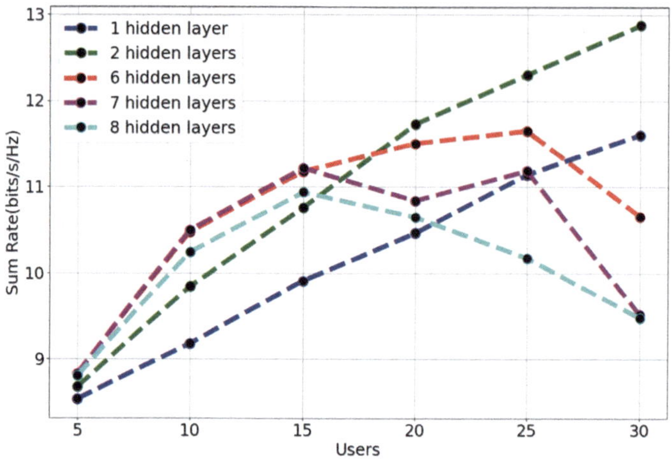

Fig. 8 Effect of depth of the neural network on the performance of the code

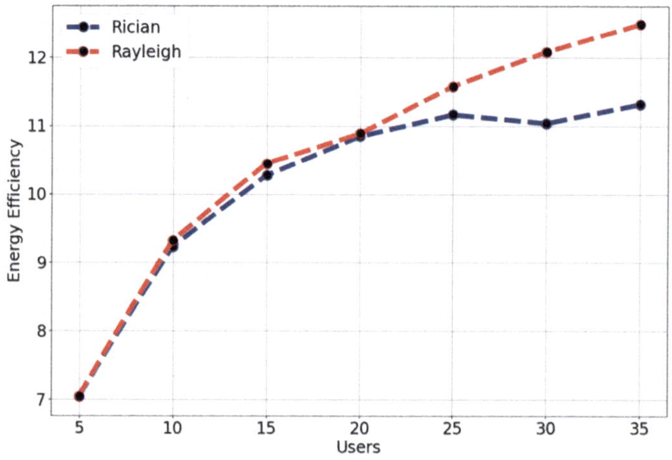

Fig. 9 Comparison of energy efficiency achieved for Rician and Rayleigh with physical layer security

Sequential quadratic programming (SQP) is implemented for comparison. It is implemented in Python from the SciPy library. The algorithm stops after 100 iterations and esp which is the step size used for numerical approximation of the Jacobian is set to $1.490e^{-08}$. We have generated 2×10^3 samples which are being used to evaluate the algorithm's performance which is randomly generated based on the channel model.

We have implemented a genetic algorithm in Python. During the implementation, we used a mutation rate of 0.4, an elite size of 40, and the algorithm ran for 10 generations. We have used a population of size 200 and the initial population is

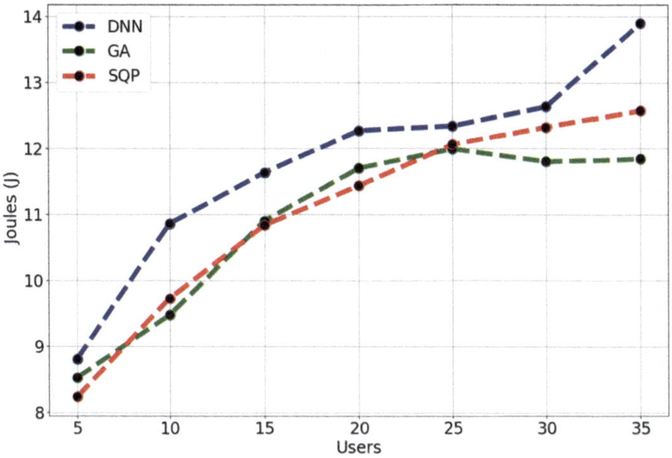

Fig. 10 Comparison of performance of DNN, SQP, and genetic algorithm for efficiency maximization problem

Table 1 Time taken by DNN, SQP, and GA to compute power profiles for 2000 samples for power efficiency problem

DNN	SQP	GA
1.6	18.4	566
1.9	33.9	612
2.1	55.8	598
2.2	79.0	750
2.3	103.7	852
2.4	134.4	921
2.7	165.0	1012

Populated with completely random solutions. From Fig. 10, we can see that the energy obtained using deep neural networks is higher than that achieved using other techniques like genetic algorithm (GA) and sequential quadratic programming (SQP) (Table 1).

6 Conclusions and Future Work

In this paper, we use a generic artificial intelligence (AI) framework for power allocation in an energy-efficient interference network, while achieving a desirable secrecy sum rate. In particular, we use a popular machine learning technique called ensemble learning to find the optimal power for in an interference network. We first simulate some results in literature like that of PCNet which is a fully connected multi-layer neural network that takes as input the channel coefficients and gives transmit power of all transmitters as the output. It adopts an unsupervised learning strategy to train by directly maximizing the system sum rate. An ensemble of these PCNets is used so that the output of each PCNet is compared to reach the global minima and therefore optimal power control is achieved in this way. We can say from the results achieved in this paper that deep learning has the potential to tackle this as well as other non-convex optimization problems that are often encountered in wireless communications.

Some of the areas that are worth further research in the future include the network generalization to different scenarios, robustness to channel estimation errors as it currently assumes ideal channel estimation, distributed power.

References

1. Mozaffari M, Lin X, Hayes S (2021) Toward 6G with connected sky: UAVs and beyond. IEEE Commun Mag 59(12):74–80
2. Huang T, Yang W, Wu J, Ma J, Zhang X, Zhang D (2019) A survey on green 6G network: architecture and technologies. IEEE Access 7:175758–175768
3. Lai L, El Gamal H (2008) The water-filling game in fading multiple-access channels. IEEE Trans Inf Theory 54(5):2110–2122
4. Haddad M, Wiecek P, Habachi O, Perlaza SM, Shah SM (2022) Fair iterative water-filling game for multiple access channels. In: Proceedings of the 25th international ACM conference on modeling analysis and simulation of wireless and mobile systems, pp 125–132
5. Haddad M, Wiecek P, Habachi O, Perlaza SM, Shah SM (2022) FEAT: fair coordinated iterative water-filling algorithm. arXiv preprint arXiv:2206.10211
6. Jiang K, Jing T, Li Z, Huo Y, Zhang F (2017) Analysis of secrecy performance in fading multiple access wiretap channel with SIC receiver. In: IEEE INFOCOM 2017—IEEE conference on computer communications. IEEE, pp 1–9
7. Shah SM, Chaitanya AK, Sharma V (2016) Resource allocation in fading multiple access wiretap channel via game theoretic learning. In: 2016 information theory and applications workshop (ITA). IEEE, pp 1–7
8. Shah SM, Sharma V et al (2017) Resource allocation in a MAC with and without security via game theoretic learning. EURASIP J Wireless Commun Netw 2017(1):1–14
9. Shah SM, Riyaz R, Wani TM, Ashraf A, Naz SF (2020) Optimal rotation angle for finite constellation over additive white Gaussian noise multiple access wiretap channel. IET Commun 14(21):3845–3852
10. Shah SM, Kumar V, Sharma V (2012) Achievable secrecy sum-rate in a fading MAC-WT with power control and without CSI of eavesdropper. In: 2012 international conference on signal processing and communications (SPCOM). IEEE, pp 1–5
11. Shah SM, Sharma V (2020) Enhancing secrecy rates in a wiretap channel. Digit Commun Netw 6(1):129–135
12. Showkat S, Dar ZB, Shah SM (2023) Achievable sum-rate of variants of QAM over Gaussian multiple access channel with and without security. arXiv preprint arXiv:2308.11405
13. Jindal N, Goldsmith A (2003) Capacity and optimal power allocation for fading broadcast channels with minimum rates. IEEE Trans Inf Theory 49(11):2895–2909
14. Han T, Kobayashi K (1981) A new achievable rate region for the interference channel. IEEE Trans Inf Theory 27(1):49–60
15. Tuninetti D (2008) Gaussian fading interference channels: power control. In: 2008 42nd Asilomar conference on signals, systems and computers. IEEE, pp 701–706

16. Wang C-X, Di Renzo M, Stanczak S, Wang S, Larsson EG (2020) Artificial intelligence enabled wireless networking for 5G and beyond: recent advances and future challenges. IEEE Wireless Commun 27(1):16–23
17. Liang F, Shen C, Yu W, Wu F (2019) Towards optimal power control via ensembling deep neural networks. IEEE Trans Commun 68(3):1760–1776
18. Luo L, Zhang J, Chen S, Zhang X, Ai B, Ng DWK (2022) Downlink power control for cell-free massive MIMO with deep reinforcement learning. IEEE Trans Veh Technol
19. Sun Y, Peng M, Zhou Y, Huang Y, Mao S (2019) Application of machine learning in wireless networks: key techniques and open issues. IEEE Commun Surv Tutor 21(4):3072–3108
20. Shen C (2018) PCNet-ePCNet. https://github.com/ShenGroup/PCNet-ePCNet
21. Kalantari A, Maleki S, Zheng G, Chatzinotas S, Ottersten B (2015) Joint power control in wiretap interference channels. IEEE Trans Wireless Commun 14(7):3810–3823
22. Gupta A, Gupta HS, Bohara VA, Srivastava A (2022) Energy resource allocation for green FiWi network using ensemble learning. IEEE Trans Green Commun Netw

An Adaptive Time Quantum-Based Efficient Round-Robin Algorithm for CPU Scheduling

Menaka Pushpa Arthur, Rizwan Mohamed Kareem, and G. Yogeshwar

Abstract The efficiency of the Round-Robin (RR) algorithm for CPU scheduling depends on its Time Quantum (TQ) value. The optimal time quantum value can reduce the average waiting time and average turnover time of the processes in the ready queue. Finding an adaptive time quantum for the processes in the ready queue is a challenging task to get an efficient performance from the round-robin algorithm. Too lengthy or too short of time quantum induces starvation or computation overhead in the process execution. TQ value is adaptive to various features of waiting processes in the ready queue such as the number of processes in the ready queue, average burst time, arrival time, and priority of the waiting processes. Hence, we have modified the round-robin algorithm by considering its performance-influencing factors to set the optimal time quantum value. Also, this proposed RR sets a dynamic time quantum for every iteration that aligns with the changing nature of the ready status properties. The performance improvement of the proposed round-robin algorithm is analyzed using implementation results. With the support of performance metrics such as average waiting time (AWT) and average turnaround time (ATT), the proposed modified round-robin algorithm is compared with other CPU scheduling algorithms and existing modified RR. The comparison results show the better performance of the proposed RR.

Keywords Round-robin algorithm · CPU scheduling algorithm · Time quantum · Process starvation

1 Introduction

The CPU scheduling algorithms in the Operating Systems (OS) select the appropriate waiting process from the ready queue and schedule the processor for its execution. First-come-first-serve (FCFS), shortest job first (SJF), shortest remaining

---

M. P. Arthur (✉) · R. M. Kareem · G. Yogeshwar
Vellore Institute of Technology, Chennai Campus, Chennai 600127, India
e-mail: menakapushpa.a@vit.ac.in

S. M. Thampi et al. (eds.), *Fifth International Conference on Computing and Network Communications*, Lecture Notes in Electrical Engineering 1219,
https://doi.org/10.1007/978-981-97-4540-1_23

time first (SRTF), and round-robin (RR) are well-known CPU scheduling algorithms in the Operating Systems. Depending upon the characteristics of waiting processes in the ready queue, different CPU scheduling algorithms perform well. CPU scheduling algorithms are majorly classified into two types, namely preemptive and non-preemptive. Under non-preemptive, the processor cannot be released from the currently running process until it has finished execution. In preemptive, the processor keeps on switching over different waiting processes during their execution. The round-robin (RR) is a preemptive CPU scheduling algorithm in the operating systems that pre-empts the process at every time quantum from the allocated processor. The optimal time quantum (TQ) value improves its efficiency with less average waiting time (AWT), average turnaround time (ATT), and the number of context switch. The RR with high TQ works like FCFS, whereas less TQ induces more context switch. Hence, the TQ value is a trade-off between high and low values to ensure the best performance of the algorithm. The TQ value should be dynamic and adaptive with respect to the current processes in the ready queue. The existing research works have identified that the round-robin algorithm with an adaptive TQ value is more efficient than a fixed TQ value.

In the state-of-the-art techniques, the TQ is updated in three ways, namely (i) process-wise (ii) iteration or round-wise, and (iii) elastic or extended time. In the first case, the TQ is updated for every process when the process gets the processor. This can assign a TQ value based on the remaining execution time of the corresponding process. The TQ update in this method is purely process-oriented. The frequency of TQ updates is very high in this method which will add reasonable delay in the overall system performance. In the second method, the TQ is updated for every iteration or round in the round-robin CPU scheduling algorithm. This TQ update is based on the current status of the ready queue such as the number of the waiting processes and their CPU burst times. When compared to first method, the additional delay is reduced, and it is equivalent to the number of iterations. The third method reduces the number of iterations by extending the assigned TQ value at the time of process execution based on its remaining burst time. If the remaining CPU burst time is marginally higher than the TQ, then the algorithm allows the process to hold the CPU until it gets over within that iteration or round. This method resolves the starvation problem in the short burst time process and reduces the number of context switches, and iterations in the RR algorithm.

The optimal TQ value can reduce the AWT and ATT along with less context switches since its cost is higher in the system process. The identification technique of the optimal TQ gets complicated with different arrival times of the process in the ready queue. Few existing works have used multilevel ready queues to classify the processes based on their properties for the enhanced round-robin algorithm. In these techniques, various adaptive TQs have been utilized at different levels of the ready queue. Adaptive TQ is recalculated when a new process enters the queue as well as when the completed process leaves the queue.

In our work, the round-robin algorithm is modified in three different ways, namely sorting, setting dynamic time quantum value, and multiple queues. Using the metrics AWT and ATT, the best strategy is analyzed and identified. The results are compared with the thump rule mentioned in the OS concept to set an optimal value for the time quantum value, i.e., 80% of the burst time of the waiting process. We observed an interesting results from this comparison. This paper is organized as follows; Sect. 2 discusses the existing works in the customization of round-robin algorithm. Section 3 explains the proposed updated round-robin algorithm, and its implementation results are shown in Sect. 4 along with the comparison of existing works. Finally, this work is concluded in Sect. 5.

2 Literature Review

The round-robin CPU scheduling algorithm has been customized with the intention of finding the best optimal value for time quantum in the existing research works. The optimal time quantum of the RR can increase its efficiency in the CPU scheduling process with respect to average waiting time (AWT), average turnaround time (ATT), and the number of context switch. In this aspect, the round-robin algorithm has been modified in three ways [1]; (i) Algorithm to sort the processes in the ready queue (ii) Technique to find the best time quantum value, and (iii) Hybrid: Find the best time quantum for the sorted ready queue. The taxonomy of the existing enhanced RR algorithm is given in Fig. 1. In most existing works [2–4], the average burst time of the processes in the ready queue is used to set the TQ value of the RR algorithm. In a few existing works, the average of the remaining burst time is used as TQ [5]. The priority-based sorting is applied in the ready queue before running the round-robin algorithm in the existing works [6, 7]. This method can reduce the process of starvation along with improved AWT and ATT of RR. The lowest burst time is used to set TQ for different priority value-based processes. For the same priority processes, the TQ is equivalent to the mean burst time of those processes.

In Neha et al. work [8], processes are categorized into three groups as per the priority; low, medium, and high. Different TQs are set for these three groups. The group with a high-priority process gets a longer TQ value than the medium and low-priority process groups. The thump rule to set the best TQ value is 80% of the burst time of the waiting process. This rule is adopted in [3, 9] along with elastic time for TQ value. The dispersion is calculated from the minimum and maximum burst time of the waiting processes in the ready queue to set the TQ value [10]. The processes are divided into four groups and applied different TQs, based on the mean burst time of the corresponding group [11]. The processes are already sorted based on the burst time after grouping. In [12], the average difference between adjacent process burst times is divided in half and then combined with precedence. After that, out of random burst time and calculated burst time, the minimum one is selected for TQ. The entry and exit of every process in the ready queue initiate a TQ update based on the current process list in the ready queue [2].

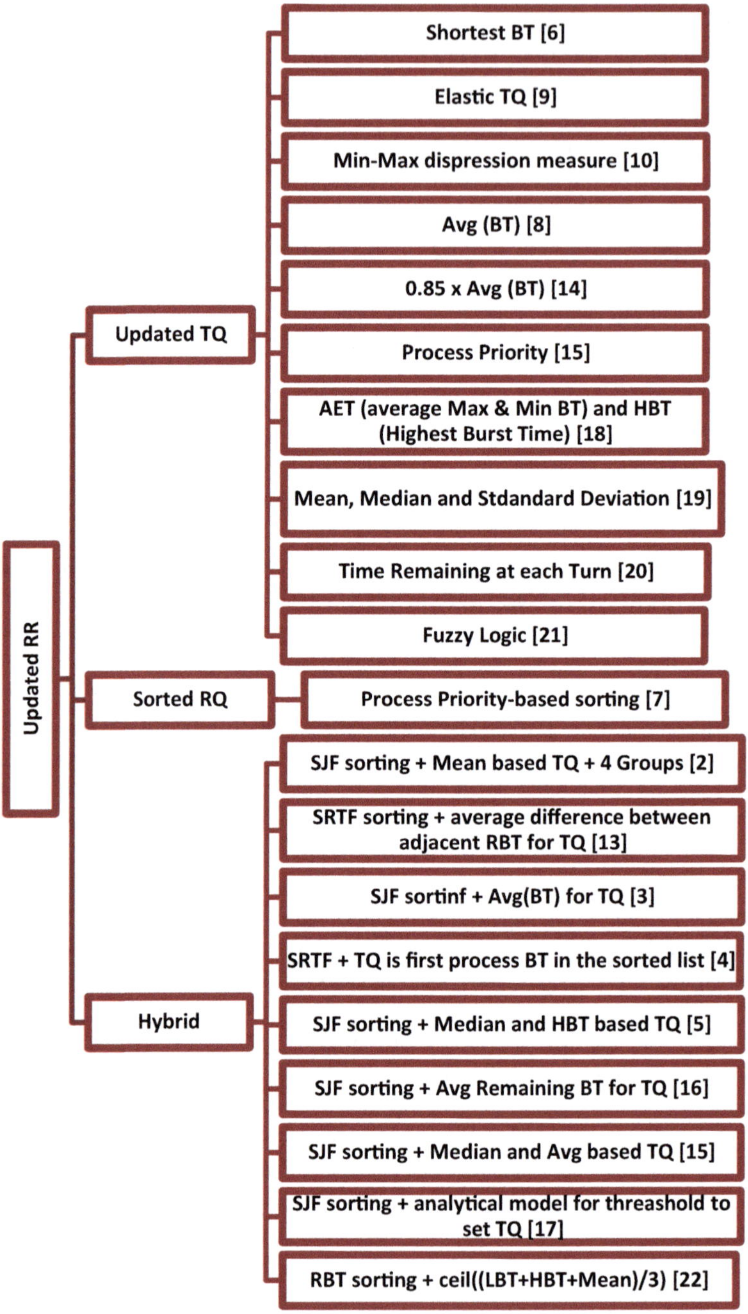

Fig. 1 Taxonomy of existing modified round-robin CPU scheduling algorithm

The work [13] sorted the process based on the remaining burst time in the ready queue before the commencement of the round-robin algorithm. The varying TQ is equivalent to the burst time of the first process in the sorted list. This list is updated as per the process completion status after every iteration. In this method, for every process, the TQ is updated. It induces additional overhead in the overall process execution. The median of burst time-based sorted ready queue processes is calculated and combined with the highest burst time to find the dynamic TQ [14]. Conditional preemption is used in this work. The TQ is set using the median and average burst times of all processes in [15]. The process is sorted as per the process burst time. For every new process arrival at the ready queue makes the update in the TQ. The intelligent TQ is identified using a fuzzy decision-making system in [16, 17] using three parameters; nice value, burst time, and recent CPU usage of all processes. Totally 27 fuzzy rules are identified on three parameters and the dynamic priority value of the process. The rules are identified using the Mamdani–Larsen inference method. In [18], the ready queue is divided into two groups such as short and long tasks based on its burst time. An analytical model is used to set the optimal TQ for two groups. The average maximum and minimum execution time and high burst time are major factors in calculating the TQ in [19]. The mean, median, and standard deviation are used in [20] to calculate the dynamic intelligent TQ for the round-robin algorithm. The maximum, minimum, and average burst time of the waiting processes are considered for TQ in the updated RR [21].

3 Proposed Round-Robin CPU Scheduling Algorithm

In our work, the round-robin algorithm is updated to generate the optimal and dynamic time quantum value for enhanced performance of the RR. Rather than using the fixed and random time quantum value for the round-robin algorithm, the adaptive TQ value is calculated based on the characteristics of ready queue processes. We updated the round-robin CPU scheduling algorithm in two ways, and results are observed for comparison, namely (i) Sorted ready queue-based round-robin algorithm with fixed time quantum and (ii) Adaptive round-robin algorithm with dynamic time quantum.

3.1 Sorted Ready Queue Round-Robin Scheduling Algorithm

The round-robin CPU scheduling algorithm is updated by considering the order of process allocation for the processor from the ready queue. In the actual round-robin algorithm, the process dispatcher assigns the processor to processes in the ready queue according to their arrival order. The time quantum is assigned as per the process arrival order. Hence, the ready queue waiting processes are sorted based on their arrival time. As per our first strategy, the ready queue waiting processes are sorted using (i) process arrival order, (ii) CPU burst time, and (iii) remaining CPU burst time of the process and then compared their performance using fixed time quantum value. The time quantum value is randomly selected. The processes are sorted within the ready queue after every iteration in the round-robin algorithm. Then using the fixed time quantum value, the processes are selected from the sorted ready queue and assigned to the processor. The three sorting algorithms are compared by varying the random fixed time quantum value for an in-depth analysis of round-robin performances with respect to AWT and ATT. The implementation results show that the burst time and remaining burst time-based sorted ready queue of the round-robin algorithm gives better results in both AWT and ATT when compared to the traditional RR algorithm, i.e., process arrival order-based sorting. This proposed method changes the order of the process in the ready queue with respect to its characteristics before assigning the processor's time quantum.

3.2 Adaptive Round-Robin Algorithm with Dynamic TQ

In our second strategy, the round-robin algorithm is customized by combining three important techniques, namely (i) sorting of ready queue processes, (ii) multiple queues, and (iii) dynamic time quantum for process categories. The process of the ready queue is sorted based on their CPU burst time. From our first strategy, we have identified that the CPU burst time or remaining CPU burst time-based sorting in the ready queue waiting process gives the same performance for RR algorithm. Hence, the CPU burst time is used to sort the ready queue processes. Then in the sorted queue, the center position is identified to divide the waiting processes into two sets such as the first half and the second half of the ready queue. These two process sets, i.e., the first and second half, are moved into two different ready queues for maintenance. In this regard, we need more than one ready queue for the proposed RR algorithm. Obviously, the first ready queue jobs are short when compared to the second ready queue jobs. The average CPU burst time of the first half and second half ready queues is calculated. Then, the percentage of processes less than the average burst time is identified. If this percentage is less than 20%, then the time quantum value is set as the CPU burst value of the first process in the queue. Otherwise, update the queue average burst time by dividing this average value by 2. Repeat the process until to set the proper value for time quantum. In certain cases, the first process has a burst time, 1. At this time, the queue average burst time is used as the time quantum value. For every iteration of the round-robin algorithm or when the new process arrives in the ready queue, the average burst time of the corresponding queue is updated which leads to a time quantum update. In this way, the time quantum value is dynamically updated for the RR algorithm. Also, different time quantum values are used for two levels of ready queues as per the characteristics of the waiting processes in such queues. The algorithm of the proposed adaptive RR algorithm is given in Fig. 2.

```
Algorithm 1: Adaptive Round-Robin CPU Scheduling Algorithm
Data: Number of Processes, CPU Burst time, arrival time and process ID
Result: Average Waiting Time (AWT) and Average Turnaround Time (ATT)
initialization;
Sort the ready queue processes using CPU burst time;
Find the center position of the sorted queue;
Move the first half process into Ready Ququel, RQ1;
Move the second half process into Ready Queue2, RQ2;
Calculate the average CPU burst time Avgl, for RQ1;
Calculate the average CPU burst time Avg2, for RQ2;
N ←Percentage of processes less than Avgl;
while Sizeof(RQ1) > 0 do
    while CPUBurstTime(FirstProcess) > 1 do
        if N < 20/100 then
            if CPUBurstTime(Firstprocess) < 1 then
                TQ ←CPUBurstTime(FirstProcess);
            else
                TQ ←Avgl;
            end
        else
            NewAvgl ← Avgl/2;
            if CPUBurstTime(Firstprocess) < 1 then
                TQ ←CPUBurstTime(FirstProcess);
            else
                TQ ←Avgl;
            end
        end
    end
end
Follow the above procedure for RQ2;
Calculate AWT1, ATT1, AWT2 and ATT2 from RQ1 and RQ2;
OverallAvgofAWT ← (AWT1+AWT2)/2 ;
OverallAvgofATT ← (ATT1+ATT2)/2 ;
```

Fig. 2 Proposed adaptive CPU scheduling algorithm

4 Results Discussion

The implementation of the updated round-robin algorithm is tested using four input test cases with the number of processes 5, 6, 8, and 10 in the environment set. The average CPU burst time of test cases is as follows: 10.8, 20.67, 25.6, and 9.6. We created the input test cases in such a way that the average burst time of the process is not dependent on a number of processes. Under two perceptions, the implementation results are well-analyzed; (i) Fixed time quantum versus different sorting algorithm in the ready queue and (ii) Evaluation of proposed adaptive TQ-based round-robin algorithm. The processes in the ready queue are sorted using three ways; first-come-first-serve (FCFS), i.e., process arrival order at the ready queue, CPU burst time (SJF), and remaining CPU burst time (SRTF). Different time quantum value is also adapted to observe the results under each input case and sorting algorithm. Totally, six fixed time quantum values, namely 2, 6, 10, 14, 18, and 22, are used in this implementation for analysis of the performance of the round-robin algorithm with respect to time quantum. The average waiting time (AWT) of the round-robin algorithm is calculated and compared under different cases to analyze the performance of the algorithm.

4.1 Analysis of Average Waiting Time for Sorted RR Algorithm

Figure 3 shows the plotting of the average waiting time (AWT) of the ready queue sorted round-robin algorithm against the average CPU burst time of all waiting processes in the ready queue. The ready queue processes are sorted using three different sorting algorithms such as FCFS, CPU burst time, and remaining CPU burst time of the process and then compared by increasing the average CPU burst time of the ready queue processes. Four diagrams are presented in Fig. 3 by changing the time quantum to 2, 10, 14, and 22. The changing pattern of the 3D surface plots of these four figures is very similar. It gives three major observations (i) Sorting based on CPU burst time and remaining CPU burst time after every iteration of the round-robin algorithm are the same AWT (ii) Sorting based on CPU burst time gives better AWT than FSFC sorting, and (iii) AWT is increasing by increasing the average CPU burst time of the ready queue processes.

The average waiting time (AWT) of the sorted ready queue round-robin algorithm is analyzed under different sorting algorithms such as FCSF, SJF, and SRTF by increasing the average CPU burst time of the ready queue waiting process (Fig. 4) and number of processes in the ready queue (Fig. 5). Under two time quantum values 2 and 14, the AWT is observed by increasing the above-mentioned factors. Figure 4 shows that the AWT increases as the average CPU burst time of the ready queue waiting processes increases. SJF and SRTF soring AWT results are the same and less than FCFS. Figure 5 shows the fluctuation in AWT as number of processes in the ready queue increases. For 5, 6, and 8 processes, AWT is less in all sorting algorithms when compared to 8 processes in the ready queue. The AWT peak value for 8 processes is triggered by the higher average CPU burst time of 8 processes than in other cases. Hence, the number of processes in the ready queue does not create a major direct impact on the AWT of the updated round-robin algorithm. Whereas, the average CPU burst time of the ready queue waiting processes creates a direct impact on the AWT. These two factors are directly proportional to each other.

The AWT of the updated round-robin algorithm presented in Fig. 6 shows better results from SJF, SRTF rather than FCFS. In all four input test cases with varying numbers of processes 5, 6, 8, and 10, the AWT decreases when the time quantum increases from 2 to 22.

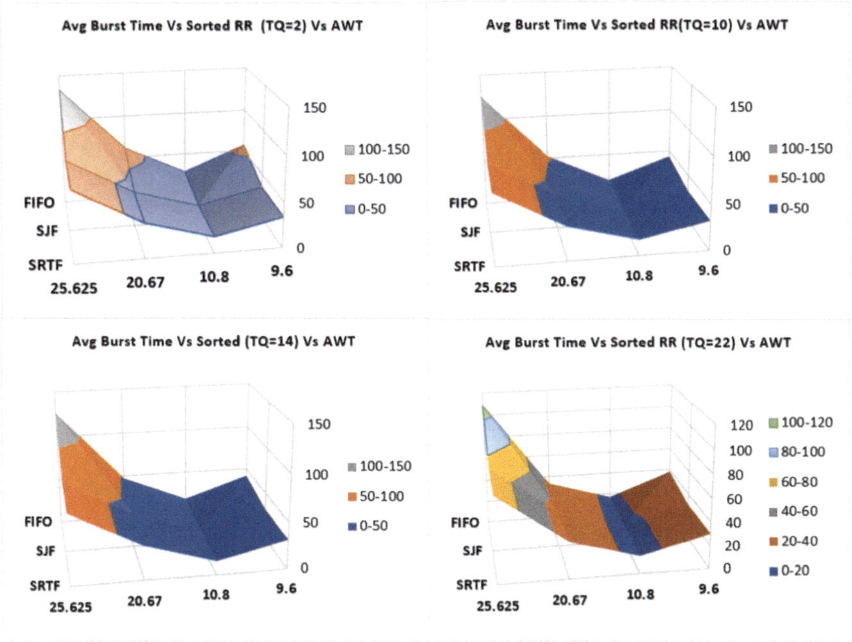

Fig. 3 3D surface graph for average burst time of RQ processes versus sorting algorithm versus average waiting time

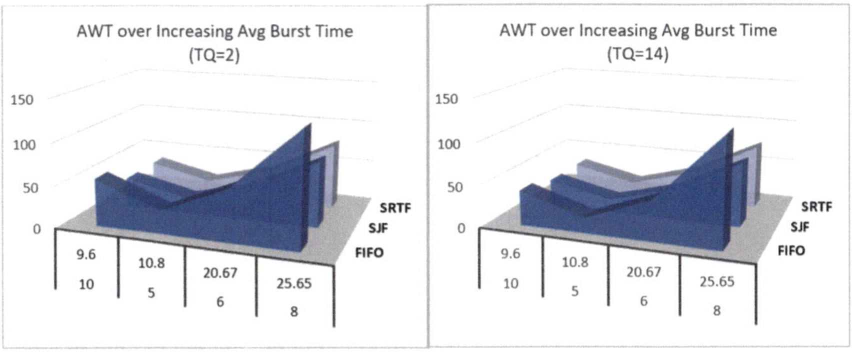

Fig. 4 Analysis of AWT by increasing average burst time of the RQ processes

4.2 Analysis of Average Turnaround Time for Sorted RR Algorithm

Figure 7 shows the plotting of the average turnaround time (ATT) of the ready queue sorted round-robin algorithm against the average CPU burst time of all waiting processes in the ready queue. The ready queue processes are sorted using three

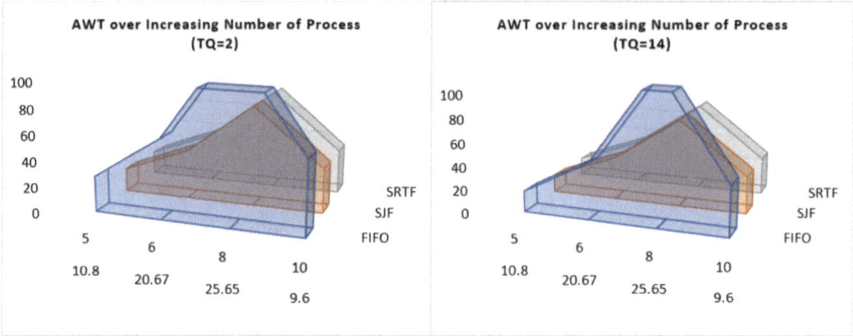

Fig. 5 Analysis of AWT by increasing the number of processes in the ready queue

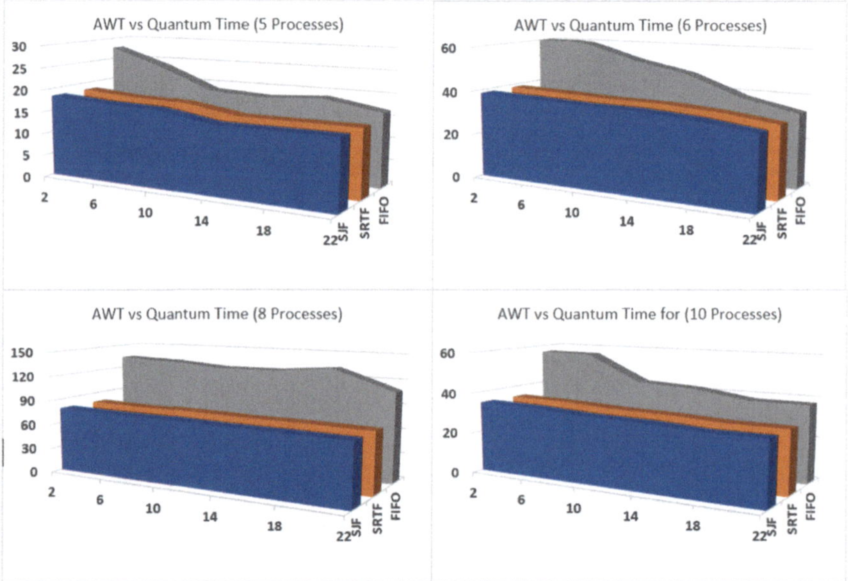

Fig. 6 Analysis of AWT by increasing the time quantum for different sorting algorithm

different sorting algorithms such as FCFS, CPU burst time, and remaining CPU burst time of the process and then compared by increasing the average CPU burst time of the ready queue processes. Two diagrams are presented in Fig. 3 by changing the time quantum to 2 and 18. The changing pattern of the 3D surface plots of these four figures is very similar. It gives three major observations: (i) Sorting based on CPU burst time and remaining CPU burst time after every iteration of the round-robin algorithm are the same (ii) Sorting based on CPU burst time gives better ATT than FSFC sorting, and (iii) ATT is increasing by increasing the average CPU burst time of the ready queue processes.

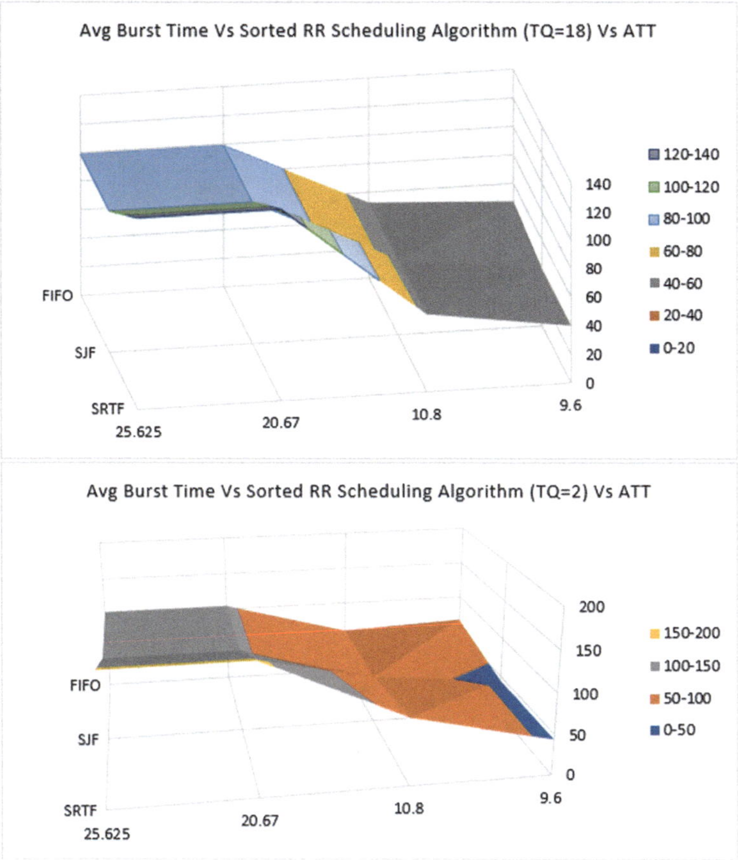

Fig. 7 3D surface graph for average burst time of RQ processes versus sorting algorithm versus average turnaround time

The average turnaround time (ATT) of the sorted ready queue round-robin algorithm is analyzed under different sorting algorithms FCSF, SJF, and SRTF by increasing the average CPU burst time of the ready queue waiting process (Fig. 8) and a number of processes in the ready queue (Fig. 9). Under two time quantum values 2 and 22, the ATT is observed by increasing the above-mentioned factors. Figure 8 shows that the ATT increases as the average CPU burst time of the ready queue waiting processes increases. SJF and SRTF soring ATT results are the same and less than FCFS. Figure 9 shows the fluctuation in ATT as the number of processes in the ready queue increases. For 5, 6, and 8 processes, ATT is less in all sorting algorithms when compared to 8 processes in the ready queue. The ATT peak value for 8 processes is triggered by the higher average CPU burst time of 8 processes than in other cases. Hence, the number of processes in the ready queue does not create a major direct impact on the ATT of the updated round-robin algorithm. Whereas,

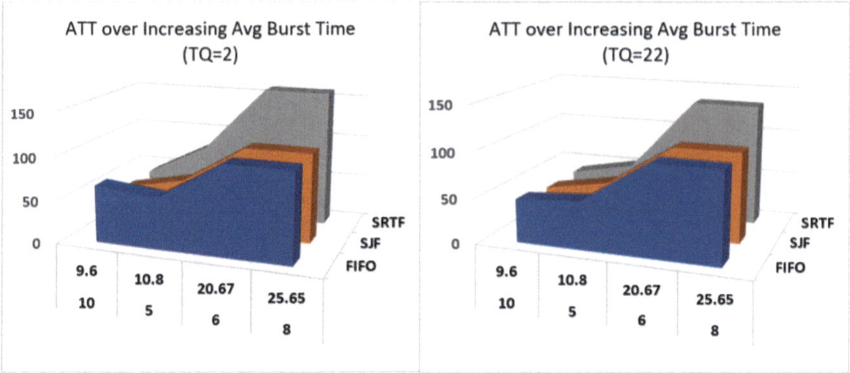

Fig. 8 Analysis of ATT by increasing average burst time of the RQ processes

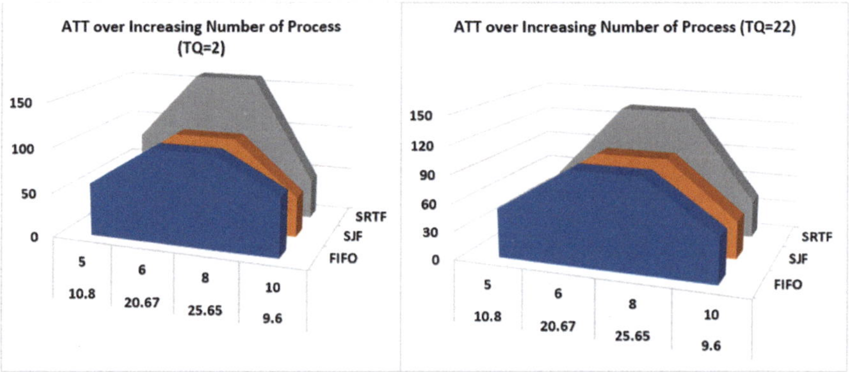

Fig. 9 Analysis of ATT by increasing the number of processes in the ready queue

the average CPU burst time of the ready queue waiting processes creates a direct impact on the ATT. These two factors are directly proportional to each other. The ATT of the updated round-robin algorithm presented in Fig. 10 shows better results from SJF, SRTF rather than FCFS. In all four input test cases with varying numbers of processes 5, 6, 8, and 10, the ATT decreases when the time quantum increases from 2 to 22.

4.3 Comparison of Sorted RQ-RR and Adaptive TQ-based RR

In this section, the proposed system is analyzed by the comparison analysis under two different perceptions, namely (i) different sorting algorithms for ready queue and (ii) other existing customized round-robin CPU scheduling algorithms. Figure 11

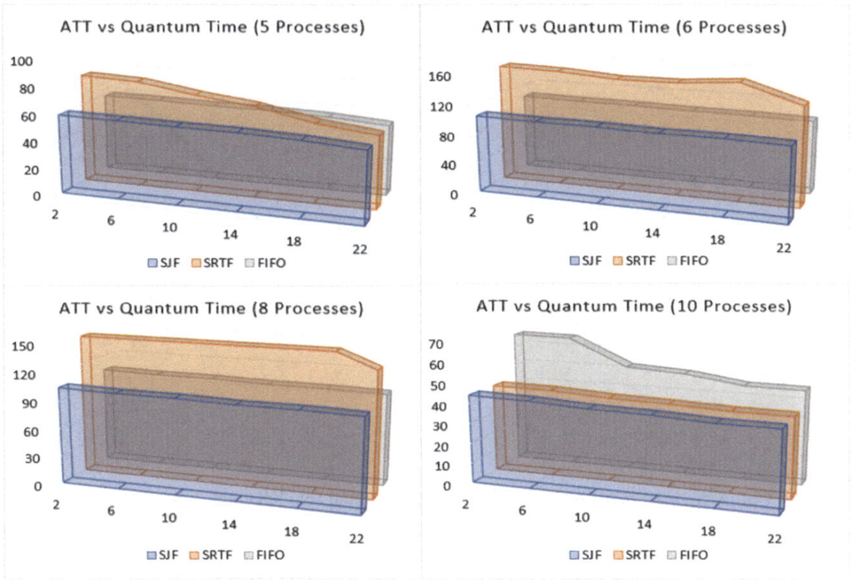

Fig. 10 Analysis of ATT by increasing the time quantum for different sorting algorithm

shows the comparison between the two different proposed methods in the round-robin algorithm with respect to average waiting time (AWT) by increasing the average CPU burst time of the ready queue waiting process. In Fig. 11, various sorting algorithms-based updated round-robin algorithm and dynamic TQ-based adaptive round-robin algorithm are compared using AWT. From this result, the adaptive RR algorithm gives better performance than the sorted ready queue-based RR algorithm. The CPU burst time-based or remaining CPU burst time-based (SJF/SRTF) sorted ready queue of the round-robin algorithm gives better AWT than the conventional RR algorithm (FCFS). The average CPU burst time of the waiting process is directly proportional to the AWT of the updated RR algorithm. Figure 12 shows the performance comparison of the proposed RR and other customized RR in the existing works with respect to AWT. The updated RR in the work [19] uses the analytical formula to set the dynamic time quantum value according to the maximum, minimum, and mean of the CPU burst time of the waiting processes. The ready queue in this method is also sorted based on the burst time. Another work [21] used for comparison analysis uses the mean value of the waiting process CPU burst time for TQ. They have compared the calculated TQ value with the threshold value of TQ. If TQ is greater than a threshold, then the updated RR retains the calculated TQ; otherwise, TQ is initialized by the threshold value. This comparison shown in Fig. 12 shows the best AWT from our proposed adaptive round-robin with dynamic TQ than the existing works [19, 21].

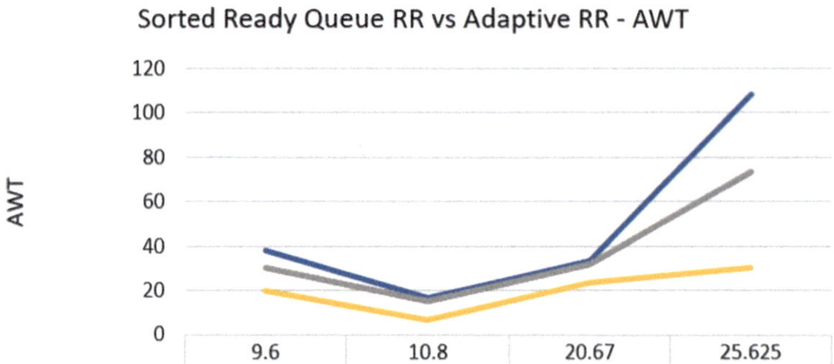

	9.6	10.8	20.67	25.625
FCFS	37.8	16.6	33.3	108.4
SJF	30.4	15.2	32	73.5
SRTF	30.4	15.2	32	73.5
Adaptive RR	19.96	6.84	23.5	30

Fig. 11 Comparison of AWT in sorted RQ-RR versus adaptive RR

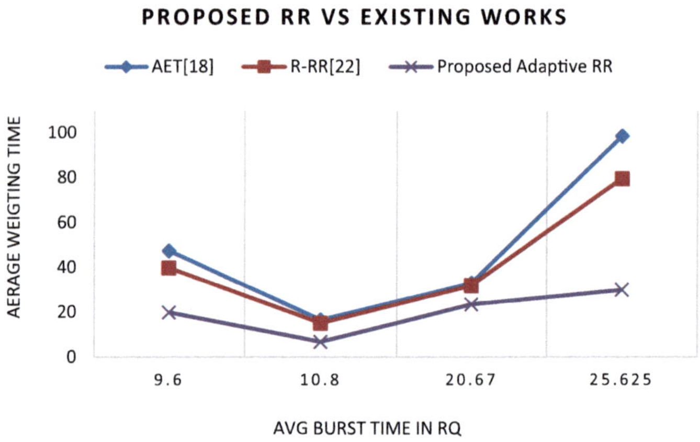

Fig. 12 Comparison of AWT in proposed and existing systems

5 Conclusion

The round-robin CPU scheduling algorithm is updated in two different ways in this work in order to improve the algorithm performance with respect to average waiting time (AWT) and average turnaround time (ATT). In the first technique, the sorted ready queue round-robin algorithm uses a different sorting algorithm in the ready queue before allocating the process to the processor as per the actual round-robin algorithm. Rather than conventional FCFS in RR, sorting based on CPU burst time and remaining CPU burst time after every iteration of RR are considered in the first strategy. The random, fixed time quantum is used for this sorted ready queue-based RR algorithm. The implementation results show that the sorting based on burst time gives better results than FCFS in the ready queue. In the second proposed technique, the adaptive round-robin algorithm finds the optimal dynamic time quantum value according to the characteristics of sorted processes in the ready queue. The comparison results between these two proposed strategies, the adaptive TQ-based round-robin algorithm, give better performance in AWT and ATT than the just sorted ready queue-based RR algorithm.

References

1. Balharith T, Alhaidari F (2019) Round robin scheduling algorithm in CPU and cloud computing: a review. In: 2019 2nd international conference on computer applications & information security (ICCAIS), Riyadh, Saudi Arabia, 2019, pp 1–7. https://doi.org/10.1109/CAIS.2019.8769534
2. Srujana R, Roopa YM, Mohan MDSK (2019) Sorted round robin algorithm. In: 2019 3rd international conference on trends in electronics and informatics (ICOEI), Tirunelveli, India, pp 968–971. https://doi.org/10.1109/ICOEI.2019.8862609
3. Jbara YH (2019) A new improved round robin-based scheduling algorithm—a comparative analysis. In: 2019 international conference on computer and information sciences (ICCIS), Sakaka, Saudi Arabia, pp 1–6. https://doi.org/10.1109/ICCISci.2019.8716476
4. Mishra R, Mitawa G (2021) Improved round robin algorithm for effective scheduling process for CPU. In: 2021 third international conference on intelligent communication technologies and virtual mobile networks (ICICV), Tirunelveli, India, pp 590–593. https://doi.org/10.1109/ICICV50876.2021.9388561
5. Sabha SU (2018) A novel and efficient round robin algorithm with intelligent time slice and shortest remaining time first. Mater Today Proc 5(5, Part 2):12009–12015. ISSN 2214-7853
6. Chandiramani K, Verma R, Sivagami M (2019) A modified priority preemptive algorithm for CPU scheduling. Proc Comput Sci 165:363–369
7. Chauhan H, Inani A (2019) Modified concept to achieve maximum efficiency of CPU scheduling algorithm. In: 2019 3rd international conference on electronics, communication and aerospace technology (ICECA), Coimbatore, India, pp 660–663. https://doi.org/10.1109/ICECA.2019.8821820
8. NEHA, ANKITA JIYANI (2018) An improved round robin CPU scheduling algorithm. IRE J 1(9). ISSN: 2456-8880
9. Farooq MU, Shakoor A, Siddique AB (2017) An efficient dynamic round robin algorithm for CPU scheduling. In: 2017 international conference on communication, computing and digital systems (C-CODE), Islamabad, Pakistan, pp 244–248. https://doi.org/10.1109/C-CODE.2017.7918936
10. Panda S, Bhoi S (2014) An effective round robin algorithm using min-max dispersion measure. Int J 4
11. Mijinyawa HM, Abdullahi SE (2017) Improved group based time quantum (IGBTQ) CPU scheduling algorithm. In: 13th international conference on electronics, computer and computation (ICECCO), Abuja, Nigeria, pp 1–6. https://doi.org/10.1109/ICECCO.2017.8333324
12. Mody S, Mirkar S (2019) Smart round robin CPU scheduling algorithm for operating systems. In: 2019 4th international conference on electrical, electronics, communication, computer technologies and optimization techniques (ICEECCOT), Mysuru, India, pp 309–316. https://doi.org/10.1109/ICEECCOT46775.2019.9114602
13. Alaa F, Zoulikha MM, Hayat B (2020) Improved round robin scheduling algorithm with varying time quantum. In: 2020 second international conference on embedded & distributed systems (EDiS), Oran, Algeria, pp 33–37. https://doi.org/10.1109/EDiS49545.2020.9296452

14. Mora H, Abdullahi SE, Junaidu SB (2017) Modified median round robin algorithm (MMRRA). In: 2017 13th international conference on electronics, computer and computation (ICECCO), Abuja, Nigeria, pp 1–7. https://doi.org/10.1109/ICECCO.2017.8333325
15. Sakshi, Sharma C, Sharma S, Kautish S, Alsallami SAM, Khalil EM, Mohamed AW (2022) A new median-average round robin scheduling algorithm: an optimal approach for reducing turnaround and waiting time. Alex Eng J 61(12):10527–10538. ISSN 1110-0168
16. Butt MA, Akram M (2015) A novel n decision-making system for CPU scheduling algorithm. Neural Comput Appl 27:1–13. https://doi.org/10.1007/s00521-015-1987-8
17. Sharma R, Goel AK, Sharma MK, Dhiman N, Mishra VN (2023) Modified round robin CPU scheduling: a fuzzy logic-based approach. In: Gunasekaran A, Sharma JK, Kar S (eds) Applications of operational research in business and industries. Lecture notes in operations research. Springer, Singapore. https://doi.org/10.1007/978-981-19-8012-1_24
18. Elmougy S, Sarhan S, Joundy M (2017) A novel hybrid of shortest job first and round robin with dynamic variable quantum time task scheduling technique. J Cloud Comp 12(6). https://doi.org/10.1186/s13677-017-0085-0
19. Das S, Mishra S, Sahu A (2019) An efficient average execution time-round-robin (AET-RR) scheduling algorithm. Int J Inf Technol 14. https://doi.org/10.1007/s41870-019-00351-5
20. Sharma PS, Kumar S, Gaur MS et al (2022) A novel intelligent round robin CPU scheduling algorithm. Int J Inf Technol 14:1475–1482. https://doi.org/10.1007/s41870-021-00630-0
21. Ullah W, Shah M (2022) A novel resilent round robin algorithm based CPU scheduling for efficient CPU utlilization, pp 41–48. https://doi.org/10.1049/icp.2022.2038
22. Oshani Bandara AM, Samantha Rajapaksha UU (2022) An improved round robin algorithm for an efficient CPU scheduling. In: 2022 2nd international conference on advanced research in computing (ICARC), Belihuloya, Sri Lanka, pp 349–354. https://doi.org/10.1109/ICARC54489.2022.9754037

Low-Cost Advanced Undetectable Hardware-Based Malicious Keyboard

Kevin Diaz and Jaya Dofe

Abstract The widespread presence of peripheral devices in households, connecting to desktops and laptops, creates a substantial security vulnerability. These devices can be manipulated at the hardware level, allowing unauthorized access to host systems. Although modern operating systems like Windows, Linux, and Mac incorporate software security measures to counteract malicious keylogging programs, these defenses can be evaded through a combination of encryption, decryption, and hardware techniques. This evasion makes them effectively undetectable at the OS level, as demonstrated in prior research. The existing literature emphasizes the considerable threat such devices pose to future users. In this study, we develop a custom malicious keyboard at a low cost, providing a practical demonstration of potential attacks. Through this research, our aim is to illuminate the inherent vulnerabilities within peripheral devices, highlighting the critical need to bolster hardware security measures to safeguard against these covert and pernicious threats.

Keywords Keylogger · Security · Privacy · Malware · Operating systems

1 Introduction

Computer security is continuously evolving today because we store a large volume of information digitally. Since this data can be sensitive, it's crucial to protect it effectively. A keylogger can be employed to monitor employees, children, or other individuals to address security weaknesses and concerns effectively. Businesses can benefit from using keyloggers to enhance transparency within their operations. Additionally, ethical hackers can utilize keyloggers to assess computer system vulnerabil-

K. Diaz · J. Dofe (✉)
Electrical and Computer Engineering, California State University, Fullerton, 800 N State College Blvd, Fullerton, CA 92831, USA
e-mail: jdofe@fullerton.edu

K. Diaz
e-mail: kevin_diaz@csu.fullerton.edu

S. M. Thampi et al. (eds.), *Fifth International Conference on Computing and Network Communications*, Lecture Notes in Electrical Engineering 1219,
https://doi.org/10.1007/978-981-97-4540-1_24

ities [1]. However, attackers frequently misuse keyloggers with malicious intent to steal sensitive information. Keyloggers have been responsible for stealing numerous credit card numbers and passwords [2, 3].

A peripheral device like a keyboard distinguishes each physical key and sends this information as a report to the controlling software, typically the operating system (OS). The software is responsible for handling these keystrokes. There are three classes of attacks known as Type I, Type II, and Type III, which target user keystrokes in a predefined user space [2]. Type I keyloggers use global data structures and run in a separate process. Type II keyloggers rely on local data structures and execute within a process, logging communication between the target process. Type III keyloggers leverage a specific attack application to turn the application into a single-based keylogger. Today, physical hardware keyloggers are placed between the keyboard and the computer's hardware. However, these are visible and easily detected through software because the USB host can monitor such devices [4]. In this work, we aim to bridge a gap by making the following contributions:

- We analyze various attack types targeting user spaces to enhance our understanding of their mechanisms and vulnerabilities.
- We develop more effective methods for accessing user spaces to target users, focusing on creating a custom hardware-based malicious keyboard capable of bypassing and manipulating the target device.
- We ensure the compatibility of our approach with virtual machines [5].
- We present a low-cost device for the end-users and researchers.

To achieve our goals, we integrate hardware and software components within a specialized chip, disguising it as a regular peripheral device to deceive the OS into believing it's safe. Our discreet keylogging device serves a dual purpose—capturing user inputs and providing a low-cost, efficient tool for cybersecurity research. Such a device is invaluable for researchers and penetration testers, as it facilitates in-depth analysis and testing of security measures and vulnerabilities.

The rest of the paper is organized as follows. Section 2 summarizes the background and related work. The proposed work and the attack model used in this work are described in Sect. 3. Experimental setup and results are presented in Sect. 4. The future directions are discussed in Sect. 5. This work is concluded in Sect. 6.

2 Background and Related Work

2.1 Keyloggers Types

Keyloggers are unlike other types of spyware or malware, such as viruses and worms. They discreetly use system resources like the CPU and memory alongside legitimate programs. They remain hidden on the system for as long as needed and are designed to perform their tasks without alerting users. Keyloggers have long been an issue in the realm of computer security, falling into two broad categories: software and hardware keyloggers [6, 7]. Software keyloggers operate by installing malicious software onto a host machine, typically exploiting vulnerabilities in software applications or operating systems [8, 9]. These types of keyloggers can often be detected and mitigated through the use of antivirus or other security software. On the other hand, hardware keyloggers exist as physical devices, usually embedded between the keyboard and the computer, intercepting and logging keystrokes. The evolution of hardware keyloggers has seen the transition from easily detectable inline devices to sophisticated, embedded components, reducing their visibility and increasing the difficulty in detection—for example, research proposed a method of recording keystrokes using a teensy board that is installed within any typical keyboard but has the downside of only being 6-key rollover, which only allows 6 keys to be pressed simultaneously [10].

2.2 Related Work

Research in hardware-based keyloggers has led to the development of methods for blending and hiding them. These methods are becoming more advanced and sophisticated over time [8]. Many studies have looked into adding logging components inside keyboards to fit them well into the printed circuit board (PCB) [11]. There are various complex methods employed to conceal hardware-based keyloggers. Research has emphasized using RISC-based microcontrollers for programming the keylogging component, ensuring discreet operation [12]. The ability of these microcontrollers to manage and log data while maintaining the appearance of normal functionality is a critical factor in the success of the keyboard design. Some core aspects of Keylogging code are mentioned below:

- Obfuscation of Code: Ensuring the keylogging code remains obfuscated is crucial to its stealth and effectiveness. We employ code obfuscation techniques to render software analysis tools ineffective for researchers to understand and analyze. As this paper shows, analysis tools failed to decipher important functions.
- Impact and Efficacy: The impact of integrating such concealed hardware-based keyloggers is profound, pointing to the evident vulnerabilities in existing security models. The efficacy of these integration methodologies was analyzed using various evaluation metrics, revealing good success rates in maintaining concealment against state-of-the-art security software, Ghidra [13]. The integration techniques' evolution shapes the landscape of hardware-based attacks, demonstrating the possibilities of exploiting even the most seemingly secure environments.
- Analytical Evaluation: Through analytical evaluation and real-world application testing, the metrics obtained have demonstrated the capability of such concealed devices to bypass detection mechanisms successfully. The analyses involved testing against well-known security software, aiming to assess the resilience and stealthiness of the integrated keyloggers.

Fig. 1 Front side of the PCB keyboard

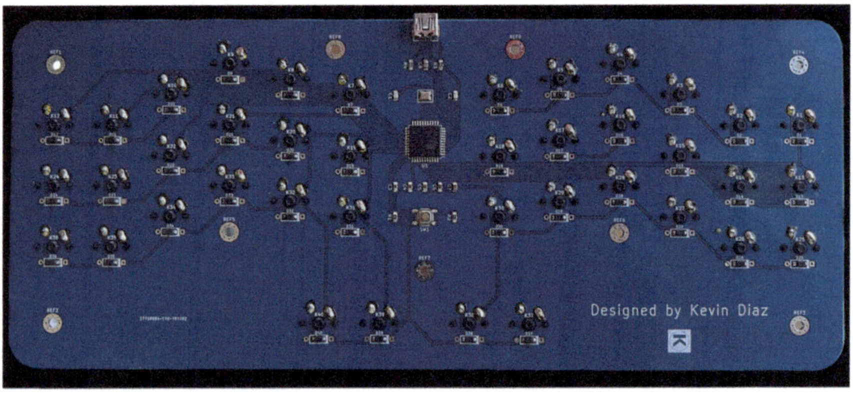

Fig. 2 Back side of the PCB keyboard

3 Our Approach

In this study, we have developed a specialized chip designed to conceal both hardware and software components, giving it the appearance of a conventional device to deceive operating system (OS). This device effectively captures user inputs and offers a cost-efficient solution for conducting cybersecurity research. To fabricate this device, we utilized the KiCad Electronic Design Automation (EDA) software [14] for board design, creating a modern hardware keylogger.

Figures 1, 2, and 3 display the front side, back side, and schematic of the proposed keyboard, respectively. We also conducted an in-depth analysis of modern operating systems, including Windows 11, Ubuntu 22.10, and macOS Ventura 13.1, to assess how the imperceptible keylogger device interacts with them. Our ultimate goal is to achieve complete detection evasion with this device.

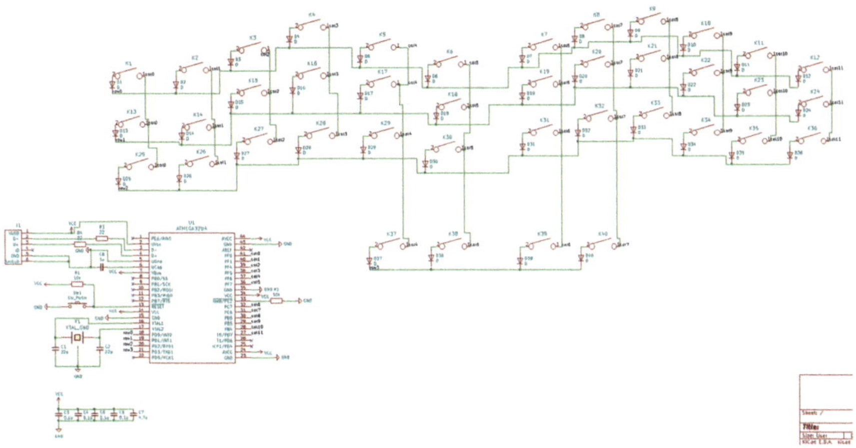

Fig. 3 Keyboard schematic design

Table 1 Windows 11 build versions (February–May 2023)

Version	Build number	KB number	Release date
Windows 11 22H2	10.0.22621.1702	KB5026372	9th May 2023
Windows 11 21H2	10.0.22000.1936	KB5026368	9th May 2023
Windows 11 22H2	10.0.22621.1555	KB5025239	11th April 2023
Windows 11 21H2	10.0.22000.1817	KB5025224	11th April 2023
Windows 11 22H2 (prod. release)	10.0.22621.1265	KB5022845	14th Feb 2023
Windows 11 21H2 (prod. release)	10.0.22000.1574	KB5022845	14th Feb 2023

Attack Model

The creation of the attack model involved careful design and advanced planning to infiltrate multiple operating systems using a custom malicious keyboard seamlessly. Specific methodologies and structured approaches are used to ensure a comprehensive assessment of the level of infiltration, the volume of data extracted, and the resilience of the attack model against security measures. We focus our attention on the following operating systems, with specific details regarding the build version of Windows 11 provided in Table 1:

- Windows 11
- Ubuntu 22.10
- macOS Ventura 13.1.

4 Experimental Analysis

4.1 Compatibility and Interface Testing

We conducted a brief analysis across multiple USB interfaces to guarantee the universal stealthiness of our proposed custom malicious keyboard across various devices and interfaces. We utilized a multi-interface device, as depicted in Fig. 4, which accommodates various USB types, including USB Type-A, USB Type-C, Micro USB, and Apple's lightning interface. Each of these interfaces is equipped with test points for the differential data lines (D+ and D−), power (VCC), and ground (GND). In the testing process, this device ensures that our keylogger device receives power and verifies the correct functioning of the D+ and D− lines. The status of D+, D−, VCC, and GND is indicated by faint blue LEDs, except for the CC light, which remains off as it pertains to the USB Type-C protocol that is not in use.

We program the device firmware to replicate the Human Interface Device (HID) profile of conventional keyboards. Since operating systems inherently trust input devices, they do not monitor these profiles once authenticated. The keylogger bypasses software barriers entirely by interfacing directly with the keyboard's hardware. It does not rely on any system-level drivers or software that could be monitored or flagged by security protocols within the OS. Also, the keylogger operates by monitoring electrical signals directly from the keyboard matrix, which are not typically observed by the OS beyond the point of the keypress interpretation.

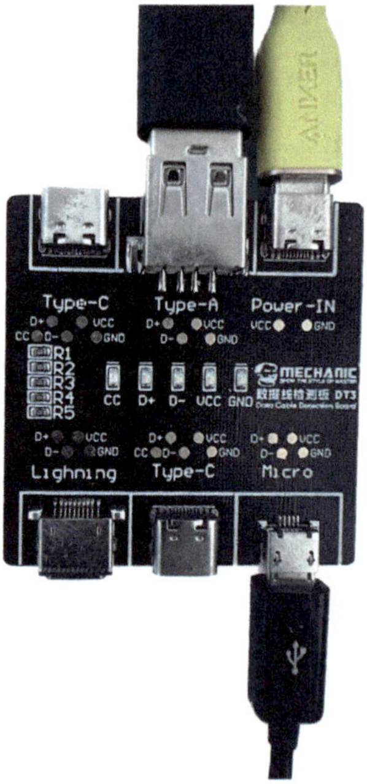

Fig. 4 Multi-interface device for USB data testing across different devices types

Table 2 Different OS attack analysis

OS type	No. of physical OS	No. of virtual OS	Success rate (%)	Average infiltration time (s)
Windows 11	40	13	100	2.6
Ubuntu 22.10	4	1	100	2.1
macOS Ventura 13.1	5	16	100	2.5

4.2 *Infiltration and Data Extraction Methodology*

Our infiltration and data extraction methods are thoroughly designed to record and store keystrokes in real-time. The captured data is then securely stored within the internal storage of the ATMEGA32U4 chip. To access the keylogger's contents, one needs to overwrite the keyboard's flash memory with the EEPROM flash file. This action allows us to read the data stored in the EEPROM. When we flash the keyboards keylogger flash file again, there is an initial sequence where the EEPROM addresses are all written with zeros, so you don't have any previous keylogged data stored.

In our research, we gathered essential metrics to evaluate the performance of our keylogging device in infiltrating systems and efficiently extracting data. We defined them as below:

Average Infiltration Time: This was calculated as the total time taken for all successful infiltrations divided by the number of successful infiltrations.

Data Extraction Efficiency: This was computed based on the volume of relevant data extracted against the total volume of extracted data, including garbage data and characters such as @.#*!.

Our test environments primarily comprised virtual machines for evaluating different Windows 11 builds, while some tests were conducted on physical systems (Table 2). It's important to note that installing the HID driver for the keyboard may extend the time required for Windows 11 machines, although this doesn't apply universally. The performance of Ubuntu and macOS systems remained unaffected as well. We have calculated the Success Rate and Average Infiltration Time for different operating systems, as detailed in Table 2, to provide insights into the effectiveness of the proposed model under various OS conditions. The average metrics are listed below:

- Number of Successfully Infiltrated Systems: We successfully infiltrated 79 systems.
- Average Time for Infiltration: The average time taken for each infiltration was approximately 2.4 s. While slight variations existed between physical and virtual machines, the overall infiltration process remained fast.
- Volume of Extracted and Stored Data: We collected 80 kB of data from all the infiltrated machines, equating to 81,920 characters logged during the operation.

- Efficiency of Data Extraction: Our data extraction process demonstrated an efficiency rate of 96.7%. This percentage represents the proportion of relevant data, specifically characters A to Z, successfully extracted. It's worth noting that some garbage data was logged due to certain factors, but our extraction process effectively filtered and retained the pertinent information.

4.3 *Security Assessment Methodology*

We employed Ghidra, an open-source reverse engineering tool developed by the National Security Agency (NSA) [13], to analyze the software components of the proposed Keylogger. This tool aided us in comprehending the Keylogger's operational capabilities and allowed us to assess its susceptibility to detection or countermeasures. As illustrated in Fig. 5, the .hex file is flashed into the ATMEGA chip in the Intel hex format. While we obtained certain bootloader information through this analysis, we did not identify any significant code beyond that as demonstrated in Fig. 6.

This approach to calculations and analyses provides good insights into the attack model's adaptability, success rate, and resilience, aiding in understanding its operational effectiveness and potential areas for refinement and enhancement.

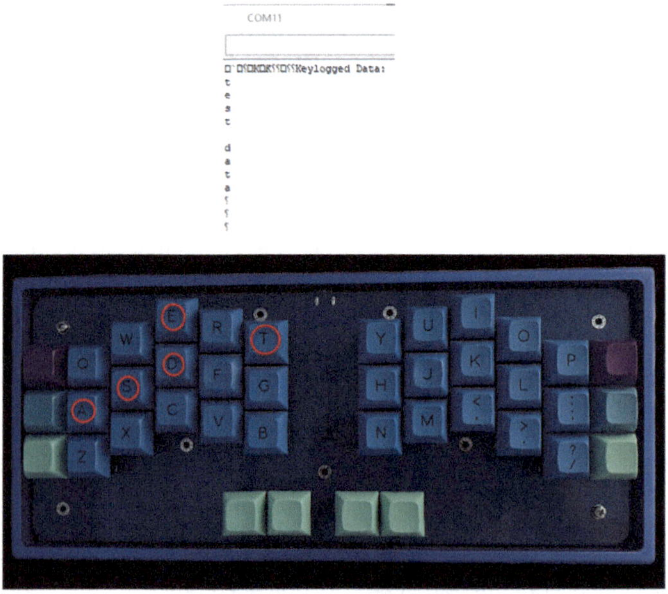

Fig. 5 Data extracted from EEPROM ("test data" is pressed and keylogged)

```
undefined2 FUN_code_042c(void)

{
  byte bVar1;
  char cVar2;
  char cVar3;
  undefined1 *puVar4;
  bool bVar5;
  bool bVar6;

  R1 = 0;
  SREG = 0;
  Ylo = 0xff;
  Yhi = 10;
  bVar1 = MCUSR;
  if ((bVar1 & 8) != 0) {
    bVar5 = DAT_mem_0231 < 0xb0 || (byte)(DAT_mem_0231 + 0x50) < (DAT_mem_0230 < 7);
    cVar2 = DAT_mem_0232 - (bVar5 + '\a');
    bVar6 = DAT_mem_0232 < 7 || (byte)(DAT_mem_0232 - 7) < bVar5;
    bVar1 = DAT_mem_0233 + 0x50;
    cVar3 = DAT_mem_0233 - (bVar6 + -0x50);
    bVar5 = DAT_mem_0233 < 0xb0;
    X = (undefined1 *)CONCAT11(cVar3,cVar2);
    if (((DAT_mem_0230 == 7 && DAT_mem_0231 == (byte)((DAT_mem_0230 < 7) + 0xb0U)) && cVar2 == '\0')
        && cVar3 == '\0') {
      DAT_mem_0230 = 0;
      DAT_mem_0231 = 0;
      DAT_mem_0232 = 0;
      DAT_mem_0233 = 0;
      MCUSR = 0;
      R25R24._0_1_ = MCUSR;
      R25R24._0_1_ = (byte)R25R24 & 0xf7;
      MCUSR = (byte)R25R24;
      SREG = (bVar5 || bVar1 < bVar6) | ((byte)R25R24 == 0) << 1 | ((char)(byte)R25R24 < '\0') << 2
             | ((char)(byte)R25R24 < '\0') << 4;
      watchdog_reset();
      R25R24._0_1_ = DAT_mem_0060;
      R25R24._0_1_ = (byte)R25R24 | 0x18;
      DAT_mem_0060 = (byte)R25R24;
      DAT_mem_0060 = 0;
      Z = (undefined1 *)0x3800;
      DAT_mem_0afe = 0x45c;
      (*(code *)0x3800)();
    }
```

Fig. 6 Ghidra analysis of hex file

5 Possible Future Directions

To further enhance the capabilities of our custom malicious keyboard, we have identified several key areas of improvement and potential enhancements:

1. **Separate Storage Chip Integration**: One of the limitations of our proposed model is its reliance on the ATMEGA32U4's in-built EEPROM storage, which restricts our data logging capacity to 1024 bytes. Future iterations of this device will incorporate a dedicated storage chip to address this limitation and provide the storage of a larger volume of keylogger data. This enhancement will expand our data storage capacity and provide the flexibility to partition and better organize the data.

2. **Wireless Capabilities**: We intend to incorporate Wi-Fi features into our keyboard to improve our device's data exfiltration capabilities. The model will be able to function as a Wi-Fi hotspot, facilitating the real-time transmission of logged data to a designated device as long as the user is close to our malicious keyboard. This wireless data transmission method enables real-time monitoring for the attacker and adds an extra layer of stealthiness, eliminating the need for any physical connection to retrieve logged data.
3. **Advanced Script Execution**: Our current device can automatically type and execute basic scripts, including actions like opening websites or running malicious payloads. We aim to improve this feature by taking inspiration from the USB Rubber Ducky [15], known for its efficiency in running complex payloads swiftly. We plan to enhance the scripting engine in future iterations of our malicious keyboard. This improvement will enable the storage and execution of more intricate and versatile scripts. Although we have conducted limited script testing, we intend to perform thorough testing to ensure seamless operation with a wider range of scripts. This enhancement will significantly expand the available attack methods, transitioning from simple keylogging to more proactive attacks like downloading and executing malicious payloads or modifying system settings.
4. **FPGA ATMEGA32U4 Architecture Cloning**: To avoid being detected by speed analysis techniques, we can clone the ATMEGA architecture and make certain modifications to the architecture to match the keystroke speed of a regular keyboard (assuming both keyboards are using the same architecture and are of the same model).

6 Conclusion

This paper closely examined hardware-based keyloggers and its effectiveness in cybersecurity. Our innovative and cost-effective approach demonstrated how microcontrollers can skillfully handle and hide logged data while pretending to be regular everyday devices that go unnoticed. We tested our method on 79 different operating systems and achieved a success of 100%. The high success rate means this technique can infiltrate various computer systems and gather significant data. While our research provides valuable insights into hardware-based keylogging, it also reminds us of the importance of staying alert and taking proactive steps to defend against such threats. As technology advances, so do the tactics of those who seek unauthorized access to personal information. We hope our findings will serve as a helpful resource for the cybersecurity community, emphasizing the importance of ongoing learning, adaptation, and progress in response to evolving threats.

References

1. Bejo SP, Kumar B, Banerjee P, Jha P, Singh AN, Dehury MK (2023) Design, analysis and implementation of an advanced keylogger to defend cyber threats. In: 2023 9th international conference on advanced computing and communication systems (ICACCS), vol 1, pp 2269–2274. https://doi.org/10.1109/ICACCS57279.2023.10112977
2. Ortolani S, Giuffrida C, Crispo B (2013) Unprivileged black-box detection of userspace keyloggers. IEEE Trans Depend Secure Comput 10(1):40–52. https://doi.org/10.1109/TDSC.2012.76
3. Ekele Victoria C, Adebiyi Ayodele A, Igbekele Emmanuel O (2023) Keylogger detection: a systematic review. In: 2023 international conference on science, engineering and business for sustainable development goals (SEB-SDG), vol 1, pp 1–6. https://doi.org/10.1109/SEB-SDG57117.2023.10124477
4. Tuscano A, Koshy TS (2021) Types of keyloggers technologies—survey. In: Kumar A, Mozar S (eds) ICCCE 2020. Springer, Singapore, pp 11–22
5. Blåfield T (2020) Different types of keyloggers: mitigation and risk relevancy in modern society. https://trepo.tuni.fi/handle/10024/122479
6. Waterson D (2021) How keyloggers work and how to defeat them. ITNOW 63(1):40–41. https://doi.org/10.1093/itnow/bwab017
7. Sagiroglu S, Canbek G (2009) Keyloggers: increasing threats to computer security and privacy. IEEE Technol Soc Mag 28(3):10–17. https://doi.org/10.1109/MTS.2009.934159
8. Sabu J, Ananthanarayanan S, Gopan A, Gowtham S, Murali S (2023) Advanced keylogger with keystroke dynamics. In: 2023 international conference on inventive computation technologies (ICICT), pp 1598–1603. https://doi.org/10.1109/ICICT57646.2023.10134044
9. Huseynov H, Kourai K, Saadawi T, Igbe O (2020) Virtual machine introspection for anomaly-based keylogger detection. In: 2020 IEEE 21st international conference on high performance switching and routing (HPSR), pp 1–6. https://doi.org/10.1109/HPSR48589.2020.9098980
10. Nasution SM, Purwanto Y, Virgono A, Alam GC (2014) Integration of kleptoware as keyboard keylogger for input recorder using teensy USB development board. In: 2014 8th international conference on telecommunication systems services and applications (TSSA), pp 1–5. https://doi.org/10.1109/TSSA.2014.7065954
11. Ramadhanty AD, Budiono A, Almaarif A (2020) Implementation and analysis of keyboard injection attack using USB devices in Windows operating system. In: 2020 3rd international conference on computer and informatics engineering (IC2IE), pp 449–454. https://doi.org/10.1109/IC2IE50715.2020.9274631
12. Fischer M, Langer F, Mono J, Nasenberg C, Albartus N (2021) Hardware penetration testing knocks your SoCs off. IEEE Des Test 38(1):14–21. https://doi.org/10.1109/MDAT.2020.3013730
13. Cybersecurity speaker series: Ghidra—beyond the code—nsa.gov. Accessed 07-10-2023. https://www.nsa.gov/Press-Room/News-Highlights/Article/Article/2958453/cybersecurity-speaker-series-ghidra-beyond-the-code/
14. KiCad EDA—kicad.org. Accessed 07-10-2023. https://www.kicad.org/
15. GitHub—hak5/usbrubberducky-payloads: the official USB rubber ducky payload repository—github.com. Accessed 07-10-2023. https://github.com/hak5/usbrubberducky-payloads

Constructing a Blind Digital Signature Scheme Based on the Discrete Logarithm Problem and the Root Problem

Van Luu Xuan, Dung Luu Hong, and Hoa Doan Van

Abstract E-government is the latest trend in many countries in which the government is using the online system for delivering government services to citizens. The need for information transmission and transactions on computer networks and the Internet is increasing, leading to requirements for ensuring the safety of information transmission, and the digital signature is used to validate contents and authorize users who are involving in the E-governance system. In some cases, blind digital signatures are used to authenticate the integrity of an electronic message and the identity of the signer, but do not allow authentication of the true origin of the signed message. Although there are many digital signature schemes currently known as: RSA, Elgamal,... the approach of improving the safety level of digital signature schemes based on the hardness of solving simultaneous difficult assumptions has still remained underdeveloped and attracted rising attentions from researchers. This paper proposes a blind digital signature scheme developed from a digital signature scheme built on the difficulty of simultaneously solving the discrete logarithm problem and the root problem. The advantage of the new scheme proposed here is that it has a higher level of security than the previously published schemes in terms of its ability to resist attacks that reveal the source of the signed message.

Keywords Digital signature algorithm · Blind digital signature · Discrete logarithm problem

V. L. Xuan (✉)
People's Security Academy, Hanoi, Vietnam
e-mail: vanlx.hvan@gmail.com

D. L. Hong
Military Technical Academy, Hanoi, Vietnam

H. D. Van
Academy of Military Science and Technology, Hanoi, Vietnam

S. M. Thampi et al. (eds.), *Fifth International Conference on Computing and Network Communications*, Lecture Notes in Electrical Engineering 1219,
https://doi.org/10.1007/978-981-97-4540-1_25

1 Introduction

The concept of a blind digital signature was proposed by D. Chaum in 1983 [1]. Blind digital signature is a type of digital signature used to verify the integrity of an digital message and the identity of the signing object, but it does not allow authentication of the identity of the object that generated the signed message. With conventional digital signatures, the signing object is also the object that generates the signed message, but with blind digital signatures, the signing object and the object creating the signed message are completely different. This is a characteristic feature of blind digital signatures and is also an important criterion to evaluate the security of this type of digital signature.

In [1–5], the authors proposed some blind digital signature schemes that are applied when it is necessary to protect the privacy of customers in electronic payment systems or the anonymity of voters in holding an online election. However, the common weakness of the above schemes is their inability to resist attacks that reveal the source of signed messages, so the applicability of these schemes in practice is very limited. The content of the article focuses on analyzing the vulnerability that can be exploited to reveal the signed message origin of a number of published blind digital signature schemes, thereby proposing to build a new safe scheme. It has the ability to keep the source of the signed message secret so that it can meet the requirements set forth by practice.

2 Weaknesses of a Blind Digital Signature Scheme

In [6], the attack method to reveal the source of the message has been shown for some blind signature schemes developed from the signature schemes RSA, DSA,... Here, we will once again illustrate the use of that method to attack Moldovyan's blind digital signature scheme. With other schemes, the attack can be performed in a completely similar way.

2.1 Moldovyan's Blind Digital Signature Scheme

This is a blind digital signature scheme proposed by Moldovyvan on the basis of development from the Belarusian digital signature standard STB 1176.2-9 [7]. The system parameters include two primes, p and q, which satisfy $q|(p-1)$ and the generator $g \in Z_p^*$ have degree q. The signer has a private key $x \in Z_q$ and a corresponding public key $y = g^x \bmod p$.

The procedure for creating a "blind" signature includes the following steps:

- Signer (A) chooses a value $k \in (1, q)$ at random and computes the value T using: $T = g^k \bmod p$, and then sends T to the requester to sign (B).

- B chooses two values τ and ε at random and computes the values $T' = T \times y^{\tau} \times g^{\varepsilon} \bmod p$, $e' = F_H(T'||M)$ and $e = (e' - \tau) \bmod q$. Here: $F_H(.)$ is a hash function, and "||" is the concatenation operator of two bit strings. Then B sends e to A.
- A calculates the value: $s = (k - x \times e) \bmod q$ and sends it to B.
- B computes the second component of the signature: $s' = (s + \varepsilon) \bmod q$. A's digital signature on the message M is a pair of (e', s').

The signature verification procedure is similar to that in STB 1176.2-9 scheme, including the following steps:

- Check if $1 < s' < q$ and $0 < e' < q$ then go to next step. Otherwise, the signature (e', s') will be rejected for validity.
- Calculation of value $T^* = g^{s'} \times y^{e'} \bmod p$.
- Calculation of value: $e^* = F_H(T^*||M)$.
- Checks if $e^* = e'$ then the digital signature (e', s') is recognized as valid. Otherwise, the digital signature (e', s') will be rejected.

2.2 *Attacking to Discover the Source of the Signed Message*

To attack the origin of a signed message M, signer A needs to store the parameter values (T, e, s) and the identity of sign requester B (ID_B) at each signing. From there, A can determine B's identity using Algorithm 1 as follows:

Algorithm 1 Attack algorithm

Input: $(M, e', s'), \{(e_i, s_i, T_i, ID_{Bi}) | i = 0, 1, 2, \ldots, N\}$
Output: ID_{Bi}
1: $m \leftarrow H(M), i = 0$
2: **select:** (e_i, s_i, T_i, ID_{Bi})
3: $\tau \leftarrow (e' - e_i) \bmod q$
4: $\varepsilon \leftarrow (s' - s_i) \bmod q$
5: $T^* = T_i \times y^{\tau} \times g^{\varepsilon} \bmod p$
6: $e^* = F_H(T^*||M)$
7: **if** $(e^* \neq e')$ **then**
8: $\quad i \leftarrow i + 1$
9: $\quad$ **goto** 2
10: **end if**
11: **return** ID_{Bi}

So, Moldovyan's blind digital signature scheme uses two values τ and ε to generate the difference between the parameters e and e' as well as between s and s'. However, the calculation of these two values as shown in step 3 and step 4 of Algorithm 1 is very simple, so Algorithm 1 has shown that determining the message source is completely possible by signer A. Thus, similar to the RSA, DSA, and Nyberg–Rueppal's blind digital signature schemes in [6], the Moldovyan's blind signature scheme is also an insecure scheme against an attack that reveals the source of the signed message.

3 Building a Blind Digital Signature Scheme that Is Resistant to Attack to Reveal the Origin of the Message

The attack algorithm that reveals the source of the signed message in [6] and Algorithm 1 shows signature schemes that "blinding" the message with a secret parameter as shown in RSA digital signature scheme, or with 2 parameters as in DSA, Nyberg–Rueppal and Moldovyan schemes,... the signer can still find the true source of the signed message (identity of the sign requester). This section proposes the development of a blind signature scheme from a basic signature scheme built on the Discrete Logarithm Problem—DLP [8–12]. The advantage of the proposed new scheme is that it also only uses two secret parameters, like in blind signature schemes like DSA, Nyberg–Rueppal, or Moldovyan,... but can resist an attack to reveal the origin of the signed message.

3.1 *Build the Basic Signature Scheme*

Basic Digital Signature Scheme

The basic signature scheme here is built based on the difficulty of the discrete logarithm problem and the root problem. This is used as the basis for developing the blind signature scheme in the next section. In [8], the authors have proposed some new types of difficult problems that can be applied to build digital signature schemes as follows:

Let a prime number $p \in Z_p^*$, for each pair of numbers $r, s \in Z_p^*$, find x satisfy following equation:

$$r^x = x^s \bmod p \tag{1}$$

Formally, we can see that, if the left side of the Eq. (1) is a constant number (C), then the proposed problem will have the following form:

$$x^s = C \bmod p \tag{2}$$

Easily, we can see that, with s, C are known numbers, the Eq. (2) is a Root Problem on Z_p (RP). Otherwise, if the right side of the Eq. (1) is a constant number, then the proposed problem will have the following form:

$$r^x = C \bmod p \tag{3}$$

Now, with r, C are known numbers, the Eq. (3) is a Discrete Logarithm Problem (DLP), solving this problem is difficult.

However, two sides of the Eq. (1) are not constant numbers here, none of them is known number, as such, algorithms for solving RP and DLP can not be applied for solving this form (1). To solve this hard problem, we must solve two hard problems

(RP and DLP) simultaneously. So that, we can assert that the new form of hard problem is proposed here is much more difficult to solve than DLP.

Key scheme of Elgamal digital signature algorithms is built based on the discrete logarithm problem:

$$y = g^x \bmod p \tag{4}$$

Here, (x, y) are private key and public key of a signer, (g, p) are domain parameters, system parameters that were created by digital authentication service provider, in which, g is generator of the group Z_p, p is a prime number. Finding private key x, from public parameters y, g and p, is a difficult problem if p is a large and strong prime number.

From (4), it can be seen that if g is also a secret parameter that is equal to x then (4) will become the form:

$$y = x^x \bmod p \tag{5}$$

We have: $y = x^x \bmod p \Leftrightarrow y^{1/x} = x \bmod p$

$$\Leftrightarrow y^{\left(x^{-1}\right)} = \left(x^{-1}\right)^{-1} \bmod p \tag{6}$$

Let $x^{-1} = X$, we see that (6) will become:

$$y^X = X^{-1} \bmod p \tag{7}$$

It's easy to see that, (7) is a special case of (1) with $a = y, b = -1$. Thus, the difficulty of solving (5) is similar to the difficulty of solving (1).

In the basic scheme, the signer's private key x is randomly chosen from the interval $(1, p)$ and the corresponding public key y is formed from x by: $y = x^x \bmod p$. Here, p is a prime number chosen so that solving the problem DLP is difficult, and g is the generator of the group Z_p^* of degree q, with $q|(p-1)$.

The basic signature scheme here includes parameter and key formation algorithms, signature generation, and checking algorithms as follows:

• *Parameters and key generation algorithm*:

The process of creating parameters and keys is done through Algorithm 2 as follows:

Algorithm 2 Parameters and key generation algorithm

Input: lp, lq - length by bits of two prime numbers p, q.
Output: p, q, x, y.
1: generate p, q: $len(p) = lp, len(q) = lq, q|(p-1)$
2: select $\alpha : 1 < \alpha < p$
3: $x \leftarrow \alpha^{(p-1)/q} \bmod p$
4: **if** $((x = 1)$ or $(x = q))$ **then**
5: go to 2
6: **end if**
7: $y \leftarrow x^x \bmod p$
8: **return** (p, q, x, y)

In which:

- $len(.)$: Function to calculate length by bits of an integer.
- p, q: System parameter/domain parameter.
- x, y: Private key and public key of the signing object.

• *Digital signature generation algorithm*:
The process of creating an digital signature is carried out through Algorithm 3 as follows:

Algorithm 3 Digital signature generation algorithm

Input: p, q, x, y, M.
Output: (r, s).
1: **select** $k : 1 < k < q$
2: $z \leftarrow x^k \bmod p$
3: $e_1 \leftarrow H(Z||M), e_2 \leftarrow H(Z||y), e_3 \leftarrow H(z||y||M)$
4: $e_4 \leftarrow (x \cdot e_3 - k \cdot e_2) \times (e_1 - e_2)^{-1} \bmod q$
5: $r \leftarrow x^{e_4} \bmod p$
6: $s \leftarrow z \cdot r^{-1} \bmod p$
7: **return** (r, s).

In which:

- M: Message must be signed.
- (r, s): Digital signature of the message M.

• *Testing Algorithm*:
The process of checking digital signature is performed through Algorithm 4 as follows:

Algorithm 4 Testing Algorithm

Input: $p, q, y, M, (r, s)$
Output: $True/False$.
1: $\overline{z} \leftarrow r \cdot s \bmod p$
2: $\overline{e_1} \leftarrow H(\overline{z}||M), \overline{e_2} \leftarrow H(\overline{z}||y), \overline{e_3} \leftarrow H(\overline{z}||y||M)$
3: $A \leftarrow r^{\overline{e_1}} \times s^{\overline{e_2}} \bmod p$
4: $B \leftarrow y^{\overline{e_3}} \bmod p$
5: **if** $(A == B)$ **then**
6:     **return** $True$
7: **else**
8:     **return** $False$
9: **end if**

In which:

- $M, (r, s)$: Message and signature to be verified.
- If $True$ is returned, the integrity and origin of M are asserted. Otherwise, if the result is $False$ then the message M is denied for origin and integrity.

On the Validity of the Basis Scheme

The correctness of the scheme is proved as follows:
Indeed, from Step 6 of Algorithm 3 it follows:

$$z = r \times s \bmod p \tag{8}$$

From Step 1 of Algorithm 4 and (8) we have:

$$\overline{z} = z \tag{9}$$

Substituting (9) into the expression in Step 2 of Algorithm 4 we get::

$$\begin{aligned} \overline{e_1} &= H(\overline{z}||M) \bmod q = H(z||M) \bmod q \\ \overline{e_2} &= H(\overline{z}||y) \bmod q = H(z||y) \bmod q \\ \overline{e_3} &= H(\overline{z}||y||M) \bmod q = H(z||y||M) \bmod q \end{aligned} \tag{10}$$

From (10) and Step 3 in Algorithm 3 deduce:

$$\overline{e_1} = e_1, \overline{e_2} = e_2, \overline{e_3} = e_3$$

Therefore, from Steps 3 and 4 of Algorithm 4, we have::

$$A = r^{\overline{e_1}} \times s^{\overline{e_2}} \bmod p = r^{e_1} \times s^{e_2} \bmod p \tag{11}$$

and:

$$B = y^{\overline{e_3}} \bmod p = y^{e_3} \bmod p \tag{12}$$

Otherwise, from Step 6 in Algorithm 3 it follows:

$$s = x^{k-e_4} \tag{13}$$

Substituting the expression in Step 5 in Algorithm 3, (13) into (11) we get:

$$\begin{aligned} A &= r^{\overline{e_1}} \times s^{\overline{e_2}} \bmod p = r^{e_1} \times s^{e_2} \bmod p \\ &= \left(x^{e_4}\right)^{e_1} \times \left(x^{k-e_4}\right)^{e_2} \bmod p \\ &= x^{e_1 \cdot e_4} \times x^{e_2 \cdot (k-e_4)} \bmod p = x^{e_1 \cdot e_4 + e_2 \cdot k - e_2 \cdot e_4} \bmod p \end{aligned} \tag{14}$$

On the other hand, from step number 4 in Algorithm 3 deduce:

$$e_4 \cdot (e_1 - e_2) \bmod q = (x \cdot e_3 - k \cdot e_2) \bmod q$$

or:

$$(e_1 \cdot e_4 + k \cdot e_2 - e_2 \cdot e_4) \bmod q = x \cdot e_3 \bmod q \tag{15}$$

From $y = x^x \bmod p$, (14) and (15), we get:

$$\begin{aligned} A &= x^{e_1 \cdot e_4 + e_2 \cdot k - e_2 \cdot e_4} \bmod p = x^{x \cdot e_3} \bmod p \\ &= (x^x \bmod p)^{e_3} \bmod p = y^{e_3} \bmod p \end{aligned} \tag{16}$$

From (12) and (16) deduce: $A = B$. This is what needs to be proved.

Security Level of the Basis Algorithm

The level of security of the base scheme can be assessed by its ability to resist some types of attacks, such as:

(a) Private key attack

The secret key attack is performed on the Key Generation Algorithm (Algorithm 2) and steps 2, 4 and 5 of the Signing Algorithm (Algorithm 3). In steps 2 and 5, the parameters r and z are public, but the parameters k and e_4 are secret. Therefore, finding x from steps 2 and 5 of the Signing Algorithm is as difficult as finding x from the key generation algorithm, which is a difficult problem that has no solution. In step 4 of Signing Algorithm 3, e_4 and k themselves are also secret parameters, so finding x from step 4 is not possible. Thus, in order to find the private key, the attacker is forced to solve the discrete logarithm problem combined with root problem using the “brute force attack" method with a computational complexity of $O(p)$.

(b) Signature forgery attack

From the Signature checking algorithm (Algorithm 4) of the proposed new scheme, a pair of (r, s) will be recognized as a valid signature with a message M if it satisfies the following condition: $r^{e_1} \times s^{e_2} \equiv y^{e_3} \bmod p$.

or:

$$\begin{aligned} & r^{H(r \cdot s \bmod p||M)} \times s^{H(r \cdot s \bmod p||y)} \\ & \equiv y^{H(r \cdot s \bmod p||y||M)} \bmod p \end{aligned} \tag{17}$$

From (17) shows, choosing first one of the two values r or s and then calculating the second value (s or r) is the second form of combination of the discrete logarithm problem and root problem. As you know, this is a type of hard problem that currently in mathematics has no other solution than the “brute force” method. Furthermore, using the hash function here, solving (17) to find (r, s) is even more difficult.

Thus, to generate a forged signature corresponding to a given message, the attacker has no choice but to randomly choose a pair of (r, s) satisfying (17), but this is actually a "brute force" attack.

3.2 Constructing Blind Digital Signature Scheme Based on the Discrete Logarithm Problem and the Root Problem

Blind Signing Algorithm

The process of creating a blind digital signature is performed through Algorithm 5 as follows:

Algorithm 5 Blind Signing Algorithm

Input: (p, q, x, y, M).
Output: (r, s).
1: **select** $k : 1 < k < q, z_a \leftarrow x^k \bmod p$
2: **select** $\alpha, \beta : 1 < \alpha < q, z_b \leftarrow z_a^\alpha \times y^{2 \cdot \beta} \bmod p$
3: $e_1 \leftarrow H(z_b||M) \bmod q$
4: $e_2 \leftarrow H(z_b||y) \bmod q$
5: $e_3 \leftarrow \alpha^{-1} \times (H(z_b||y||M) - \beta \times (e_1 + e_2)) \bmod q$
6: $e_4 \leftarrow (x \cdot e_3 - k \cdot e_2) \times (e_1 - e_2)^{-1} \bmod q$
7: $r_a \leftarrow x^{e_1} \bmod p$
8: $s_a \leftarrow z_a \times r_a^{-1} \bmod p$
9: $r \leftarrow r_a^\alpha \times y^\beta \bmod p$
10: $s \leftarrow s_a^\alpha \times y^\beta \bmod p$
11: **return** (r, s)

In which:

- Steps 1, 6, 7, 8 are performed by the signer (A).
- Steps 2, 3, 4, 5, 9, 10 are signed by the requester (B).

Signature Validation Algorithm

The process of checking blind digital signatures is performed through Algorithm 6 as follows:

Algorithm 6 Signature Validation Algorithm

Input: $p, q, y, M, (r, s)$.
Output: $True/False$.
1: $\overline{z} \leftarrow r \times s \bmod p$
2: $\overline{e_1} \leftarrow H(\overline{z}||M) \bmod q$
3: $\overline{e_2} \leftarrow H(\overline{z}||y) \bmod q$
4: $\overline{e_3} \leftarrow H(\overline{z}||y||M) \bmod q$
5: $A \leftarrow r^{\overline{e_1}} \times s^{\overline{e_2}} \bmod p$; $B \leftarrow y^{\overline{e_3}} \bmod p$
6: **if** $(A == B)$ **then**
7: **return** $True$
8: **else**
9: **return** $False$
10: **end if**

On the Validity of Newly Proposed Scheme

The correctness of the scheme is proven as follows:

Indeed, from step 8, 9, 10 of Algorithm 5 and step 1 of Algorithm 6 we have:

$$\begin{aligned}\overline{z} &= r \times s \bmod p = (r_a)^\alpha \times (s_a)^\alpha \times y^\beta \times y^\beta \bmod p \\ &= (r_a \times s_a)^\alpha \times y^{2\beta} \bmod p = z_a^\alpha \times y^{2\beta} \bmod p\end{aligned} \tag{18}$$

From step 2 of Algorithm 5 and (18) we have:

$$\overline{z} = z_b \tag{19}$$

Substituting (19) into the expressions in steps 2, 3, 4 of Algorithm 6 we get:

$$\begin{aligned}\overline{e_1} &= H(\overline{z}||M) \bmod q = H(z_b||M) \bmod q \\ \overline{e_2} &= H(\overline{z}||y) \bmod q = H(z_b||y) \bmod q \\ \overline{e_3} &= H(\overline{z}||y||M) \bmod q = H(z_b||y||M) \bmod q\end{aligned} \tag{20}$$

From steps 3, 4, 5 of Algorithm 5 and steps 2, 3, 4 of Algorithm 6, deduce:

$$\overline{e_1} = e_1, \overline{e_2} = e_2, \overline{e_3} = (\alpha \cdot e_3 + \beta \times (e_1 + e_2)) \bmod q$$

Therefore, from step 5 and step 6 of Algorithm 6 we have:

$$A = r^{\overline{e_1}} \times s^{\overline{e_2}} \bmod p = r^{e_1} \times s^{e_2} \bmod p \tag{21}$$

and:

$$B = y^{\overline{e_3}} \bmod p = y^{\alpha \cdot e_3 + \beta \cdot (e_1 + e_2)} \bmod p \tag{22}$$

On the other hand, from step 8 of Algorithm 5 it follows:

$$s_a = x^{k-e_4} \bmod p \tag{23}$$

Substituting the expression in steps 7, 9, 10 in Algorithm 6 and the expression (23) into (21) we have:

$$\begin{aligned} A &= r^{e_1} \times s^{e_2} \bmod p = r_a^{\alpha \cdot e_1} \times y^{\beta \cdot e_1} \times s_a^{\alpha \cdot e_2} \times y^{\beta \cdot e_2} \bmod p \\ &= (x^{e_4} \bmod p)^{\alpha \cdot e_1} \times (x^{k-e_4} \bmod p)^{\alpha \cdot e_2} \times y^{\beta (e_1 + e_2)} \bmod p \\ &= x^{\alpha \cdot e_1 \cdot e_4} \times x^{\alpha \cdot e_2 \cdot (k - e_4)} \times y^{\beta \cdot (e_1 + e_4)} \bmod p \\ &= x^{\alpha \cdot (e_1 \cdot e_4 + e_2 \cdot k - e_2 \cdot e_4)} \times y^{\beta \cdot (e_1 + e_2)} \bmod p \end{aligned} \tag{24}$$

On the other hand, from step 6 of Algorithm 5 it follows:

$$e_4 \times (e_1 - e_2) \bmod q = (x \times e_3 - k \times e_2) \bmod q$$

or:

$$(e_1 \times e_4 + k \times e_2 - e_2 \times e_4) \bmod q = x \times e_3 \bmod q \tag{25}$$

From $y = x^x \bmod p$, (24) and (25) we get:

$$\begin{aligned} A &= x^{\alpha \cdot (e_1 \cdot e_4 + e_2 \cdot k - e_2 \cdot e_4)} \times y^{\beta \times (e_1 + e_2)} \bmod p \\ &= x^{\alpha \cdot x \cdot e_3} \times y^{\beta \cdot (e_1 + e_2)} \bmod p \\ &= (x^x \bmod p)^{\alpha \cdot e_3} \times y^{\beta \cdot (e_1 + e_2)} \bmod p \\ &= y^{\alpha \cdot e_3} \times y^{\beta \cdot (e_1 + e_2)} \bmod p \\ &= y^{\alpha \cdot e_3 + \beta (e_1 + e_2)} \bmod p \end{aligned} \tag{26}$$

From (22) and (26) deduce: $A = B$. This is what needs to be proved.

Security Level of the Newly Proposed Algorithm

To determine the source of the message, the signer A needs to store the value of the parameters $(k, z_a, r_a, s_a, e_1, e_2, e_3)$ at each sign, along with the identifiers of the person requesting the signature: $ID_B = \{ID_{Bi})|i = 0, 1, 2, \ldots, N\}$.

If the value of the pair (α, β) generated by B can be known, the signer A can determine the relationship between $\{M, (e, s)\}$ with ID_{Bi}, i.e. the source of the message can be determined using one of the following algorithms:

Algorithm 7 Algorithm to determine source of the message

Input: $\{(z_{ai}, r_{ai}, s_{ai}, k_i, ID_{Bi})|i = 0, 1, 2, \ldots, N\}, M, (r, s), \alpha.$
Output: ID_{Bi}.
1: $i = 0$
2: **select**: $(z_{ai}, r_{ai}, s_{ai}, k_i, ID_{Bi})$
3: $z \leftarrow r \times s \bmod p$
4: $e_1 \leftarrow H(z||M) \bmod q$
5: $e_2 \leftarrow H(z||y) \bmod q$
6: $e_3 \leftarrow \alpha^{-1} \times H(z||y||M) \bmod q$
7: $e_4 \leftarrow (x \times e_3 - k_i \times e_2) \times (e_1 - e_2)^{-1} \bmod q$
8: $r_a^* \leftarrow x^{e_4} \bmod p$
9: **if** $(r_a^* \neq r_{ai})$ **then**
10: $i \leftarrow i + 1$
11: goto 2
12: **end if**
13: $s_a^* \leftarrow z_{ai} \times (r_a^*)^{-1} \bmod p$
14: **if** $(s_a^* \neq s_{ai})$ **then**
15: $i \leftarrow i + 1$
16: goto 2
17: **end if**
18: **return** ID_{Bi}

Or:

Algorithm 8 Algorithm to determine source of the message

Input: $\{(e_{1i}, e_{2i}, e_{3i}, ID_{Bi})|i = 0, 1, 2, \ldots, N\}, M, (r, s), \alpha.$
Output: ID_{Bi}.
1: $i = 0$
2: **select**: $(e_{1i}, e_{2i}, e_{3i}, ID_{Bi})$
3: $z \leftarrow r \times s \bmod p$
4: $e_1^* \leftarrow H(z||M) \bmod q$
5: **if** $(e_1^* \neq e_{1i})$ **then**
6: $i \leftarrow i + 1$
7: goto 2
8: **end if**
9: $e_2^* \leftarrow H(z||y) \bmod q$
10: **if** $(e_2^* \neq e_{2i})$ **then**
11: $i \leftarrow i + 1$
12: goto 2
13: **end if**
14: $e_3^* \leftarrow \alpha^{-1} \times H(z||y||M) \bmod q$
15: **if** $(e_3^* \neq e_{3i})$ **then**
16: $i \leftarrow i + 1$
17: goto 2
18: **end if**
19: **return** ID_{Bi}

Or:

Algorithm 9 Algorithm to determine source of the message

Input: $\{z_{ai}, k_i, ID_{Bi} | i = 0, 1, 2, \ldots, N\}, M, (r, s), \alpha.$
Output: ID_{Bi}.
1: $i = 0$
2: **select**: (k_i, z_{ai}, ID_{Bi})
3: $z \leftarrow (z_{ai})^{\alpha}$
4: $e_1 \leftarrow H(z||M) \bmod q$
5: $e_2 \leftarrow H(z||y) \bmod q$
6: $e_3 \leftarrow \alpha^{-1} \times (H(z_b||y||M) - \beta \times (e_1 + e_2)) \bmod q$
7: $e_4 \leftarrow (x \times e_3 - k_i \times e_2) \times (e_1 - e_2)^{-1} \bmod q$
8: $r_a \leftarrow x^{e_4} \bmod p$
9: **if** $(r^* \neq r)$ **then**
10: $\quad i \leftarrow i + 1$
11: $\quad$ goto 2
12: **end if**
13: $s_a \leftarrow z_{ai} \times r_a^{-1} \bmod p$
14: $s^* \leftarrow s_a^{\alpha} \bmod p$
15: **if** $(s^* \neq s)$ **then**
16: $\quad i \leftarrow i + 1$
17: $\quad$ goto 2
18: **end if**
19: **return** ID_{Bi}

To find the pair (α, β), A must solve the system of equations as follows:

$$\begin{cases} r = (r_{ai})^{\alpha} \times y^{\beta} \bmod q \\ s = (s_{ai})^{\alpha} \times y^{\beta} \bmod q \end{cases} \tag{27}$$

However, solving (27) will not guarantee the desired (α, β) pair. Because for a $\{M, (r, s)\}$ and each (r_{ai}, s_{ai}) belongs to: $\{(r_{ai}, s_{ai}) | i = 0, 1, 2, \ldots, N\}$ will always find a pair (α, β). Therefore, it is not feasible to apply the above algorithms to attack to reveal the message origin for this scheme.

4 Conclusion

From the weak analysis of some published blind digital signature schemes, the paper proposes a new blind digital signature scheme developed from the basic signature scheme built based on the difficulty of the discrete logarithm and the root problem, this new scheme has a higher level of security than known schemes in terms of resistance to the attack of revealing the source of the signed message. This is an important factor that allows the proposed new scheme to be feasible in real-world applications.

References

1. Chaum D (1983) Blind signature systems. In: Advances in cryptology, Crypto' 83. Plenum Press, p 153
2. Chaum D, Fiat A, Naor M (1990) Untraceable electronic cash. In: Advances in cryptology, Crypto' 88, LNCS 403. Springer, Heidelberg, pp 319–327
3. Chaum D (1989) Privacy protected payment. SMART CARD 2000, Elsevier Science Publishers B.V., pp 69–93
4. Ferguson N (1993) Single term off-line coins. In: Advances in cryptology, Eurocrypt'93, LNCS 765. Springer, Heidelberg, pp 318–328
5. Chaum D, den Boer B, van Heyst E, Mjolsnes S, Steenbeek A (2005) Efficient offline electronic checks. In: Advances in cryptology, Eurocrypt'89, LNCS 434. Springer, Heidelberg, pp 294–301
6. Giang NT, Thai NV, Dung LH (2014) The blind digital signature scheme is built on the root problem. J LQD Universite 5(10):102–114
7. Moldovyan NA (2011) Blind collective signature protocol. Comput Sci J Moldova 19, 1(55)
8. Elgamal T (1985) A public key cryptosystem and a signature scheme based on discrete logarithms. IEEE Trans Inf Theor IT-31(4):469–472
9. Xuan VL, Van HD, Hong DL (2022) Constructing digital signature algorithm based on a new key scheme. In: Fong S, Dey N, Joshi A (eds) ICT analysis and applications. Lecture Notes in Networks and Systems, vol 314. Springer, Singapore
10. GOST R 34.10-94: Russian Federation Standard, Information Technology (1994) Cryptographic data security, produce and check procedures of electronic digital signature based on asymmetric cryptographic algorithm. Government Committee of the Russia for Standards (in Russian)
11. Dung LH, Duc TM, Van LX (2018) A new method for constructing digital signature schemes base on difficulty of the integer factorization and discrete logarithm root problems the Z_n. In: Fundamental and applied IT research conference, pp 1–9
12. Moldovyan NA (2008) Digital signature scheme based on a new hard problem. Comput Sci J Moldova 47(2):163–182

LSTM-Based Deep Learning Model for Energy Management

Shridatha Hegde, Suneeta V. Budihal, and Saroja V. Siddamal

Abstract Smart home devices and its applications have a great potential in successfully predicting the electricity consumption across households. The proposed framework comprises of a hardware setup for data acquisition and software setup for predicting the energy consumption. As an initial process, the energy consumption is measured and stored in a database with the help of PHP via the ESP 32 microcontroller. The AWS cloud service is used for users to view the consumption in real time. Data acquired is utilized by the server to run LSTM neural networks to predict possible future usage. The accuracy check is performed by comparing the predicted future value with the actual consumption. The proposed framework provides 98% accuracy using LSTM in training data and 99% in testing data.

Keywords Electricity · Dataset · LSTM · Deep learning

1 Introduction

It was necessary to track electricity consumption in order to forecast electricity usage with accuracy. The demand for energy depends on weather condition, occupancy, and the kind of machinery and appliances used. The dependence on a large number of elements has increased the complexity of forecasting techniques. For effective distribution, accurate estimation of electricity usage is crucial. Applying every factor that influences power usage, however, may result in a complex forecasting model that is unstable and unpredictable. As a result, time-series solutions are the primary focus of data-driven solutions for predicting electricity usage. Electricity use is a temporal characteristic. As a result, there exist methods for developing the model to forecast electricity use using time series. Time-series analysis is used to find solutions when past data is available, as it displays time-dependent fluctuations.

S. Hegde · S. V. Budihal (✉) · S. V. Siddamal
KLE Technological University, Hubballi, India
e-mail: suneeta_vb@kletech.ac.in
URL: http://www.kletech.ac.in

S. M. Thampi et al. (eds.), *Fifth International Conference on Computing and Network Communications*, Lecture Notes in Electrical Engineering 1219,
https://doi.org/10.1007/978-981-97-4540-1_26

Two methods RNN and LSTM are presented in this paper. To forecast daily, trimester, and monthly electricity usage, the RNN and LSTM were utilized. In comparison with previous models, both models have demonstrated a reduction in root mean square error. The studies were done to try and anticipate the electricity use of both individual homes and blocks of homes.

2 Literature Survey

In this paper, authors presented an Internet of things (IoT) for energy management in a city and corresponding simulations. The system uses Message Queuing Telemetry Transport (MQTT) and Representational State Transfer (REST)-based communication models [1]. The authors discussed building an intelligent energy management on IoT and large data platform using data analytics and scalable storage. Hardware architecture of Home Energy Management System (EMS) consists of RFID reader, microcontroller, temperature, humidity, and current sensors. EMS authorizes customers to monitor and control the device and generation of online bill through mobile application [2]. The system is designed by RFID tags on the household effects. When an user leaves, the system records the movement of household effects. It consists of four pairs of IR sensors, touch panel, and database server [3]. Smart EMS has been designed and developed using Raspberry Pi. In addition, the system includes LPG gas sensor and it can be used in gas leakage-detecting equipment. Here, Raspberry is connected to modulator-demodulator (MODEM) to access and sends e-mails to consumer [4]. Authors discussed predicting electricity usage for house appliance. They used principal component analysis (PCA) to resolve data dimensionality. They also used ML algorithms such as SVR and ANN with backpropagation. Authors also suggested that SVR model can generate better results with respect to accuracy [5].

Authors used linear regression (LR) and SVR for electricity consumption forecasting. Authors used one-year dataset. They obtained a success of 85.7%. Authors conducted a review of important methods of ML. Authors suggested that good results can be obtained using K-Means, LR, RF, SVR [6]. Authors predicted energy consumption of individual household using deep learning approach. They used state explainable autoencoder. Architecture of this model consists of projector and predictor [7]. Authors used two approaches, RNN and LSTM. They used publicly available London smart meter dataset. Authors conducted experiment for predicting a block of electricity consumption [8]. The authors used different ML regression algorithms such as support vector linear, decision tree, LASSO, logistic, Naive Bayes for smart city application. Authors suggested that big data should be required for accuracy in ML algorithm [9]. Authors used DRNN for predicting electricity consumption using LSTM network. Authors achieved an average RMS error of 0.1. Authors considered a dataset of 600 days for predicting electricity consumption [10].

Author used two models, Deep Q learning and DPG for energy optimization using DRL. Authors collected data from advanced metering infrastructure (AMI). Authors achieved minimization of cost or flattening of the net profile of energy [11]. Authors used hybrid prediction approach on evolutionary deep learning to predict energy consumption for building. Authors selected a random dataset for training. Authors suggested that evolutionary deep learning predicted better performance than compared conventional prediction method [12]. Authors suggested that genetic algorithm and swarm optimization algorithm are better for predicting energy needs. The results of this paper illustrated that the accuracy and F-measure reached 96.8% and 97.4%, respectively [13]. Authors used improved radial 7 basis function neural network for prediction of electricity consumption. Authors suggested that is this proposed approach several weather parameters can be used as the input data [14]. Authors used a X12 model and seasonal and trend decomposition using loss decomposition of forecasting the monthly electricity usage methods. Authors discussed the relationship between electricity consumption and season. Authors discussed that forecasting method by combining STL and ARIMA model performs better in comparison with traditional forecasting models [15].

Author presented novel DNN architecture to implement forecasting on an hourly basis. They used CNN components to Know rich characterization. They also used LSTM and RNN algorithms. Authors had dataset for 3 years of hourly data in north China [16]. Authors presented a novel method to find time series using DL. This allows parallelization with adaption to large data context. This methodology has been tested on the ten-minute frequency sampling rate [17]. The authors proposed a novel DNN for short-term forecasting. Experiment results on a big dataset containing hourly loads of method, supporting accurate load forecasting to help power companies to save power and minimize wastage [18]. The authors used ANN and SVM to perform accuracy of automation of switchover in the residential EMSm and SVM performed better than ANN. This can save global energy crisis and change in climatic condition [19]. Smart Home EMS system includes LPG gas sensor (MQ6) and Raspberry Pi and it can be used in gas leakage detecting equipment in consumer. In this, Raspberry is connected to MODEM to access and sends e-mails to consumer [20].

3 Proposed Methodology

The proposed design is discussed in this section, which includes steps for setting up various tools required for the proper functioning of the system. This section also gives an insight of the system framework used in this paper. Systems design is needed while defining the architecture, product design, modules, interfaces, and data to satisfy specified requirements. The proposed system framework for the problem statement as demonstrated in Fig. 1 is discussed. The two layers of the framework perform various functions as shown in Fig. 1. The data collection and control layer consist of non-invasive current sensors and an ESP32 controller. The sensors collect the usage of current data (in Amp), and the controller performs energy consumption data calculation (in KWh) and stores in a cloud service database. The database has raw data and converted data from the sensors. The energy management based on context-awareness layer administrates databases (DBs), with a frontend web interface for real-

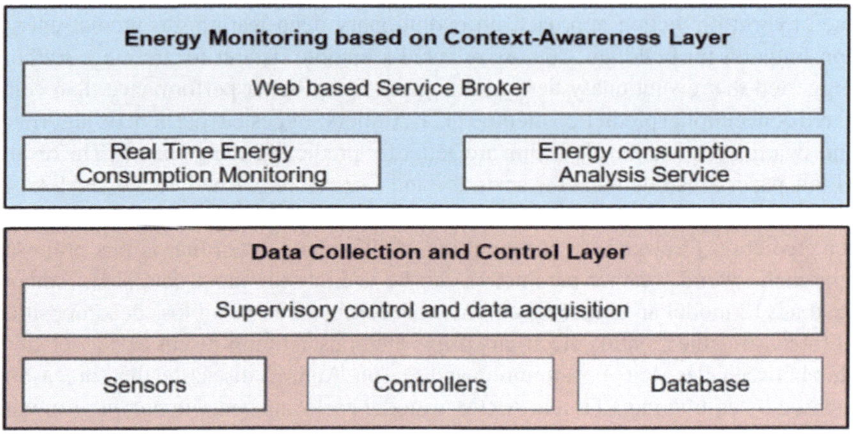

Fig. 1 Proposed energy monitoring block diagram with the first layer based on context-awareness and the second layer as data collection control

time data visualization for the user. Also with the employment of deep learning neural network-based model, analyzing the past and present energy consumption predicts future energy consumption. The layer also provides enhanced energy management strategies and forecasts and outcomes and performance metrics.

3.1 Data Acquisition Implementation

It implies how the data is being collected from the sensor node and how the collected data is transmitted to the database.

Setting up a Sensor Node

A non-invasive current sensor is used to calculate the amount of current owing through a live wire from mains using the concept of, Hall effect. The sensor is a capable of reading the current values upto 30 A. The sensor may produce output voltage when no current is own through the sensor. System needs to know this effect in order to properly calculate the actual current flow.

Data Acquisition and Storage

In this paper, a ESP32 client is built where an HTTP POST requests to a PHP script to insert sensor readings into a MySQL database, as shown in Fig. 2 for data acquisition.

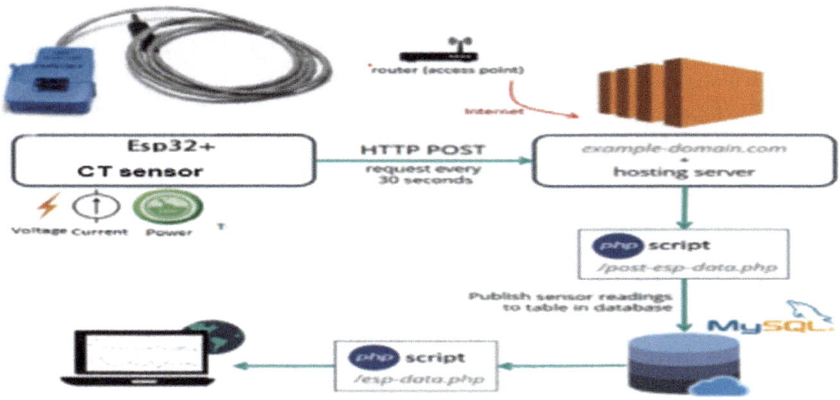

Fig. 2 Data acquisition system with server as infrastructure

Data Analytics and Forecasting

Here data preprocessing, exploratory data analysis, and LSTM neural network models are discussed.

3.2 Data Preprocessing

The reference dataset from Kaggle contains measurements of power consumption in a household with one-minute sampling rate over a period of 4 years as shown in Fig. 3.

Attribute considered:

1. date: date in format dd/mm/yyyy
2. time: time in format hh:mm:ss
3. globalactivepower: household global minute-averaged active power
4. globalreactivepower: household global minute-averaged reactive power (in KW)
5. voltage: minute-averaged voltage (in V)
6. globalintensity: household global minute-averaged current intensity (in A)
7. totalconsumption: combined household consumption from three sub-meterings (in KWh)

The dataset was resampled from minute to hourly frequency for better pattern analysis and seasonal trends. The columns of features 'globalactivepower', 'totalconsumption' along with time were considered for the prediction model.

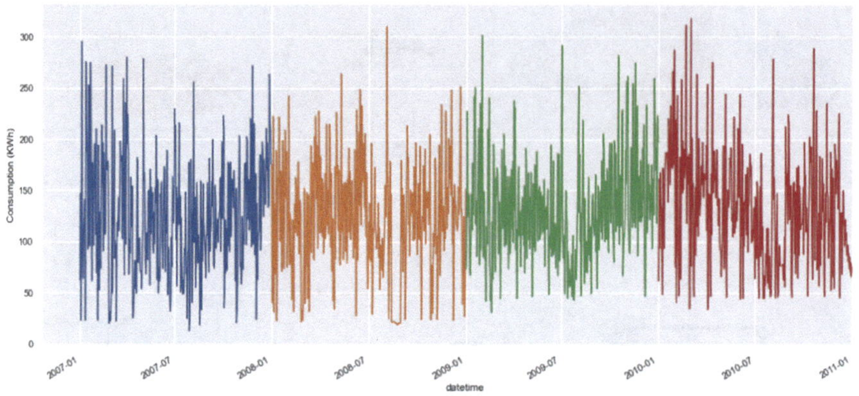

Fig. 3 Overview of dataset visualization with the selected feature as consumption (in kWh) and timestamp set as an index.

```
Augmented Dickey-Fuller Test:
ADF test statistic              -4.278833
p-value                          0.000483
# lags used                     20.000000
# observations                1421.000000
critical value (1%)             -3.434960
critical value (5%)             -2.863576
critical value (10%)            -2.567854
Strong evidence against the null hypothesis
Reject the null hypothesis
Data has no unit root and is stationary
```

Fig. 4 Stationary check using Augmented Dickey–Fuller Test shows that 'p-value' obtained (0.000483) was less than 0.05, and hence it rejects the Null Hypothesis; data has no unit root and is stationary.

Exploratory Data Analysis

Finding interesting relationships between variables, detecting outliers or anomalous events, and better understanding patterns within the data are all aided by the exploratory analysis. To check whether the data used was stationary or not, Augmented Dickey–Fuller Test was conducted on the data. The test results further proved that data is indeed stationary as the 'p-value' obtained (0.000483) was less than 0.05, and hence it rejects the Null Hypothesis, data has no unit root and is stationary as shown in Fig. 4.

3.3 LSTM Neural Networks

Long short-term memory networks (LSTMs) are a unique class of RNN that can recognize long-term dependencies. The algorithmic state for sequential data is represented by RNNs. Internal memory helps it remember the input. For sequence models, it is discovered that past values—rather than features—become the determining

Layer (type)	Output Shape	Param #
lstm (LSTM)	(None, 30, 60)	14880
lstm_1 (LSTM)	(None, 60)	29040
dense_1 (Dense)	(None, 1)	61

```
Total params: 43,981
Trainable params: 43,981
Non-trainable params: 0
```

Fig. 5 LSTM model summary

element for the forecast for the following day. LSTMs offer an advantage over traditional feed-forward neural networks because of their ability to recall certain patterns selectively for a long time. The LSTM architecture consists of four components:

1. Memory Gate—With gate interactions, it resembles a conveyor belt that runs parallel to the entire chain. The gate is a component that controls how information is added to and removed from the LSTM.
2. The forget gate is in charge of erasing data from the cell state. Information that is no longer necessary for the LSTM to comprehend the concepts is eliminated. The network's performance is improved as a result.
3. The input gate is in charge of adding information to the state of the cell. By doing this, it is ensured that the cell will only receive pertinent information and not duplicate data.
4. Output gate—This device is responsible for extracting the pertinent information from the current cell state and displaying it as an output.

The daily resampled dataset contains 1460 samples. 80% of this data is split into train and the remaining 20% as test data to validate the accuracy of the model. The train and test data were normalized using MinMax Scaler, which converts the entire data into the range from 0 to 1. The scaled data is converted into a NumpPy array. The TensorFlow takes input in the form of 3D tensors. The batch size is the parameter considered for future prediction along with timesteps. The sequential model is imported, and two hidden layers of LSTM are added with each layer having 60 input neurons. A dropout layer is added, which helps to deactivate the layers whose output is zero. A dense layer with 'relu' activation function is included, which acts as the output layer of the model as shown in Fig. 5.

3.4 Cloud Services

Saving Weights as Model File

As discussed, the LSTM architecture is used for training the model. The model is saved as a h5 file so that it can save the program's state data to disk and carry on where it left off, when restarted. It is a structured format to save model's weights and configuration in a single file.

AWS EC2 Instance

An EC2 instance is a virtual server in Amazon's Elastic Compute Cloud (EC2) for running applications on the Amazon Web Services (AWS) infrastructure. Users can run application programs in the computing environment using the EC2 service. AMI is a template that includes the software configuration needed to launch the instance, including the operating system, application server, and applications. It is an Ubuntu 20.04, Linux-based OS. A public DNS will be generated, which would be used in the next steps. A new security is created with network traffic set to all and accessible for anyone from anywhere and link this to the instance through network interfaces and permissions.

SSH Client

After creating and launching an EC2 instance, to access it an SSH client is needed. With the help of public DNS of instance and username along with private key, one can access remotely with SSH client. Putty SSH client is used, which is a free open-source serial console and network application for connecting and working with the command line interface of the Ubuntu server.

Secure Copy Protocol

WinSCP is an SFTP, FTP, Amazon S3, and SCP client for Windows that is free and open source. A safe method of transmitting computer files between two remote hosts or between a local host and a remote host is the safe copy protocol (SCP). The Secure Shell (SSH) protocol serves as its foundation. The private key is used with WinSCP to copy all the model files into the server.

Real-Time Visualization

The frontend for the real-time graphs is done with php, html, javascript with MySQL acting as backend, which provides data from it's table.

(1) access database through .php script;
(2) check and verify connection with database;
(3) create chart using HTML5-based JavaScript to generate interactive graphs on html web pages.

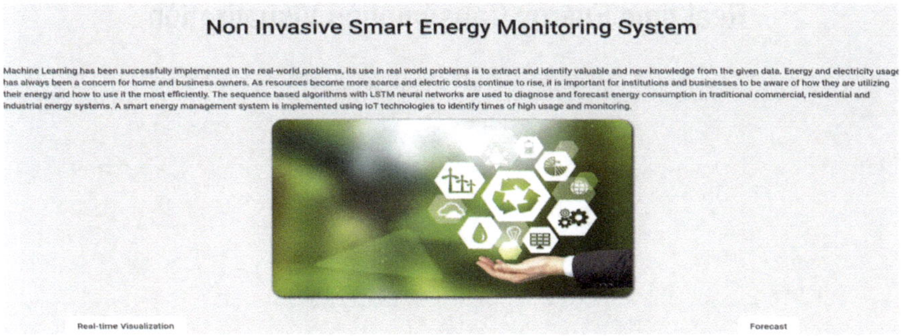

Fig. 6 An interactive-based graphical interface acting as homepage

3.5 User Interface

UI is the top layer where human users can interact with the application. A Web-based graphical user interface to visualize electricity consumption in real time and forecast based on the past consumed units is developed as shown in Fig. 6. Flask Web framework is an API of Python that allows us to build up web applications. Flask does not include a database abstraction layer. This is used to create a web API that would display the prediction in a HTML page using POST method request, which means the model will take input from the user through HTML and returns the predicted value. By defining the app route the HTML page is rendered.

4 Results and Discussions

The results achieved, the optimization used to achieve it and analysis of the results are discussed in this section. It is a procedure that is executed iteratively by comparing various solutions till an optimum solution is found. A well-optimized system is low power consuming, with an optimum operation coverage, cost-effective, and compact.

Hardware optimization: Efficient use of power is a concern that should not be neglected. ESP32 uses 240mA in active mode, this small number can be problematic in the long run, thus to control the power consumption 5 configurable power modes can help in power management of the overall system. The different modes are Active Mode, Modem sleep mode, Light sleep mode, Deep sleep mode, and Hibernation sleep mode. For the setup convenient mode is the light sleep mode. In light sleep mode digital peripherals, CPU and RAM are clock-gated. The CPU is halted by turning of its clock pulses in light sleep mode, although the RTC and ULP-co-processor remain functioning. This uses less power than the modem's sleep mode, which uses roughly 0.8 mA.

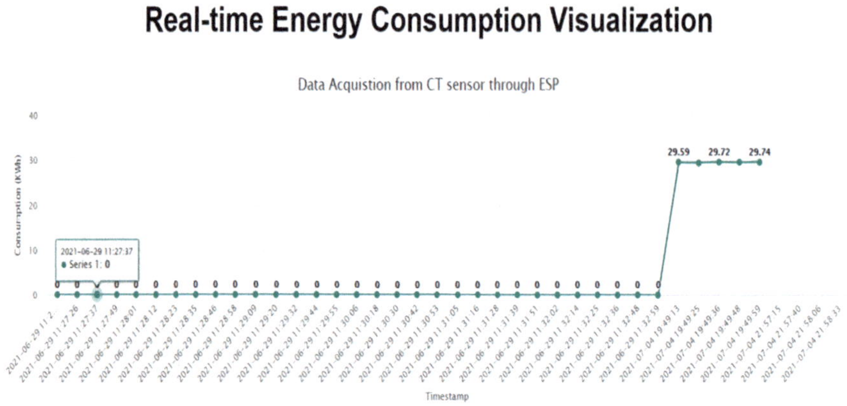

Fig. 7 Dashboard for real-time visualization made with creative chart, which displays data acquired from sensor (consumption value in kWh)

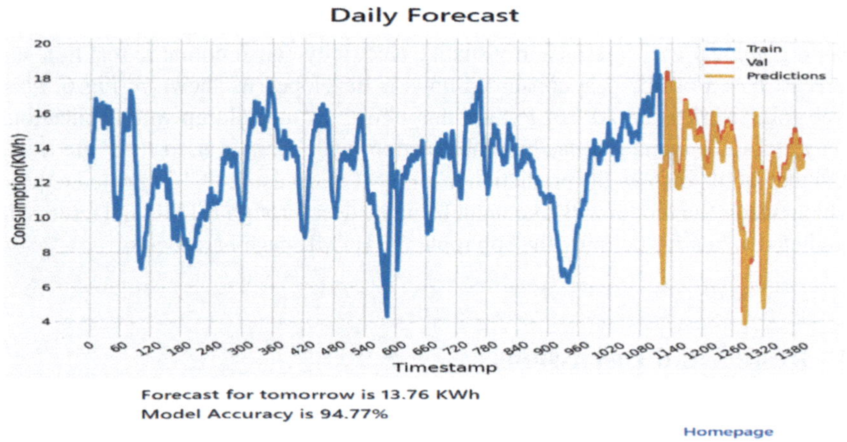

Fig. 8 Plot of forecast model with train, validation, and predicted data values rendered from LSTM model

Deep learning model optimization: In this section, improving the accuracy of the model employing hyper-parameter tuning is discussed. A hyper-parameter is a parameter whose value is used to control the learning process. Hyper-parameter optimization finds a couple of hyper-parameters that yield an optimal model, which minimizes a pre-defined loss function on given independent data. List of hyper-parameters taken into consideration:

1. Batch Size—Number of data feeding in batches for the model.
2. Learning Rate—Parameter used to control how much to change the model in response to the estimated error each time the model weights are updated.
3. Epoch—The number of passes the entire training dataset has completed.

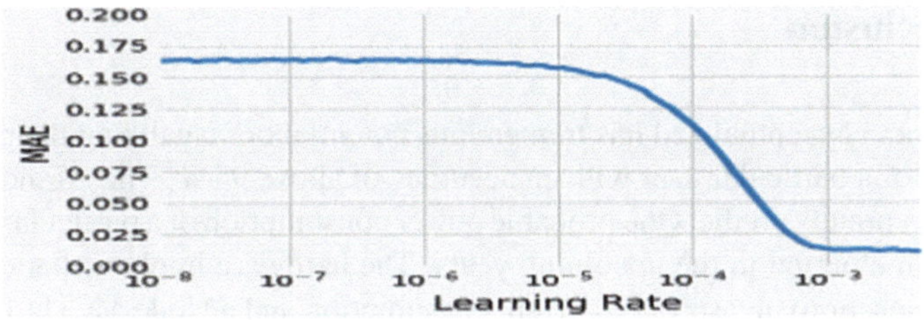

Fig. 9 Learning rate scheduler is used with over 100 epochs, for selecting optimal learning rate

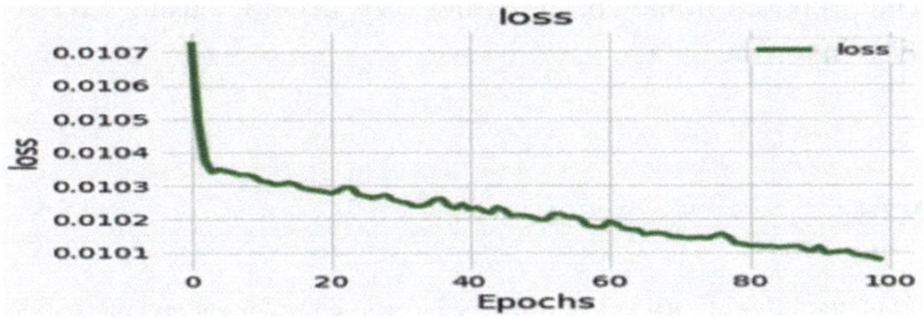

Fig. 10 Optimal learning rate of 0.002 to train the model with 120 epochs to reduce the MAE with much better convergence to improve model accuracy

Table 1 Model accuracy comparison

Model considered	Accuracy in %
ARMA	57
PROPHET	59.42
LSTM	**95**

4. Optimizer—Update the various parameters that can reduce the loss with less effort. Adam optimizer is used as it helps to identify global minima quickly
5. Loss function—It is a method of evaluating how well specific algorithm models the given data. The usage of MAE gradient is large, which can lead to missing minima at the end of training. Huber loss helps to identify and correct this. Figure 7 shows the dashboard for real-time visualization made with creative chart, which displays data acquired from sensor (consumption value in kWh).

The literature survey of different papers and publications yielded the decision matrix for selecting the best-suited system for proof of concept. Various prediction models, including both statistical and ML, were taken into consideration for appropriate trade-off. The usage of the deep learning approach with LSTM architecture resulted in better forecast accuracy of the model. Plot of forecast model for train, validation, and predicted data values rendered from LSTM is shown in Fig. 8. Figures 9 and 10 show the optimal learning rate of 0.002 which is used to train the model with 120 epochs to reduce MAE with much better convergence to deliver improvement in model accuracy. The proposed framework provides 98% accuracy using RNN in training data and 99% in testing data as shown in Table 1.

5 Conclusion

The product conceptualized has tremendous potential of visualizing the power consumption of a particular user with an accuracy of above 95 %. The AI model implemented accurately predicts the probable power consumption of a particular place during the same period in the upcoming years. The hardware implementation requires a current sensor to measure the current consumption and a Node MCU to accept the values from the sensor and send the generated data to the database. The proposed work focuses on models, which can be used to predict electricity consumption for a house. The proposed framework provides 98% accuracy using LSTM in training data and 99% in testing data.

References

1. Brundu FG, Patti E et al (2016) IoT software infrastructure for energy management and simulation in smart cities. IEEE Trans Ind Inf 13(2):832–840. ISSN 1551-3203
2. Al-Ali et al (2017) A smart home energy management system using IoT and big data analytics approach. IEEE Trans Consumer Electronics 426–434
3. Mandava S, Gudipalli A (2018) Analysis of home energy management system using IoT. Int J Pure Appl Math 118(18):3957–3969
4. Lakshmi P et al (2016) Smart home energy management system based on Raspberry Pi using IOT. IOSR J Electronics Commun Eng 11(6):20–25
5. Rasel RI et al (2019) Predicting electric energy use of a low energy house: a machine learning approach. In: International conference on Electrical, Computer and Communication Engineering (ECCE), pp 1–6
6. Gonzalez-Briones A et al (2019) Machine learning models for electricity consumption forecasting: a review. In: 2nd international conference on Computer Applications Information Security (ICCAIS)
7. Jin-Young K, Sung-Bae C (2019) Electric energy consumption prediction by deep learning with state explainable autoencoder. Energies 12(4):739
8. Sachdeva et al. Machine learning approaches in smart cities. In: International conference on innovative computing and communications, vol 471, pp 1–9
9. Mohapatra B. Machine learning approaches to smart city. ACCENTS Trans Image Process Comput Vis 5(14):2455–4707
10. Arasteh H et al (2016) Iot-based smart cities: a survey. In: Proceedings of the EEEIC, pp 2–7
11. Mocanu E et al (2016) Iot-based smart cities: a survey. In: International conference on environment and electrical engineering, Florence, Italy, pp 2–7
12. Li W et al (2021) A comprehensive survey on machine learning-based big data analytics for IoT-enabled smart healthcare system. Mob Netw Appl 26:234–252
13. Cai H et al (2019) Predicting the energy consumption of residential buildings for regional electricity supply-side and demand-side management. Nature Protocols. Print ISSN: 1754-2189
14. Nguyen DH, Tung Nguyen A (2019) A machine learning-based approach for the prediction of electricity consumption. In: 12th Asian Control Conference (ASCC), Kitakyushu, Japan, pp 1301–1306
15. Gunay ME (2016) Forecasting annual gross electricity demand by artificial neural networks using predicted values of socio-economic indicators and climatic conditions: case of Turkey. Energy Policy 90:92–101

16. Ozawa A, Furusato R, Yoshida Y (2016) Determining the relationship between a households lifestyle and its electricity consumption in Japan by analyzing measured electric load profiles. Energy Build 119:200–210
17. Torres JF et al (2017) Deep learning-based approach for time series forecasting with application to electricity load division, ES-41013 Seville, Spain
18. Balouji E, Salor O (2017) Classification of power quality events using deep learning on event images. Gazi University Ankara, Turkey
19. Krishna Prakash N, Prasanna Vadana D (2017) Machine learning based residential EMS. Department of Electrical and Electronics Engineering
20. Ferreira R, Jacobs SIS, Bean CP (1963) Monthly energy consumption forecast: a deep learning approach. Academic, New York, pp 271–350

Inter-frame Video Forgery Detection—A PCA-Based Approach

Wincy Abraham, B. H. Shekar, and Bharathi Pilar

Abstract This research paper explores the utilization of Principal Component Analysis (PCA) for the detection of video forgeries which encompasses the identification of duplicated frames, deleted frames, and inserted frames within the video content. PCA, a potent dimensionality reduction technique widely employed across various domains, is explored for its adaptability to video data. The approach involves preprocessing video frames, extracting principal components, and establishing a PCA-based feature space. The primary objective of this method is to identify irregularities and inconsistencies within video content, which may signify potential forgery. The proposed methodology is rigorously tested using a diverse dataset, with a focus on evaluating the performance of the PCA-based forgery detection system. The results show promising accuracy and precision, underscoring the potential effectiveness of PCA in video forgery detection. We have undertaken a comparative analysis with some of the state-of-the-art techniques. The findings from our observations demonstrate that the proposed method outperforms some established techniques in the detection of video forgery.

Keywords Inter frame · Video forgery · Eigenvalue · Principal Component Analysis

1 Introduction

Video forgery encompasses the process of altering or fabricating video content in a manner designed to deceive viewers into perceiving it as authentic. Such manipulations of video content can serve various malicious purposes, including the dis-

W. Abraham (✉)
Assumption College, Changanassery, Kerala, India
e-mail: wincya@gmail.com

B. H. Shekar
Mangalore University, Mangalagangothri, Karnataka, India

B. Pilar
University College, Mangalore, Karnataka, India

S. M. Thampi et al. (eds.), *Fifth International Conference on Computing and Network Communications*, Lecture Notes in Electrical Engineering 1219,
https://doi.org/10.1007/978-981-97-4540-1_27

semination of misinformation, propaganda, or the creation of false evidence. The widespread availability of video editing tools has significantly simplified the creation of deceptive video content and emphasized the increased demand for robust video forgery detection techniques. With the advancement of AI and machine learning techniques, creating realistic deepfake videos has become increasingly accessible, raising serious concerns in the digital age. The implications of video forgery are multifaceted and far-reaching. Under the serious implications, developing robust and reliable video forgery detection techniques is crucial to combat the spread of deepfake videos and maintain the integrity and trustworthiness of digital media. Researchers, policymakers, and technology developers need to work together to address the challenges posed by video forgery and develop countermeasures to mitigate its potential negative impacts.

Video forgery can be inter-frame forgery or intra-frame forgery. This paper proposes a technique to detect inter-frame forgeries in videos using Principal Component Analysis (PCA). PCA is a widely used dimensionality reduction technique that can be employed for video forgery detection. PCA aims to transform high-dimensional data into a lower-dimensional space while retaining the most important information. PCA achieves this by identifying and computing new orthogonal variables, known as principal components, which are linear combinations of the original features. Eigenvalues and eigenvectors play a fundamental role in Principal Component Analysis (PCA) [1]. They are the key components that allow PCA to identify the principal components. In video forgery detection, PCA can be used to analyze and compare video frames' visual features, seeking inconsistencies that may indicate potential forgeries.

The following section delves into existing works within the field. Section 3 deals with the proposed methodology and Sect. 4 details the experimentation using various available datasets. Section 5 concludes the paper.

2 Literature Review

Numerous studies conducted over the years have aimed to detect various forms of video forgery. In the following section, we highlight key studies that have made notable contributions to this field.

Al-Sanjary et al. [2] focus on passive blind methods for detecting cloning forgery, source camera identification, and splice forgery. Kaur et al. [3] analyze inter-frame forgery detection, intra-frame forgery detection, object-based mechanisms, and pixel-based approaches. Papinwar et al. [4] review video forgery detection using watermarking techniques, while Nilesh et al. [5] discuss video authentication methods that detect region duplication and frame duplication. These papers provide insights into different approaches and algorithms used for video forgery detection. In [6] spatial analysis methods analyze the spatial properties of the video, such as the brightness, contrast, and texture of the images. These methods can be used to detect forgery techniques that involve changing the brightness, contrast, or texture of the images.

Hue, Saturation, Value (HSV) color histogram difference between adjacent frames is used by Zhao et al. [7] as a similarity measure and a matching recheck is done based on tampering positions. Frame insertion, copying, and deletion can be detected using this method. Experimental results shown in the paper indicate that the model is capable of doing detection with precision, recall, and accuracy of 98.07%, 100%, and 99.01%, respectively. The model, therefore appears to perform better than the existing state-of-the-art methods. In the method proposed by Akumba et al. [8] mean and standard deviation of the correlation coefficient between adjacent frames are used as the features to perform classification to detect video forgery. They claim an accuracy of 100% for the VIFFD dataset [9] and self-created manipulated video dataset. In [10] authors use a statistical correlation feature for detecting duplication forgery in digital video. The statistical correlation of the frames in the video content through textural features is used to identify frame duplication. The paper [11] presents an analysis of forgery detection techniques like inter-frame forgery detection and intra-frame forgery detection.

A comparative analysis of new techniques in this direction is performed and brings out the difficulties and opportunities in video forgery detection. Correlation consistency of Entropy coded (DistrEn2D and MSE2D) frames is used by Nitin et al. [12] for forgery detection. They use two-dimensional distribution entropy (DistrEn2D) and bi-dimensional multiscale entropy (MSE2D) in their work. Experimentation is done on the original data collected from SULFA [13], REWIND [14], and VTL [15] and are used in the experimentation after performing the necessary forgery operations in order to create the forged dataset. The overall detection accuracy is claimed to be 96.6%. In [16] the authors use the relationship between the adjacent frames using the correlation coefficient to detect video forgery due to frame insertion and frame deletion. They use the inter-frame correlation distance between the frames, calculate the minimum distance score, and statistical features, and compute upper-bound, and lower-bound threshold and sigma coefficient for the identification of forgery location. Inter-frame forgery operations like frame shuffling, frame insertion, and deletion are detected and localized by Fadl et al. [17] using the temporal average and the universal image quality index. 15 tampered videos were developed for testing and are made publicly available. There exist methods that use DCT (Discrete Cosine Transform) based features for forgery detection as in [18] where the authors use it for image forgery detection. In [19] authors use a DCT-based feature (FCAN-DCT) that captures the spatio-temporal properties for video forgery detection.

Various methods which can do video forgery detection have been explored. Some of them offer promising results. But we identify the various constraints faced by the above methods like colossal time complexity, lack of appropriate dataset, etc. The problem of finding an effective inter-frame video forgery detection technique which can overcome the above mentioned challenges is taken care of here. We propose a method that primarily focuses on addressing inter-frame forgery, specifically frame insertion, duplication, and deletion. The proposed method is designed to effectively handle various types of inter-frame forgeries and can classify a video as authentic or forged. The experimentation is performed on data available for the purpose.

3 Proposed Methodology

3.1 Principal Component Analysis (PCA)

PCA is used to reduce the dimensionality of video frames by identifying the principal components, which are orthogonal vectors that capture the most significant variations in the data. These principal components represent the directions along which the data varies the most, allowing us to project the high-dimensional video frames onto a lower-dimensional subspace while preserving the most important information. PCA works by identifying the principal components of a dataset in a way that maximizes the variance along these components. Eigenvalues and eigenvectors play a fundamental role in Principal Component Analysis (PCA). Below we list their roles in PCA:

Eigenvalues (λ): Eigenvalues are scalars (numbers) associated with a square matrix. For a square matrix A, an eigenvalue λ is a number such that when A is multiplied by an eigenvector v, the same vector scaled by λ is obtained:

$$Av = \lambda v \quad (1)$$

In other words, the matrix A only stretches or compresses the eigenvector v by λ without changing its direction.

- Eigenvalues represent the amount of variance captured by each principal component. They indicate the importance or significance of each principal component in explaining the data's variability.
- When PCA performs eigenvalue decomposition on the covariance matrix of the data, it computes a set of eigenvalues. Each eigenvalue corresponds to one principal component.
- The eigenvalues are sorted in descending order. The higher the eigenvalue, the more variance is explained by the corresponding principal component.
- Retaining the top-k principal components (those associated with the highest eigenvalues) allows PCA to preserve a substantial amount of the original data's variance while achieving dimensionality reduction.

Eigenvectors (v): Eigenvectors are nonzero vectors that remain in the same direction after multiplication by a matrix A. These vectors are associated with eigenvalues and can be found by solving the eigenvalue Eq. (1) for each eigenvalue.

- Eigenvectors represent the directions in which the data varies the most. They provide the basis for the new lower-dimensional space in which the data will be projected during dimensionality reduction.
- Each eigenvector corresponds to a principal component, and they are orthogonal to each other. This orthogonality ensures that the principal components capture independent patterns in the data.

- The first eigenvector (associated with the highest eigenvalue) represents the direction of maximum variance in the data. It is the first principal component. The second eigenvector (associated with the second-highest eigenvalue) represents the direction of the second most significant variance and forms the second principal component, and so on.
- The eigenvectors provide the transformation matrix that projects the original data onto the lower-dimensional space defined by the selected principal components.

3.2 *Feature Extraction Using PCA*

In the proposed approach, PCA is employed for video forgery detection. Initially, the preprocessing steps prepare the video data for PCA analysis. Subsequently, PCA identifies the principal components that capture the most significant variations in the video frames, enabling feature extraction and the potential detection of video forgery.

- **Data Preprocessing**: Before applying PCA, the video frames are converted to grayscale to reduce the processing effort without sacrificing efficiency. The data matrix is normalized or standardized to remove any potential scaling differences that could affect the PCA results.
- **Covariance Matrix Calculation**: The covariance matrix of the feature data is computed. The covariance matrix describes the relationships and correlations between different visual features across the frames.
- **Eigenvalue Decomposition**: PCA performs eigenvalue decomposition on the covariance matrix. This process computes the eigenvalues and eigenvectors of the covariance matrix, representing the principal components of the data.
- **Selection of Principal Components**: The eigenvectors are ranked based on their corresponding eigenvalues. The top principal components, which capture the most significant variations in the video data, are retained.
- **Forgery Detection**: In the PCA feature space, the authenticity of the video can be assessed based on the distribution of top eigen values and eigen vectors in the various frames of the video. Any significant deviations or inconsistencies within the feature space might indicate potential forgery. Thresholds or statistical measures can be defined to classify videos as authentic or potentially forged.

Our proposed method relies on the utilization of the foremost principal component for processing, involving the largest eigenvalue ($\lambda 1$) and the corresponding eigenvector ($v1$). It is noteworthy that the number of retained principal components (k) significantly influences the quality of the reduced representation. However, as the first principal component captures the maximum variance and holds the most dominant pattern or structure within the image, we have exclusively employed the first principal component in our calculations. To establish the PCA feature space representation of the video, Principal Component Analysis (PCA) is performed on the preprocessed and normalized video frame.

In the context of a video, the first principal components (PC1) of its frames depict the most prominent spatio-temporal pattern present in the video. In a video with a constant frame rate and minimal abrupt movement, PC1 exhibits a gradual pattern of change. Any sudden deviations in this pattern may indicate a potential video forgery, which is utilized for feature extraction via PCA. Despite PC1 consisting of both the eigenvalue and the corresponding eigenvector, we solely utilize the eigenvalue for feature extraction. Specifically, we compute the difference in eigenvalues between adjacent frames as a feature. If $\lambda 1$ and $\lambda 2$ are the eigenvalues of the first principal component of the adjacent video frames $F1$ and $F2$, their difference is found as $\text{diff1} = \lambda 1 - \lambda 2$ and each such calculated difference for all adjacent frames is stored in an array. Further, the calculated difference values are analyzed and the one which deviates most from the rest of the values in the array is considered as a suspected outlier. It is confirmed by checking whether its z-score exceeds a threshold. Threshold is empirically set as 3. The z-score (standard score) of an element in an array is a measure of how many standard deviations that element is away from the mean of the array. It is calculated as

$$z = \frac{x - \mu}{\sigma} \tag{2}$$

where z is the z-score of the element, x is the value of the element to standardize, μ is the mean (average) of the array and σ is the standard deviation of the array.

If an outlier can be detected in the array the video is considered as forged. Moreover, to enhance the precision of detection, a feature based on the intensity variations among the frames is employed. In the temporal domain of the video, we anticipate a consistent change in the average intensity of each frame, while any variation may signify forgery. The steps involved in the classification process are detailed below.

1. Extract the frames in the video.
2. Perform gray-scale conversion of the frames.
3. Find the spatial average of the intensity of each frame and store it as diffAvgSequence for further processing.
4. Apply eigenvalue decomposition on the covariance matrix of each frame by setting the number of components to 1.
5. Store the eigenvalue of the first principal component for further processing.
6. Find the difference between the value found in (5) of each frame with its immediate next neighbor and use this difference sequence for forgery detection.
7. Consider the difference sequence in (6) and call it diff-sequence for further processing.
8. Among the top ten values in diff-sequence if any value appears repeatedly report forgery due to frame duplication and go to (13).
9. Consider the maximum value in the diff-sequence as an outlier and confirm it by checking whether its z-score value is larger than 3 in a sub-sequence of diff-sequence starting from 3 positions to the left of the outlier and ending at 3 positions to the right of the outlier.
10. If (9) confirms the maximum value in diff-sequence as an outlier, indicate forgery and the location of forgery and go to (13).

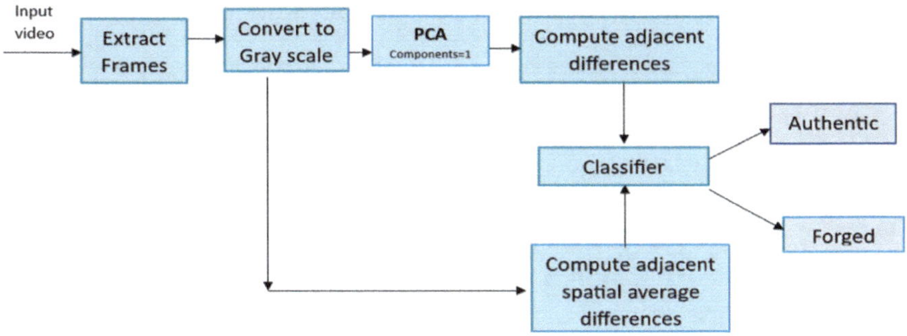

Fig. 1 Block diagram of the proposed classifier

11. Repeat (8) by considering diffAvgSequence as diff-sequence if not already done.
12. Mark the video as Authentic.
13. Stop.

Feature extraction and classification based on the extracted features by setting the threshold to 3 is depicted in the block diagram (Fig. 1). The difference sequences carry the required information for classification. As a two-level detection of outlier is performed, accuracy is considerably improved. Hence, some of those undetected using eigenvalue differences during the first attempt get detected by the frame intensity difference sequence. Figure 2 shows the pattern of change at the forgery locations in the eigenvalue difference sequence at the left and pixel intensity difference sequence at the right in copy, delete, insert, and no forgeries, respectively.

In the case of frame duplication, the same difference value appears many times in the sequence as seen in Fig. 2, the detection of which results in accurate classification as frame duplication forgery. It may be noted that the proposed method detects and localizes forgery irrespective of the type of forgery. It can detect whether a video is forged or authentic with the accuracy mentioned in the next section.

4 Experimentation

The methodology described above is experimented with for detecting inter-frame video forgeries like frame deletion, frame insertion, and frame duplication (copy). The same methodology without any change in the approach or threshold value is applied for the detection of all types of inter-frame forgeries. We use the common performance evaluation metric 'accuracy' to evaluate the effectiveness of the model and precision and recall are used when they are needed for comparison of performance.

$$\text{Accuracy} = \frac{\text{Number of Correct Predictions}}{\text{Total Number of Predictions}}$$

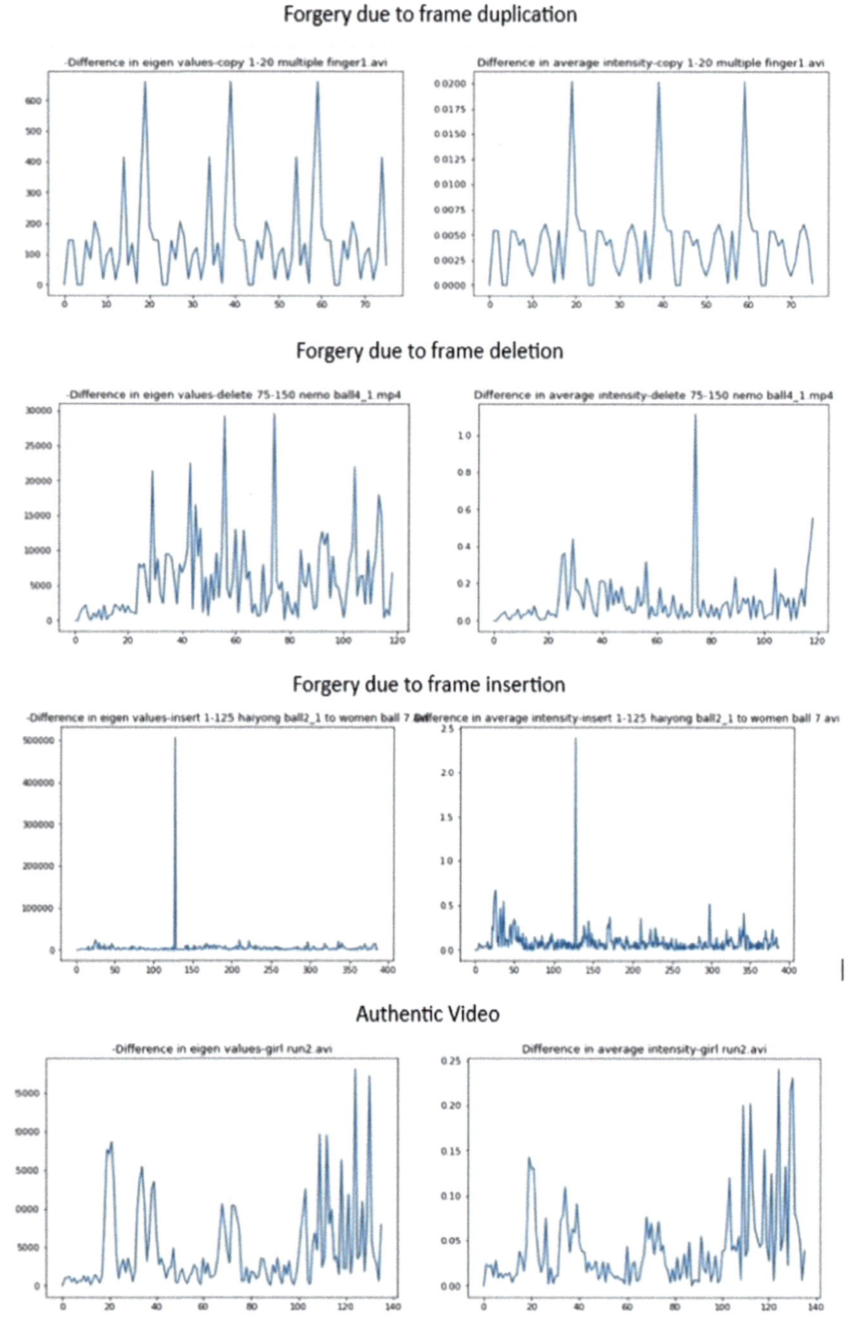

Fig. 2 Difference sequences

$$\text{Precision} = \frac{\text{True Positives}}{\text{True Positives} + \text{False Positives}}$$

$$\text{Recall} = \frac{\text{True Positives}}{\text{True Positives} + \text{False Negatives}}$$

where Number of Correct Predictions = True Positives + True Negatives.
True Positive (TP): A true positive occurs when the model correctly predicts a positive outcome, and the actual outcome is indeed positive.
True Negative (TN): A true negative occurs when the model correctly predicts a negative outcome, and the actual outcome is indeed negative.
False Positive (FP): A false positive occurs when the model incorrectly predicts a positive outcome, but the actual outcome is negative.
False Negative (FN): A false negative occurs when the model incorrectly predicts a negative outcome, but the actual outcome is positive.

4.1 Dataset

Extensive experimentation is performed using the datasets described in Table 1 and performs better than many state-of-the-art methods. Experimentation with each of the datasets is described below.

VIFFD dataset [9] In this dataset containing 90 forged videos, 30 videos are forged due to frame duplication (copy), another 30 videos are forged due to frame deletion and the remaining 30 are forged due to frame insertion. In experimentation, the model is tested with all the forged and authentic videos. The classification to authentic and forged is done as shown in Table 2. The overall accuracy of forgery detection and localization is specified there.

Table 1 Datasets used in the experimentation

Name of the dataset	Total No. of videos	Forged	Authentic
VIFFD [9]	120	90	30
Surveillance video dataset [17]	15	15	0
TDTVD dataset [15]	226	210	16
Inter-frame forgery dataset [20]	48	32	16

Table 2 VIFFD dataset—experimentation results

Video type	Accuracy
Forged	71.11
Authentic	93.33

Table 3 Surveillance video dataset—experimentation results

Forgery type	No. of videos	Accuracy
Shuffling	5	80
Delete	5	80
Insert	5	100

Table 4 Surveillance video dataset—comparison with other methods

Method	Precision	Recall
Fadl et al. [17]	0.976	0.98
Proposed method	1	0.866

The highlight of this approach is that it can detect authentic images correctly with high recall and accuracy.

Surveillance video dataset The authors of [17] created this dataset and made it publicly available. So, the proposed method is tested on this dataset and the result is shown in Table 3. It must be noted that the proposed method sets the threshold for classification empirically, irrespective of the dataset and irrespective of the type of forgery. Hence the same algorithm works well for all the datasets considered, and for all types of forgery with no change in the methodology or threshold. In the paper [17], we see that high performance is exhibited for forgery detection and localization, by using separate thresholds for each type of forgery. The proposed approach may leave some forgery undetected (false rejection) but mostly doesn't do any false acceptance. It fails to detect one shuffling and one delete forgery because the detected outliers could not exceed the threshold. The method detects and localizes all other forgeries present in the videos. Performance comparison in terms of the overall precision and recall irrespective of the type of forgery is shown in Table 4.

TDTVD dataset This temporal domain tampered video dataset [15] is based on Event/Object/Person (EOP) removal or modification and Smart Tampering/Multiple Tampering (ST). There are 40 EOP videos each for frame deletion, insertion and duplication. The 120 ST videos have 30 videos each in the same categories with multiple tampering at multiple locations. The 16 original videos are from the SULFA dataset. Table 5 shows the results of the experimentation and the comparisons with methods which use the same dataset are shown in Table 6.

Inter-frame forgery dataset This dataset is created by Fadl et al. [20], and is also used for experimentation of their approach for video forgery detection. The original video folder is copied from the TDTVD dataset.

Among the 32 forged videos, there are forgeries due to insertion, copy, deletion and shuffling. Table 7 shows the details of the videos, and the result of experimentation using the proposed method.

The experimentation detailed above reveals the fact that the proposed method can detect frame insertion and duplication forgery with high accuracy while the method is found not to perform well on the frame deletion forgery. A lower performance is

Table 5 TDTVD dataset—result of experimentation

Forgery type	No. of videos	Accuracy in %
EOP copy	40	97.5
EOP delete	40	85
EOP insert	40	97.5
Original	16	81.25
ST copy-10	10	90
ST copy-20	10	90
ST copy-30	10	90
ST delete-10	10	40
ST delete-20	10	70
ST delete-30	10	80
ST insert-10	10	100
ST insert-20	10	100
ST insert-30	10	100

Table 6 TDTVD video dataset—comparison of accuracy with other methods

Method	Frame deletion	Frame duplication	Frame insertion	Multi frame deletion	Multi frame duplication	Multi frame insertion
CCCoGV [21]	80.00	82.5	95.00	76.67	87.50	93.33
Optical flow model [22]	65.00	87.5	87.50	60.00	86.67	86.67
Proposed method	85.00	97.50	97.50	63.33	90.00	100

Table 7 Inter-frame forgery dataset—result of experimentation

Forgery type	No. of videos	Accuracy in %
Copy	10	90
Delete	5	80
Insert	5	100
Shuffling	9	100
Copy and shuffling	3	66.66
Original	16	81.25

seen in the case of delete forgery because the frame brightness and intensity pattern don't show a significant deviation even after deletion in some deletion forgeries. It can identify the genuineness of authentic images with high accuracy.

5 Conclusion

This paper presents an efficient and simple method for inter-frame video forgery detection which can detect and localize video forgery using Principal Component Analysis and frame intensity analysis of adjacent frames. The change happening to the sequence of adjacent frames, when the original sequencing is altered is found by extracting features from the first principal component of the frames and using the arithmetic difference between adjacent frame intensities of the video. It is found to be a simple and efficient method, especially for detecting frame insertion and duplication forgeries. The method may be modified to incorporate intra-frame forgery detection too. We discuss the strengths and limitations of the approach and compare it with other forgery detection methods. Real-world applications and challenges of deploying PCA for video forgery detection are also explored. This research contributes to the growing body of knowledge in combating video forgery and offers insights into the effectiveness of PCA in enhancing digital media authenticity.

References

1. GreeksforGreeks. Principal component analysis. https://www.geeksforgeeks.org/principal-component-analysis-pca/
2. Al-Sanjary OI, Sulong G (2015) Detection of video forgery: a review of literature. J Theoret Appl Inf Technol 74(2):207–220
3. Kaur J, Kumar R (2020) Analysis of various video forgery detection types techniques. Int Res J Eng Technol (IRJET)
4. Papinwar SR (2016) Forgery detection in video using watermarking: a review. Int J Comput Sci Inf Technol 7(1):270–274
5. Nilesh G, Gajanan G (2016) Review on video forgery detection. Int J Res 3:619–633
6. Jin JS, Xu C, Xu M, Hyun DK, Lee MJ, Ryu SJ, Lee HY, Lee HK (2013) Forgery detection for surveillance video. In: The era of interactive media. Springer, New York, NY. https://doi.org/10.1007/978-1-4614-3501-3-3
7. Zhao Z-M, Wang D-N, Lu R-K (2018) Inter-frame passive-blind forgery detection for video shot based on similarity analysis. Multimed Tools Appl. https://doi.org/10.1007/s11042-018-5791-1
8. Akumba BO, Iorliam A, Agber S, Okube EO, Kwaghtyo KD (2021) Authentication of video evidence for forensic investigation: a case of Nigeria. J Inf Secur 12:163–176. https://doi.org/10.1007/s11042-018-5791-1
9. Nguyen XH, Hu J (2020) VIFFD—a dataset for detecting video inter-frame forgeries. Mendeley Data 5. Multimed Tools Appl. https://doi.org/10.17632/r3ss3v53sj.5
10. Sowmya KN, Basavaraju HT, Lohitashva BH, Chennamma HR, Aradhya VNM (2019) Similarity analysis of residual frames for inter frame forgery detection in video. In: ICICC 2019. Advances in intelligent systems and computing, vol 1034, p 20. https://doi.org/10.1007/978-981-15-1084-7
11. Kumar V, Singh A, Kansal V, Gaur M (2020) A comprehensive analysis on video forgery detection techniques. In: Proceedings of the international conference on innovative computing & communications (ICICC), 29 Mar 2020. Available at SSRN: https://ssrn.com/abstract=3563382 or https://doi.org/10.2139/ssrn.3563382

12. Shelke NA, Kasana SS (2022) Multiple forgeries identification in digital video based on correlation consistency between entropy coded frames. Multimed Tools Appl 28:267–280. https://doi.org/10.1007/s00530-021-00837-y
13. Bestagini P, Milani S, Tagliasacchi M, Tubaro S (2013) Local tampering detection in video sequences. In: 2013 IEEE 15th international workshop on multimedia signal processing (MMSP)
14. Video trace library. http://trace.eas.asu.edu/
15. Panchal HD, Shah H (2020) Video tampering dataset development in temporal domain for video forgery authentication. Multimed Tools Appl 79:33–34. https://doi.org/10.1007/s11042-020-09205-w
16. Kumar V, Gaur M (2022) Multiple forgery detection in video using inter-frame correlation distance with dual-threshold. Multimed Tools Appl. https://doi.org/10.1007/s11042-022-13284-2
17. Fadl S, Han Q, Li Q (2018) Surveillance video authentication using universal image quality index of temporal average. In: 17th international workshop IWDW 2018, Jeju Island, Korea, 22–24 Oct 2018
18. Shekar BH, Abraham W, Pilar B (2023) Local DCT-based deep learning architecture for image forgery detection. In: Tistarelli M, Dubey SR, Singh SK, Jiang X (eds) Computer vision and machine intelligence. Lecture notes in networks and systems, vol 586. Springer, Singapore. https://doi.org/10.1007/978-981-19-7867-8-37
19. Wang Y, Peng C, Liu D, Wang N, Gao X (2023) Spatial-temporal frequency forgery clue for video forgery detection in VIS and NIR scenario. IEEE Trans Circuits Syst Video Technol. https://doi.org/10.1109/TCSVT.2023.3281475
20. Fadl S, Han Q, Qiong L (2020) Exposing video inter-frame forgery via histogram of oriented gradients and motion energy image. Multidimens Syst Signal Process 31:1365–1384. https://doi.org/10.1007/s11045-020-00711-6
21. Wang Q, Li Z, Zhang Z, Ma Q (2014) Video inter-frame forgery identification based on consistency of correlation coefficients of gray values. J Comput Commun 2:51–57. https://doi.org/10.4236/jcc.2014.24008
22. Wang W, Jiang X, Wang S, Wan M, Sun T (2013) Identifying video forgery process using optical flow. Int Work Digit Watermark 22:244–257

Meta-Learning-Based Multimodal Fake News Detection

M. S. Varun, Ishaan Bhattacharjee, Pratheek Karike Venkatesha, A. Sathvik, S. Pramath, and Bhaskarjyoti Das

Abstract In the contemporary digital era, spreading false information via social media and Internet platforms poses a significant challenge. Not only the data has become increasingly multimodal, lack of sufficient labeled data due to prohibitive labeling costs and data privacy issues poses an additional challenge. An approach such as meta-learning can help address this issue. This work showcases meta-learning with scarcely labeled data for the detection of multimodal fake news and provides a performance comparison of the meta-learning approach with the fully supervised learning approach. We leverage the power of Model-Agnostic Meta-Learning (MAML) and Fast Context Adaption via Meta-Learning (CAML) in this work. The results are encouraging as these meta-learning models achieve heightened accuracy and speed even with insufficient labeled data for multimodal fake news detection.

Keywords Meta-learning · Multimodal · Fake news · MAML · CAML

1 Introduction

In today's digital age, disinformation, such as fake news, propaganda, and misinformation, such as rumors, have become a significant concern. This phenomenon has weakened faith in trustworthy sources and has other far-reaching implications in society, economy, and geopolitics. Therefore, the ability to detect disinformation and misinformation has become extremely important. However, 'disinformation research' has struggled to keep up with the rapid pace with which disinformation is evolving and growing. Das et al. [2] have examined multi-contextuality in disinformation as a key challenge faced by disinformation researchers. A disinformation campaign encompasses contextual factors such as publisher, knowledge, network, content, temporal,

M. S. Varun · I. Bhattacharjee · P. K. Venkatesha · A. Sathvik · S. Pramath
Department of Computer Science and Engineering, PES University, Bengaluru, Karnataka, India

B. Das (✉)
Department of Computer Science and Engineering in AI and ML, PES University, Bengaluru, Karnataka, India
e-mail: bhaskarjyoti01@gmail.com

S. M. Thampi et al. (eds.), *Fifth International Conference on Computing and Network Communications*, Lecture Notes in Electrical Engineering 1219,
https://doi.org/10.1007/978-981-97-4540-1_28

and spatial aspects. The challenge of multi-contextuality gets further exacerbated by the insufficiency of available labeled data. Since disinformation like fake news is heterogeneous considering emerging events, there is never enough labeled data to facilitate early detection of emerging fake news. This makes semi-supervised learning the preferred option in modern disinformation research. Meta-learning [6, 19] that enables few-shot learning can be an alternative to a semi-supervised approach.

It is well-established [4, 8] that a picture is more effective than a pure text message on social media. The ever-increasing popularity of multimodal content [1] presents an additional challenge. There are numerous challenges with the presently available multimodal disinformation datasets [2], i.e., lack of explainability of labels, differing annotation schemes, the small size of the datasets, etc. An approach such as meta-learning is the most appropriate to solve the issue caused by the small size of multimodal datasets as meta-learning algorithms are far more effective in producing results with small datasets than general machine learning algorithms which is leveraged in this study to achieve a high accuracy on a very small dataset while also being very efficient by taking much lesser time in training the models.

The work investigates a novel meta-learning-based approach for multimodal fake news and disinformation detection to address the twin challenge of lack of labeled data and multi-modality of content. The remainder of this article is structured as follows. Section 2 provides the necessary background for the techniques used in this work. Section 3 discusses the dataset used. Section 4 discusses the methodology adopted in this work. In Sect. 5, the results are discussed. Finally, Sect. 6 concludes the article by summarizing the findings of this work and future steps.

2 Existing Work

2.1 Multimodality in Disinformation Research

Multimodality in disinformation research is relatively unexplored [1]. There are two aspects of disinformative content, i.e., factuality and harmfulness. Factuality can be best assessed with access to knowledge, but in the absence of available knowledge, machine learning models typically rely on characteristic footprints in the content itself. Successful detection of disinformation in multimodal content requires not only an individual assessment of each of these aspects but also requires semantic consideration of these modalities in tandem. There is a semantic challenge involved in making meaning of the two modalities in question, i.e., image and text as any of them in isolation may not necessarily deliver the intended meaning. Satirical content, frequently observed in multimodal memes, falls into this category, where the divergence between modalities actively contributes to the final conveyed meaning.

Hence, the existing work in multimodal disinformation research can be fit into one of these strategies of learning models:

- Supervised learning models relying on the fusion [7] of modalities are used. There are different ways to achieve this fusion, i.e., early, late, and hybrid. Nevertheless, the strategy assumes that fusing modalities will capture the necessary information for downstream tasks. Typically, in this strategy [15], transformer-based embeddings such as BERT for text and CNN-based models for the image data are used.
- Supervised learning models that rely on techniques of cross-attention [12, 13, 17] are used to deliver an improved fusion.
- Supervised learning approach using pre-trained visual linguistic models [19] is used.
- Supervised learning models actively utilizing the semantic gap [16, 18] between modalities in addition to the modalities themselves are used to address cases such as memes, satire, parody, false connection.
- Based on the fact that disinformation is multi-contextual, few recent works have adopted a graph neural network-based approach to fuse knowledge context and content context from text and image [11] for downstream machine learning tasks. This graph-based approach is typically based on semi-supervised learning.

Nevertheless, none of the previously employed strategies has integrated meta-learning as an approach to tackle the issue of lack of labeled data for multimodal fake news detection.

2.2 Meta-learning in Disinformation Research

As compared to hand-engineered feature-based machine learning followed by deep learning that is based on the model as well as feature learning, meta-learning today brings in a learning paradigm [5] based on feature, model, and algorithm learning. Meta-learning addresses the challenge of learning new concepts without sufficient labeled data. The meta-learning algorithms can be classified into categories [6], including metric-based, model-based, and optimization-based techniques.

With disinformation research itself being a new field, there is very little existing work using meta-learning for disinformation detection for multimodal data. Recently, Salem et al. [14] have used meta-learning on a dataset of modest size for detecting fake news about the Syrian war. However, this work does not use multimodal data. To our knowledge, at the time of writing this article, meta-learning-based multimodal fake news detection has not been attempted so far.

3 Dataset

We use an existing dataset named Fakeddit [10], which showcases the fusion of image and text data extracted from Reddit. This is a fine-grained multimodal fake news dataset where fake news is not only false information [9] but also includes other categories such as satire, misreporting. This dataset has a vast repository of more than 1 million samples collected across more than 20 subreddits. Due to computational resource constraints, the work described in this article is on a sampled subset of about 56,000 data entries with 6-way labeling, each representing unique aspects of the content. This dataset was made by stratified sampling based on the label to ensure that the sampling ensures an equal distribution of labels and no bias is induced into the dataset by means of any particular labels having disproportionately higher frequencies in the original dataset. The challenge of conducting a multimodal 6-way classification on this diverse dataset lies in its potential to find the relation between textual and visual elements. The six classes are as follows:

- *Label 0—True*: True information is truthful and based on facts. This also indicates a complete match between the image and the caption in terms of content.
- *Label 1—Satire/Parody*: This category includes content that satirizes genuine contemporary content or information that makes it untrue.
- *Label 2—Misleading Content*: This category includes information that has been purposefully modified in order to deceive the audience.
- *Label 3—Imposter Content*: There are two subreddits in this category that feature bot-generated content and are trained on a significant number of other subreddits.
- *Label 4—False Connection*: This category's submission photos do not adequately support their written descriptions.
- *Label 5—Manipulated Content*: Content purposefully altered by manual picture editing or other means.

The challenge of conducting a multimodal 6-way classification on this dataset lies in its potential to find the relation between textual and visual elements (Fig. 1).

4 Methodology

In the work described, we compared the use of the meta-learning approach with 20% training data and 80% test data with a traditional supervised learning approach with 80% training and 20% test data.

In the supervised learning approach, two distinct approaches for multimodal fusion in classification tasks are presented:

- This approach uses a traditional fusion model and does not use openAI CLIP embeddings, i.e., a powerful transformer-based architecture.
- This approach leverages the powerful Contrastive Language-Image Pretrained (CLIP) model. This model is trained on paired multimodal data.

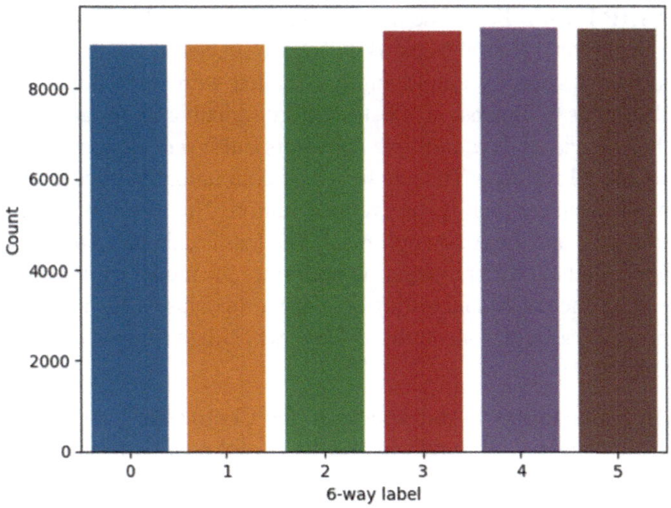

Fig. 1 Dataset analysis graph

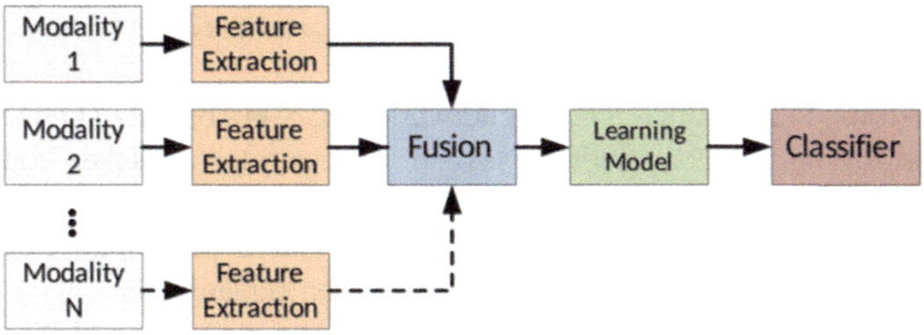

Fig. 2 Working of fusion

Our meta-learning approach uses two algorithms that focus on meta-optimization as meta-optimization as an approach works well for related tasks in a multi-class dataset.

- The first approach uses Model-Agnostic Meta-Learning (MAML) with openAI CLIP Embeddings.
- The second approach uses Fast Context Adaptation via Meta-Learning (CAVIA) with openAI CLIP Embeddings.

4.1 Non Meta-Learning Approach

The traditional non-meta-learning models are built using a fusion strategy as shown in Fig. 2.

Without CLIP Embeddings The model is based on deep learning techniques, utilizing a Convolutional Neural Network (CNN) to process text and image data effectively. The model starts by employing pre-trained word embeddings, specifically GloVe embeddings of dimension 300, to capture semantic information from the textual data. For the visual data, multiple convolutional layers along with fine-tuning a ResNet-101 model are used. The late fusion of both modalities occurs through concatenation, allowing for joint feature representation. The model is trained using the CrossEntropyLoss and the Adam optimizer. The loss is calculated using the negative log-likelihood, and the backpropagation optimizes the model parameters.

During experimentation and testing on real-world datasets, the model achieves an average validation accuracy stands at 65.26%, while the testing accuracy is reported to be 71.85%.

With CLIP Embeddings In this approach, the Contrastive Language-Image Pre-training (CLIP) model [19], renowned for its exceptional performance in handling multimodal data.

This approach involves several key steps namely—dataset preparation, obtaining text, and image embeddings using the CLIPTokenizerFast and CLIPProcessor, respectively, defining a custom ClassificationModel, and evaluating its accuracy. The text data is tokenized using the CLIPTokenizerFast, and the image data is processed with the CLIPProcessor to extract embeddings.

Our custom model utilizes a neural network consisting of fully connected layers in addition to multiple convolutional layers. This architecture takes both text and image embeddings as inputs to predict category labels. To train the model, we use the CrossEntropyLoss and Adam optimizer. The model's performance is evaluated on another testing dataset to assess its generalization capabilities.

The model achieves an impressive average validation accuracy of 82.41% and a testing accuracy of 82.88%. The CLIP embedding-based approach performs better as shown in Fig. 3.

4.2 Meta-Learning-Based Approaches

Model-Agnostic Meta-Learning (MAML) with CLIP Embeddings Model-Agnostic Meta-Learning (MAML) [3] is one of the meta-optimization algorithms that tries to achieve quick adaptation to new tasks using limited data. Unlike traditional machine learning methods that optimize for a single task, MAML employs a two-step optimization process (as shown in Fig. 4) that learns a good initialization for model parameters, facilitating efficient adaptation to diverse tasks.

In this model, CLIP embeddings have been used as input features. We first preprocessed the dataset to obtain CLIP embeddings for both textual and image data. We utilized the CLIP model to obtain semantically meaningful embeddings by projecting images and their corresponding textual descriptions into a shared embedding space. A custom dataset class is designed to handle the CLIP embeddings alongside their

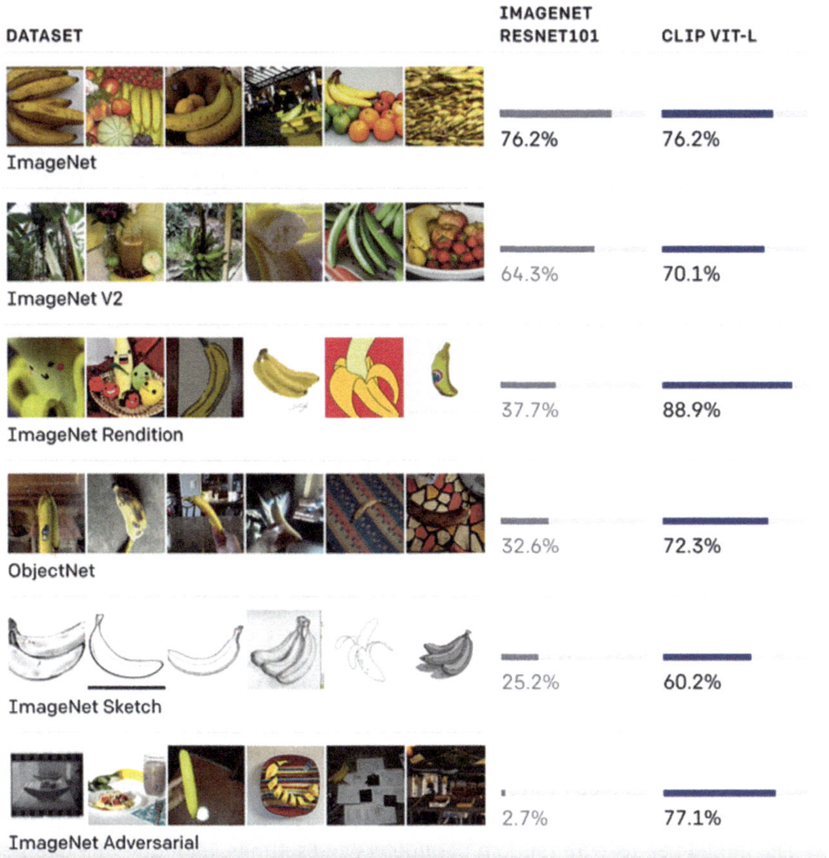

Fig. 3 Comparison of CLIP with the ResNet-101 model

corresponding class labels for efficient training and evaluation, enabling seamless integration with the MAML model. The core MAML algorithm was implemented with the training process encompassing a two-step optimization loop. In the inner loop, the model's parameters were updated on small batches of data from each task, generating task-specific parameters that improve performance on individual tasks. The outer loop utilized these task-specific parameters to update the meta-learner's parameters, facilitating better generalization across tasks.

During training, the MAML model was exposed to multiple tasks, each containing less number of training samples. In this case, we completely flipped the amount of data to be used for training to 20%. The dataset was split into training, validation, and testing sets in order to assess the perfomance of the MAML model. The model was trained on the training set, and its hyperparameters were tuned using the validation set. The testing accuracy obtained for few-shot learning tasks was an impressive 75.1%, showcasing the ability of the model to generalize accurately even with limited

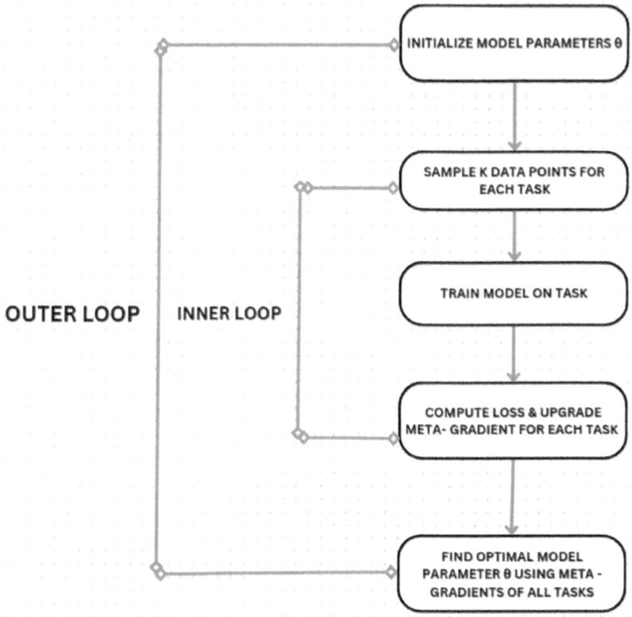

Fig. 4 Working of MAML algorithm

training data. Additionally, the model also took significantly less time for training and testing as a consequence of the efficiency of the algorithm and the minimal training required.

The model begins with random parameters and settings, including the number of training samples, epochs, and learning rates. It processes input data through a linear transformation, adjusting parameters and converting predictions into probabilities using softmax.

During training, the model undergoes an inner loop to minimize the cross-entropy loss between predictions and actual outcomes for specific tasks. In the outer loop (meta-training), it adapts its core parameters based on the knowledge gained from various tasks, guided by cross-entropy loss on a validation set.

Fast Context Adaptation via Meta-learning (CAVIA) with CLIP Embeddings

$$\phi_i = \phi_0 - \alpha \nabla_\phi \mathcal{L}_{\mathcal{T}_i}(f_{\phi_0}, \theta) \tag{1}$$

$$\theta = \theta - \beta \nabla_\phi \sum_{\mathcal{T}_i \sim p(\mathcal{T})} \mathcal{L}_{\mathcal{T}_i}(f_{\phi_i}, \theta) \tag{2}$$

In this approach, an improved version of the MAML algorithm, i.e., Fast Context Adaptation via Meta-Learning [20] (abbreviated as CAVIA or CAML) has been used. The core aspect of this method uses the same idea of MAML as shown in Eqs. 1 and 2, i.e., the concept of 2 loops (an inner loop and an outer loop) to find a better

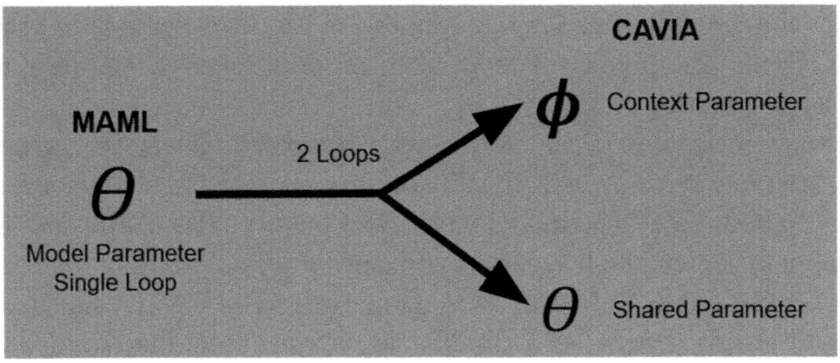

Fig. 5 CAVIA optimization

initial parameter overall. However, it differs from MAML as instead of using a single parameter, it splits up the model parameter into 2 as shown in Fig. 5. One of them is a Context Parameter denoted by ϕ (phi) which is a task-specific parameter updated on the inner loop and is specific to each task. The second one is the shared parameter θ (theta) which is shared across tasks and updated in the outer loop to find the optimal model parameter.

In this work, CAVIA meta-learning model has been used with CLIP embeddings as input features from the multimodal dataset. The model architecture consists of 2 parameter matrices, theta and phi. These matrices are learned during training and used for classification. In the forward pass, the model takes an input tensor having multiple values based on the batch size, and a transformation is applied to make sure each of the values is transformed to the required output dimension. The 'softmax' function is applied to it to get the probability distribution over the output classes.

For training, we use 2 steps an inner loop and an outer loop.

1. **Inner-loop training**: In the course of inner-loop training, the model undergoes training using the cross-entropy loss variant. Given the predicted probabilities from the softmax output, the custom loss function computes the logarithm of the reciprocal of the likelihood associated with the ground-truth labels. Adam, with an adjusted learning rate alpha, is the optimizer used for inner-loop training.
2. **Outer loop training**: The model's parameters are updated using a meta-gradient calculated from a few shot examples during the outer-loop training, also referred to as meta-learning. The model is trained using a limited number of examples from each class, simulating few-shot learning. The 'num_samples' hyperparameter determines how many samples are used for meta-learning.

During meta-learning, the model samples a few-shot dataset from the training data and computes the meta-gradient. The meta-gradient is obtained as the average gradient of the few-shot examples' loss concerning the model's theta parameters. This gradient represents how the model's theta parameters should change to adapt to new, unseen contexts. The phi matrix is updated using context-aware variational inference. The update is driven by two terms: context_loss, which encourages similarity between the theta and phi matrices, and kl_divergence, which measures the divergence between theta and phi to incorporate uncertainty into the adaptation process. After the model has undergone training, it becomes capable of predicting for new data points. The predict method takes an input tensor x and returns the class predictions as a numpy array.

For the specific case of our model and its accompanying data, the learning rate and meta-learning rate were adjusted manually through trial and error to get an average accuracy of about 76%. The model worked on merely 20% of the data for training and was able to achieve high accuracy with such minimal training. Furthermore, the efficiency of CAVIA over MAML was demonstrated as CAVIA finished the entire training and testing process in a significantly shorter time than MAML while still retaining a high testing accuracy.

5 Results and Discussion

Table 1 compares the performance of traditional fusion model meta-learning-based models. With in regards to the overall performance of the three main types of model for each label, there is no clear winner between CAVIA and MAML though CAVIA trains much faster than MAML. The fusion model scores better than the meta-learning models as it is trained with 80% of data. Considering the lack of labeled data (20% of data), the meta-algorithms perform really well.

On analyzing the accuracy per label across all models some trends begin to appear. At first glance, there is a trend of the accuracy of predicting label 4 being the highest compared to other labels across all models. Label 4 stands for False connection which means that the image and text do not match up in terms of context or in simple terms that the text associated with the image seems wrong. We hypothesize that the reason for this label having the highest accuracy is due to the simple reason that even for a human it is fairly easy to understand something of this sort where the image and text don't match up. The model does not need to do any kind of deeper analysis of the text or image to determine its veracity or any kind of message given by it as it would for any of the other labels. It needs to simply check if the context matches for both image and text and it is a much simpler process than for any of the other labels hence the model always has a high accuracy for label 4.

Conversely, we see that the accuracy is comparatively low each time for labels 0 and 3 which stands for True and Imposter content. We hypothesize that these two

Table 1 Label accuracy for different models

	Label					
	0	1	2	3	4	5
Fusion (%)	74.55	83.75	79.99	77.17	94.50	88.30
MAML (%)	67.94	74.21	73.19	59.34	89.43	84.44
CAVIA (%)	69.09	73.79	72.03	66.50	93.07	80.64

Table 2 Ablation models

Method used	Accuracy (%)
CLIP image only	67.90
CLIP text only	66.72
MAML image only	64.5
MAML text only	62.62
CAVIA image only	64.04
CAVIA text only	62.59

are related. Due to how good image faking by bots has become, the model finds it hard to classify images and we believe it often classifies images of label 0 as 3 and vice versa due to this.

Ablation Studies

A general trend seen in the ablation studies shows how the lack of one of the modalities reduces the accuracy both in the case of the image and the text models. This further signifies the importance of having multimodal data and supports earlier given hypotheses of image and text complimenting each other to improve accuracy particularly in the field of disinformation detection. While here again we can see the meta models have performed well despite the far limited dataset and comapred ot the CLIP models the general trend of both of the models having a decreased accuracy due to the fact that both the modalities are seperated and connot complement each other is an important observation here (Table 2).

Weakness of the Model

The mispredictions of the four models have been analyzed. The analysis reveals that all four models suffer from their inability to comprehend the subtle semantics, if any. This is not something that meta-learning or fusion strategy can address. To address this, one possible strategy can be to harness the knowledge context that the presented models have not used. A few examples are listed.

In Fig. 6, it appears that the CLIP model made an incorrect prediction due to its inability to analyze different parts of an image separately from others. In this particular case, the model identified the image as a false connection to a circus, which is incorrect. Although the image primarily consists of candy, a human observer can identify additional aspects such as the folded dollar bill and the arrangement of objects, which suggest a parody. In Fig. 7, the model has once again made a wrong prediction due to its limited imagination in envisioning situations. It interpreted the title too literally, and since there are more than two individuals present in the image, it has labeled the content as misleading. However, it is typical for humans to use symbolic images that may not precisely match every element mentioned in the title.

In Fig. 8, the model may classify this as bot-generated content due to the unusual appearance of the image when compared to other human-designed objects. However, the image is actually a design of ships inspired by Star Wars, which may give the

Fig. 6 An image of what looks like candy and other materials which are a dollar note and a plastic cover, and candy being arranged in a specific pattern with the caption given in the data set as circus-candy. Output from CLIP fusion model: actual label (1), predicted label (4)

Fig. 7 In this image we get to see a picture of what looks like a theme park entrance called LEGOLAND with people moving through the gates with the caption being about a physical altercation happening in the a particular place denoted by the image. CLIP MAML model: actual label (0), predicted label (2)

false impression that the US government is creating military spaceships in the style of the movie. This can cause confusion for the model as the image falls into both categories.

Fig. 8 Here we take an example of a picture from the dataset with some sort of spaceship in outer space with the caption denoting its some sort of government space fleet. CLIP CAVIA model: actual label (2), predicted label (3)

In general, the model appears to favor text and therefore struggles to function correctly without a clear text label. Additionally, because of this emphasis on text, it often fails to comprehend humor and clever wordplay, which can result in multiple errors. The model also struggles to understand abbreviations and slang that humans commonly use.

6 Conclusion and Next Steps

The lack of labeled data is a key concern in this domain due to cost and privacy issues. This makes semi-supervised learning and meta-learning important strategies in disinformation research. With content becoming increasingly multimodal, meta-learning needs to play a role in multimodal disinformation detection. The work described in this article is a step in that direction as it successfully showcases the efficacy of such an approach as compared to traditional supervised deep learning.

While the developed model performed well in accuracy, it was biased toward text and struggled to recognize comedy, smart wordplay, and slang. As evident from analyzing a few failure cases, meta-learning addresses a limited amount of labeled data but may not necessarily address the subtle semantics in a multimodal meme, satire, or false connection. The semantic complexity of such content requires the help of additional context, such as external knowledge. As a next step, we plan to enhance the adopted approach by incorporating knowledge context. While the work described in this article adopts an optimization-oriented meta-learning, we also plan to investigate the model-based and metric-based approaches of meta-learning.

References

1. Alam F, Cresci S, Chakraborty T, Silvestri F, Dimitrov D, Martino GDS, Shaar S, Firooz H, Nakov P (2021) A survey on multimodal disinformation detection. arXiv preprint arXiv:2103.12541
2. Das B et al (2023) Multi-contextual learning in disinformation research: a review of challenges, approaches, and opportunities. Online Soc Netw Media 34:100247
3. Finn C, Abbeel P, Levine S (2017) Model-agnostic meta-learning for fast adaptation of deep networks. In: International conference on machine learning. PMLR, pp 1126–1135
4. Hameleers M, Powell TE, Van Der Meer TG, Bos L (2020) A picture paints a thousand lies? The effects and mechanisms of multimodal disinformation and rebuttals disseminated via social media. Polit Commun 37(2):281–301
5. Hospedales T, Antoniou A, Micaelli P, Storkey A (2021) Meta-learning in neural networks: a survey. IEEE Trans Pattern Anal Mach Intell 44(9):5149–5169
6. Huisman M, Van Rijn JN, Plaat A (2021) A survey of deep meta-learning. Artif Intell Rev 54(6):4483–4541
7. Lahat D, Adali T, Jutten C (2015) Multimodal data fusion: an overview of methods, challenges, and prospects. Proc IEEE 103(9):1449–1477
8. Li Y, Xie Y (2020) Is a picture worth a thousand words? An empirical study of image content and social media engagement. J Mark Res 57(1):1–19
9. Molina MD, Sundar SS, Le T, Lee D (2021) "Fake news" is not simply false information: a concept explication and taxonomy of online content. Am Behav Sci 65(2):180–212
10. Nakamura K, Levy S, Wang WY (2019) r/Fakeddit: a new multimodal benchmark dataset for fine-grained fake news detection. arXiv preprint arXiv:1911.03854
11. Qian S, Hu J, Fang Q, Xu C (2021) Knowledge-aware multi-modal adaptive graph convolutional networks for fake news detection. ACM Trans Multimed Comput Commun Appl (TOMM) 17(3):1–23
12. Qian S, Wang J, Hu J, Fang Q, Xu C (2021) Hierarchical multi-modal contextual attention network for fake news detection. In: Proceedings of the 44th international ACM SIGIR conference on research and development in information retrieval, pp 153–162
13. Sachan T, Pinnaparaju N, Gupta M, Varma V (2021) SCATE: shared cross attention transformer encoders for multimodal fake news detection. In: Proceedings of the 2021 IEEE/ACM international conference on advances in social networks analysis and mining, pp 399–406
14. Salem FKA, Al Feel R, Elbassuoni S, Ghannam H, Jaber M, Farah M (2021) Meta-learning for fake news detection surrounding the Syrian war. Patterns 2(11)
15. Segura-Bedmar I, Alonso-Bartolome S (2022) Multimodal fake news detection. Information 13(6):284
16. Singhal S, Dhawan M, Shah RR, Kumaraguru P (2021) Inter-modality discordance for multimodal fake news detection. In: ACM multimedia Asia, pp 1–7
17. Song C, Ning N, Zhang Y, Wu B (2021) A multimodal fake news detection model based on crossmodal attention residual and multichannel convolutional neural networks. Inf Process Manag 58(1):102437
18. Zhou X, Wu J, Zafarani R (2020) Safe: similarity-aware multi-modal fake news detection (2020). arXiv preprint 200304981, p 2
19. Zhou Y, Ying Q, Qian Z, Li S, Zhang X (2022) Multimodal fake news detection via clip-guided learning. arXiv preprint arXiv:2205.14304
20. Zintgraf L, Shiarli K, Kurin V, Hofmann K, Whiteson S (2019) Fast context adaptation via meta-learning. In: International conference on machine learning. PMLR, pp 7693–7702

A Certificate-Based Device Authentication System for an Enterprise Network

Aleena Terese George, Hrishikesh Rajendra Neve, and N. Muraleedharan

Abstract In modern networks and systems, due to the increase of security breaches using device compromise, it is inevitable to identify the device before connecting to the organization's networks. Existing approaches such as physically unclonable functions (PUF) that use the devices' digital fingerprint for authentication, require hardware-level implementation. Hence, it is difficult to implement in a generic environment such as an organizational network that may need to provide access to the devices including Bring Your Own Devices (BYOD). This paper proposes a certificate-based device authentication system for an enterprise network. The proposed system can authenticate both the organizational devices and BYOD by automatically issuing the certificate. It reduces the overhead related to certificate management such as issuance, revocation, renewal, etc., and helps to minimize human interventions, therefore increasing overall efficiency. By implementing the proposed approach, an extra layer of security can be provided to the organization's network.

Keywords Device authentication · PKI · Digital certificate · BYOD

1 Introduction

In this age of digital transformation, safeguarding information has become crucial for ensuring privacy, confidentiality, and trust. It is essential to authenticate the identity of a device before allowing access to a system or service. Device authentication methods that depend on the digital fingerprint of the device have limitations to implement in a generic network environment such as an organization network. Hence, the public key

A. T. George (✉) · H. R. Neve · N. Muraleedharan
Centre for Development of Advanced Computing (C-DAC), Bangalore, India
e-mail: aleenatg@cdac.in

H. R. Neve
e-mail: hrishineve@cdac.in

N. Muraleedharan
e-mail: murali@cdac.in

S. M. Thampi et al. (eds.), *Fifth International Conference on Computing and Network Communications*, Lecture Notes in Electrical Engineering 1219,
https://doi.org/10.1007/978-981-97-4540-1_29

infrastructure (PKI)-based device authentication, which relies on the use of digital certificates and public key cryptography, can be considered to authenticate devices.

The PKI framework constitutes a certificate authority (CA) responsible for issuing the digitally signed certificate and maintaining the repository of issued certificates. When the CA certifies a device, it verifies the device's identity and grants a certificate, signing with the private key of the CA. The CA has its certificate and public key, which it publishes, and a private key, which is securely maintained. Before providing the requested access to devices, the resource-granting entity uses the root certificate of the CA to verify signatures that the certificate authority has made. This certificate ensures that the device information is correct and that the public key belongs to that device [1].

1.1 Advantages of PKI-Based Authentication

The PKI-based authentication system has many advantages compared to the traditional password-based authentication system. The advantages of PKI-based authentication systems are

- Enhanced Security: Device authentication using PKI is safer than device authentication using more conventional techniques. Using a trusted CA ensures that the device's identity is validated, and digital certificates are considerably harder to forge than passwords or PINs.
- Non-repudiation: PKI-based device authentication offers non-repudiation, which means that once a user does an action, they cannot later deny their involvement, as their digital certificate is an undeniable proof of their identity.
- Centralized Management: Device identities can be centrally managed with PKI-based device authentication. Administrators will find it simpler to control device access to various systems and applications as a result.
- Scalability: The scalability of PKI-based device authentication makes it ideal for large organizations with numerous users.
- Interoperability: Devices can more easily access various systems using the same digital certificate due to the interoperability of PKI-based device authentication across various systems and applications.

A certificate-based authentication system for the devices is proposed in this paper. Some of the use-cases of the proposed system are

- Bring Your Own Device (BYOD): Identification and authentication of the device brought by the user can be implemented using the proposed approach. The short-term certificate can be issued to the device while it is connecting to the organization network [2, 3].
- Internet of Things (IoT): Device authentication is an essential requirement for IoT security. The proposed approach can be used to issue the certificate to IoT device [4, 5].

- Zero Trust: The proposed approach can be used to authenticate the devices for a zero trust network access system [6].

In this paper, we propose a certificate-based device authentication system for an enterprise network. The proposed system can authenticate both the organizational devices and BYOD by automatically issuing the certificate. As it allows only the authenticated device to connect to the networks, the security issues introduced by the unauthorized device can be kept in check. It also reduces the overhead related to certificate management such as issuance, revocation, renewal, etc., and helps to minimize human interventions, therefore increasing overall efficiency. The novelty of the paper is the two-stage authentication approach using the media access control (MAC) address from the network traffic and PKI-based certificate for the device authentication.

The organization of the paper is as follows: Sect. 2 describes the background and related works of device authentication and Sect. 3 explains the proposed system, its components and their functions in detail. Section 4 explains the device authentication process used in the proposed system and its implementation details. The challenges related to the implementation of proposed system is discussed in Sect. 5. Conclusion and future works are shown in Sect. 6.

2 Background and Related Work

This section discusses the fundamental aspects of device authentication and surveys the relevant works in the field. We explore the background of device authentication to establish a contextual foundation. Additionally, we review related works, examining various methodologies and frameworks that have shaped the landscape of certificate-based device authentication.

2.1 Background

The most popular kind of authentication is password-based [7], although this method is susceptible to brute-force and password-guessing attacks. A tangible token, like a smart card or a USB device, is used in token-based authentication to verify the identity of the user. A special identifying number found on the token is used to confirm the user's identity. Because the token cannot be easily copied, token-based authentication is more secure than password-based authentication. However, it necessitates that the user carries the token around, which might not always be practical [8, 9]. Utilizing biological traits to verify the user is known as biometric authentication. These traits could include voice recognition, facial recognition, iris scanning, or fingerprints. Because each person's biological traits are distinctive, biometric authentication is extremely safe. It calls for specialized hardware and software, though, and their

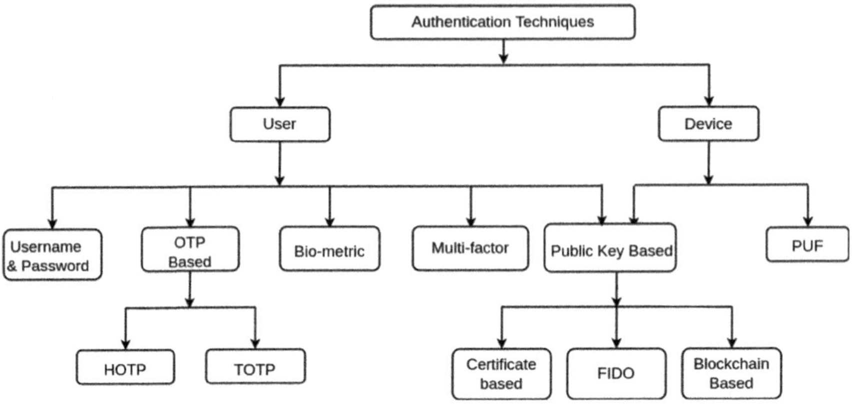

Fig. 1 Authentication techniques

implementation can be expensive. Considering how sensitive biometric data is, it also creates concerns with data privacy [10].

Authentication Techniques Different authentication techniques are available to authenticate the user and devices. The taxonomy of the common authentication techniques is depicted in Fig. 1.

User Name and Password In this method, the user must provide a secret password, this password will be converted into a hash value which is then compared with a hashed value of the password that is saved on the server [7]. The user is given access to the system or service if the two match. Despite being popular and easy to use, password-based authentication has certain security flaws. Brute-force attacks can be used to steal or guess passwords [9].

Biometric Biometric authentication uses unique physical or behavioral characteristics of a person to verify their identity. This includes techniques such as voice recognition, facial recognition, iris scanning, or fingerprint recognition. Biometric authentication provides a high level of security because these characteristics are distinct to each individual. However, because sensitive biometric data must be protected, its deployment demands specialized hardware and software and raises questions regarding data privacy [10].

OTP Based An OTP, or one-time password, is a temporary and dynamically generated numerical code used for authentication. Each OTP is valid only for a brief period, usually, just one login session or a limited time frame, making it difficult for attackers to reuse or guess the code and hence it adds an extra layer of security. There are different types of OTPs, such as HMAC-based OTP (HOTP) and time-based OTP (TOTP).

- HMAC-based OTP (HOTP): A shared secret key and a counter value are the fundamental elements of the OTP generation technique known as HOTP, or HMAC-based one-time password. The same secret key is maintained by both the server and the client. The counter value is increased with each authentication request, and the HOTP algorithm uses this counter value and the secret key to generate a unique OTP by utilizing a cryptographic hash function (HMAC). The OTP can be correctly validated as long as the server and client are in synchronization and the counter value is properly managed [11].

- Time-based OTP (TOTP): Another OTP generating technique that increases security by adding the current time as an additional parameter is TOTP, or time-based one-time password. Similar to HOTP, TOTP demands that the server and client share a secret key. However, TOTP uses the current time as part of the calculation instead of a counter. The server and client's device must be synchronized to the same time. A time-sensitive OTP is created by applying a cryptographic hash function to the secret key and the current time [12].

Multi-factor Authentication To increase security, multi-factor authentication combines two or more distinct authentication techniques. For example, before providing access, require both a password (something the user knows) and a fingerprint scan (something the user is).

Public Key-Based Digital certificates and public-private key pairs are used in the public key infrastructure (PKI) system to authenticate users and encrypt data. PKI-based authentication provides strong encryption and security features.

- Certificate-Based: Digital certificates are issued by a trusted certificate authority. These certificates contain the user's public key and other information, providing a secure way to verify the user's identity [1].
- Fast IDentity Online (FIDO): FIDO is an open standard for secure authentication that aims to eliminate passwords. Public key cryptography is used in FIDO-based authentication to identify users without disclosing any private data to the service provider. FIDO authentication procedures often use public key cryptography and biometrics. FIDO incorporates many protocols and techniques, including FIDO2 and FIDO U2F (Universal 2nd Factor) [13].

Physical Unclonable Function (PUF) Another technique for device authentication is physical unclonable functions (PUFs). PUFs create a distinctive digital fingerprint for each device by using its unique physical properties. The fingerprint is unique and unclonable since the physical characteristics are inherent to the device. By creating a fingerprint specifically for a device and comparing it to a reference fingerprint that has been stored, PUFs can be used for authentication. The device is authenticated if the fingerprints match [5].

Blockchain-Based Blockchain-based device authentication entails compiling a digital register of all the networked devices. Each device receives a unique identification number, and the blockchain stores the authentication information for each device [14]. Each time a device wants to join the network, its authentication information is verified against the blockchain.

Authentication Protocols The common protocols used for authentication are explained below.

- Lightweight Directory Access Protocol (LDAP): LDAP is a standardized protocol used for accessing and managing directory information services. It provides a standardized method to query and modify directory entries and acts as a foundation for centralizing user authentication, keeping user profiles, and managing resources in network environments [15].
- Remote Authentication Dial-In User Service (RADIUS): RADIUS is a networking protocol that offers centralized authentication, authorization, and accounting (AAA) services for network and remote access devices. It is widely used to enable safe user authentication for network resources including Wi-Fi networks and virtual private networks (VPNs). The RADIUS server verifies a user's credentials when they attempt to connect to a network and either gives or denies access based on preset policies [16].
- Security Assertion Markup Language (SAML): SAML is an XML-based protocol for sharing authentication and authorization information between parties, often between an identity provider (IdP) and a service provider (SP). Single Sign-On (SSO), which allows users to access different applications and services with a single set of credentials, is made possible using SAML. The IdP creates a SAML assertion with authentication data when a user logs in to one application, and sends it to the SP to provide access without requiring the user to log in again [17].
- OpenID Connect (OIDC): The OpenID Connect (OIDC) authentication protocol, which is based on OAuth 2.0, enables safe and convenient user authentication for web and mobile apps. It enables users to sign in to one application with their credentials from an identity provider and then use those credentials to access other connected applications without having to enter them again. Applications can use the identification data provided by OIDC in the form of JSON web tokens to confirm the user's identity and acquire required attributes [18].

2.2 Related Works

In the dynamic field of device authentication, researchers have devoted considerable effort to addressing the increasing challenges of securing access to electronic devices. Research in this field has concentrated on creating reliable, flexible, and easy-to-use authentication solutions that can successfully negotiate the complex balance between security and usability. A device authentication technique using Message Queue Telemetry Transport Protocol (MQTT) was introduced by Bhawiyuga et al. [19] which replaces the traditional password and username-based authentication. However since MQTT only has one channel for authentication, an attacker may simply obtain the credentials by wireless sniffing.

Ch et al. [20] introduced an authentication system that uses a smart card reader (RFID reader) or a biometric sensor. This technique is prone to disclosure attacks. A blockchain-based secure device authentication and key agreement system is proposed in [21]. Enhanced trust is provided by this mechanism, but using it will make deployment, development, and scaling challenging. To reduce the risk of authentication key

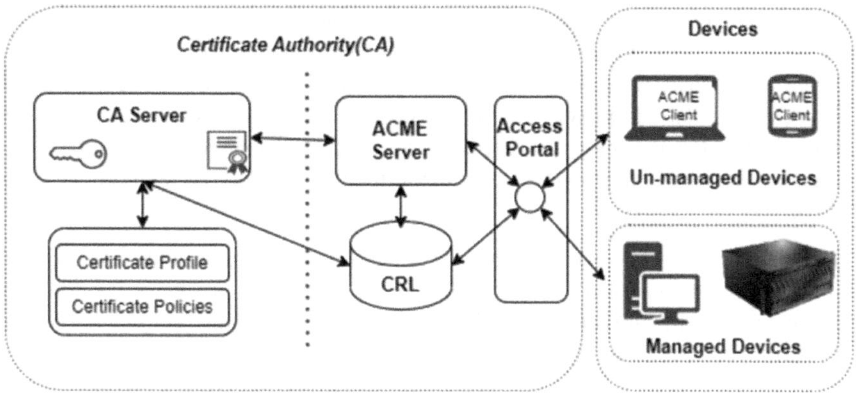

Fig. 2 Components of certificate-based device authentication system

exposure and the load on the authentication server, a PUF-based authentication system was introduced by Yoon et al. [22]. This method has scalability concerns, higher cost, and complex implementation.

3 Proposed System

The proposed system uses certificates to authenticate devices. An organizational Certificate Authority (CA) needs to be deployed and configured to issue, manage, and revoke digital certificates used for the devices. The major components of the proposed system are depicted in Fig. 2.

3.1 Certificate Authority (CA)

The Certificate Authority issue, manage, and revoke digital certificates. The CA ensures that the information included in a certificate is accurate and confirms the identification of certificate applicants. The CA can be configured as a public CA or private CA. The public CA is a trusted third party which issues the publicly trusted certificate. The private CA is an enterprise-level CA that can generate the certificates for the organization-level usage. In the proposed system, we are using the private CA to issue the certificates to the devices. The private CA generates and maintains a self-signed certificate that will act as the root of the chain of trust.

Depending on the size of the organization, the Registration Authority (RA) or subordinate CA may exist in an organization [23]. The RA collaborates with the CA to confirm the device identities. The RA could also be in charge of overseeing the device enrollment procedure. In the proposed system, depending on the number of users and devices, the RA role may exist. For a small organization, the CA can verify the device requests and issue the certificates to the device. However, for a bigger organization, the CA can issue the certificate to the RA. The RA needs to verify the device request and upon verification, it can issue the certificate to the devices. For example, the department of a bigger organization can play the role of the subordinate CA and it can verify the devices in the department and further issue the certificate to them.

Certificate Policy The use of digital certificates in a particular context is governed by the certificate policy, a collection of regulations and best practices. In addition to guidance for certificate use and maintenance, the policy may also contain regulations for certificate issue, validity, and revocation.

Key Management The creation and storage of cryptographic keys for encryption and decryption are the responsibility of the key management system. The system shall have mechanisms for key generation, distribution, backup, key recovery, and key destruction.

Certificate Repository Digital certificates are kept and given out to users and programs that require them by the Certificate Repository.

Certificate Revocation List (CRL) The list of certificates that have been revoked by the CA is known as the Certificate Revocation List. Clients can check the status of a certificate before depending on it by using the CRL, which the CA publishes.

Automated Certificate Management Certificate management including the issuance, renewal, and revocation is the major challenge in the deployment of the certificate-based authentication system. We have addressed this issue using the Automated Certificate Management Environment (ACME) protocol [24] that automates the certificate management process without human interventions. ACME is a protocol that allows one to automatically obtain and manage certificates. Enabling automation eases the load of manual management on certificate issuance, renewal, and revocation. In order to request and get a certificate, ACME uses client-server communication. ACME streamlines the certificate renewal process by giving clients an interface to request certificate renewals and enabling the CA to issue renewed certificates automatically. This automated procedure makes managing certificates quick and safe by helping to keep credentials valid and up to date.

Access Portal The access portal component in the proposed system acts as an interface between the device and the certifying authority. The devices that need to connect to the network need to connect with the access portal. Upon verification of the requests, the access portal forwards the request as per the defined policy.

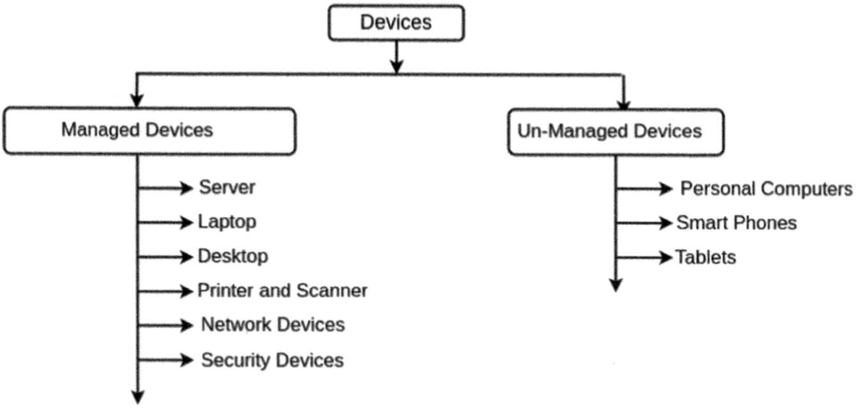

Fig. 3 Device category

3.2 *Device*

In the proposed authentication system, we are categorizing the devices into two types, i.e., managed devices and unmanaged devices. Figure 3 summarizes the device category used in the proposed system and the details of them are explained below.

Managed Devices Managed devices refer to computers, servers, and other devices that are under the control and management of an organization. These devices are owned by the organization and are used for specific business purposes. In the proposed system, we assume that the organization maintains the list of all managed devices in a searchable database which can be accessed and verify the device's existence. Managed devices are configured, maintained, and secured by the organization, and may have specific policies and procedures in place for their use. The validity of the certificate issued to a managed device depends on the organization's policy. However, typically the validity of the managed device certificate shall be in years.

Unmanaged Devices The unmanaged devices, on the other hand, are devices that are not under the control or management of the organization. These devices are typically owned by individual employees or contractors and may be used for personal or business purposes. The unmanaged devices are not typically configured, maintained, or secured by the organization. Hence access to these devices needs to be restricted. Once it is authenticated, the access permission of an unmanaged device is limited for a short duration. By limiting the access duration of an unmanaged device, we will be able to control the indefinite use of a device in the network. It is implemented using by configuring the validity of the certificate. For example, the validity of the certificate issued to an unmanaged device shall be configured for one day.

The access portal with ACME client will be intelligently designed to categorize devices into managed and unmanaged groups. This categorization allows for efficient filtering, ensuring that the appropriate CSRs are sent to the CA.

4 Device Authentication Using Certificates

A trusted certificate authority (CA) signs a special digital certificate that is assigned to the device for device authentication. It delivers its digital certificate to the server when the device tries to connect to a network or server. Using the CA's public key, the server uses the digital certificate to verify it, and if successful, it allows access to the device. It offers a safe way to make sure that only approved devices can access a network or a server. Additionally, it offers a way to guarantee the security and privacy of the data sent between the device and the server. The steps to generate a device certificate are as follows:

1. Registration of the device at the registration authority (RA).
2. Validation of device credentials by RA. (Using a database of the asset management team of the organization with the credentials of the new device, if credentials are valid.)
3. Based on the information filled in the configuration file, a certificate signing request (CSR) will be generated.
4. Issue of certificate by the certificate authority (CA) to the device and updating of the certificate in the certificate database.
5. The issued certificate will be sent to the device.

The set of events that take place when a device logs into the network as in our system is explained in Algorithm 1.

4.1 Implementation Details

The implementation details of core components of the proposed system is explained next.

Device Certificate The following parameters [25] were used to derive the device certificate:

(i) Country.
(ii) State or Province Name.
(iii) Organization Name.
(iv) Organization Unit.
(v) Validity of the Certificate.
(vi) Key Algorithm used.
(vii) Key size.
(viii) Email Address.
(ix) MAC Address.

In our system, the device certificates are generated using an extended key usage [26]. An extended key, commonly referred to as an Extended Key Usage extension (EKU extension), is a field in a digital certificate that describes the intended uses for

Algorithm 1: Device authentication using certificates

```
Device enters the network
Lookup the device MAC in the asset list
if (MAC found) then
    Device Type ← Managed
    if Valid Device certificate then
        Extract MAC from the request traffic
        Read the MAC from certificate
        Compare the MAC from traffic and certificate
        if MAC matching then
            Access Allowed
            Update the logs
        else
            Access Denied
            Update the logs
        end
    else
        Access Denied
        Update the logs
    end
else
    Device Type ← UnManaged
    Extract MAC from the request traffic
    Initiate the registration process
    if Successful Registration then
        Generate a public-private key pair for the device.
        Request a digital certificate from a trusted Certificate Authority (CA) using the
          device's public key and configuration file
        Receive the signed certificate from the CA and store it securely
        Update database and log
    else
        Access Denied
        Update the log
    end
end
```

the public key it contains. The extended key usage extension indicates the certificate is a device certificate.

We have configured the client certificate with the EKU extension with OID value 1.3.6.1.5.5.7.3.2 which represents the client authentication. To uniquely identify the device, we have added the media access control (MAC) address in the 'SerialNumber' field of 'Subject' in the certificates. To ensure that a device is approved to access the network, each one will have a device certificate. Before each login attempt, the user and device certificates will both be checked. Access to the network is only permitted with a valid device certificate. By using the MAC address in the device certificate, we achieve unique device identification. Moreover, when a device connects to the network, the MAC address can also be identified using the request packet originating from the device. Figures 4 and 5 show the sample certificate issued to a managed

Fig. 4 Sample certificate for managed device

device and an unmanaged device, respectively. From Fig. 4 we can observe that the validity of the issued managed device certificate is one year. Similarly, Fig. 5 shows the validity as one day since it is the certificate issued for an unmanaged device.

Automatic Certificate Management Using ACME In the proposed system, we used the Automated Certificate Management Environment (ACME) protocol [24] to automate the certificate management process without human intervention. The access portal can effectively incorporate the ACME protocol to streamline the certificate issuance process since the access portal manages the connection between the clients and the server. This centralization of ACME implementation within the access portal ensures a more organized and manageable approach. The access portal, acting as the client interface, becomes the focal point for handling ACME-related operations, including certificate signing requests (CSRs) and certificate management. In addition to the access portal, the server must also be configured with ACME capabilities.

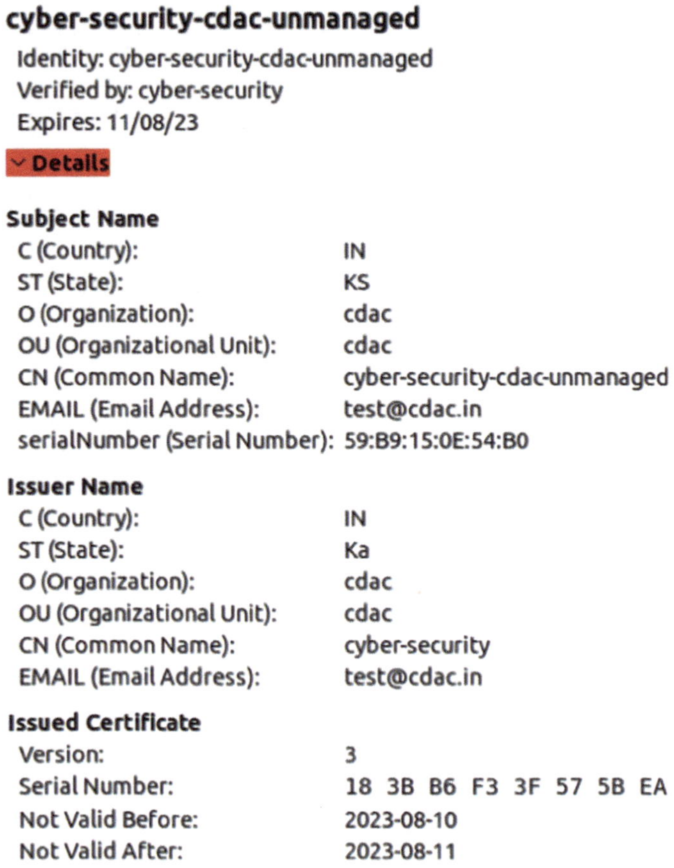

Fig. 5 Sample certificate for unmanaged device

CAs can be used and set up along with the ACME server. The CA is configured to generate the self-signed root and intermediate CA which is linked with the ACME PKI that would process the CSR, revoke, and manage client requests received through the ACME client. The private CA is implemented using 'step-ca' open source solution [27]. The 'acme' provisioner available in the 'step-ca' is configured to support the ACME features.

We have used the 'certbot' as the ACME client for our implementation. The ACME client should register itself initially with the configured server and an authoritative key pair is generated. The CA generates a DNS or HTTP challenge to demonstrate the control of the client over the requested domain or service running on a device to which the CSR belongs. This is done by the server sending a random generated text which the client has to sign. On approval of the challenge by the CA, the CSR

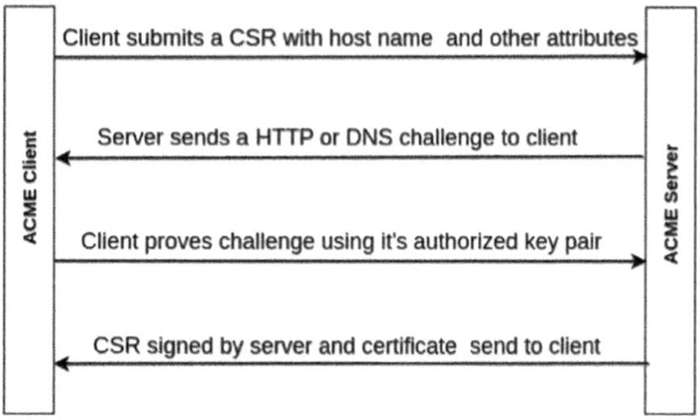

Fig. 6 ACME client-server communication

is signed, and the certificate is sent to the client automatically. The same process can be implemented for the revocation and renewal of the certificates. The ACME client-server communication sequence is depicted in Fig. 6.

5 Challenges and Vulnerabilities

Although ACME automates the process of obtaining, renewing, and revoking certificates, service interruptions may result from incorrect setups or missed certificate renewals. Maintaining a safe and functional authentication system depends on ensuring seamless and timely certificate renewal.

The safekeeping of the private keys connected to the certificates presents another difficulty. Compromised private keys can result in unwanted access and possible exploitation of private data if they are not sufficiently secured. Therefore, to protect the integrity of the certificate-based authentication system, strong key management procedures are necessary. Moreover, if a CA is compromised or issues certificates without proper verification, it could undermine the trustworthiness of the entire authentication system.

Therefore, in order to ensure that an ACME certificate-based device authentication system is resilient against potential vulnerabilities, it is imperative that issues associated with certificate lifecycle management, key protection, CA security, and secure communication channels are addressed.

6 Conclusion and Future Works

A certificate-based device authentication system for an enterprise network is discussed in this paper. The proposed system uses a two-layered approach for device authentication that uses the public key-based certificate and MAC address derived from the network traffic to confirm the device's identity. The system authenticates the existing devices in the organization and BYOD devices connecting to the network. Depending on the policy configured, access to the BYOD system can be restricted using the proposed approach. To reduce the overhead related to certificate management such as issuance, revocation, renewal, etc., the system provides an automated certificate management using ACME protocol and minimizes the human interventions. As future work, we would like to incorporate the certificate-based user authentication to the system.

Acknowledgements The authors would like to express their sincere gratitude to the Ministry of Electronics and Information Technology, Government of India, for providing financial support and resources that made this research project possible.

References

1. Grassi P, Garcia M, Fenton J (2017) Digital identity guidelines: revision 3. National Institute of Standards. https://doi.org/10.6028/NIST.SP.800-63-3
2. Downer K, Bhattacharya M (2015) BYOD security: a new business challenge. In: IEEE international conference on smart city/SocialCom/SustainCom (SmartCity). https://doi.org/10.1109/SmartCity.2015.221
3. Eke CI, Anir AN (2021) Bring Your Own Device (BYOD) security threats and mitigation mechanisms: systematic mapping. In: International conference on computer science and engineering (IC2SE), Padang, Indonesia. https://doi.org/10.1109/IC2SE52832.2021.9791907
4. Karthikeyan S, Patan R, Balamurugan B (2019) Enhancement of security in the internet of things (IoT) by using X.509 authentication mechanism. In: Recent trends in communication, computing, and electronics, Singapore
5. Siddiqui Z, Gao J, Khan MK (2022) An improved lightweight PUF-PKI digital certificate authentication scheme for the internet of things. IEEE Internet Things J 9(20):19744–19756. https://doi.org/10.1109/JIOT.2022.3168726
6. Rose S, Borchert O, Mitchell S, Connelly S (2020) Zero trust architecture. National Institute of Standards and Technology. https://doi.org/10.6028/NIST.SP.800-207
7. Zhao Y, Li S, Jiang L (2018) Secure and efficient user authentication scheme based on password and smart card for multiserver environment. Secur Commun Netw
8. Khan HZU, Zahid H (2010) Comparative study of authentication techniques. Int J Video Image Process Netw Secur (IJVIPNS) 10(04)
9. Wang X, Yan Z, Zhang R, Zhang P (2021) Attacks and defenses in user authentication systems: a survey. J Netw Comput Appl 188:103080. https://doi.org/10.1016/j.jnca.2021.103080
10. Yang W, Hu J, Wang S, Wu Q (2018) Biometrics based privacy-preserving authentication and mobile template protection. Wirel Commun Mob Comput. https://doi.org/10.1155/2018/7107295
11. M'Raihi D, M'Raihi D, Hoornaert F, Naccache D, Bellare M, Ranen O (2005) HOTP: an HMAC-based one-time password algorithm
12. M'Raihi D, Rydell J, Pei M, Machani S (2011) TOTP: time-based one-time password algorithm
13. Czeskis A, Lang J (2015) FIDO NFC protocol specification v1.0, vol 17, pp 872–888. FIDO Alliance Proposed Standard
14. Yang X et al (2022) Blockchain-based secure and lightweight authentication for internet of things. IEEE Internet Things J 9(5):3321–3332. https://doi.org/10.1109/JIOT.2021.3098007
15. Sermersheim J (2006) Lightweight directory access protocol (LDAP): the protocol
16. Rubens A, Rigney C, Willens S, Simpson W (2000) Remote authentication dial in user service (RADIUS)

17. Campbell B, Mortimore C, Jones M (2015) Security assertion markup language (SAML) 2.0 profile for OAuth 2.0 client authentication and authorization grants
18. Sakimura N, Bradley J, Jones M, De Medeiros B, Mortimore C (2014) OpenID connect core 1.0. The OpenID Foundation, p S3
19. Bhawiyuga A, Data M, Warda A (2017) Architectural design of token based authentication of MQTT protocol in constrained IoT device. In: 11th international conference on telecommunication systems services and applications (TSSA), Lombok
20. Ch R, Lakshmi B, Krishna PN, Sai Rama Krishna C, Sree AS (2021) Implementation on RFID based device control and authentication system. In: IEEE international conference on mobile networks and wireless communications (ICMNWC)
21. Shen M et al (2020) Blockchain-assisted secure device authentication for cross-domain industrial IoT. IEEE J Sel Areas Commun 38(5):942–954
22. Yoon S, Kim B, Kang Y, Choi D (2020) PUF-based authentication scheme for IoT devices. In: International conference on information and communication technology convergence (ICTC), pp 1792–1794
23. Prodanovic R, Vulic I, Tot I (2019) A survey of PKI architecture. In: Knowledge based sustainable development
24. Barnes R, Hoffman-Andrews J, McCarney D, Kasten J (2019) Automatic certificate management environment (ACME)
25. Boeyen S, Santesson S, Polk T, Housley R, Farrell S, Cooper D (2008) Internet X.509 public key infrastructure certificate and certificate revocation list (CRL) profile
26. https://www.openssl.org/docs/man3.1/man1/openssl-verification-options.html
27. Step-CA. Easily manage certificates for production workloads. Available at: https://smallstep.com/certificates/. Accessed 24 Nov 2023

Analyzing the Correlation of Smart Contract Vulnerability and Readability: A Machine Learning Approach

K. M. Sameera, P. Vinod, K. A. Rafidha Rehiman, Poonam Kumari, Harsh Raj, Rinki Kumari, Aman Kumar, and Ritika Mittal

Abstract Blockchain's rapid growth is revolutionizing sectors like finance and healthcare, providing universal computational access for diverse, decentralized applications (DApps). It also facilitates the creation of DApps, which integrates smart contracts on the blockchain with a user-friendly front-end interface. This research aims to assist researchers and developers in analyzing the smart contract's code readability level and vulnerability types. In this work, we collect and analyze smart contracts from the "Etherscan" platform to demonstrate the tool's effectiveness. To the author's knowledge, this is the first work examining the association between the readability and vulnerability of solidity smart contracts. To examine the degree of readability of smart contracts, we employ a set of readability matrices that the previous research demonstrated related to code readability. Our suggested method evaluates the smart contract automatically and calculates a set of metrics for the level of code readability. Furthermore, We propose a machine learning model to create an automated system that discerns between readable and non-readable smart contracts. The result shows that XGBoost outperformed other classifiers with an impressive 98.856% F-Measure and 98.858% accuracy. Additionally, we utilize the *Slither* to analyze security issues in smart contracts. Afterward, we examined the most prevalent and critical vulnerability types discovered in each smart contract. Also, it is observed that there is a correlation between non-readability and vulnerability. Our analysis revealed that 72.6% of non-readable files exhibit vulnerabilities, showing that files with high vulnerability are characterized by low readability.

K. M. Sameera · P. Vinod · K. A. Rafidha Rehiman (✉) · P. Kumari · H. Raj · R. Kumari · A. Kumar · R. Mittal
Department of Computer Applications, Cochin University of Science and Technology, Kochi 682022, Kerala, India
e-mail: rafidharehimanka@cusat.ac.in

K. M. Sameera
e-mail: sameerakm@cusat.ac.in

P. Vinod
e-mail: vinod.puthuvath@unipd.it

P. Vinod
Department of Mathematics, University of Padua, 35121 Padua, Italy

S. M. Thampi et al. (eds.), *Fifth International Conference on Computing and Network Communications*, Lecture Notes in Electrical Engineering 1219,
https://doi.org/10.1007/978-981-97-4540-1_30

Keywords Blockchain · Smart contracts · Code readability · Vulnerability

1 Introduction

In recent years, blockchain technology has experienced rapid growth due to its decentralized and tamper-resistant characteristics [1]. The development of blockchain technology has bestowed universal computational convenience upon a broad spectrum of DApps spanning multiple domains, including healthcare, finance, and more [2]. The immutable nature of blockchain ensures that they cannot be altered once smart contracts are deployed [3]. This feature provides fairness and trustworthiness to transactions and makes blockchain technologies more appealing to developers. Furthermore, it facilitates DApps creation [4], which integrates smart contracts on the blockchain with a user-friendly front-end interface. Ethereum, an open-source blockchain platform, supports smart contracts using the cryptocurrency ether. These self-executing code blocks enhance security by automating agreements, reducing the need for intermediaries. Despite their numerous advantages, creating standardized smart contracts presents challenges for blockchain developers in analysis, testing, maintenance, and development [5].

Smart contract developers prioritize optimizing their code and usually design using unconventional software development processes; they often deviate from traditional object-oriented programming coding. This affects gas consumption (the computational cost required for executing operations on the Ethereum blockchain) and resource usage, primarily due to the high execution costs. Code readability is a fundamental aspect of software development, dramatically influencing developers' ability to understand and maintain code efficiently. Furthermore, it significantly affects program comprehension and enhances maintainability, portability, and reusability. More specified codes have a low possibility of causing errors and hence reduce security vulnerabilities. Readable codes ensure portability, maintainability, and reusability and reduce the complexity of the source code. Recent research indicates that code optimization can potentially lead to a decrease in code readability and understandability [6]. Recent research has revealed that code reuse is prevalent in smart contracts, with approximately 26% of contract code blocks being reused [7]. Consequently, when the code exhibits low readability, it not only hampers the understandability and maintainability of smart contracts but also acts as a barrier to further code reuse.

In the meantime, smart contracts play an essential role in maintaining the security of blockchain platforms. Their security has consistently posed a substantial challenge within the blockchain domain. The development of smart contracts can introduce vulnerabilities, which, in turn, may result in severe consequences, including financial losses. The open accessibility for anyone to create and deploy smart contracts on Ethereum has led to a considerable accumulation of similar contracts through code reuse and propagating vulnerabilities. Once a smart contract is deployed, it becomes immutable, meaning that any changes or modifications to the code are no longer possible. This immutability is a fundamental characteristic of smart contracts on blockchain platforms and maintains the integrity, transparency, and trustworthiness of transactions and agreements executed through them. Reusing code within smart contracts can magnify substantial risks, including security breaches and resource inefficiency. It encompasses a range of vulnerabilities, including Reentrancy, UncheckedCall, Suicidal vulnerability, overflow errors, logic flaws, and others [8–11]. Mitigating these issues requires a rigorous process of code review and comprehensive testing.

Based on prior research, we conducted empirical research to explore the relationship between smart contract readability and vulnerability. We have utilized 2186 smart contracts from the Etherscan platform to showcase the tool's capability to calculate readability metrics accurately and conduct vulnerability analysis. This study represents the first quantitative and qualitative investigation into whether non-vulnerable smart contracts are readable. Additionally, we developed models to assess smart contract readability using various machine learning algorithms. To conduct our investigation, we employed a tool that automatically evaluates the source code of Solidity[1] smart contracts and computes a set of code readability metrics discussed in iSCREAM [12]. Furthermore, we leveraged machine learning models to create an automated system for distinguishing readable from non-readable smart contracts. Despite the absence of labeled data for metric validation, we manually labeled the readability of the smart contracts. Our findings revealed that the XGBoost model outperformed other machine learning models, achieving an impressive test F-Measure of 98.858%. Following this, we utilized Slither's static analysis tool to detect vulnerabilities within each smart contract, analyzing the frequency of occurrence for various vulnerabilities in each contract and identifying the absence of specific vulnerabilities. By prioritizing these vulnerabilities, we provided developers and security experts with valuable insights into the security issues of the contracts. Additionally, we conducted an examination of the most prevalent and critical vulnerability types discovered in each smart contract. Finally, we assessed the correlation between readability and vulnerability. Our analysis has provided insights indicating that smart contracts possess a notably low level of readability, and concurrently, they exhibit a high degree of vulnerability. In summary, the primary contributions of this paper are as follows:

- We developed a system that automatically assesses Solidity smart contracts' readability characteristics. We provided labeled classifications for each contract, distinguishing them as readable or non-readable.
- We propose a machine learning model to create an automated system that discerns between readable and non-readable smart contracts.
- We utilized Slither's static analysis tool to detect vulnerabilities within each smart contract and analyze the frequency of occurrence for various vulnerabilities in each contract.
- We examined each smart contract's most prevalent and critical vulnerability types. This process involved a detailed analysis of the vulnerabilities in determining their potential impact and exploitability.

[1] https://solidity.readthedocs.io/.

- We analyzed the correlation between smart contract vulnerability and readability to determine whether vulnerable files were readable.

Paper structure. The remaining portion of this paper is organized as follows. The related works were made in Sect. 2. Section 3 describes the proposed approach for vulnerability and readability analysis. Section 4 presents the experimental evaluation, and Sect. 5 provides the conclusion and future directions.

2 Related Work

This section delves into the literature concerning code readability and vulnerabilities in smart contracts.

2.1 Code Readability

Code readability is the most prominent attribute for developing a software product and writing source code. More specified codes have a low possibility of causing errors and enhancing maintainability, portability, and reusability. Numerous research studies have suggested various methods and approaches for evaluating the readability of different programming languages, such as Java, Python, and more [13, 14]. Buses and Weimer introduced the measures for determining code readability within a code snippet [15]. This study developed a code readability model using 100 code snippets from various open-source projects. The model comprises 25 structural code metrics. To validate it, 120 computer science students rated the readability of the code snippets on a scale of 1 to 5. The results indicate that the model is often more effective at predicting readability than human assessors and strongly correlates with software quality metrics like code changes and defect reports. This research underscores the significance of code readability in software development and suggests that structural code metrics play a crucial role in assessing it. Scalabino et al. [13] introduced textual features to enhance code readability models, aiming to encompass both syntactic and textual aspects of source code. Involving over 600 code snippets assessed by around 5k participants, their research replicated Buse and Weimer's study. They demonstrated that a comprehensive model combining all features correlated with FindBugs warnings. These findings highlight how the newly proposed textual features complement the existing structural characteristics. In the context of code readability, combining both sets of features demonstrated significantly improved accuracy compared to using structural features alone. Farokey et al. [16] uncovered inconsistencies between cohesion, coupling, complexity, and readability metrics and their practical interpretation. They proposed utilizing tools for structural metrics, refactoring, and style issues to enhance code readability. Their research examined 548 readability enhancements reported by developers across 63 software projects, revealing the inadequacy of current readability models. To improve readability models in the future, they advocated considering tools that compute additional structural metrics and detect refactorings and style problems. Canfora et al. [12] introduced iSCREAM, a tool designed to automate the Solidity smart contract source code analysis. iSCREAM calculates the same code features demonstrated by Buse and Weimer to be associated with code readability. This assessment used a diverse sample of 90 functions of varying sizes extracted from multiple smart contracts. Vacca et al. [17] presented an empirical study that explores the relationships between various metrics and gas consumption, as well as between different pairs of metrics. Their investigation revealed that smart contracts typically demonstrate lower readability than conventional software regarding metrics like the frequency of parentheses, inline comments, and blank lines. Furthermore, they observed significant correlations between specific readability metrics, such as the average length of identifiers, the average count of keywords, and gas consumption.

2.2 Smart Contract Vulnerability

A smart contract is a self-executing computer program that enforces agreement terms upon meeting specific conditions, eliminating the need for intermediaries to enhance transaction speed, security, and cost-effectiveness [18]. Demir et al. [19] discussed vulnerabilities in smart contracts and categorized them based on their nature. The paper also addresses deployment challenges, library code reuse through delegate calls, assembly operation vulnerabilities, re-entrancy risks, compiler version problems, exception handling issues, and unsafe coding practices. Khan et al. [20] explored different vulnerabilities in smart contracts and categorized them based on their specific domains of problematic operations. They address various vulnerabilities, classifying them into inter-contractual, contractual, integer-related, gas-related, transactional, deprecated, and randomization vulnerabilities. Perez et al. [21] identified a range of vulnerabilities in smart contracts, including Re-Entrancy, Unhandled Exceptions, and more. Their paper emphasizes that despite the significant number of reported vulnerable contracts, these vulnerabilities have seen minimal exploitation. They examined just 0.29% of the 3 million smart contracts in their study, leading to a loss of 9,066 ETH.

Summary: In the existing literature, it is evident that the prior studies have yet to explore the relationship between the readability and vulnerability of smart contracts. Furthermore, a noticeable gap exists in the literature where automated systems for distinguishing between readable and non-readable smart contracts are absent. Furthermore, there is a gap in the existing literature concerning an in-depth examination of the most prevalent and critical vulnerability types identified within each smart contract.

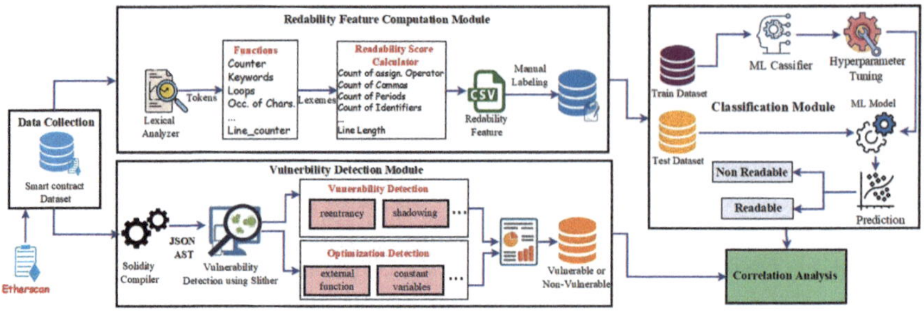

Fig. 1 Proposed architecture

3 Proposed Methodology

Enhancing code readability is paramount in software development and source code writing. Well-structured and clear code minimizes the likelihood of errors and mitigates security vulnerabilities. This research quantifies code readability and scrutinizes vulnerabilities within Solidity smart contracts. As illustrated in Fig. 1, the proposed architecture comprises five key components: the data collection module, the readability feature computation module, the vulnerability detection module, the classification module, and the correlation analysis module. The readability feature computation module calculates a readability matrix for each smart contract, while the vulnerability detection module examines smart contract vulnerabilities using the Slither tool. Various machine learning models are employed within the classification module to automatically assess smart contracts' readability. Moreover, using the Pearson correlation coefficient, the correlation analysis module calculates the correlation between readability and vulnerability in smart contracts.

3.1 Dataset Collection and Labeling

In this experiment, we collected 2186 solidity smart contracts from Etherscan[2], a decentralized smart contracts platform for Ethereum where developers publish the smart contract source code. We performed the manual annotation, where one Ph.D. student, with guidance from two experienced professors in smart contracts and blockchain, categorized contracts as "1" for readable and "0" for non-readable ones. To enhance the reliability of the annotation process, the labeling of the same data was undertaken by a single annotator three times. The Fleiss' Kappa [22] was then calculated to quantify the level of disagreement. We obtained a Kappa value of 0.86, signifying nearly perfect agreement. Our dataset includes only statistically

[2] https://etherscan.io/.

significant samples of smart contracts, with a confidence level of 99% and an error margin of less than 3%. To train the model, we utilized 80% of the labeled data, while the remaining samples were evenly split between the validation and testing sets.

3.2 Redability Feature Computation Module

The primary purpose of the readability computation module is to establish readability metrics for each smart contract, which serve as distinctive features in the classification process. This module automatically assesses and calculates readability scores for individual functions within the smart contract. It comprises two key components: the Lexical Analyzer and the Readability Metric Calculator. The analysis process for each function within the targeted smart contract unfolds as follows.

Lexical Analyzer: The lexical analyzer divides the raw input file into tokens, and these tokens are subsequently processed to extract the necessary results known as lexemes. This extraction involves the use of various functions such as *counter(), keyword(), loops(), comments(), occurrence_of_characters(), count_parentheses(), number_count(), and Line_counter().*

Readability Metric Calculator: The extracted lexeme is utilized for the calculation of the readability score, taking into account a comprehensive set of 19 different metrics, as detailed in Table 1. Each metric, designed to assess code readability, provides an average or maximum value across all lines, irrespective of the code block. For instance, in the context of calculating assignment metrics related to the number of assignment operators, our method computes both average and maximum values for each lexeme. Moreover, for the occurrence of any character and identifier metrics, we compute the number of occurrences of the identifier and the character that most commonly occurs in the smart contract.

1. Average Value (Avg): The total occurrence of lexemes divided by the total number of lines in smart contract functions.
2. Maximum Value (Max): The occurrence of lexemes in each line in the smart contract functions.

After processing all the functions in Solidity smart contracts, the system stores the outcome in a tabular format. Subsequently, it exports these results as a CSV file for enhanced readability and comprehension.

3.3 Classification Module

In this study, we utilized six ML classifiers: Logistic Regression (LR), Support Vector Classifier (SVC), K-Nearest Neighbor (KNN), Decision Tree (DT), Random Forest (RF), and XGBoost for classifying smart contracts into readable or non-readable

Table 1 Set of metrics for readability score calculation

ID	Redability metric	Avg	Max	Correlation
1	Assignment	✓		N
2	Commas	✓		N
3	Periods	✓		N
4	Comparison	✓		N
5	Spaces	✓		N
6	Blank lines	✓		P
7	Keywords	✓	✓	N
8	Operators	✓		P
9	No of Identifiers	✓	✓	N
10	Identifier Length	✓	✓	N
11	Occurrence of any identifier		✓	N
12	Indentation	✓	✓	N
13	Loops	✓		N
14	Conditionals	✓		N
15	Occurrence of any character		✓	N
16	Parenthesis	✓		N
17	Comments	✓	✓	P
18	Number of numbers	✓	✓	N
19	Line length	✓	✓	N

The checkmark indicates the value considered for the readability score calculation. In the final column, the symbol 'N' denotes a negative correlation, and 'P' signifies a positive correlation with code readability, as found in prior research [15]

categories. Each model has its own set of hyperparameters that require optimization to achieve optimal results. To achieve this optimization, we employed hyperparameter tuning techniques using GridSearchCV to identify the most suitable parameter values for the classifiers.

3.4 Vulnerability Detection Module

To assess the Smart contract's vulnerability, initially compile the Ethereum-based solidity smart contract using the solidity compiler with version 0.4.0. After a successful compilation, the solidity compiler generates a Solidity Abstract Syntax Tree and directs it as input to the Slither for vulnerability detection. Slither[3] is a Solidity static analysis framework that recognizes 78 vulnerabilities, finds code optimizations the compiler misses, and summarizes the contract information [23]. That assists Smart contract developers in detecting vulnerabilities and improving code comprehension. The result is generally recorded in JSON format. To examine the different vulnerabil-

[3] https://github.com/crytic/slither.

ities in the smart contract, we extract the information from the description component of the JSON file containing the list of vulnerabilities. Furthermore, the vulnerabilities for each Solidity file function are recorded in a CSV file. Additionally, when Slither found a vulnerability, we labeled the contract "vulnerable"; otherwise, we labeled it "non-vulnerable".

3.5 Correlation Analysis

We use the Pearson correlation coefficient (r) to evaluate the correlation between vulnerability and readability, computing it through the Eq. (1), as defined in [24].

$$r = \frac{\sum_{i=1}^{n}(X_i - \bar{X})(Y_i - \bar{Y})}{\sqrt{\sum_{i=1}^{n}(X_i - \bar{X})^2 \sum_{i=1}^{n}(Y_i - \bar{Y})^2}} \tag{1}$$

where r is the Pearson correlation coefficient between parameters X and Y, n is the number of samples for parameters X, and Y, X_i, and Y_i are individual sample points indexed with i, $\bar{X}$ is the sample mean of X_i, and $\bar{Y}$ is the sample mean of Y_i.

4 Experimental Evaluation

In this section, we discuss the experimental setup and the effectiveness of the proposed method. In our experiments, we employed a system featuring a 10-core i9 CPU, 32 GB of RAM, and an NVIDIA GDDR5X Quadro P2000 GPU, configuring the simulation using Python 3.

4.1 Evaluation Matrix

We assessed the proposed models and their influence on the test samples by employing the following set of evaluation metrics.

1. Accuracy: Accuracy is the correctness of a classification model, expressed as the ratio of correctly predicted instances to the total instances in the dataset, calculated using Eq. (2).

$$\text{Accuracy} = \frac{\text{TP} + \text{TN}}{\text{TP} + \text{FN} + \text{FP} + \text{TN}} \tag{2}$$

   TP denotes the number of samples correctly predicted as the true class, and TN indicates the number of samples correctly predicted as the negative class. FP

represents the samples incorrectly predicted as the true class, and FN indicates the number of samples predicted as a different class by the model.

2. F-Measure: The F-Measure measures a test's accuracy and can be seen as a weighted average of precision and recall.

$$\text{F-Measure} = 2 * \frac{\text{Precision} * \text{Recall}}{\text{Precision} + \text{Recall}} \tag{3}$$

Here, Precision measures true positive predictions to all positive predictions, while Recall indicates the model's ability to identify all actual positive instances. Due to the skewed nature of the dataset, we also consider the F-Measure to assess the performance of the proposed model.

4.2 *Result and Discussion*

In this section, we delve into the results obtained for the proposed model. We launch a Graphical User Interface (GUI) with the buttons `Choose file` and `Choose folder` for the smart contracts readability score calculation. Users can choose a specific smart contract by clicking the `Choose file` button to browse the file and then the `OK` button in the pop-up message. The system computes and displays the smart contract's readability metrics score in tabular form, as shown in Fig. 2. In particular, the button `Choose folder` allows the user to select a directory that contains a set of solidity files and calculate the readability metrics for each function(s) included in the solidity file. Once the `Export CSV` option is activated, the results are exported to a CSV file for easier interpretation and readability.

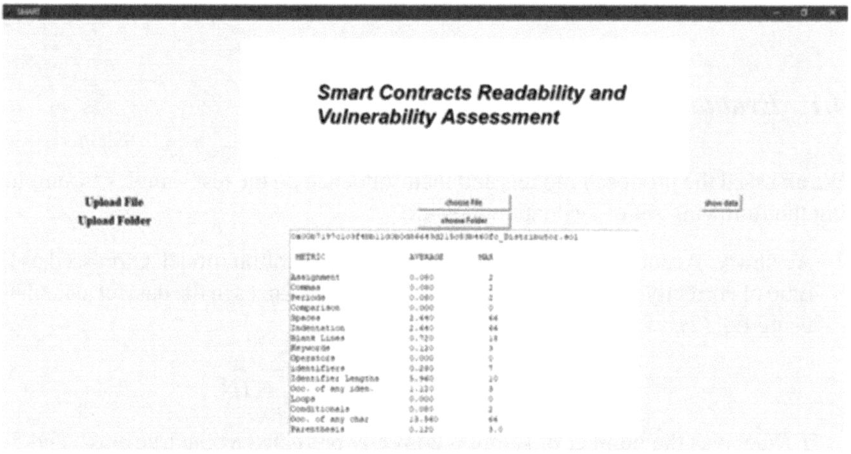

Fig. 2 Visualizing the readability score for one smart contract using GUI

Table 2 Hyperparameter used for each ML model

Model	Hyperparmeter values
LR	C=0.1, penality=none
KNN	metric=manhattan, n_neighbors=15, weights=distance
SVC	C=1, class_weight='balanced', gamma=1
DT	criterion='entropy', max_depth=10, max_features=0.8
RF	class_weight='balanced_subsample', max_features='auto', n_estimators=2000
XGBoost	base_score=0.5, booster='gbtree', gamma=3.2, learning_rate=0.15, max_depth=6, min_samples_leaf=2, min _samples_split=2

Table 3 Performance of the machine learning model

Model	Accuracy (%)	F-measure (%)
LR	92.690	92.593
KNN	96.347	96.319
SVC	98.17	98.16
DT	97.48	97.477
RF	98.17	98.17
XGBoost	**98.858**	**98.856**

Highlighted the best result in bold

We first conducted hyperparameter tuning to optimize and determine the best parameter values for all the ML models, as illustrated in Table 2. Subsequently, the performance of various machine learning models is comprehensively presented in Table 3. Our analysis reveals that XGBoost notably outperformed the other classifiers, achieving an impressive F-Measure of 98.856% and an accuracy of 98.858%. We employ the Slither tool for vulnerability detection, which allows the smart contract source code to complete the compilation process. Only 2070 were examined for vulnerability analysis because 116 of the 2186 files had compilation issues. We execute Slither on each `.sol` file and generates a JSON file for each solidity file. We extracted information from the description component of the JSON file containing the list of vulnerabilities. We recorded it in a CSV file to investigate the potential vulnerabilities in the smart contract.

Furthermore, among the different vulnerabilities, we divide the vulnerabilities into three different categories based on the values of *Impact* and *Confidence*. The *Impact* denotes the impact of the issues in the contract, and *Confidence* represents your confidence in the analysis. The permissible values for *Impact* are *High, Medium* and *Low*. The allowed values for *Confidence* are *High, Medium*, and *Low*. The vulnerabilities are categorized as *High* when the *Impact* and *Confidence* is either *High* or *Medium*, as shown in Table 4. The *Medium* vulnerabilities categories include the vulnerability if both *Impact* and *Confidence* levels are *Medium*. The *Low* vulnerabilities categories include the vulnerability if the *Impact* is low. Table 5 illustrates the 31 distinct vulnerabilities in the highly vulnerable category.

Table 4 The number of vulnerabilities detected in smart contracts at various vulnerability levels

Vulnerability type	Impact	Confidence	Number of contract
High	H or M	H or M	1055
Medium	M	M	319
Low	L	L or M	172

H: High, M: Medium, L: Low

Table 5 Set of 31 distinct vulnerabilities in the highly vulnerable category

Confidence	Impact	Vulnerabilities
High	High	V_1:abiencoderv2-array; V_2:arbitrary-send-erc20; V_3:array-by-reference; V_4:incorrect-shift; V_5:multiple-constructors; V_6:name-reused; V_7:protected-vars; V_8:public-mappings-nested; V_9:rtlo; V_{10} :shadowing-state; V_{11} :suicidal; V_{12} :uninitialized-state; V_{13} :uninitialized-storage; V_{14} :unprotected-upgrade
Medium	High	V_{15}:arbitrary-send-erc20-permit; V_{16}:arbitrary-send-eth; V_{17}:controlled-array-length; V_{18}:controlled-delegatecall; V_{19}:delegatecall-loop; V_{20}:msg-value-loop; V_{21}:reentrancy-eth; V_{22}:storage-array; V_{23}:unchecked-transfer; V_{24}:weak-prng
High	Medium	V_{25}:domain-separator-collision; V_{26}:enum-conversion; V_{27}:erc20-interface; V_{28}:erc721-interface; V_{29}:incorrect-equality; V_{30}:locked-ether; V_{31}:mapping-deletion

Solidity vulnerabilities are weaknesses or flaws in smart contracts written in the Solidity programming language. Exploitable by malicious actors, these vulnerabilities can lead to unintended behavior or fund theft. In Fig. 3, we present an analysis of high-level vulnerabilities within smart contracts. This illustration highlights that out of the examined vulnerabilities, 12 distinct types have not been reported in any of the smart contracts. Among the vulnerabilities detected, the most significant ones include *unprotected-upgrade*, *arbitrary-send-erc20-permit*, *weak-prng*, *enum-conversion*, and *storage-array*.

Table 4 overviews the number of smart contracts identified within each vulnerability category. The last column in this table shows the count of contracts with various identified vulnerabilities. The study revealed that some contracts were associated with multiple vulnerability types. For example, if a contract exhibited high and moderate vulnerability, we categorized it as a highly vulnerable group. The results showed that 1055 smart contracts had high vulnerability, 319 fell into the medium vulnerability category, and 172 were classified as having low vulnerability. Furthermore, Slither detected optimizations in 1962 files and summarized information for all 2070 contracts utilized in this study. By prioritizing these vulnerabilities, we offered developers and security experts more precise information for addressing the most critical issues. Additionally, it is noteworthy that among the analyzed smart contracts, a total of 524 contracts exhibited a notable absence of high, medium, or low vulnerabilities, indicating a particularly robust level of security.

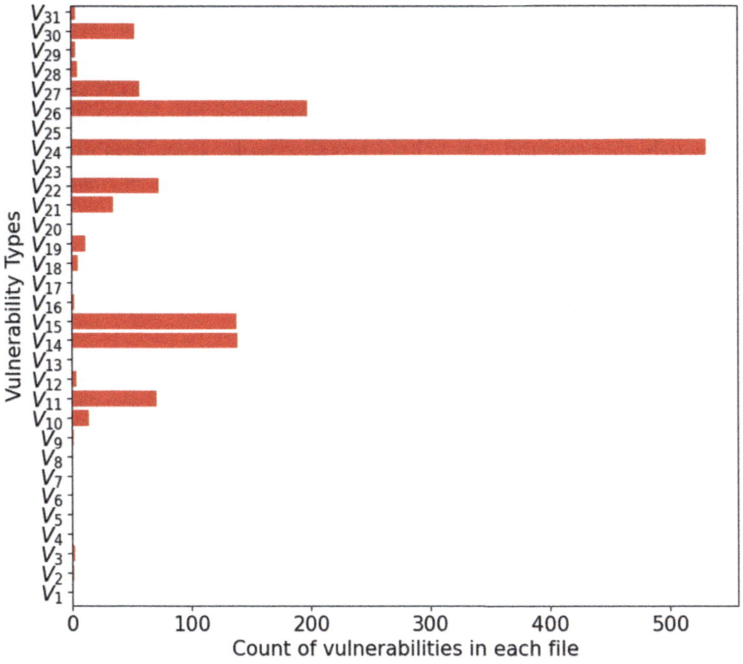

Fig. 3 Analysis of various vulnerabilities in smart contracts, with a particular focus on high-risk vulnerabilities

Correlation Analysis: In scrutinizing the relationship between readability and vulnerability, our initial method employed predictions produced by an XGBoost model. We subjected a test dataset, constituting around 20% of the entire dataset (comprising 438 samples), to this analysis. We then scrutinized whether the machine learning model's output effectively discerns the readability status of files. Following that, we computed a correlation coefficient of the output generated by the classification module for the same solidity file to detect the presence of vulnerabilities, yielding a value of −0.208 using Eq. (1). The findings revealed a negative correlation between non-readability and vulnerability. After carefully analyzing the test dataset for correlating the vulnerability and readability, 72.6% of non-readable files were identified as vulnerable, while 12.6% of readable files exhibited vulnerabilities. Our study indicated that files with lower readability are more likely to exhibit vulnerabilities.

5 Conclusion

Code readability is the most prominent attribute for developing a software product and writing source code. More specified codes have a low possibility of causing errors and hence reduce security vulnerabilities. The proposed tool examines a smart contract's source code to calculate a set of metrics, which can be demonstrated for checking the readability of smart contracts using 19 metrics. We also study the vulnerabilities in this smart contract and group them into different categories. We observed a negative correlation between file readability and vulnerability, indicating that files with higher readability tend to be less susceptible to vulnerabilities. In future research, we plan to assign weights for each vulnerability category and evaluate the vulnerabilities based on these values.

References

1. Zheng Z, Xie S, Dai H-N, Chen W, Chen X, Weng J, Imran M (2020) An overview on smart contracts: challenges, advances and platforms. Future Gener Comput Syst 105:475–491
2. Saini A, Zhu Q, Singh N, Xiang Y, Gao L, Zhang Y (2020) A smart-contract-based access control framework for cloud smart healthcare system. IEEE Internet Things J 8(7):5914–5925
3. Tolmach P, Li Y, Lin S-W, Liu Y, Li Z (2021) A survey of smart contract formal specification and verification. ACM Comput Surv (CSUR) 54(7):1–38
4. Wang Y, Li Z, Gou G, Xiong G, Wang C, Li Z (2020) Identifying DApps and user behaviors on ethereum via encrypted traffic. In: Security and privacy in communication networks: 16th EAI international conference, SecureComm 2020, Washington, DC, USA, October 21–23, 2020, proceedings, Part II vol 16. Springer International Publishing, pp 62–83
5. Vacca A, Di Sorbo A, Visaggio CA, Canfora G (2021) A systematic literature review of blockchain and smart contract development: techniques, tools, and open challenges. J Syst Softw 174:110891
6. Zarir AA, Oliva GA, Jiang ZM, Hassan AE (2021) Developing cost-effective blockchain-powered applications: a case study of the gas usage of smart contract transactions in the ethereum blockchain platform. ACM Trans Softw Eng Methodol (TOSEM) 30(3):1–38
7. Chen X, Liao P, Zhang Y, Huang Y, Zheng Z (2021) Understanding code reuse in smart contracts. In: 2021 IEEE international conference on software analysis, evolution and reengineering (SANER). IEEE, pp 470–479
8. Perez D, Livshits B (2021) Smart contract vulnerabilities: vulnerable does not imply exploited. In: 30th USENIX security symposium (USENIX Security 21), pp 1325–1341
9. Singh A, Parizi RM, Zhang Q, Choo K-KR, Dehghantanha A (2020) Blockchain smart contracts formalization: approaches and challenges to address vulnerabilities. Comput Secur 88:101654
10. Nicola A, Bartoletti M, Cimoli T (2017) A survey of attacks on ethereum smart contracts (sok). In: Principles of security and trust: 6th international conference, POST 2017, held as part of the European joint conferences on theory and practice of software, ETAPS 2017, Uppsala, Sweden, April 22–29, 2017, proceedings vol 6. Springer Berlin Heidelberg, pp 164–186
11. Samreen NF, Alalfi MH (2021) A survey of security vulnerabilities in ethereum smart contracts. arXiv preprint arXiv:2105.06974
12. Canfora G, Di Sorbo A, Fredella M, Vacca A, Visaggio CA (2021) iSCREAM: a suite for smart contract readability assessment. In: 2021 IEEE international conference on software maintenance and evolution (ICSME). IEEE, pp 579–583
13. Scalabrino S, LinaresVásquez M, Oliveto R, Poshyvanyk D (2018) A comprehensive model for code readability. Journal of Software: Evolution and Process 30(6):e1958
14. Yang Y, Xia X, Lo D, Bi T, Grundy J, Yang X (2022) Predictive models in software engineering: challenges and opportunities. ACM Trans Softw Eng Methodol (TOSEM) 31(3):1–72
15. Buse RPL, Weimer WR (2009) Learning a metric for code readability. IEEE Trans Softw Eng 36(4):546–558
16. Fakhoury S, Roy D, Hassan A, Arnaoudova V (2019) Improving source code readability: theory and practice. In: 2019 IEEE/ACM 27th international conference on program comprehension (ICPC). IEEE, pp 2–12

17. Vacca A, Fredella M, Di Sorbo A, Visaggio CA, Canfora G (2022) An empirical investigation on the trade-off between smart contract readability and gas consumption. In: Proceedings of the 30th IEEE/ACM international conference on program comprehension, pp 214–224
18. Zou W, Lo D, Kochhar PS, Le X-BD, Xia X, Feng Y, Chen Z, Xu B (2019) Smart contract development: challenges and opportunities. IEEE Trans Softw Eng 47(10):2084–2106
19. Demir M, Alalfi M, Turetken O, Ferworn A (2019) Security smells in smart contracts. In: 2019 IEEE 19th international conference on software quality, reliability and security companion (QRS-C). IEEE, pp 442–449
20. Khan ZA, Namin AS (2020) Ethereum smart contracts: vulnerabilities and their classifications. In: 2020 IEEE international conference on big data (Big data). IEEE, pp 1–10
21. Perez D, Livshits B (2019) Smart contract vulnerabilities: does anyone care?. arXiv preprint arXiv:1902.06710 pp 1–15
22. Kılıç S (2015) Kappa testi. J Mood Dis 5(3):142–144
23. Feist J, Grieco G, Groce A (2019) Slither: a static analysis framework for smart contracts. In: 2019 IEEE/ACM 2nd international workshop on emerging trends in software engineering for blockchain (WETSEB). IEEE, pp 8–15
24. Cohen I, Huang Y, Chen J, Benesty J, Benesty J, Chen J, Huang Y, Cohen I (2009) Pearson correlation coefficient. Noise reduction in speech processing, pp 1–4

Hardware Security in Evolving FinTech Landscape

Biswajit Bhowmik, Joshitha Reddy Dongala, Kruthika K. Sudhama, Reshma Tresa Antony, and K. K. Girish

Abstract The assimilation of technology into the financial sector, often referred to as FinTech, has brought about a significant transformation. This shift has not only widened the scope of financial inclusivity but has also fundamentally reshaped the contours of financial solutions delivered. As FinTech solutions continue to empower individuals with greater control over their finances through mobile banking, digital wallets, and advanced data analytics, the security of these innovations becomes paramount. While software security has traditionally received more attention, this paper underscores the significance of hardware security, which serves as the foundational infrastructure for software security measures. It delves into the factors used to evaluate hardware security and outlines various categories of hardware attacks. A case study, focusing on point-of-sale (PoS) systems, exemplifies the importance of hardware security in FinTech. Ultimately, this research contributes to a comprehensive understanding of the evolving FinTech landscape and its implications for both financial inclusion and cybersecurity.

Keywords FinTech · Hardware security · Hardware attack · Cybersecurity · PoS

B. Bhowmik (✉) · J. R. Dongala · K. K. Sudhama · R. T. Antony · K. K. Girish
Maharshi Sushrut CAS Lab, BRICS Laboratory Department of Computer Science and Engineering, National Institute of Technology Karnataka, Surathkal, Mangalore 575025, India
e-mail: brb@nitk.edu.in

J. R. Dongala
e-mail: joshithareddydongala.191cs126@nitk.edu.in

K. K. Sudhama
e-mail: kruthikaksudhama.191cs224@nitk.edu.in

R. T. Antony
e-mail: reshmatantony.191cs149@nitk.edu.in

K. K. Girish
e-mail: girishkk.217cs003@nitk.edu.in

S. M. Thampi et al. (eds.), *Fifth International Conference on Computing and Network Communications*, Lecture Notes in Electrical Engineering 1219,
https://doi.org/10.1007/978-981-97-4540-1_31

1 Introduction

The financial industry has experienced a profound shift due to the integration of technology, often referred to as FinTech. This transformation has led to a broader embrace of financial inclusion. Through the growth of mobile banking, digital wallets, and online payment platforms, previously unbanked or underbanked individuals now enjoy access to fundamental financial services, enabling them to save, engage in transactions, and establish financial security [1, 2]. Subsequently, to stay competitive and align with evolving customer demands, established financial institutions are progressively engaging in partnerships with FinTech startups or crafting their own FinTech solutions.

In the ever-changing landscape of the financial industry, innovative solutions are constantly being developed, encompassing both software and hardware components [3]. FinTech companies and financial institutions are constantly developing and improving software applications, platforms, and digital services that empower customers with greater control over their finances. These software innovations include mobile banking apps, digital wallets, investment management platforms, and sophisticated data analytics tools. On the other hand, hardware innovations are also contributing significantly to the financial sector's evolution. Point-of-sale (PoS) devices, ATMs, secure card readers, and other hardware components are becoming more advanced and secure. Figure 1 shows the synergy of software and hardware innovations in FinTech.

The synergy of software and hardware in the financial sector underscores the critical need for both components' security. Software security often garners more attention in cybersecurity discussions due to vulnerabilities and increasing cyber threats. However, it's imperative to recognize that hardware security is equally, if not more, essential, as it underpins and shapes software security. This paper explores hardware security, examining factors for assessing hardware security and catego-

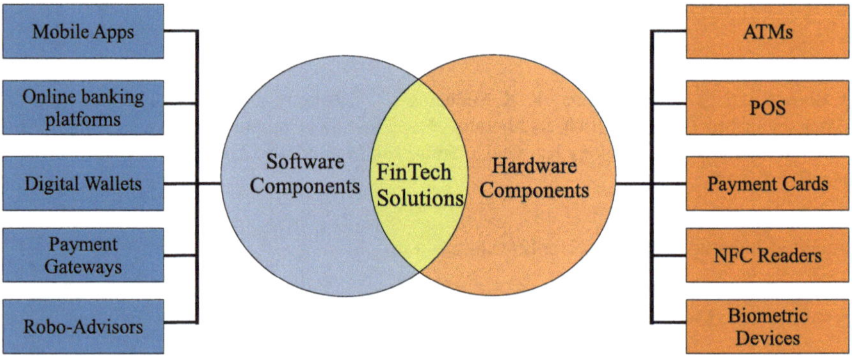

Fig. 1 Software and hardware synergy in FinTech

rizing hardware attacks. Additionally, it uses point-of-sale (PoS) systems as a case study to illustrate the importance of hardware security in the FinTech domain.

The rest of the paper is organized as follows: Sect. 2 describes hardware security landscape. Section 3 factors in assessing security of hardware devices. Section 4 elaborates on hardware attacks. Section 5 describes hardware security in FinTech. Section 6 illustrates a case study on PoS system. Section 7 concludes the paper.

2 Hardware Security Landscape

Emerging from cryptographic engineering, hardware security encompasses methods to safeguard a product's supply chain, hardware design, access control, secure multi-party computation, and key storage. Hardware security modules (HSMs) are tangible devices that handle cryptoprocessing and key management for secure authentication [4]. Its historical roots extend to the 1800s, with applications in military contexts for protecting semiconductors from radiation-induced single-event effects (SEEs). These security measures eventually found application in consumer products, giving rise to hardware attacks, leading to today's established security standards [5].

Modern security efforts are primarily focused on safeguarding or breaching operating systems. While software alone falls short in offering comprehensive system security, hardware-based security provides performance and power efficiency benefits. However, encryption complexity and data processing challenges remain, with many encryption algorithms implemented in hardware. Hardware security issues entail higher costs compared to software problems, with risks increasing as the product development progresses. Figure 2 illustrates the increase in hardware security risks throughout the manufacturing process.

Hardware attacks are on the rise, exploiting existing vulnerabilities in hardware. Users' adoption of general security practices and manufacturers' use of HSMs and crypto-processors are vital for enhancing hardware security. Hardware security finds applications in various domains, including finance, healthcare, and defense. Despite limited research, the field is gaining traction as more vulnerabilities are exposed.

3 Assessing Security of a Hardware Device

The importance of hardware security is equivalent to that of software security. When evaluating the safety of a hardware device, it is crucial to consider both its inherent vulnerabilities and potential sources of risk, which encompass running code and the device's networked data input/output (I/O). While constructing a hardware device, several essential security objectives should be considered to protect the device, its data, and the users [6]. These objectives are essential for maintaining the device's integrity, confidentiality, and availability, as depicted in Fig. 3.

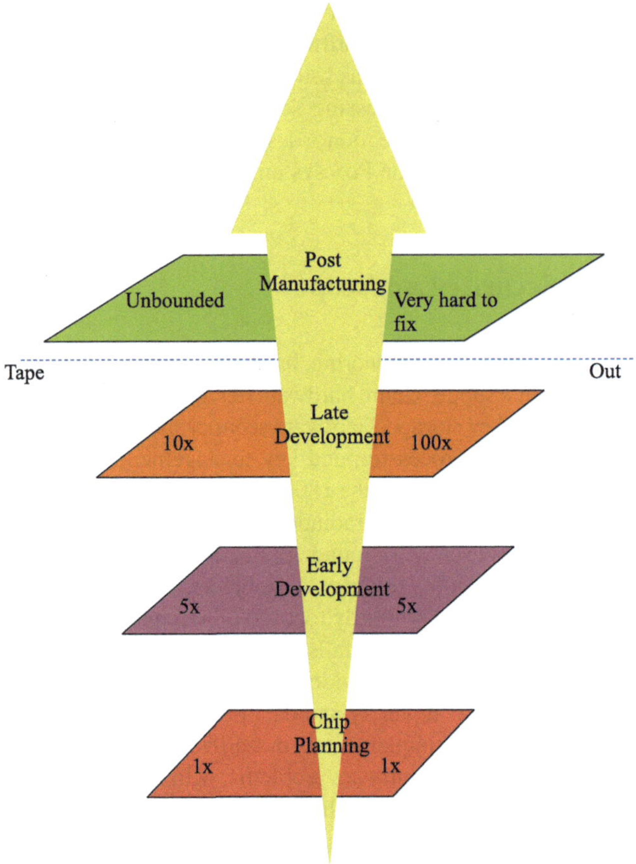

Fig. 2 Increase of hardware risk in manufacturing

Here are the key security objectives in the process of constructing a hardware device:

- Availability—Ensure that the hardware device is available for legitimate users when they need it. Protection against physical damage, environmental factors, and distributed denial of service (DDoS) attacks is essential to maintain availability.
- Integrity—Ensuring the integrity of the hardware and its data is crucial. This means that data remains unaltered and trustworthy. To achieve integrity, hardware components should be designed and manufactured to resist tampering, and cryptographic mechanisms can be used to verify data integrity.
- Confidentiality—Protecting the confidentiality of data is a primary objective. This involves ensuring that unauthorized users cannot access sensitive information stored on the device. Techniques such as encryption and access control mechanisms are commonly used to achieve confidentiality.

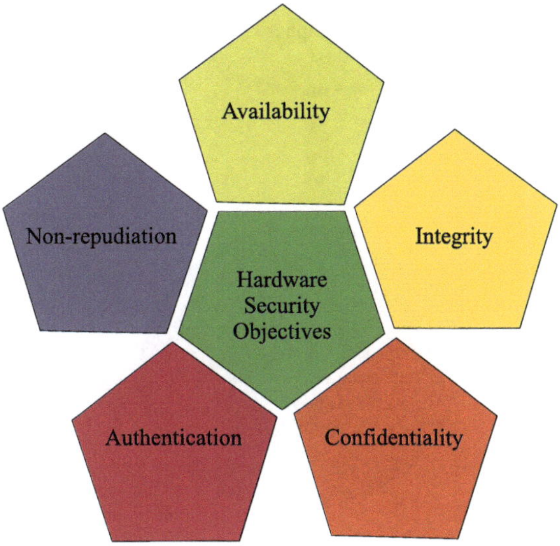

Fig. 3 Hardware security objectives

- Authentication—Verify the identity of users, devices, or components trying to access the hardware. Strong authentication mechanisms, such as passwords, biometrics, or multi-factor authentication, should be implemented to prevent unauthorized access.
- Non-repudiation—Assurance that both the sender and the recipient of the information are given proof of delivery and the sender's identity, so neither can subsequently deny processing the data. Digital signatures and audit logs are common tools for achieving non-repudiation.

4 Hardware Attacks

Although software attacks receive more attention, hardware attacks are equally damaging, often being a cybercriminal specialty. These attacks, although more complex than their software counterparts, introduce security concerns, including common passwords, outdated firmware, and lack of encryption. As companies aim to enhance user experience through innovative technologies, they inadvertently expose data to vulnerabilities, as detailed in this section on various hardware attacks [7].

4.1 Side-Channel Attack

Side-channel attacks exploit discrepancies in the electrical emissions data from a computer's monitor or hard drive, potentially revealing information such as on-screen

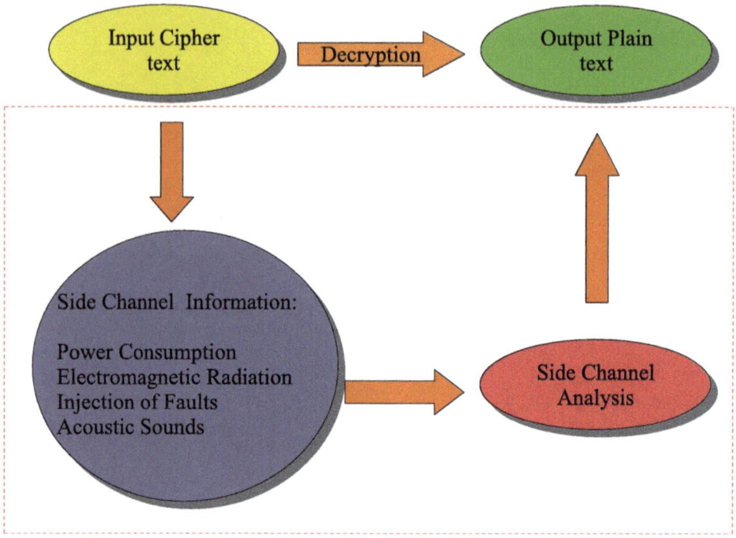

Fig. 4 Side-channel attack

content or the power consumption of different hardware components. Through the measurement of these hardware emissions, cybercriminals can steal sensitive data, including cryptographic keys [8]. Figure 4 provides a general overview of how a side-channel attack can be employed to decrypt confidential information.

4.2 *Rowhammer Attack*

Exploiting a bug in Dynamic RAM modules produced since 2010, Rowhammer attacks repeatedly access or "hammer" DRAM memory cells, releasing electric charges that flip bits from 0 to 1 and vice versa. This vulnerability grants cybercriminals full system security privileges and allows them to bypass security sandboxes [9]. In modern Rowhammer attacks, users aggressively read or write to different memory locations to induce bit flips in specific physical memory areas, as depicted in Fig. 5.

4.3 *Evil Maid Attack*

The term "evil maid" was coined to represent the scenario of a malicious individual attempting to gain access to electronic devices left unattended in a hotel room. In this type of attack, the cybercriminal physically breaches unguarded hardware devices in order to obtain sensitive information [10], as depicted in Fig. 6.

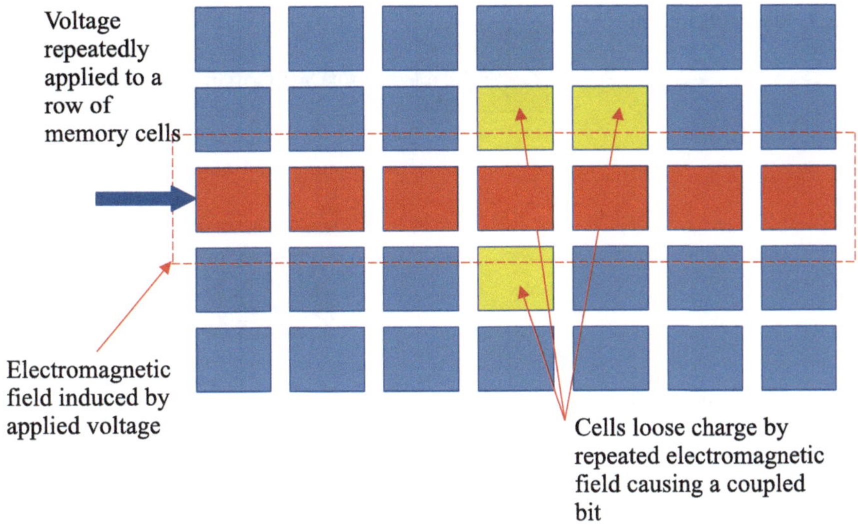

Fig. 5 Rowhammer attack

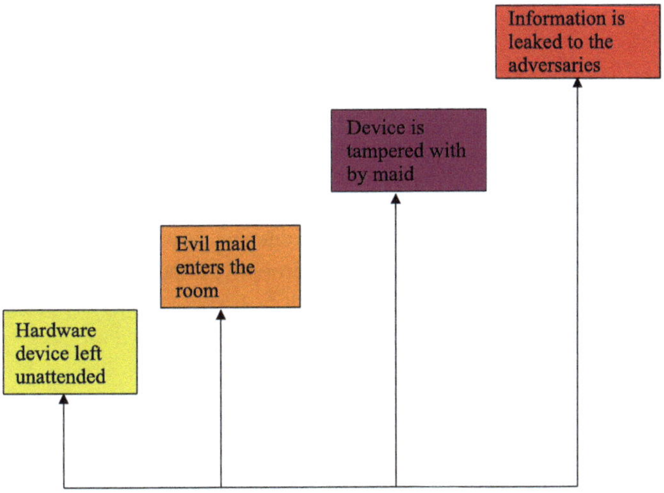

Fig. 6 Evil maid attack

4.4 Man-in-the-Middle Attack

Cybercriminals breach hardware security to carry out man-in-the-middle (MITM) attacks, introducing malicious software or exploiting vulnerabilities to intercept and manipulate data packets [11]. MITM attacks involve attackers infiltrating a communication between a user and an application, either eavesdropping or impersonating a participant, as depicted in Fig. 7.

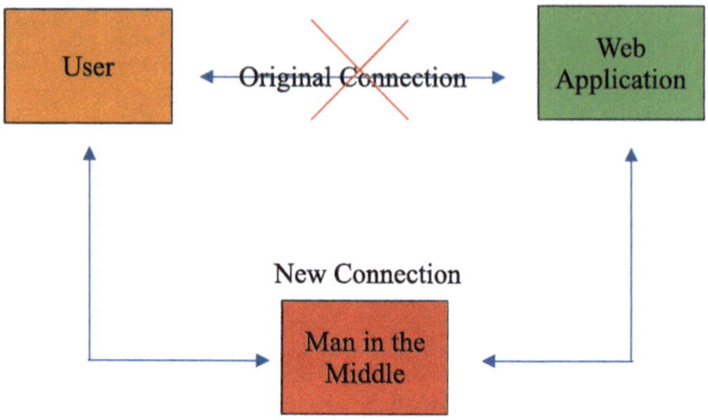

Fig. 7 Man-in-the-middle attack

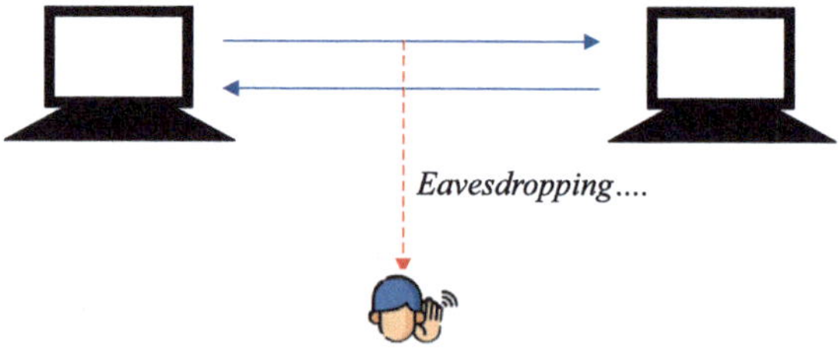

Fig. 8 Eavesdropping attack

4.5 Eavesdropping Attack

A covert attack takes place when confidential data, such as credit card information and passwords, is transmitted between devices. These attacks are highly effective for cybercriminals as they occur without triggering any notifications on insecure networks [12]. Figure 8 illustrates this eavesdropping attack.

4.6 Miscellaneous Attacks

There are several other methods to simplify the task for hackers seeking to compromise hardware, which include:

- Triggering Fault Attack: Attackers often employ this tactic by deliberately introducing hardware defects into the system to disrupt the device's normal operation.
- Counterfeit Hardware Attack: This type of supply chain attack involves the sale of counterfeit or unauthorized equipment to businesses, creating opportunities for hackers to gain unauthorized access through a hidden entry point.
- Timing Attack: The cybercriminals analyze the time it takes for the cryptosystem to respond to different inputs, using this information to deduce and exploit various algorithms and functions.

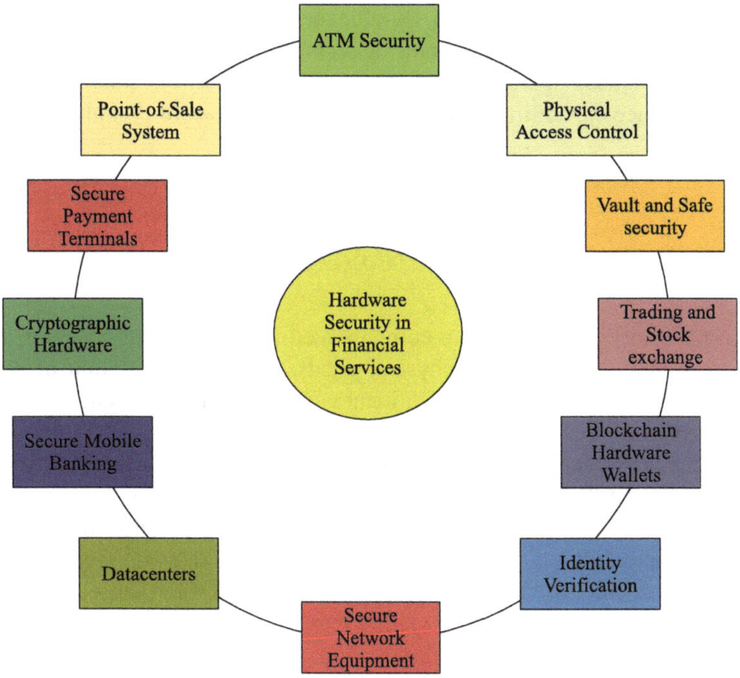

Fig. 9 Hardware security in financial services

5 Hardware Security in FinTech

Financial institutions are prime targets for cyberattacks due to the sensitive nature of the data they handle. Therefore, ensuring their security is of significant importance, and hardware security plays a vital role in this regard. Figure 9 outlines key domains in financial sector benefitting from hardware security.

The HSMs represent a groundbreaking innovation in hardware security, especially for financial institutions. HSMs are physical devices designed to enhance the security of sensitive data [13]. They are used for generating cryptographic keys and supporting encryption, decryption, and authentication processes in applications. While they have their primary application in financial technology, they are also found in defense and anti-piracy fields. There are two primary types of HSMs based on their application:

1. General Purpose HSMs: Used in Public Key Infrastructures, crypto wallets, and other applications involving simple yet sensitive data.
2. Transaction and Payment HSMs: Specifically designed to safeguard sensitive transaction data, including credit card information.

The HSMs address the challenge of obtaining cryptographic randomness for keys while reducing CPU load, primarily motivated by the imperative need for security amid growing software attacks [14]. By offloading key generation and secret storage to a dedicated device, an added layer of security is achieved. HSMs are fortified with security-focused operating systems and minimal network interfaces, ensuring high-performance capabilities, such as generating millions of digital signatures per second.

6 Case Study: Point-of-Sale (PoS)

A point-of-sale (PoS) system is a computerized process commonly used in retail stores to streamline the sale of physical goods, serving as the point of purchase and payment settlement. Unlike traditional cash registers, PoS systems efficiently record, store, and analyze transaction data, enhancing customer service by eliminating redundant tasks [15]. They replace cash registers, handling tasks like calculating totals, verifying transactions, and recording details, with some allowing post-payment adjustments. When integrated with online stores, PoS systems centralize business operations. These systems comprise hardware components like tablets and scanners, distinct from standalone point-of-sale software that requires installation [16].

6.1 PoS Architecture

A secure PoS system comprises two primary components: a terminal used by the retailer for transaction processing and a PIN Entry Device (PED) used by customers for PIN input [17]. Typically, sensitive information is encrypted by either the PED or a remote server as it passes through the terminal, ensuring security. In financial institutions, the PED communicates with a remote server to verify the card and cardholder using a security PIN, and it includes internal security measures to detect tampering. The most sensitive data within the PED includes the cardholder's PIN and the cryptographic key for encryption, which are protected by well-known algorithms like Rivest–Shamir–Adleman (RSA) and Data Encryption Standard (DES). The key's security is critical, stored in the PED's SRAM or registers, requiring the security core to detect unauthorized access or tampering. Figure 10 provides an overview of a PoS system's components.

6.2 Hardware Attacks on PoS

When considering attacks on the PED, they fall into two main categories: peripheral security and core security [18]. Peripheral attacks involve various methods of PIN exposure, such as eavesdropping on physical signals during transactions using

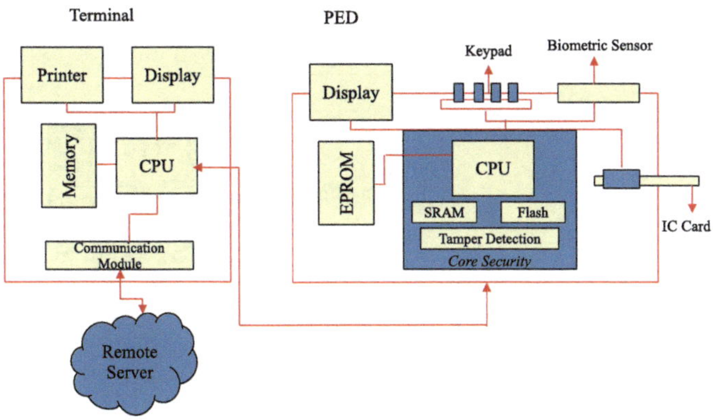

Fig. 10 PoS architecture

Fig. 11 Peripheral attacks on PED

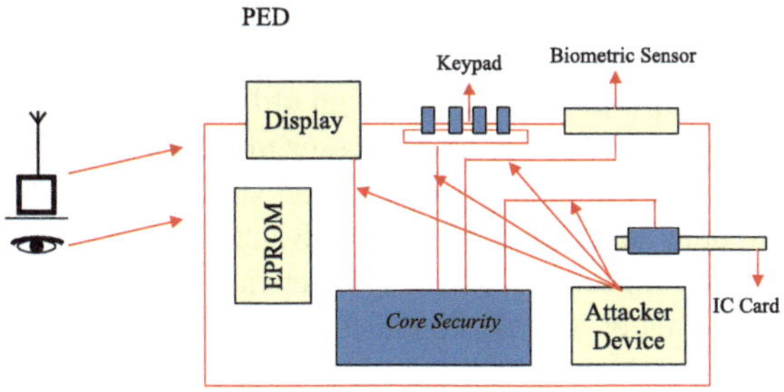

deceptive devices, remotely capturing PINs, passive relays on contactless payment terminals, and accessing the PED's internal components for potential exploitation. These attacks exploit vulnerabilities like unencrypted data transmission, electromagnetic radiation, sound monitoring, and power consumption variations [19, 20].

On the other hand, core security concerns encompass more advanced attack methodologies, such as differential power analysis (DPA) attacks, electromagnetic analysis, timing-based attacks, and differential fault analysis, among others, as detailed in [18]. Figures 11 and 12 visually represent peripheral attacks and core security attacks on PEDs, respectively.

In order to establish effective peripheral protection, it is imperative for the hardware designer to take certain measures. These include implementing a robust privacy shield that envelops the keypad to a significant extent and preventing the monitoring of electromagnetic radiation [21]. When PINs are entered and accompanied by tones, it is essential to ensure that these tones are uniform and indistinguishable for all digits. Additionally, incorporating sensors that can activate alarms in the event of tampering with the PED or any attempts to open it is crucial. When these alarms are triggered, the device should initiate a data erasure process and shut down[22].

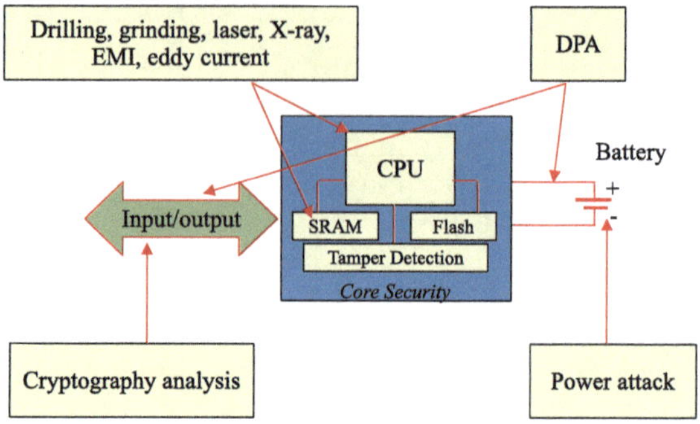

Fig. 12 Core attacks on PED

6.3 *Related Works*

The security of PoS systems is a well-researched topic, with a focus on safeguarding the smart card reader and WiFi interface, which handle sensitive data. Sahana et al. [23] propose a secure design utilizing an i.MX6 processor to oversee smart card reader operations. This processor manages data encryption and decryption through a divided secret key stored in the smart card reader and key controller. The WiFi module, supporting WLAN and Bluetooth functions, provides reliable 2.4GHz short-range connectivity, allowing independent activation for both WLAN and Bluetooth.

Solms et al. [24] conducted an experimental study on transaction receipt data misuse. They found that both merchant and client copies of transactions can contain sensitive card data, including Primary Account Numbers (PAN) and expiration dates. This data can potentially be exploited for online shopping fraud and identity theft. They also highlighted the risk of One-Time Passwords (OTPs) being accessed without device unlocking.

Contactless card payments have evolved into mobile proximity payments using near-field communication (NFC) technology, but they face security challenges, notably passive relay attacks. To mitigate these risks, Biader et al. [25] propose the KerNeeS protocol, enabling mutual authentication between NFC phones and PoS machines through an Authentication Server (AS). The AS shares cryptographic keys for mutual authentication. This six-step protocol provides logical proof of the authentication processes, but its effectiveness depends on the integrity of the Authentication Server, which, if compromised, poses a single point of failure.

Contactless card security faces passive relay threats, countered by Nezhad et al. [26] innovative OPay solution. OPay leverages NFC device and terminal alignment during legitimate transactions, rejecting misaligned card transactions, and enhancing security. It also restricts attempts to prevent multiple attacks. Though portable X-ray scanners could theoretically compromise the system, health risks and conspicuousness in public spaces offer defense.

7 Conclusion

As technology reshapes finance and enhances financial inclusion through diverse solutions, safeguarding both software and hardware components becomes paramount. While software security is often at the forefront of cybersecurity discussions, this paper highlights the foundational importance of hardware security. By focusing on hardware security factors and delineating attack categories, it underscores the need for comprehensive security measures. The case study of point-of-sale (PoS) systems demonstrates the real-world implications of hardware security in FinTech. Furthermore, this paper offers valuable insights and a roadmap for a more secure financial services environment.

References

1. Girish KK, Bhowmik B (2023) Recent advancements and challenges in fintech. In: 2023 14th International Conference on Computing Communication and Networking Technologies (ICCCNT), pp 1–7
2. Arner DW, Buckley RP, Zetzsche DA, Veidt R (2020) Sustainability, fintech and financial inclusion. Eur Bus Organization Law Rev 21:7–35
3. Guez G (2017) Why hardware-based design security is essential for every application, White Paper
4. Jin Y (2015) Introduction to hardware security. Electronics 4(4):763–784
5. Han J, Kim S, Kim T, Han D (2019) Toward scaling hardware security module for emerging cloud services. In: Proceedings of the 4th workshop on system software for trusted execution, pp 1–6
6. Kazemi Z, Fazeli M, Hely D, Beroulle V (2020) Hardware security vulnerability assessment to identify the potential risks in a critical embedded application. In: 2020 IEEE 26th international symposium on On-Line Testing and Robust System Design (IOLTS), IEEE, pp 1–6
7. Moein S, Gebali F, Traore I (2014) Analysis of covert hardware attacks. J Convergence 5(3)
8. Standaert F-X (2010) Introduction to side-channel attacks. In: Secure integrated circuits and systems, pp 27–42
9. Qiao R, Seaborn M (2016) A new approach for rowhammer attacks. In: 2016 IEEE international symposium on hardware oriented security and trust (HOST), IEEE, pp 161–166
10. Knapp AC (2022) Firmware security in IoT: should smart homes be afraid of evil maids? PhD thesis, University of Colorado Colorado Springs
11. Mallik A (2019) Man-in-the-middle-attack: Understanding in simple words. Cyberspace: Jurnal Pendidikan Teknologi Informasi 2(2):109–134
12. Maggi F, Volpatto A, Gasparini S, Boracchi G, Zanero S (2011) A fast eavesdropping attack against touchscreens. In: 2011 7th international conference on Information Assurance and security (IAS), IEEE, pp 320–325
13. Samir N, Gamal Y, El-Zeiny AN, Mahmoud O, Shawky A, Saeed A, Mostafa H (2019) Energy-adaptive lightweight hardware security module using partial dynamic reconfiguration for energy limited internet of things applications. In: 2019 IEEE International Symposium on Circuits and Systems (ISCAS). IEEE, pp 1–4
14. Pricope R et al (2017) Hardware and software technologies used in the financial industry. Int J Inf Secur Cybercrime (IJISC) 6(1):33–40
15. Kumar S, Bhowmik B (2023) Covid-19 waves and their impacts to society. In: 2023 IEEE Guwahati Subsection Conference (GCON), pp 1–5

16. Lee H-J, Lee Y, Won D (2014) Protection profile for pos (point of sale) system. In: Mobile, ubiquitous, and intelligent computing: MUSIC 2013. Springer, Heidelberg, pp 495–500
17. Samaranayake C, Achchige RK, Shanaz T, Ranasinghe A, Senarathne A (2019) Enhanced secure solution for pos architecture. In: 2019 International Conference on Advancements in Computing (ICAC). IEEE, pp 109–115
18. Caldwell T (2014) Securing the point of sale. Comput Fraud Secur 2014(12):15–20
19. Bhowmik B, Gagan N (2023) Design of a fault-tolerant pseudo-3d routing. In: 2023 IEEE international test conference India (ITC India), pp 1–6
20. Prathyusha MR, Bhowmik B (2023) Iot-enabled smart applications and challenges. In: 2023 8th International Conference on Communication and Electronics Systems (ICCES), pp 354–360
21. Obi EN (2023) Impact of criminal acts on point of sales (pos) business operations in Karu local government area of Nasarawa state. KIU J Soc Sci 9(2):119–127
22. Kredina A (2021) Transformation of fintech: impact of pos and atm on non-cash payments. Eurasian J Econ Bus Stud 2(60):16–32
23. Sahana V, Latha H, Bhandari R (2018) Design of secure smart card reader and wi-fi interface for point of sale terminal. In: 2018 3rd IEEE international conference on Recent Trends in Electronics, Information & Communication Technology (RTEICT). IEEE, pp 1341–1345
24. Von Solms S (2015) An investigation into credit card information disclosure through point of sale purchases. In: 2015 Information Security for South Africa (ISSA). IEEE, pp 1–8
25. Ceipidor UB, Medaglia CM, Marino A, Sposato S, Moroni A (2012) Kernees: a protocol for mutual authentication between nfc phones and pos terminals for secure payment transactions. In: 2012 9th international ISC conference on information security and cryptology. IEEE, pp 115–120
26. Mehr Nezhad M, Hao F (2021) Opay: an orientation-based contactless payment solution against passive attacks. In: Annual computer security applications conference, pp 375–384

Evolving Trends in Ransomware: Inherent Advanced Persistent Threat

Sandhya Addetla and Rajalakshmi Pachamuthu

Abstract Cyber-attacks are ever growing and ransomware being one of the major threat to the safety of cyber users. The number of Ransomware attacks has increased considerably more so after the COVID-19 pandemic. Ransomware is targeting businesses, governments and critical infrastructure rather than individual cyber users due to huge monetary gains from bigger targets. This paper presents an overview of ransomware and why ransomware should be perceived as an Advanced Persistent Threat (APT). The current trends of Ransomware are described. In this paper, various ransomware attacks that occurred recently are discussed, the state of the art techniques for detection of Ransomware, guidelines for prevention are also presented in this paper. Future work based on the predictions of ransomware trends is also presented.

Keywords Advanced persistent threats · TTPs · Ransomware · Extortion · Mass exploitation · Zero day threats

1 Introduction

Ransomware attacks are on the rise targeting many sectors including critical sectors like transportation, health, communication, energy, banking & finance and defense sectors. The ransomware may not be considered only as a simple attack for extortion of victims for funds. They might be part of more targeted attacks like advanced persistent threats (APTs), considering the increase in their attacks on the critical sectors of the nation. The ransomware groups, instead of only demanding ransom from organizations to give back their sensitive data, are now opting for multi-extortion methods, wherein instead of only encrypting the sensitive data, a copy of data is also being exfiltrated. The victims are threatened that the sensitive data would be released

S. Addetla (✉) · R. Pachamuthu
Indian Institute of Technology, Hyderabad, India
e-mail: AI20RESCH14001@iith.ac.in

R. Pachamuthu
e-mail: raji@ee.iith.ac.in

S. M. Thampi et al. (eds.), *Fifth International Conference on Computing and Network Communications*, Lecture Notes in Electrical Engineering 1219,
https://doi.org/10.1007/978-981-97-4540-1_32

in the public domain if the ransom is not paid. This stolen data is never deleted by the groups, in fact, it is used further for cyber espionage on targeted organizations. Hence it is important to consider ransomware as part of an advanced persistent threat and should not be taken lightly. Security analysts should focus on the possibilities of Ransomware being part of a larger threat landscape and if it is attributable to APT actors, to mitigate the possible sophisticated attacks that could cause catastrophic damage to the nation's security. The objective of this paper is to emphasize on this concept that, in view of the evolution in the ransomware techniques, there is a need to consider these attacks as inherent APTs.

Ransomware is a type of cyber security attack in which cyber criminal demands payments from the victims to provide access to their critical data. The ransom transactions are done in crypto currency to maintain anonymity. There are three categories of ransomware: (i) Locker ransomware—only locks access to the system, it does not delete or encrypt the files. (ii) Crypto Ransomware—encrypts user's sensitive data and demands ransom to provide decryption key. (iii) Scareware—does not do any harm to the victim's system, it only exploits their fear [1]. There are three encryption schemes for crypto ransomware (i) Symmetric—the encryption key embedded in Ransomware. This type of encryption is vulnerable to reverse engineering. (ii) Asymmetric—it takes long time to encrypt large files. (iii) Hybrid encryption—This uses a combination of symmetric and asymmetric encryption [2]. It first creates random symmetric key, after the symmetric key encrypts all data, public-private key is generated by C&C server. Public key is sent to ransomware and the symmetric key is encrypted. This is a most effective approach and hardest one to decode [3]. The phases in a ransomware attack are given in Fig. 1 [4]. The ransomware can be automated or human operated. The automated attacks usually spread across the network like virus, through emails, etc., whereas human operated ransomware are not self propagating. The cyber criminals explore security misconfigurations, vulnerabilities in a system, gain access, escalate privileges, deploy ransomware, and exfiltrate data [5]. The ransomware attacks have been severely effecting many critical sectors like health care, transportation, energy, communications etc., causing disruption to the services. Severe financial losses, compromise to national security and reputational damages occur due to these attacks. Security firms and researchers have been working on designing effective solutions for real-time detection of ransomware attacks. This paper presents the need to consider ransomware as advanced persistent threat (APT) in Sect. 2, some of the recent ransomware attacks have also been presented. Section 3 presents the changing trends in ransomware. Section 4 presents the state of the art detection techniques proposed by researchers and the preventive measures to be considered for mitigating such attacks. This section also highlights the solutions to maintain cyber resilience. Section 5 presents the challenges faced by researchers and security analysts in devising solutions for ransomware detection. Section 6 presents the predictions for ransomware and solutions to be designed in future. Section 7 is the conclusion.

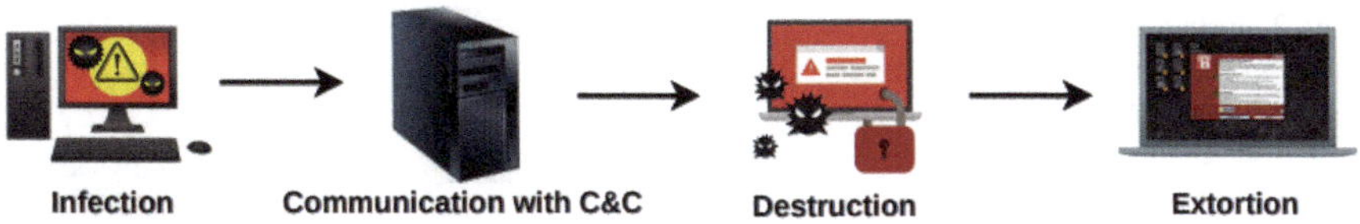

Fig. 1 Attack phases of ransomware [4]

2 Ransomware as Advanced Persistent Threat (APT)

A Malware is any malicious software that intentionally harms a computer. Ransomware is a type of malware which makes the victim's data inaccessible and demands ransom for providing the access to data back. On the other hand, APT is a multi-stage attack which is state-sponsored and very targeted attack causing catastrophic damages to a nation's security. There is an overlap of features among these three types as shown in Fig. 2. The APT attacks have a life cycle with many stages represented in Fig. 3. The attacks are executed in six phases [6]: (i) **Reconnaissance**: attacker tries to understand the target network, its weaknesses, through passive monitoring and social engineering (ii) **Initial compromise**: gaining initial access to the target network. In this, spear-phishing, attachments with malware, and zero-day vulnerabilities can be used. (iii) **Command & Control**: establishes connection with the command and control server, and forwards additional information about the system configurations, etc., which will help in planning next moves. (iv) **Lateral movement and privilege escalation**: The attacker tries to move laterally in the network to find vulnerable systems and escalates privileges. (v) **Identification of critical assets**: Through lateral movement and privilege escalation, the attacker tries to identify the critical assets in the network which has most of the sensitive data. (iv) **Exfiltration of data**: Once the data is identified, it is then exfiltrated by establishing connections to multiple levels of C&C servers. The attackers also clear their tracks after exfiltration to ensure that no traces are left on the target network. As shown in Fig. 1, the ransomware also has similar the phases of access gain to the network, lateral movement, privilege escalation, and data exfiltration with main goal of extortion. However, encryption of data and demand of ransom are not part of advanced persistent threats. Ransomware should be considered as an inherent APT as it also has the potential for destruction and espionage. Also, several ransomware are managed by APT groups. APT groups use ransomware to raise fund for their operations. In some cases, ransomware is used as a distraction from a more severe attack on critical systems. The TTPs used to deploy ransomware either for APT or a cybercrime group are mostly identical. Both types of attacks exploit vulnerabilities to gain access of target network, compromise system credentials, and lateral movement. Organizations must focus on these common areas to improve their defense mechanism. Paying ransom would not be the end of the story if there is a blend in ransomware and APT groups. In cases where APT groups are involved, most of the ransomware attacks would be multi-extortion cases were data also is exfiltrated.

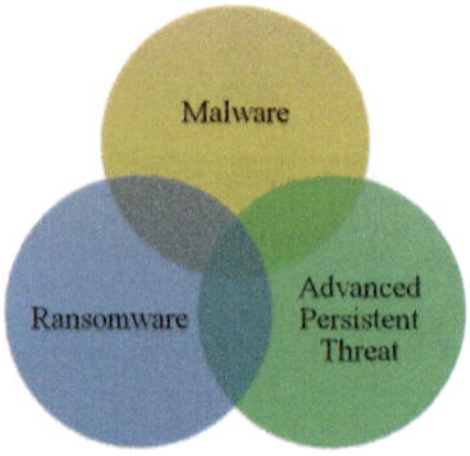

Fig. 2 Relation between malware, ransomware, and advanced persistent threats

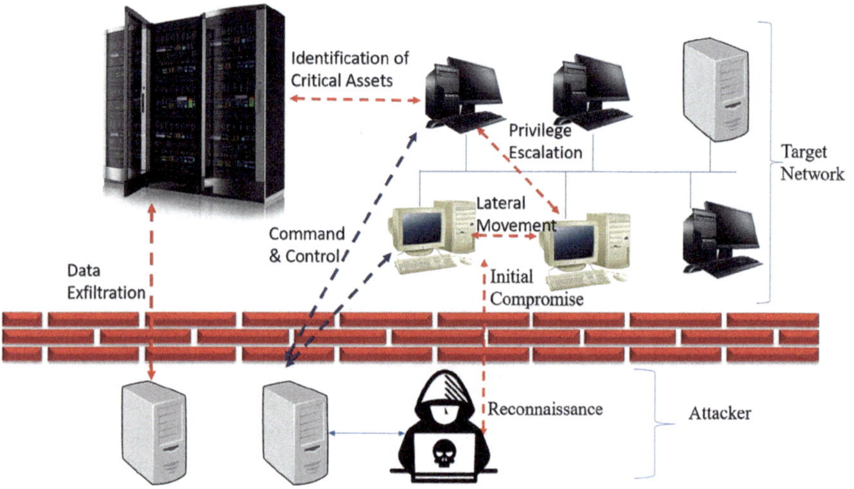

Fig. 3 Various phases of APTs

This is of grave concern where cyber insurance providers state that act of war is not covered [7]. Some of the recent ransomware attacks occurred on critical sectors like health, transportation, and energy sectors are listed in Table 1.

In some cases, the ransomware, based on its TTPs can be attributable to APT groups. Ransomware multi-extortion cases where data theft happens and ransom is demanded for not publishing the data, the data may not be deleted/destroyed after payment of ransom by the victim. It would be used for the advantage of threat actors got causing severe impairment to security and defenses of a nation [7].

3 Changing Trends of Ransomware

The Ransomware Threat Landscape is evolving and there is a need to understand the ransomware to be able to develop solutions for detection. Following are some of the trends in ransomware observed in last few years:

Table 1 Some of recent ransomware attacks

Period	Victim	Ransomware group	Sector	Effect
Aug 2023	Lankan Govt cloud [8]	Not known	Government	5000 email accounts data encrypted
Jul 2023	Port Nagoya, Japan [9]	Lockbit 3.0	Transportation	Cargo services disrupted for two days
Jun 2023	Eisai, pharmaceutical company [10]	Not known	Health	Services have been taken offline
May 2023	ABB, Swiss company [11]	Black Basta	Industry	Exfiltration of data. Disruption of operations
May 2023	MCNA dental [12]	Lockbit	Health	9M patients data stolen
May 2023	Swiss government Xplain software solutions [13]	Play & NoName	Government	Private, confidential data, financial and taxation details were stolen. All stolen data was published
Apr 2023	German arms company: Rheinmetall [14]	Black Basta	Manufacturing	Sensitive data posted on leak sites
Apr 2023	Harvard pilgrim health care [15]	Not known	Health	Exfiltration of sensitive data. Systems disrupted
Apr 2023	Insurance information Bureau of India [16]	Russian malware	Finance	Data inaccessible due to encryption of 30 servers
Jan 2023	Des Moines Public Schools, Iowa [17]	Not known	Education	Personal information of 6700 individuals was exposed
Apr 2022	Oil India [18]	Russian malware	Energy	Services partially restored after 10 d
Feb 2022	Bridgestone [19]	Lockbit 2.0	Manufacturing	SSNs, bank information, etc. of customers stolen
Feb 2021	St. Margaret health [20]	Not known	Health	Complete closure of operations in Peru & Spring Valley in Jun 2023

1. **Encryption-less attacks**. Ransomware basically involves encryption of user data rendering inaccessible and demanding ransom to provide access to data. However, the trends have changed. There is double-extortion used, in which the data with user is encrypted and exfiltrated before encryption. This way, the attacker has additional leverage and can still demand ransom to make the victim's sensitive data public even though the users might have restored their services. Further, only exfiltration of data is also being used, which is called encryption-less attack. With this, massive data theft from users is observed. However, it causes less disruption to services. A combination of encryption-less attacks with the supply chain vulnerabilities will cause even severe damages, as they generally target organizations with less security controls and mass exploitation can be carried out with single attack vector [21].
2. **Disbanding of Ransomware groups**. Many ransomware groups are shutting down and coming up in a new way. Conti Ransomware was shut in 2022, however, strong connections of Conti with other ransomware brands (Karakurt, Diavol, SchoolBoys, TommyLeaks, BlackBasta, Royal, etc.) are observed [21].
3. **Shift in ransomware design**. The malware has migrated from C/C++ to Golang and Rust Programming languages with cross platform compatibility. Figure 4 shows the details of programming languages used by various ransomware.
   It is more time consuming for reverse engineering and many third party libraries are being included for encryption. There is migration observed from symmetric to asymmetric encryption techniques (RSA & Elliptic Curve Cryptography) in the malware. Also, increased obfuscation techniques are used, which has made it difficult for detection of ransowmare [21].
4. **Mass exploitation** Presence of vulnerability in one product is leading to extortion campaigns against many organizations, stealing their data, exploiting them and threatening to post sensitive data on leak sites, if ransom is not paid. The zero-day vulnerability identified in MoveIt File transfer software was exploited by UNC4857 (merged into FIN11) in May 2023 to target many organizations. This software was widely used for secure file transfer. The vulnerability of SQL Injection present in the MoveIt software (CVE-2023–34362 assigned) was utilized to enabled privilege escalation and resulted in gaining unauthorized access to resources. Further, the vulnerability was leveraged to plant webshell LEMURLOOT into the vulnerable systems. This webshell was further exploited to exfiltrate data which was uploaded by users for data transfer [22].
5. **Ransomware as a Service.** The threat groups make ransomware tool-kits available on the dark web for sale. They also provide support for the software and negotiations. Availability of such services enables any person to buy the tools and initiate an attack. There is no time required to spend on developing the software. Easy access to such destructive tools is enabling even script kiddies leverage attacks.

Ransomware Family	C/C++	Golang	Rust
Hive		✓	✓
BlackCat/ALPHV			✓
BlackByte	✓	✓	
RansomExx	✓		✓
BianLian		✓	
Snatch		✓	
Nokoyawa	✓		✓
Agenda/Quilin		✓	✓
Luna			✓

Fig. 4 Shift in programming languages in ransomware [21]

4 Detection and Prevention Techniques of Ransomware

4.1 Detection Techniques for Ransomware

Many researchers have been exploring plethora of methodologies for detection of ransomware. Usman et al. in [23] have proposed a supervised learning ensemble method for detection of ransomware. Static analysis (includes permissions, content & network features) and Dynamic analysis (includes system logs, memory & CPU usage) on malware samples from android applications is carried out. Further, the malware is categorized as ransomware or non-ransomware if both static and dynamic ensemble analyzers classify it as ransomware. Another supervised machine learning model, random forest with gain ratio technique for feature selection and detection was proposed by Ban et al. [24] for ransomware detection. Jurijs et al. in [25] propose Bayesian Neural Network with approximate posterior distribution resulting in spike and slab distributions. This handles missing/ sparse data, scales to variation in number of parameters and prevents overfitting of data. In this method, the low-level events are represented as MITRE ATT&CK TTPs for better generalization to unknown ransomware. Yahye et al. in [26] propose supervised machine learning techniques for detection of ransomware and also leverage enhanced minimum redundancy maximum relevance filter that can remove the noisy features for better extraction of ransomware behavioral features. Abdullahi et al. in [27] have analyzed the parameters like CPU usage, memory usage, read and write operations of applications (malware and benign software) to determine if they are ransomware or not. Further, the details are input to ML models (decision tree, random forest, neural network) with threshold values assigned for different type of data. Random Forest gives the best performance for ransomware detection. However, the supervised learning-based methods heavily rely on the datasets on which the model is trained and fail to capture the dynamic nature of the Ransomware characteristics. Grant et al. in [28] carryout dynamic analysis of how ransomware affects windows server-based services. It was observed that the domain services were not affected. Farnood et al. in [29] have brought out a method based on user data in shared storage, entropy of data and carries out dynamic & static analysis of malware for real-time detection of ransomware on smart phones. Their method also ensures zero data loss in such attacks. Te-Min et al. in [30] have used network packet analysis for ransomware detection. Works identifying the network environment and base lining of normal traffic. Further, the disk i/o operations, memory usage are monitored for anomalies. System behavior is analyzed when LooCipher ransomware is executed. Yamany et al. in [31] use absolute jaccard index on static features of ransomware binaries to identify similarities between different samples. It is also observed that usage of Import Address Table feature gave better results than any other feature. Khan et al. in [32] propose an architecture DNAact-Ra in which Digital DNA sequence is generated for selected features through MOGWO & BCS algorithms and further classified as ransomware or non-ransomware.

4.2 Prevention of Ransomware

The following are some of the solutions to prevent ransomware attacks on networks:

1. **Update and patch all systems**. The security of a network is as good as its weakest link. As seen in many of the ransomware attacks described above, the presence of systems with outdated software is one of the easy entry for the attackers to gain access into the network. Hence, it is absolutely necessary to update the systems operating system with patches, software, and antivirus. The network administrators and the security analysts of an organization should ensure that the relevant patches for all the software and fixes for the vulnerabilities present are installed in a timely manner.
2. **Have effective defense mechanisms**. Instead of relying on a single security solution, organizations should have a defense in depth approach. Using multiple security strategies which include physical security, access restrictions, perimeter defense, encryption, multi-factor authentication, host-based security solutions and solutions for anomaly detection have to be included in the network and all logs have to be monitored by the security analysts for timely action.
3. **Zero trust network architecture**. Trust nothing is the security trend that has to be followed by every security administrator of a network. Whether a user is connected to the network within or from outside the premises of a network, the security protocol for accessing the systems should be same. Usage of multi-factor authentication, role-based access and logging of all events in a network are mandatory. The zero trust architecture can be implemented in existing networks also, hence migrating to the zero-trust architecture should be considered for effective network security.
4. **User training and awareness**. The Ransomware attacks use phishing, spear-phishing, emails with malicious links to gain access into the network, after iden-

tifying the vulnerable systems. Educating the users on cyber hygiene will prevent them to be cautious in clicking on any suspicious links or attachments. All organizations should conduct regular cyber awareness programs to all the employees to avoid such incidents of unauthorized access of attackers to network.

5. **3–2–1–1 strategy for backup of critical data**. To be prepared to recover from ransomware attack and to minimize the effects of such attacks, organizations need to have regular backup of all data. The concept of 3–2–1–1 backup for all critical data to be considered to have minimum disruption of services due to such attacks. The 3–2–1–1 strategy of backup is to have 3 copies of data backup (One original and two copies). Out of these "*3*" data copies, at least "2" different storage media (Like SAN, Magnetic tape) should be used. "*1*" copy of the data has to stored in remote location from the main data backup. The other "*1*" in the strategy signifies immutability of the data. The data as write- once-ready-many-times. The data cannot be deleted or altered by anyone. Hence that copy of data would be protected and helps quick recovery from ransomware attacks [33].
6. **Improve security posture**. Proper analysis of the ransomware incidents will enable organizations to understand the potential security issues that lead to such attacks. Based on these details, the security posture of their networks can be tightened by introducing solutions/ policies & rules that would improve detection of any intrusions. The security of a network is not a one time activity, there has to be continuous update on the settings to prevent attacker from entering the network and moving laterally.
7. **Do not pay ransom**. Paying of ransom to get the data back is discouraged, as it would make the attackers to further involve in similar attacks on different organizations. Also, even after paying ransom, it is not guaranteed that the attacker would give the data back and would not keep the sensitive data for future. Organizations also need to perform cost benefit analysis if it is worth paying ransom to get back the data.

4.3 Solutions for Cyber Resilience

Cyber-attacks are on a rise affecting large and small organizations. It is of utmost importance for the organizations to recover from such attacks and continue with business operations with minimum downtime. National Institute of Standards and Technology (NIST) has proposed a framework for cyber resilience with five core functionalities as shown in Fig. 5 which include **identify**—understanding the organization in terms of its assets, people, systems, data and capabilities, **protect** —implementing safeguards to minimize the risk due to cyber security event, **detect**—identification of occurrence of a security event, **respond**—action taken to contain the impact of detected cyber security event, and **recover**—maintain plans for resilience and to restore the services which were impaired due to cyber security event [34]. Following are some of the solutions for organizations to build cyber resilience:

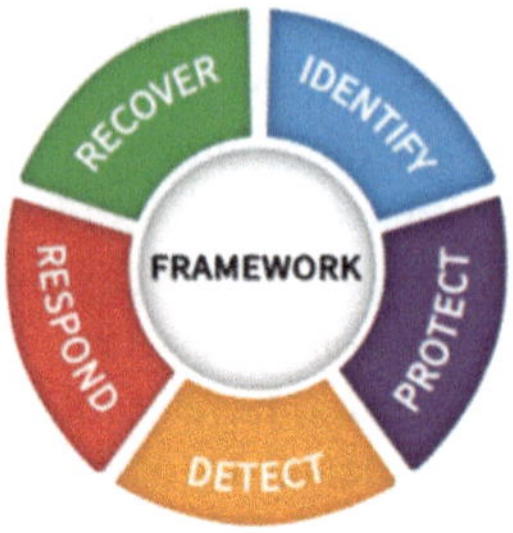

Fig. 5 NIST framework for cyber resilience

1. **Maintain fundamental cyber hygiene**. It is pertinent to mention that adhering to minimum security standards can protect a network against 99% of the attacks. Minimum security standards like multi-factor authentication, regular updates to system software, regular data backups, least privilege access can protect the networks from most of the attacks [35].
2. **Collaborations**. Cyber poverty line is a minimum standard for security from any kind of attacks. Collaborations between public and private sectors, Academia and research organizations should be considered for building policies, integration of technologies in security domain, which will ultimately aide the organizations to rise above the cyber-poverty line. Productive relations among all stakeholder in an organization would aide in improvement of resilience and defense [35].
3. **Innovative security leveraging AI**. AI has the potential to update the security landscape by advancements in analyzing humongous security data on daily basis for a safer system. Large Language Models can be leveraged to automate and augment cyber defense modules like threat intelligence, monitoring, response, etc. [35].

5 Challenges

Ransomware attack is carried out by threat actors leaving minimum scope for the security analysts to identify their TTPs and mitigate such attacks. Some of the challenges faced in detection and attribution of the ransomware are as follows:

1. **Non-traceability of the attack**. The attacks can be executed anonymously through a third party by an attack group. In such cases, arriving at the origin of the attack and attribution would be difficult. The payments made for the ransomware attacks are also in cryptocurrency, so, tracing the attacker is difficult [36].
2. **Asymmetric or hybrid encryption**. usage of asymmetric or hybrid encryption will render it difficult to find the key and decrypt the data [2].
3. **Reverse engineering**. There is a shift observed in the programming languages used for development of ransomware. Because of this, reverse engineering of the malware has become time consuming. Even though the code of ransomware has

been reverse engineered, due to the change in the design and encryption techniques, the reverse engineered code may not be relevant or helpful in identifying the parameters of new ransomware [21].

4. **Ransomware as service**. Availability of ransomware as a service enables any person to be able to execute a ransomware attack. Even the initial access to networks are also made available on dark web as a service along with ransomware toolkit [21].
5. **Data security not guaranteed**. As the cyber criminals are using data exfiltration or multi-extortion methods, sensitive data of the victim is available with the cyber criminal. Hence even if ransom is paid by the victim, it cannot be guaranteed that the sensitive data of the victim will be deleted/ destroyed permanently by the cyber criminal [7].

6 Future Work

We have seen that the ransomware attacks have increased on corporates, government organizations and various critical infrastructure instead of individuals as it is very lucrative for the cybercriminals. Though many detection methods have been formulated by reverse engineering the Ransomware code, many ransomware attacks are left undetected due to ever changing ransomware patterns. Some of the future extortions have been predicted by security experts. Few of the ransomware extortions would be compromise of cloud environments, novel means like Search Engine optimization poisoning, callback phishing can be used for initial access gain to a network, extortion without encryption would increase, politically motivated incidents would rise (would destabilize the critical infrastructure considering it as an APT) [7]. Future research should be in the direction considering occurrence of such extortions. Security solutions leveraging AI should be considered for automation of cyber defense against ransomware.

7 Conclusion

Considering the ever growing attack patterns, it is important for every organization to be proactive to keep their systems up to date in every department. Ensure regular user awareness campaigns, maintain backups of all critical data, install ransomware detection methods into the network and not to pay any ransom to the attacker in case any such attack takes place. The security analysts need to focus more on the detection methods for ransomware where in, considering the ransomware as APT would be necessary as it follows most of the steps of the APT and is also supported by APT groups. Solutions to detect the ever changing patterns of the ransomware in a timely manner are extremely essential, as data extortion is also involved in it, which could be utilized by cyber criminals to their advantage.

References

1. Gómez-Hernández J, Álvarez-González L, García-Teodoro P (2018) R-locker: thwarting ransomware action through a honeyfile-based approach. Comput Secur 73:389–398
2. Cicala F, Bertino E (2022) Analysis of encryption key generation in modern crypto ransomware. IEEE Trans Dependable Secure Comput 19:1239–1253
3. Beaman C, Barkworth A, Akande T, Hakak S, Khan M (2021) Ransomware: recent advances, analysis, challenges and future research directions. Comput Secur. 111:102490. https://europepmc.org/articles/PMC8463105
4. Oz H, Aris A, Levi A, Uluagac A (2022) A survey on ransomware: evolution, taxonomy, and defense solutions. ACM Comput Surv 54:1–37. https://doi.org/10.1145/3514229
5. Daniel Simpson J (2023) What is ransomware?. https://learn.microsoft.com/en-us/security/ransomware/human-operated-ransomware
6. Abdullayeva F (2021) Advanced persistent threat attack detection method in cloud computing based on autoencoder and softmax regression algorithm. Array 10:100067
7. Unit 42 Ransomware and Extortion Report (2023) (PaloAlto, 2023). https://start.paloaltonetworks.com/2023-unit42-ransomware-extortion-report
8. Antoniuk D (2023) Sri Lankan government loses months of data following ransomware attack. https://therecord.media/sri-lanka-loses-months-of-government-data-in-ransomware-attack
9. Kass D (2023) Japanese port of nagoya resumes operations after 2-day Russian ransomware attack. https://www.msspalert.com/news/japanese-port-of-nagoya-resumes-operations-after-2-day-russian-ransomware-attack
10. Arghire I (2023) Pharmaceutical giant Eisai takes systems offline following ransomware attack. https://www.securityweek.com/pharmaceutical-giant-eisai-takes-systems-offline-following-ransomware-attack/
11. Group ABB provides details about IT security incident (2023) https://new.abb.com/news/detail/103405/abb-provides-details-about-it-security-incident
12. Karabus J (2023) Criminals spent 10 days in US dental insurer's systems extracting data of 9 million. https://www.theregister.com/2023/05/31/mcna_breach/
13. Toulas B (2023) Swiss government warns of ongoing DDoS attacks, data leak. https://www.bleepingcomputer.com/news/security/swiss-government-warns-of-ongoing-ddos-attacks-data-leak/
14. Martin A (2023) German arms company Rheinmetall confirms Black Basta ransomware group behind cyberattack. https://therecord.media/rheinmetall-confirms-black-basta-ransomware-group-behind-cyberattack
15. Toulas B (2023) Harvard pilgrim health care ransomware attack hits 2.5 million people. https://www.bleepingcomputer.com/news/security/harvard-pilgrim-health-care-ransomware-attack-hits-25-million-people/
16. Natani S (2023) Insurance information Bureau of India hit by ransomware attack: Russian hackers demand $250,000 as ransom. https://www.the420.in/indian-insurance-information-bureau-hit-by-ransomware-attack-russian-hackers-demand-250000-as-ransom/?swcfpc=1
17. Gatlan S (2023) Iowa's largest school district confirms ransomware attack, data theft. https://www.bleepingcomputer.com/news/security/iowas-largest-school-district-confirms-ransomware-attack-data-theft/
18. PTI Russian malware planted from Nigeria in OIL cyber attack police (2022) https://www.theweek.in/wire-updates/national/2022/04/22/cal8-as-oil-cyber-hacker.html
19. Vijayan J (2023) Bridgestone CISO: lessons from ransomware attack include acting, not thinking. https://www.darkreading.com/ics-ot/bridgestone-ciso-lessons-ransomware-attack-acting-thinking
20. Vijayan J (2023) Illinois hospital closure showcases ransomware's existential threat. https://www.darkreading.com/attacks-breaches/illinois-hospital-closure-ransomware-existential-threat
21. Gross B, Singh N (2023) Insights into the latest ransomware trends. Zscaler

22. Nader Zaveri E (2023) Zero-day vulnerability in MOVEit transfer exploited for data theft. https://www.mandiant.com/resources/blog/zero-day-moveit-data-theft
23. Ahmed U, Lin J, Srivastava G (2022) Mitigating adversarial evasion attacks of ransomware using ensemble learning. Comput Electr Eng 100:107903
24. Khammas B (2020) Ransomware detection using random forest technique. ICT Express 6:325–331
25. Nazarovs J, Stokes J, Turcotte M, Carroll J, Grady I (2022) Radial spike and slab Bayesian neural networks for sparse data in ransomware attacks
26. Abukar Y, Koçer B, Huda S, Al-rimy B, Hassan M (2020) A system call refinement-based enhanced minimum redundancy maximum relevance method for ransomware early detection. J Network Comput Appl 167:102753
27. Arabo A, Dijoux R, Poulain T, Chevalier G (2020) Detecting ransomware using process behavior analysis. Procedia Comput Sci 168:289–296
28. McDonald G, Papadopoulos P, Pitropakis N, Ahmad J, Buchanan W (2022) Ransomware: analysing the impact on windows active directory domain services. Sensors 22:953. https://doi.org/10.3390/s22030953
29. Faghihi F, Zulkernine M (2021) RansomCare: data-centric detection and mitigation against smartphone crypto-ransomware. Comput Network 191:108011
30. Liu T, Kao D, Chen Y (2020) LooCipher ransomware detection using lightweight packet characteristics. Procedia Comput Sci 176:1677–1683
31. Yamany B, Elsayed M, Jurcut A, Abdelbaki N, Azer M (2022) A new scheme for ransomware classification and clustering using static features. Electronics 11:3307
32. Khan F, Ncube C, Ramasamy L, Kadry S, Nam Y (2020) A digital DNA sequencing engine for ransomware detection using machine learning. IEEE Access 8:119710–119719
33. Arcserve (2023) How to protect against ransomware with a 3–2–1–1 strategy. https://www.arcserve.com/blog/how-protect-against-ransomware-3-2-1-1-strategy
34. Barrett M (2018) Framework for improving critical infrastructure cybersecurity version 1.1, NIST cybersecurity framework. https://doi.org/10.6028/NIST.CSWP.04162018
35. Microsoft Threat Intelligence (2023) Microsoft digital defense report 2023. (Microsoft threat intelligence). https://www.microsoft.com/en-us/security/security-insider/microsoft-digital-defense-report-2023
36. Ilker KARA, Aydos M (2020) Cyber fraud: detection and analysis of the crypto-ransomware. In: 2020, 11th IEEE annual ubiquitous computing, electronics & mobile communication conference (UEMCON). New York, NY, USA, pp 0764–0769. https://doi.org/10.1109/UEMCON51285.2020.9298128

Machine Learning Techniques for Credit Card Fraudulent Transaction Detection

M. A. Ram, S. Geetha, and U. Vignesh

Abstract Fraud detection has been a serious problem in a variety of industries such as banking, medicine, insurance, and many others. Fraudulent activities have increased along with the number of online transactions made using various payment methods, including credit/debit cards, PhonePe, Gpay, Paytm, etc. Fraudsters have also developed very sophisticated escape routes so they can make their willing successful. To avoid this kind of problems, a secure system is required for authentication and preventing customers from fraud. Nowadays, fraud detection has become a difficult task because no system is faultless and there is always a vulnerability. Therefore, machine learning algorithms for fraud detection are very helpful in reducing fraud. In this paper, we propose an efficient machine learning approach employing models like XGBoost, AdaBoost, and ensemble models to detect fraudulent credit card transaction. The proposed model for credit card fraud detection exhibits superior performance than the existing systems.

Keywords Deep learning · Classification · Fraud detection · Machine learning · Regression

M. A. Ram · S. Geetha · U. Vignesh (✉)
School of Computer Science and Engineering, Vellore Institute of Technology (VIT), Chennai, Tamil Nadu, India
e-mail: vignesh.u@vit.ac.in

S. M. Thampi et al. (eds.), *Fifth International Conference on Computing and Network Communications*, Lecture Notes in Electrical Engineering 1219,
https://doi.org/10.1007/978-981-97-4540-1_33

1 Introduction

In a credit card system, when someone other than the account proprietor uses a credit card or account information for an unauthorized transaction, this is known as credit card fraud (CCF). A credit card that has been lost, stolen, or fraudulently manufactured may be the source of fraud. Due to the rise in online purchasing, card-not-present fraud—or the use of your credit card number in e-commerce transactions—has also increased in frequency. The expansion of e-banking and online payment environments has led to an increase in fraud, such as CCF, which causes billions of dollars in annual losses. CCF detection has emerged as one of the key objectives in this era of digital payments. Machine learning (ML) models have been applied in several research to address a wide range of problems. Deep learning (DL) algorithms have been used in a variety of fields, including home automation, banking, insurance, cellular networks for mobile devices, healthcare fraud detection, malware identification for video surveillance, and the prediction of heart disease.

In this study, we investigate the practical use of ML, in particular DL algorithms, to recognize credit card frauds in the banking sector. The training process and detection phase using machine learning models are shown in Figs. 1 and 2. Numerous solutions might be offered by a vast body of research. As a result, the issue of class inequality has not yet been resolved to the best of our knowledge. By including additional layers for feature extraction and the categorization of credit card transactions as fraudulent or not, we suggest changing the DL algorithm of the CNN model.

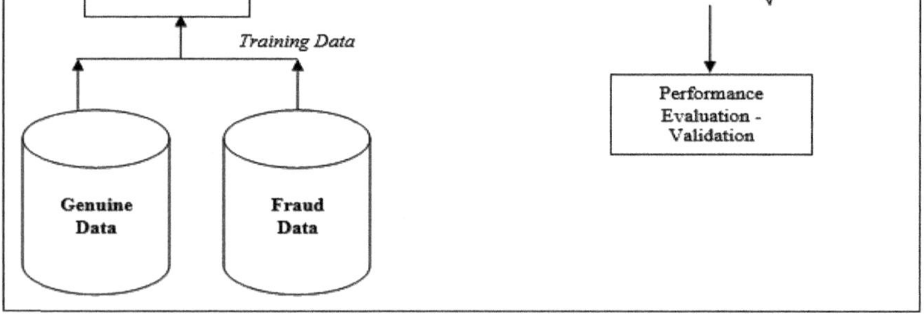

Fig. 1 Training phase

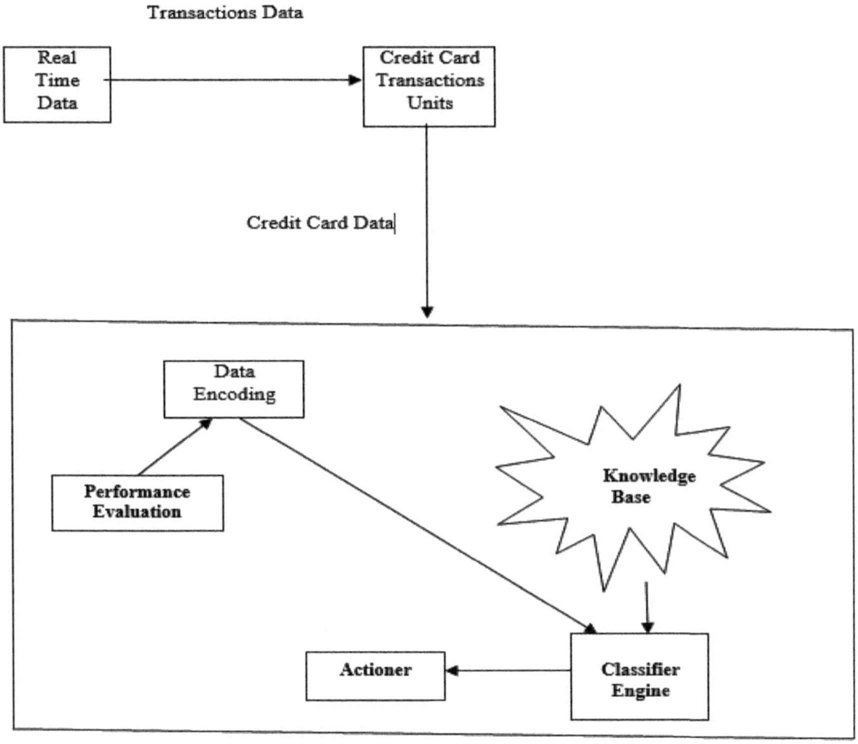

Fig. 2 Detection phase

2 Related Work

In a comprehensive exploration of fraud detection models, Zareapoor and team evaluated neural network, Bayesian network, SVM, and *K*NN based on accuracy, speed, and cost. Bayesian network demonstrated exceptional speed and accuracy, while neural network showed good performance with medium accuracy. *K*NN exhibited good speed and medium accuracy, whereas SVM lagged with low speed and medium accuracy. Notably, all models incurred high costs. Awoyemi's study compared *K*NN, Naïve Bayes, and logistic regression for credit card fraud detection. Different distributions were sampled, revealing Naïve Bayes as the top performer with 97.5% accuracy in the 10:90 distribution. In another distribution (34:66), *K*NN excelled with 97.9% accuracy, while logistic regression improved to 54.8% accuracy [1].

Jain and team utilized SVM, ANN, and *K*NN for credit card fraud detection, employing metrics like accuracy, precision, and false alarm rates for comparison. ANN achieved high accuracy (99.71%) and precision, with a minimal false alarm rate. SVM demonstrated good accuracy but higher false alarm rates, and *K*NN showed solid accuracy and precision [2]. Gupta's team focused on economically related fraudulent transactions, emphasizing Naïve Bayes, which displayed outstanding performance with 80.4% accuracy and a 96.3% area under the curve [3].

Tanouz's research proposed ML-based classification algorithms for handling strongly imbalanced datasets. Logistic regression and Naïve Bayes both scored 95.16%, random forest achieved 96.77%, and decision tree reached 91.12% across various evaluation measures [4]. Sahin and Duman used four SVM methods for credit card fraud detection, scoring high in training (99.87%) but lower in testing (83.02%) [5].

Itoo and group implemented logistic regression, Naïve Bayes, and *K*-nearest neighbors, with logistic regression leading with 91.2% accuracy, Naïve Bayes at 85.4%, and *K*-nearest neighbor at 66.9% accuracy [6]. Credit card frauds are easy and friendly targets to reduce it, the design develops a novel fraud detection method for Streaming Transaction Data, with an objective, to analyze the past transaction details of the customers and extract the behavioral patterns [7] and the different approaches of clustering concept with various applications visualization are mentioned [8].

3 Proposed Methodology

In the research experiments conducted, even the complete and consistent dataset, efficient algorithms, and appropriate evaluation metrics, a machine learning model may still have a lower accuracy, resulting in inefficient detection of fraudulent transactions. This is due to the efficacy of the model's training. Consequently, the machine learning model suggests a system in which the individual algorithms are trained on the dataset using an OOB score validation method. In the process of ensemble learning, generated models are combined and trained further. The primary difficulty is not in acquiring extremely precise base models; rather, the difficulty resides in acquiring base models with a variety of flaws. For instance, when using ensembles for classification, high accuracies can be obtained even if the accuracy of the base classifier is low, so long as distinct base models misclassify different training examples. This is feasible due to the fact that distinct base models may misclassify distinct training examples.

Due to high variance, there is a danger of overfitting for extended tree-depths in machine learning models such as random forest. Out of the bag (OOB) score is a validation technique used specifically for random forest to prevent overfitting of the model by minimizing variance. OOB score is computed on data that was not necessarily used in the analysis of the model, as opposed to validation score, which is computed on data that was used in the analysis of the model. Before training the models, however, a portion of the initial training dataset is withdrawn and set aside for the validation score calculation. In addition, the OOB score is computed using a subset of the bootstrap training dataset's DTs that exclude the OOB sample. In contrast, the validation score is calculated by employing all of the ensemble's DTs. As stated previously, the OOB score is calculated using only a subset of the available DTs. As a result, the general aggregation effect during packaging is diminished. In terms of calculating the score, therefore, validation conducted on the entire collection of DTs is preferable to validation performed on a subset of DTs. Occasionally, the dataset is insufficiently enormous, making it impractical to reserve a portion of it for validation. Therefore, the OOB score is a beneficial instrument to have in situations where we do not have a very large dataset but would like to use the entire thing as the training dataset. There are several machine learning algorithms that prove its efficiency in forming the proposed deep learning model.

3.1 XGBoost Classifier

Extreme gradient boosting, which is abbreviated as XGBoost, is a scalable and distributed machine learning library. The algorithm gradient-boosted decision tree (GBDT) is utilized. It is the most widely used machine learning library for solving problems such as regression, classification, and ranking, and it supports parallel tree boosting. The gradient boosting decision trees (GBDT) is an ensemble learning technique for decision trees that are used for classification and regression. The "bagging" operation of random forest reduces variance and overfitting, whereas the "boosting" operation of GBDT reduces bias and underfitting.

XGBoost is a highly accurate, scalable implementation of gradient boosting that stretches the limits of computing capacity for boosted tree algorithms. It was created predominantly for the purpose of enhancing the efficacy of machine learning models and accelerating their computational speed. XGBoost constructs trees in tandem, as opposed to sequentially as GBDT does. It employs a level-wise approach, examining gradient values and using these partial sums to evaluate the quality of divides at each potential split in the training set. This is done in order to obtain the finest potential results. The proposed research on machine learning models to distinguish fraudulent from non-fraudulent transactions employs OOB scores to improve accuracy and achieve the best outcome. Without the OOB score, the model resulted around 98.97% but there has been a spike in the accuracy to 99.03% once the model is validated using OOB score.

3.2 AdaBoost-Decision Tree Classifier

The concept behind boosting algorithms is to first create a model based on the data that will be used for training and then to create a second model that corrects the errors that the first model produced. This process is repeated until errors are reduced to an acceptable level and an accurate prediction can be made regarding the dataset. AdaBoost, also referred to as adaptive boosting, is a technique utilized by machine learning. This technique is categorized as an ensemble method. Most frequently, AdaBoost employs the technique known as decision trees with one level, also known as decision trees with a single split. These trees are also referred to as decision stumps.

The objective of this method is to build a model by designating equal weights to each of the input data points. After that, it assigns incorrectly classified points a greater weight in the overall calculation. Therefore, in the next model, all the points with increased weights will be given greater significance. It will continue training models until a lower error is returned, after which it will cease. At the end of the process, when all the classifiers constructed during the iterations are asked to vote for the target of a new observation, some trees will have a heavier vote than others. This is the primary distinction between AdaBoost and bagging methods such as random forests. AdaBoost considers the weight of each classifier as it is constructed, so this is the case. These are the species that have consistently demonstrated the highest levels of accomplishment across all iterations.

The most efficient application of AdaBoost is to enhance the efficacy of decision trees applied to binary classification problems. Every machine learning algorithm's efficacy can be enhanced by employing the available AdaBoost algorithm. It is most effective when used with students of lower intelligence. When applied to a classification problem, these models achieve a degree of precision that exceeds what would be expected by random chance. Recently, it has begun to be referred to as discrete AdaBoost because it is used for classification as opposed to regression.

3.3 Ensemble Classifier

The prediction performance of a single tree-based algorithm is frequently outperformed by an ensemble technique that integrates multiple tree-based algorithms into a single model. The underlying principle of the ensemble model is that a collection of less capable learners can be combined to generate a more capable learner, thereby improving the model's accuracy. Noise, variation, and bias are the three primary contributors to the difference between actual and predicted values when attempting to forecast a target variable using any machine learning method. The use of ensemble reduces the severity of these issues. The essential concept underlying the ensemble model is that a collection of less capable learners can be combined to generate a more

Table 1 Characteristics of dataset

S. No.	Feature	Description
1.	Time	The number of seconds that passed between this transaction and the dataset's initial transaction
2.	Amount	Transaction amount
3.	Class label	Binary class labels 1 and 0 for non-fraudulent and fraudulent transactions
4.	V1, V2, …, V28 attributes	The features are the results of PCA transformations in order to protect user information and confidential details

capable learner, thereby improving the model's accuracy. When attempting to predict the value of a target variable using any machine learning technique, noise, variance, and bias are the three most common contributors to the discrepancy between actual and predicted values.

4 Experimental Results

The dataset contains transactions that a cardholder completed over the course of two days in September 2018. In all, there were 284,807 transactions, and 492, or 0.172%, of them were fraudulent. Principal component analysis is used to apply the main component analysis to the bulk of the dataset's characteristics since exposing a consumer's transactional information is seen as a concern of confidentiality (PCA). In the relevant literature, PCA is a common and commonly used method for lowering the dimensionality of such datasets, improving interpretability while minimizing information loss. It accomplishes this by producing fresh, uncorrelated variables that maximize variance one after the other. The dataset has been split into two components testing dataset and training dataset with a distribution 20:80 for the experiment to conduct (Table 1).

The performance measures of ensemble classifier with the other machine learning models are shown in Table 2. The ensemble learning model in the proposed system is a combination of the XGBoost, random forest, and AdaBoost algorithms. The random forest classifier, AdaBoost classifier, and XGBoost classifier are loaded after the sampled X_train and Y_train datasets have been imported. EnsembleVoteClassifier is implemented so that classifiers can be executed. The accuracy metric used in the research is calculated based on the following parameters such as TP, TN, FP, FN stands for True Positive, True Negative, False Positive, and False Negative, respectively. Accuracy is used to measure the efficacy of data recovery and processing in the evidence domain as shown in Fig. 3.

Table 2 Performance measures

Model	Accuracy	Precision	Recall	$F1$-score	Average	Kappa Cohen	Sensitivity	Specificity
Random forest all features	0.994154	0	0	0	0.001703	− 0.00242	0.99585	0
Random forest ($n = 27$) with OOB_9	0.993908	0.200957	0.865979	0.326214	0.174253	0.324346	0.994126	0.865979
Random forest ($n = 35$) with OOB_12	0.995225	0.247839	0.886598	0.387387	0.219926	0.385752	0.99541	0.886598
AdaBoost simple classifier	0.993171	0.182609	0.865979	0.301616	0.158364	0.299646	0.993388	0.865979
AdaBoost ($n = 15$) with OOB_10	0.993171	0.181223	0.85567	0.299099	0.155313	0.297124	0.993405	0.85567
XGBoost classifier	0.995453	0.257485	0.886598	0.399072	0.228479	0.397482	0.995639	0.886598
XGBoost classifier parameter tuning	0.989765	0. 0.121	1.0.810	0.213225	0.100223	0.21089	0.995058	0.886598
XGB with OOB4	0.990309	0.137161	0.886598	0.237569	0.1218	0.235314	0.990486	0.886598
Ensemble classifier with all models	0.996664	0.321839	0.865979	0.469274	0.278934	0.467953	0.996887	0.865979

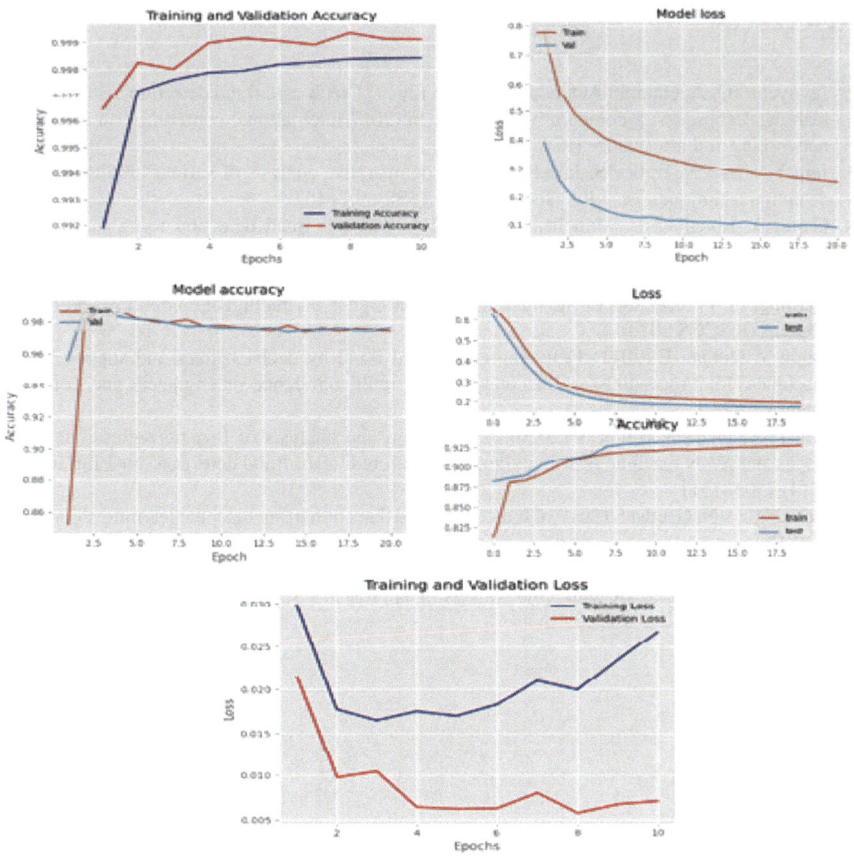

Fig. 3 Performance graphs of the proposed—ensemble model with all classifiers. **a** Training versus validation accuracy; **b** training versus validation loss

5 Conclusion

The performance of ML methods varies for each individual business case. The type of input data is a dominant factor that drives different ML methods. In this research, algorithms ensembled with boosting techniques have been implemented to accurately detect fraudulent transactions from credit card data. For detecting CCF, the number of features, number of transactions, and correlation between the features are essential factors in determining the model's performance. For optimization, the parameters of the XgBoost classifier are tuned which increases the accuracy. There will also be a focus on improving performance metrics like recall rate, precision, and F-measure which in turn can improve the accuracy and the robustness of the model to detect fraudulent transactions.

References

1. Awoyemi JO, Adetunmbi AO, Oluwadare SA (2017) Credit card fraud detection using machine learning techniques: a comparative analysis. In: 2017 international conference on computing networking and informatics (ICCNI)
2. Jain Y, Tiwari N, Dubey S, Jain S (2019) A comparative analysis of various credit card fraud detection techniques. Int J Rec Technol Eng 7(5S2):402–407
3. Gupta A, Lohani MC, Manchanda M (2021) Financial fraud detection using naive Bayes algorithm in highly imbalance data set. J Discrete Math Sci Cryptogr 24(5):1559–1572
4. Sailusha R, Gnaneswar V, Ramesh R, Rao RR (nd) Credit card fraud detection using machine learning. In: Proceedings of the international conference on intelligent computing and control systems (ICICCS 2020)
5. Sahin Y, Duman E (2011) Detecting credit card fraud by decision trees and support vector machines. In: Proceedings of the international multiconference of engineers and computer scientists, vol 1
6. Itoo F, Meenakshi, Singh S (2020). Comparison and analysis of logistic regression, Naïve Bayes and *K*NN machine learning algorithms for credit card fraud detection. Int J Inf Technol 13(4):1503–1511
7. Dornadula VN, Geetha S (2019) Credit card fraud detection using machine learning algorithms. Proc Comput Sci
8. Vignesh U, Parvathi R (2020) 3D visualization and cluster analysis of unstructured protein sequences using ARCSA with a file conversion approach. J Supercomput

Secure Cloud-Based Management System for Legal Firms

Ayushi Soumya, Bhumika Nayak, Deepthi Dayanand, V. B. Vaishnavi, and Venkatesh Prasad

Abstract Legal firms face mounting pressure to adapt to technology-induced transformations in the corporate landscape. To address this challenge, we propose a strategic approach centered on a customized Software-as-a-Service (SaaS) cloud model tailored to the specific needs of small-scale legal enterprises. This approach offers compelling advantages, including scalability, modularity, cost-efficiency, enhanced availability, reduced hardware expenses, and strengthened security measures. Leveraging diverse cloud computing resources, the proposed solution facilitates real-time collaboration, communication, and data administration within legal organizations. It incorporates automated functionalities for generating, negotiating, and executing legal contracts, documents, and templates, along with streamlined mechanisms for storing and retrieving contract-related documents. To ensure data confidentiality and security, the system employs advanced tools such as the Advanced Encryption Standard (AES) algorithm and Amazon Quantum Ledger Database (QLDB) to protect Personal Identifiable Information (PII). This project represents a convergence of visionary concepts and tangible execution, seamlessly integrating microservices, robust security features, and cloud-based deployment. By addressing the inherent requirements of legal technology in corporate entities, fostering user confidence, and embracing the benefits of cloud technology, this project holds significant long-term implications for the adoption of cloud-based management systems by legal firms. Empirical evidence supports the growing need for legal entities to collaborate with law firms that fully leverage technology. Over the next three years, legal departments will increasingly seek partnerships with firms that can provide cloud-based solutions tailored to their specific needs.

Keywords Cloud computing · Software-as-a-Service · Legal technology · Advanced Encryption Standard (AES)

A. Soumya · B. Nayak · D. Dayanand (✉) · V. B. Vaishnavi · V. Prasad
Department of Computer Science Engineering, PES University, Ring Road Campus, Bengaluru 560085, India
e-mail: deepthi.bhd@gmail.com

S. M. Thampi et al. (eds.), *Fifth International Conference on Computing and Network Communications*, Lecture Notes in Electrical Engineering 1219,
https://doi.org/10.1007/978-981-97-4540-1_34

As the legal landscape undergoes a rapid transformation driven by technological advancements, legal firms face the pressing challenge of integrating these advancements into their operations. Our paper introduces a groundbreaking approach to cloud-based management systems meticulously designed for legal enterprises, addressing this critical need. Our legal technology strategy aims to revolutionize the legal practice by seamlessly merging Software-as-a-Service (SaaS) principles with cloud computing capabilities, ushering in an era of unparalleled efficiency, security, and accessibility.

At the heart of this strategy lies the development of a bespoke cloud model tailored to the unique demands of small-scale legal entities. This novel paradigm encompasses a suite of features designed to transform how legal firms manage their workflows, including the seamless integration of essential microservices that are indispensable for modern legal technology operations. Recognizing the paramount importance of security in fostering user trust, our strategy employs cutting-edge encryption algorithms and robust key management services, ensuring the unwavering protection of sensitive data. This tailored approach, coupled with the integration of essential microservices and an unwavering emphasis on security, distinguishes our solution from generic cloud solutions.

Leveraging the inherent benefits of cloud infrastructure, our approach provides unparalleled scalability and flexibility, addressing the intricacies of modern legal practice with finesse. Our holistic approach encompasses microservices, stringent security protocols, and seamless cloud deployment, with the ultimate goal of navigating the challenges posed by the technology-driven corporate environment.

Our proposed model, synergizing the potential of SaaS, cloud computing, AES encryption, and Amazon Quantum Ledger Database (QLDB), offers a promising avenue to enhance efficiency, security, and productivity for legal practitioners in this era of dynamic technological change. This groundbreaking strategy represents a forward-looking solution poised to empower legal firms to thrive amidst the ongoing technological revolution.

By synergizing the potential of SaaS, cloud computing, AES encryption, and Amazon Quantum Ledger Database (QLDB), our proposed model offers a promising avenue to enhance efficiency, security, and productivity for legal practitioners in this era of dynamic technological change. This strategy represents a forward-looking solution poised to empower legal firms to thrive amidst the ongoing technological revolution.

1 Preliminaries

1.1 Blockchain Technology and Amazon QLDB

Blockchain technology is a revolutionary distributed ledger system that has redefined how data is securely recorded and shared across a decentralized network. It ensures transparency, immutability, and trust in digital transactions, making it a transformative force in various industries [1].

Amazon QLDB, offered by Amazon Web Services (AWS), is a fully managed ledger database service that combines the reliability of traditional databases with the transparency and security of blockchain technology. QLDB provides an immutable and cryptographically verifiable ledger for tracking changes to data over time, making it an ideal choice for applications requiring a tamper-proof and transparent audit trail.

1.2 Cryptography

Cryptography serves as the cornerstone of the system, ensuring the confidentiality and security of legal documents and sensitive data. Cryptography involves the use of encryption algorithms to safeguard valuable information.

The system must handle plaintext (the original data), keys (used for encryption and decryption), and ciphertext (the encrypted data, unreadable without the appropriate decryption key).

The cryptographic processes can employ symmetric and/or asymmetric keys. Symmetric key encryption entails the use of the same key for both encrypting and decrypting data, while asymmetric encryption involves a pair of public and private keys for these operations.

1.3 Symmetric Encryption Using AES Algorithm

Symmetric encryption, utilizing AES (Advanced Encryption Standard), employs a single key for both encryption and decryption processes. Effective key management is essential for secure key exchange among authorized parties, ensuring exclusive access to legal documents. Secret keys must be generated using a secure RNG adhering to industry standards like FIPS 140-2, ensuring randomness and resistance to attacks [2]. Robust key protection mechanisms, including secure storage and transmission, are crucial. Additionally, selecting appropriate encryption modes (e.g., ECB, CBC) and padding schemes (e.g., PKCS7) tailored to the legal firm's specific use case and threat model is vital [3].

2 Background

Moving to the cloud offers secure, remote accessibility with the advantage of more reliable, efficient, and collaborative workflow-often with less cost. Cloud is also easier to set up and use than older, on-premise legal software. Yet, many small-scale law firms still haven't embraced cloud-based solutions and are still following the traditional methods. In the years to come, however, cloud computing will become almost ubiquitous among law firms of all sizes. Firms that embrace these technologies sooner will position themselves ahead of their competitors. The larger law firms have already started embracing the five new technology advancements, that is, generative AI, multiverse, cloud-native platforms, embedded data and analytics, and cybersecurity. Therefore, small-scale legal firms need assistance in doing the same by providing them with similar technology through Software-as-a-Service product.

Many trends are changing in the legal industry. It is being said that by 2025, in-house legal departments will increase their spend on technology by threefold and will replace 20% of their generalist lawyers. More than 70% of the work will be automated and the surveys show that more than a quarter of their profit would go to non-specialist technology providers.

3 Problem Statement

The problem statement entails developing a legal technology strategy that is adaptive to the changes in the corporate environment and technological advancements to address the challenges faced by small-scale law firms in responding to technology-driven changes.

3.1 Existing Systems

The legal industry is undergoing a rapid transformation driven by technological advancements, commonly referred to as LegalTech. This surge in LegalTech solutions aims to enhance efficiency, streamline processes, and improve overall productivity within law firms. However, traditional legal technology approaches often fall short in addressing the critical aspects of data security and trust, particularly in the context of sensitive legal documents and confidential client information due to limited scalability, costly on-premises infrastructure, vulnerability to cyberattacks, and fragmented data storage [4]. Cloud-based legal technology solutions offer enhanced security, scalability, cost-effectiveness, and centralized data management. Advanced encryption algorithms and blockchain technology provide additional layers of security for legal data management [5]. By embracing cloud-based platforms and leveraging advanced security measures, legal firms can enhance efficiency, streamline processes, and improve productivity without compromising data privacy and security, both of which are of tantamount importance in the field of LegalTech.

4 Proposed System

Industry 5.0 lies at the intersection of resilient, sustainable, and human-centric applications. The usage of cloud solutions enables resilience and sustainability. An efficient application tailored to the needs of the customers can be built to make the product human-centric. A combination of these factors would result in a product that lives up to the standards of Industry 5.0 in this era of emerging technology (Fig. 1).

Our proposed model offers a promising path toward heightened efficiency, security, and productivity for small-scale legal practitioners. This project combines innovative ideas with practical implementation by integrating microservices, strong security measures, and cloud-based deployment. This project emerges as a comprehensive solution ready to reshape the market for cloud-based management systems for law firms by solving the inherent needs of legal tech corporate organizations, enhancing user confidence, and embracing the benefits of cloud technology.

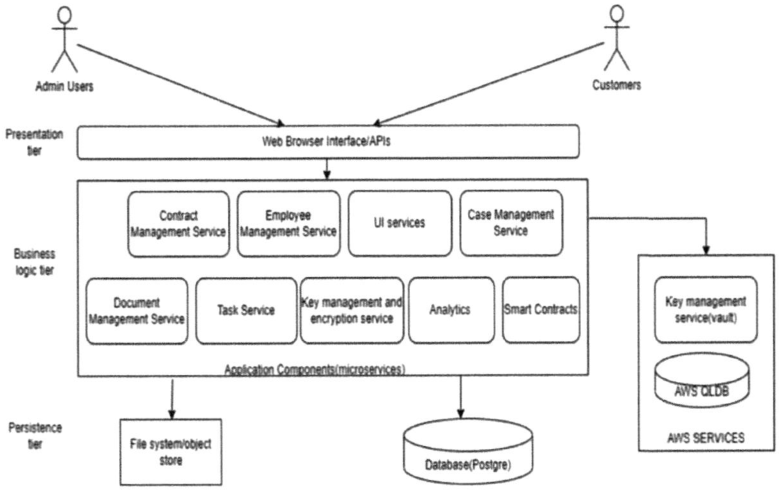

Fig. 1 Proposed system architecture

Traditional legal technology systems are often outdated, inflexible, and insecure, making them ill-equipped to handle the growing demands of modern legal practice. Our proposed system, on the other hand, is a groundbreaking cloud-based platform that is specifically designed for the needs of small-scale legal entities. It is scalable, secure, and easy to use, and it integrates essential microservices to streamline legal workflows. Our system also employs cutting-edge encryption algorithms to protect sensitive data, and it leverages the power of cloud computing to provide unparalleled flexibility and scalability.

5 Implementation

Underscoring the significance of remote work technology, the 2023 Future Ready Lawyer Survey Report revealed that 89% of lawyers consider technology that supports remote work to be crucial [6]. This growing preference for remote work necessitates a legal technology solution that empowers legal practitioners to work effectively from anywhere. Our proposed system stands at the forefront of this trend, providing a comprehensive and secure cloud-based solution tailored to the specific needs of legal firms.

Furthermore, the 2022 Future Ready Lawyer Survey Report indicated that 63% of lawyers anticipate increasing their investment in software to support legal work [7]. This growing investment highlights the growing recognition of technology as a key driver of efficiency and productivity in the legal industry. Our innovative system aligns with this trend, offering a novel approach to cloud-based legal technology that addresses the unique demands of modern legal practice.

The implementation procedure is divided into three separate production stages, each of which addresses crucial system components. The first layer, the persistence tier, deals with the essential problem of data storage. File storage will be used to store the data, and a PostgreSQL image downloaded via Docker will also be used.

The platform's administrative microservices are accommodated in the following business logic layer, which is crucial to the platform's smooth operation. These microservices enable data management and carry out critical responsibilities to guarantee the platform's reliable functioning. The layer is strengthened further by encryption algorithms that uphold core encryption principles while also improving data security.

The third tier, which serves as the system's interaction with its users, is the last presentation layer. Our cloud-based management system's tools and functions are accessible to legal practitioners through well-designed APIs and a web browser interface. The user experience is improved, and legal professionals are given the tools they must leverage technology to its full potential.

To host web applications on an Amazon EC2 instance in AWS, a virtual server through the AWS Management Console is created. Once the instance is running, it is connected remotely. All the necessary software components are set up, and the web application code is deployed. The domain and DNS settings are configured to point to the instance's IP address. The web application is secured by managing security groups, and firewall rules, and installing an SSL certificate for HTTPS. The instance's performance is monitored, and backup and data management practices are implemented. Load balancing and auto-scaling are used to maintain traffic. This approach provides flexibility, scalability, and control over the web hosting infrastructure, allowing the web applications to run with ease in the AWS cloud environment [8].

5.1 AES

The related databases are hosted on RDS. Amazon RDS (Relational Database Service) offers AES-256 encryption for securing data at rest. This encryption can be enabled when creating a new RDS instance or applied to an existing one. During setup, the AWS Key Management Service (KMS) or personalized KMS key can be used for managing encryption keys. Once enabled, RDS automatically encrypts database storage volumes and backups transparently. Applications accessing the database do so in its decrypted form without requiring manual decryption. Effective key management through AWS KMS ensures the security and compliance of your encrypted data. This feature adds a crucial layer of protection to the database infrastructure, safeguarding sensitive information.

5.2 QLDB

Amazon Quantum Ledger Database (QLDB) can be seamlessly integrated into a web application to provide an immutable and transparent ledger of data changes. A QLDB ledger is set up and data schemas that match the application's data model are defined. The QLDB driver can be integrated into the web application's backend code to record data changes as immutable journal entries in the ledger. QLDB's query capabilities are used to retrieve historical and current data and ensure data integrity through cryptographic verification features. QLDB's built-in audit capabilities are leveraged for tracking all changes to data and implementing access controls, and smart contracts are used for complex business logic. Monitoring, backup, and integration with other AWS services can further enhance the web application's capabilities, making QLDB a valuable asset for applications requiring secure, verifiable data histories.

6 Conclusion and Future Work

In conclusion, cloud-based management solutions have become crucial resources for law firms, providing a variety of advantages like improved accessibility, scalability, cost-effectiveness, and data security. These solutions allow legal professionals to cooperate easily, streamline their processes, and get crucial information at any time. The cloud-based solutions have made it possible for legal firms to remain competitive in an environment that is becoming more remote and digital. When adopting cloud-based management systems, legal companies must carefully assess their particular requirements, compliance needs, and security issues and select the strategy that most closely reflects their goals and duties. With the appropriate approach, cloud-based technologies can enable legal practitioners to improve client service and streamline their workflows.

Future work in cloud-based management systems for legal firms includes advancements in security measures, integration of AI and automation, improved collaboration tools, enhanced regulatory compliance features, data analytics and insights, mobile and remote access enhancements, improved interoperability, scalability, and cost-efficiency optimization, user experience enhancements, and comprehensive training programs. These developments aim to make cloud systems more secure, efficient, and user-friendly while meeting evolving legal industry needs and compliance standards.

References

1. Mendki P (2021) Securing cloud native applications using blockchain. In: 2021 12th international conference on information and communication systems (ICICS), Valencia, Spain, 2021, pp 419–423. https://doi.org/10.1109/ICICS52457.2021.9464583
2. Dharangan B, Praveen J, Selva Chandru M, Rajagopal S, Jegajothi B (2022) Secure cloud-based E-health system using advanced encryption standard. In: 2022 3rd international conference on electronics and sustainable communication systems (ICESC), Coimbatore, India, pp 642–646. https://doi.org/10.1109/ICESC54411.2022.9885501
3. Yu L, Li Y, Xia X (2008) Image encryption algorithm based on self-adaptive symmetrical-coupled toggle cellular automata. In: 2008 congress on image and signal processing, Sanya, China, pp 32–36. https://doi.org/10.1109/CISP.2008.48
4. Flittner M, Scheuermann JM, Bauer R (2017) ChainGuard: controller-independent verification of service function chaining in cloud computing. In: 2017 IEEE conference on network function virtualization and software defined networks (NFV-SDN)
5. Ananda Kumar KS, Metan J, Mali SS, Sirisha S, Sahana TR (2023) Ensuring the security of cloud based information using blockchain. In: 2023 international conference on data science and network security (ICDSNS)
6. Wolters Kluwer Legal & Regulatory (2023) Future ready lawyer survey report. Wolters Kluwer Legal & Regulatory, 25 Oct 2023 [Online]. Available: https://www.wolterskluwer.com/en/know/future-ready-lawyer-2023#download. Accessed: 30 Nov 2023

7. Wolters Kluwer Legal & Regulatory (2022) The Wolters Kluwer future ready lawyer. %Kluwer Legal & Regulatory, 1 Sept 2022 [Online]. Available: https://www.wolterskluwer.com/en/know/future-ready-lawyer-2022#download. Accessed: 30 Nov 2023
8. Soumya A, Nayak B, Dayanand D, Vaishnavi VB, Prasad V (2023) Literature review of approaches in cloud-based management systems for legal firms. In: Nanda SJ, Yadav RP (eds) Data science and intelligent computing techniques. SCRS, India, pp 593–609. https://doi.org/10.56155/978-81-955020-2-8-54

Amalgamation of Divergent Logs for Detection of Advanced Persistent Threats in Cyber Threat Analysis

Sandhya Addetla and Rajalakshmi Pachamuthu

Abstract Cyber world is penetrating all domains of technology, and it has become increasingly evident that digitization poses threat to security and stability of a country. The attacks launched by cyber criminals have compounded, with sophisticated and hard to detect attack techniques, challenging existing security measures. Advanced persistent threats (APTs) are nation-state attacks targeting critical information infrastructure systems. The threat posed by such attacks is catastrophic to enterprises. The dynamic and evolving nature of APTs makes them unique, and security solutions for their detection are required to be adaptive and dynamic to the changing nature of these attacks. This chapter analyzes APT attacks from different perspectives and presents recent methodologies that were proposed for detection of APTs. The concept study of amalgamation of signature-based logs and anomalous logs for effective APT detection is presented in this paper.

Keywords Advanced persistent threats · TTPs · Attribution · Amalgamation of logs · Deep learning methods

1 Introduction

Critical information infrastructure (CII) of countries is at risk from state-sponsored cyber threat actors, which lookout incessantly for opportunities to find vulnerable strategic systems. They induce attacks sneakily, upholding access and pilfering strategic data, causing a risk to national security. The increasing reliance on digital technology for automation has expanded threat platform. The main threat to CII is from advanced persistent threats (APTs), as their covert nature makes them extremely difficult to be perceived, posing a challenge to security specialists. The existing defensive mechanisms work with different detection techniques like signature-based

S. Addetla (✉) · R. Pachamuthu
Indian Institute of Technology, Hyderabad, Kandi, India
e-mail: AI20RESCH14001@iith.ac.in

R. Pachamuthu
e-mail: raji@ee.iith.ac.in

S. M. Thampi et al. (eds.), *Fifth International Conference on Computing and Network Communications*, Lecture Notes in Electrical Engineering 1219,
https://doi.org/10.1007/978-981-97-4540-1_35

and anomaly-based. APTs are advanced in terms of the attacker's capabilities. They know when to perform an attack, it could be leveraging an old or a new technique. APTs stay in target network for extended periods of time. They do not give traditional systems a scope to detect them as they are spread over a large time frame. Attackers are patient and wait for the right time when security watch is minimal to execute their next steps of attack. It poses a severe threat to strategic agencies which deal with facilities and data of national interest. Any compromise to such critical systems would lead to irrecoverable damage. Unlike many attacks that are single staged, APT attacks are multi-staged and have a life cycle as represented in Fig. 1. APTs are executed in six phases [1]: (i) **Reconnaissance**: attacker understands target network, its weaknesses, through passive monitoring and social engineering. (ii) **Initial compromise**: gains initial access to target network leveraging methods like spear-phishing and malware attachments. (iii) **Command and Control**: establishes connection with Command and Control server, forwards information about system configurations, etc., which help in planning next moves. (iv) **Lateral movement and privilege escalation**: moves laterally in network, finds vulnerable systems and escalates privileges. (v) **Identification of critical assets**: Through lateral movement and privilege escalation, attacker tries to identify critical assets in network having sensitive data. (iv) **Exfiltration of data**: after identification, data is exfiltrated by establishing connections to multiple levels of C&C servers, further, clear their traces in the network. The evolution of new technologies is opening new eras for attackers to utilize advanced attack patterns against target CII. ChatGPT is one such technology having potential in Cyber security field [2]. There are risks involved as ChatGPT can be used by attackers to automate planning and execution of attacks, making it easier to carryout sophisticated, coordinated attacks at a much faster pace. This would make it more difficult for organizations and individuals to defend against such attacks. Many such digital platforms of Large Language Models are being used by attackers to generate malware codes, enabling even individuals to carryout automated attacks. To improve upon defensive mechanisms, it is important to understand APTs and their detection technologies in vogue. This paper covered introduction to APTs in this section. Sections 2 and 3 present the attribution and detection techniques, respectively. Section 4 presents challenges in APT detection, Sect. 5 covers concept of amalgamation of logs for improved detection of APTs. Section 6 presents future work and conclusion.

2 Attribution

APTs are targeted attacks and there is always a motive behind such attack. Generally, governments also support such groups. When an APT attack occurs, the motive behind such attack, which could be political, strategic or economic, needs to be identified. The process of identifying which organization/person/group or nation is behind such attack and what could be the motive of such attack is called attribution. Analyzing the modus operandi of attack helps gaining insights into the Tactics, Tech-

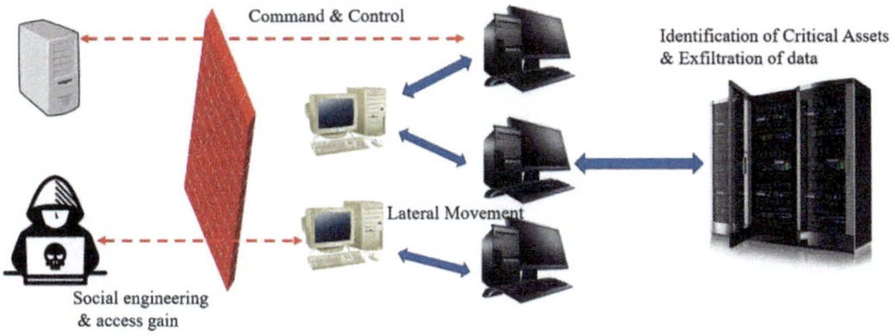

Fig. 1 Phases of advanced persistent threats

niques and Procedures (TTPs) to enable attributing attack to a nation or group. There are more than 150 APT groups identified globally [3]. Security firms like Kaspersky, Crowdstrike, Mandiant and FireEye carry out a thorough analysis of such attacks to identify the groups involved in them. These APT groups are given numbers (APT1, APT23, etc.) as well as names (e.g., 'Fancy Bear' indicates a Russian group). For attribution, security firms initially collect logs and Indicators of Compromise (IoCs) from different devices of victim network. TTPs leveraged by attackers may be repetitive for a group. Standard frameworks aid security analysts in the characterization of IoCs [4]. Cyber threat actors prefer using proven attack methods than try a completely new method that may fail.

APT groups even use ransomware TTPs as cover to hide more insidious activities [5]. A ransomware attack in April 2022 on Oil India Limited, Assam, has created huge financial losses for the company [6]. Hackers have attacked critical energy infrastructure of Tata Power in October 2022 by exploiting a vulnerability in Boa web server [7]. Information collected was released on dark Web forum. Because of evolution in TTPs of APT groups, continuous evolution of attribution techniques is need of the hour. Rosenberg et al. in [8] used Deep Neural Networks for attribution of nation-state malware. Naveen et al. in [9] have proposed DL architecture for attribution based on reports obtained from various sources of Cyber Threat Intelligence. Pajouh et al. have proposed clustering model based on multi-view fuzzy consensus for attribution of APT malware [10].

Different models exist for attribution leveraging IoCs, one of the models used is diamond model [11]. This model has four components—capabilities, infrastructure, adversary and victim—identified for attacks (with two axes—technical and socio-political axis). With the IoCs, information of capabilities (TTPs used) and infrastructure can be obtained (technical axis). The other two components are adversary and victim (socio-political axis) which enable attribution to a particular nation, group, or organization.

Another model for attribution is 4C model—collect, cluster, charge, communicate. Figure 2 shows different levels of attribution. By collecting all evidences/IoCs and clustering them into different intrusion sets, technical aspect of attribution can be identified. Moving further, charge phase requires using some hypothesis to conclude on the person, organization or nation. In communication phase, even though details of

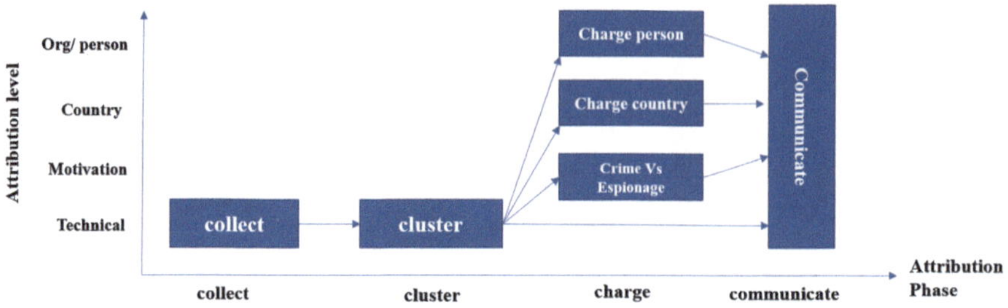

Fig. 2 4C model [11]

initiator of attack are known, full information may not be revealed for various socio-political reasons [11]. Another framework which has been widely used for attribution is MICTIC framework. This considers not only technical aspects, but other details like cyber espionage and sabotage. The aspects of this framework are Malware (M), Infrastructure (I), Control Servers (C), Telemetry (T), Intelligence (I) and Cui bono (C). An extension to the MICTIC framework is presented by Brandao in [12]. It considers social engineering and internal leaks as extension to MICTIC framework. Both these aspects consider the non-technical components of attacks, like trying to find out who all of the firms are in contact with attacker, knowing if attacker have collaborated with any employees. Such information will aide in better attribution of attackers. FireEye came up with ATOMIC model for attribution. This model uses the method of document clustering. Applying this to structured documents of each group, similarities among the groups can be evaluated, ultimately giving a holistic similarity measure. TF-IDF method brings out the unique malware. Another method of cosine similarity can also be used for evaluating the APT groups [13].

3 Detection of APTs

Considering plethora of data to be analyzed for detection of APTs, techniques leveraging machine learning are essential for design of efficient security solutions. This section puts forward miscellany of emerging solutions proposed by researchers in the last few years. Summary of research works is provided in Table 1.

Game theory approaches are in vogue for APT detection. Huang et al. in [14] propose multi-stage game using iterative algorithm for computation of Bayesian Nash Equilibrium which predicts policies of attackers and defenders providing proactive defense-in-depth strategies. Rass et al. in [15] explore usage of various classes of matrix games as defense tools against APTs handling uncertainties of system state, adversarial incentives and disagreement of experts in risk assessments. Kumar et al. in [16] provide a comprehensive view of models, objectives and features of game theory solutions, decision-making with optimal resource utilization along with APT behavior. Moothedath et al. in [17] brought out that APT has certain data-flow and

Table 1 Detection methodologies for advanced persistent threats

Category of models	Author and research work	Methodology used
Game theory	Huang et al. [14]	Leverage iterative algorithm to compute Bayesian Nash equilibrium for prediction
	Rass et al. [15]	Leverage various classes of matrix games as defense tools
	Kumar et al. [16]	Comprehensive view of models, objectives and features of game theory solutions
	Moothedath et al. [17]	Dynamic Information Flow Tracking (DIFT)
	Zhu et al. [18]	Multi-agent deep reinforcement learning (MADRL) model enables creation of strategies impromtu
Supervised learning models	Laurenza et al. [19]	Malware Triage through one-class classifier
	Alrehaili et al. [20]	Hybrid model leveraging SAE-LSTM and CNN-LSTM
	Li [21]	Extended kill chain model considering internal attacks as well. Leverages LSTM-based RNN and RRCF
	Shang et al. [22]	Explore the concept of different malware sharing same features. Leverage LSTM, CNN and Gradient boost
	Xuan et al. [23]	Combination of deep learning architecture and attention network. CNN-LSTM and ATTENTION network are leveraged
	Cho et al. [24]	Combination of MLP, CNN and LSTM
Unsupervised and semi-supervised	Hofer-Schmitz et al. [25]	Pearson correlation and Local Outlier Factor method are leveraged
	Zimba et al. [26]	Semi-supervised learning models—Shared Nearest Neighbor (SNN) and K-Nearest Neighbors (KNN) considered
	Abdullayeva [1]	Autoencoder model and softmax regression
	Alsanad et al. [27]	Propose clustering algorithm—SVM with radial basis function for APT detection
Graph-based techniques	Anjum et al. [28]	Provenance graph-based supervised APT detection framework
	Han et al. [29]	Provenance graph for anomaly detection
	Cho et al. [30]	Leverage deep learning graph network to analyzing behavior profile of a malware
	Ayoade et al. [31]	Anomaly detection leveraging provenance graph and online metric learning DL method
Deception-based techniques	Oluoha et al. [32]	Detailed survey on the deception-based cyber security defenses for network threats including APTs
	Baksi et al. [33]	HMM-based APT defense wherein Indicators of Compromise are used as features for detection
	Amin et al. [34]	Leverages HMM. Has two phases which include predicting attack path and estimating required decoy nodes in the predicted path
	Zhan et al. [35]	Hardening of FTP service as defense against APTs

control-flow commands during system interaction. With Dynamic flow tracking, information flow can be tracked. This works by tagging sensitive information as suspicious, tracking all tagged flows and analyzing these traps. Interaction of APT and Dynamic Information Flow Tracking (DIFT) is modeled as a nonzero-sum multistage game of two players. Solution of this game is an optimal defense policy which maximizes the probability of APT detection. Each stage of APT is considered as a stage of game. Zhu et al. in [18] leverage novel sampling approach on a multi-agent deep reinforcement learning (MADRL) method. This is a multi-defender, multi-attacker game model in which defenders can create strategies impromptu and allow other defenders to know about it, which enable faster and efficient learning strategies to defenders. Researchers have developed many supervised learning models for detection of APTs. Laurenza et al. in [19] leverage Malware Triage (analysis) through one-class (Isolation Forest) classifier, over their earlier method of Random Forest Classifier. The performance improved over their earlier model, as re-training the model requires only one-class inclusion. Alrehaili et al. in [20] use hybrid model with SAE-LSTM for feature extraction and time correlation and CNN-LSTM for capturing spatial and temporal relations. Anomaly detection during testing is achieved by using separate data files for each stage of attack. Performance is found to be better than individual deep learning methods. Li in [21] has presented an extended Kill chain model in which internal attacks are also considered. LSTM-based RNN is used for correlation of events in time, RRCF anomaly detection algorithm is used for detection of anomalies. Shang et al. proposal in [22] is based on sharing same features of malware by different attack groups. Flow level and packet level statistics are generated for all network flows. Further, feature extraction (LSTM, CNN) and classification (Gradient boost) are done for detection of APTs. Xuan et al. in [23] propose an efficient APT detection model which is combination of deep learning (DL) architecture with CNN-LSTM and attention network. In this method, before classification, data is analyzed by ATTENTION network. Output of which is then classified as APT or non-APT. Cho et al. in [24] propose a method utilizing combination of DL models (MLP, CNN and LSTM) which perform main task of extracting IP features from network flow and classifying these IPs as APT IP or normal IP.

Researchers are using unsupervised and semi-supervised learning models for APT detection. Hofer-Schmitz et al. in [25] used feature set selection by Pearson correlation for obtaining improved detection capability. Best feature set was selected with a Local Outlier Factor method. Impact of selection of feature set on detection performance was evaluated. Zimba et al. in [26] have proposed a semi-supervised learning model using Shared Nearest Neighbor (SNN) and K-Nearest Neighbors (KNN) considering the complex network characteristics. Abdullayeva in [1] proposed an autoencoder model that learns features and leverages softmax regression to classify data as multiclass data. Here, autoencoder also transforms data into feature set, reducing the volume of data. Alsanad et al. in [27] demonstrated use of clustering algorithms for detection of APTs. On the basis of their evaluation, SVM with radial basis function gave the best results in efficiency and accuracy with limited labeled data.

Table 2 Datasets created for APTs

Author and research work	Dataset created
Myneni et al. [36]	DAPT2020
Coulter et al. [37]	APT-Executable
Medenou et al. [38]	Customized dataset containing all phases of APTs
AL-Saraireh et al. [39]	Dataset based on TTPs and IoCs of APTs at all stages
Liu et al. [40]	SCVIC-APT-2021

Graph-based techniques are also in vogue for APT detection. Anjum et al. in [28] present Provenance graph-based supervised APT detection framework. They capture causality by leveraging a system provenance graph and explanation for predictions given by BNN Model, which achieved high detection performance. Han et al. in [29] use provenance graph for anomaly detection, performance of which evolves by understanding long-term system behavior. Cho et al. in [30] propose a model to identify APT on a system by analyzing behavior profile of a malware using a DL graph network. The model performs two tasks—building malware behavior profiles and detecting malware using graphs based on profiles. Ayoade et al. in [31] propose model for anomaly detection which finds execution traces of host systems and extracts features through provenance graph. Uses online metric learning DL method to minimize distance between similar classes (attacks) and maximize distance between dissimilar classes (malicious and benign traffic). Works for single stage attack detection. There are also deception-based cyber security defense methodologies, few of which are [32–35].

The persistent problem of availability of datasets has become a major obstacle for researchers to prove their concepts. Few researchers created datasets, as detailed in Table 2, that contain APT attack patterns that researchers can leverage worldwide to verify performance of their frameworks/algorithms. Myneni et al. in [36] present a custom dataset DAPT2020 created by simulating APT behavior at all stages in a cloud network for 5 days. Coulter et al. in [37] present APT-Executable (APT-EXE) dataset that can adapt malware samples (to retrieve system-driven access and usage structure and footprints of functionality interaction) to a target domain, enabling improved detection of APTs. Medenou et al. in [38] have created a customized dataset depicting all phases of APT kill chain to support military operations. Al-Saraireh et al. in [39] built an APT dataset by conducting attacks at all stages of APTs, based on Tactics, Techniques and Procedures (TTPs) and Indicators of Compromise (IoCs) of APTs. Liu et al. in [40] introduce a dataset, SCVIC-APT-2021, which includes attack patterns in all stages of APT. Different attack methodologies were leveraged to create attacks pertaining to each stage.

Though copious research works propose solutions for APT detection, there are certain challenges in development and implementation of these solutions.

4 Challenges of APT Detection

Machine learning/deep learning techniques for advanced security solutions face many challenges as listed below to effectively prove in real-world security applications:

- **Datasets**. There are very limited publicly available datasets which depict complete APT attack, and it is also very imbalanced dataset as the percentage of APT traffic in normal traffic is very limited.
- **Evolving Tactics, Techniques and Procedures (TTPs)**. Attackers may try to change their methodology with each of their attacks. It is important to stay updated on latest TTPs used by these attackers and improve model accordingly for effective detection of APTs.
- **Usage of Zero-Day Attacks**. Leveraging zero-day attacks in attack chain makes APTs extremely unlikely to be detected, as traditional security solutions cannot identify zero-day attacks because they rely on databases of recognized threats.
- **Extended Time Frame**. APTs are carried out over a very long time frame, even for years together. Correlation of logs extended over such a long time frame is a very challenging task. The current dataset available does not depict such long time frames of attack time-lines.
- **Performance in Real-Time Systems**. Though the ML/DL models for detection of APTs perform well in lab setup, in real-time scenarios, rate of false positives and false negatives is high compared to test bed results. Both false positives and false negatives severely affect functionality of network security tools in protecting critical information infrastructure.

5 Amalgamation of Divergent Logs for APT Detection

Various detection techniques explored by researchers recently have been listed in Sect. 3. However, detection techniques are not limited to the above. Approaches for detection of advanced persistent threats encounter some of the challenges listed in Sect. 4. A lot of research is ongoing to improve the performance of existing methods and to come up with new algorithms and techniques. Solutions for arresting APT attacks require adapting to dynamic changes in threat landscape. Advanced persistent threats are attacks carried out by various strategically motivated threat actors, which, when executed, may cause severe impact on the National Security. The difficulty in detecting APTs is due to the fact that they operate in a covert way and also spread over a long period of time. APT attacks are executed in a phased manner. Such features of APTs make them extremely difficult to be perceived and render the best security solutions ineffective. One category of solutions considers APTs utilizing the known attack patterns. These solutions work on datasets that are labeled. Most open datasets available can be utilized for this type of models. The solutions focus mainly on one stage of the APT attack which is C&C stage or data exfiltration stage, which

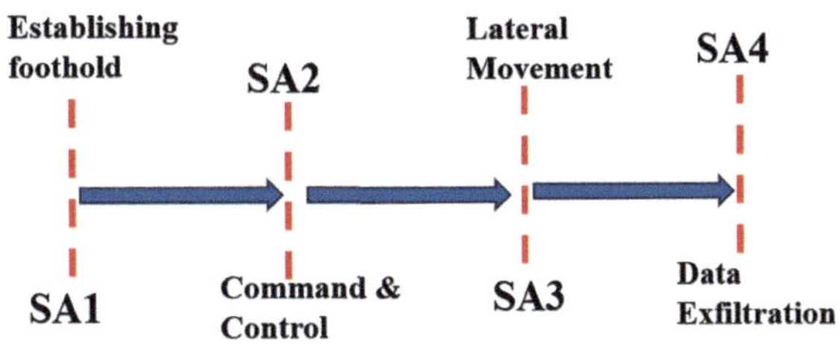

Fig. 3 APT attack chain of known attacks

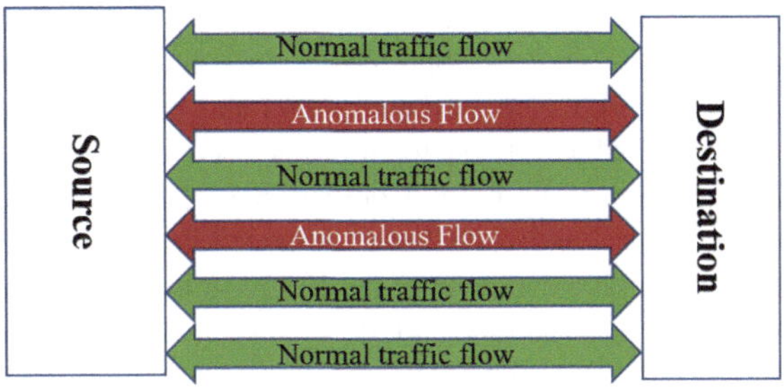

Fig. 4 Attack identification from unknown attacks

may use known attack patterns. They presume that attack signatures are available for different stages of the APT attack chain. As shown in Fig. 3 at all stages of APTs (establishing foothold, lateral movement, command and control, data exfiltration), it is assumed that signature-based attacks are executed. SA1, SA2, SA3, SA4 are individual signature-based attacks used in different stages of APT attack. In such methodologies, Graph-based techniques, tagging or time series analysis are used for identification of attack chain that could be a potential APT attack. The entire path, including individual attacks, forms an APT attack chain. The linkage of events/logs in this attack chain is obtained by analyzing network traffic or network logs. Further, established frameworks support labeling of the attack patterns based on their TTPs, which aid in identification of potential APTs. However, these solutions do not work in real-time scenario for detection of complete APT attack, and they may not use all signature-based attacks in the complete attack chain.

The other category of solutions is based on anomalous traffic present in the entire network traffic flow. These solutions focus on methodologies to detect anomalies in normal traffic flow. For this, normal traffic pattern of the network is to be understood and certain parameters like flow duration, no. of bytes transmitted, date and time of transmission, etc., would be analyzed against set threshold values to identify anomalies. The PCAP files or traffic flow are used as datasets in this category of solutions. They rely on clustering methods to detect potential APTs. The normal traffic flow and anomalies are shown in Fig. 4.

These anomalies would be considered to be part of a potential APT by verifying against threshold values of a few parameters like flow duration, no. of bytes transmitted, date and time of transmission, etc., Various ML/DL algorithms are used for this analysis. The anomalies are considered for building security solutions to identify attacks like zero-days. Zero-day exploits are used to execute cyber attacks by various threat actors, causing irrecoverable damages to organizations, including complete shutdown of facilities [41]. APTs may leverage zero-day attacks as well as known attacks to exploit the target and extract required information. Something that makes zero-day exploits even more dangerous is that advanced cyber criminal groups use them strategically. Threat actors save these zero-day exploits to be leveraged in compromising high-value targets. This aids in increased duration of exploiting the target network, as the vulnerability present in the system has not been discovered [42]. The presence of vulnerabilities makes critical infrastructure a facile target for cyber attacks. Few of the attacks on CII are Colonial pipeline attack in the USA; supply chain attacks on Solar Winds and attacks on power systems and atomic power plants in India. Imported equipment and technology are leveraged in many systems, and knowledge of these technologies acts as a key to compromising critical systems [43]. Solutions of anomaly detection consider only traffic patterns which are outliers or anomalies and zero-day attacks which are not detectable by traditional security solutions. However, this type of solution works well for detection of single stage of APT and zero-day attacks in specific. A comprehensive solution which could identify the complete APT attack chain (which may include both signature-based and anomalous attack patterns) is not feasible with such techniques. The ML/DL solutions which are based on only known patterns or only unknown patterns do not suffice for detection of complex and long duration attacks like APTs. There is a need to design and develop solutions which leverage both known and unknown attack patterns simultaneously in identifying the APT Kill Chain. It is evident that an attacker may use known vulnerabilities to exploit a strategic network. Similarly, they may also use unknown vulnerabilities to exploit the same system in any other phase of the entire APT life cycle. Unless a solution is conceived which considers known attack patterns in conjunction with unknown patterns, the difficulty in detecting APTs will prevail. In this paper, the concept of amalgamation of divergent logs is proposed, which will help in design and development of efficient cyber security solutions for detection of advanced persistent threats. To realize the concept of amalgamation of logs, the following steps need to be followed:

1. Install anomaly detection security module in a network.
2. Install signature-based security detection system like Firewall, IPS, SIEM in the same network.
3. Gather logs from both the security solutions.
4. Identify common features from different types of logs.
5. Merge the log files based on time to generate an integrated file containing coalition of both anomalous- and signature-based logs.

Attackers inducing APTs may use well-known attack patterns and unknown attack patterns/zero-day attacks as part of APT attack chain to exfiltrate data from target network. Hence, relying on defensive solutions that are based on only signatures or only anomalous traffic would not meet the requirement of detection of potential ATP attacks. There are security solutions which provide both these types of detection mechanisms, but an analysis based on combination of these logs is not available for potential threat detection. As described in the steps to be followed and as shown

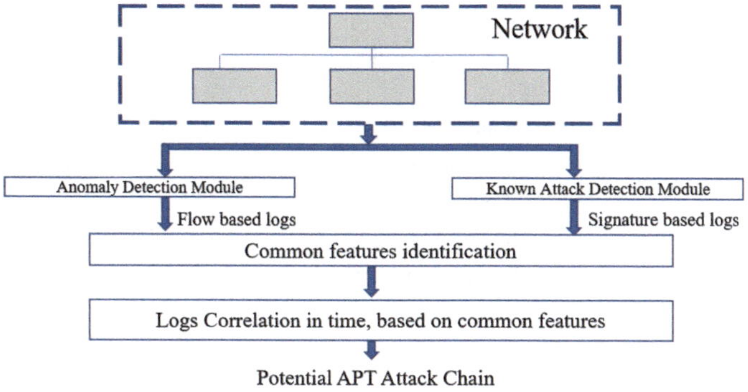

Fig. 5 Amalgamation of divergent security logs

in Fig. 5, any network needs to be integrated with security solutions that include signature-based (comparison with known IOCs or attack signatures) and anomalous traffic detection (alert on suspicious behavior) mechanisms. It is to be noted that with such combined security solutions, we will be able to capture the logs from internal network as well as external network as the security solutions installed would be available to gather logs from all devices inside the network and also security solutions used as perimeter defense enables to capture logs of all ingress and egress traffic. When considering combined logs, since these logs will include internal traffic (from the systems and hosts within the network) as well as external traffic logs (network traffic monitored) in network, important phases of APTs like access gain, command and control, lateral movement, data exfiltration can be captured through these logs and it will also minimize the possibility of non-identification of potential APTs due to presence of diverse attack patterns.

For combining logs, common features in both types of logs need to be analyzed. Signature-based logs have around main 17 features (*Time, Source IP Address, Destination IP Address, Device Name, Destination MAC address, Source MAC address, Message, Fragment, TCP, Action Taken, Deviceoutboundinterface, Deviceinbound-interface, Protocol, Destination IP Address, Source IP Address, Device Host, Device External ID, etc.*). Anomalous logs have around 36 main features (*IP Protocol, IP Protocol Version, TCP Flags, Application ID, Application Name, Application Protocol, Broken Flow, Direction, Destination IP Address, Destination IP Address Locality, Destination IP Address City, Destination IP Address Country, Latitude and Longitude of Destination, Destination MAC address, Destination Port Number, Destination Port Name, Flow End Time, Flow Locality, Flow Start Time, In Bytes, In Packets, Ingest Time stamp, Out Bytes, Out Packets, Source IP Address, Source IP Address Locality, Source IP Address City, Source IP Address Country, Latitude and Longitude of Source, Source MAC Address, Source MAC Name, Source Port Number, Source Port Name, Total Number of Bytes, Duration of Flow, Total Number of Packets, etc.*). The features which are available in both types of logs and represent the same parameters are considered as common features. The significant common

Table 3 Identification of common features

Anomalous logs features	Signature-based logs features	Fields considered for correlation
Source IP	Source IP	Source IP
Destination IP	Destination IP	Destination IP
Protocol	Protocol	Protocol
Source MAC address	Source MAC address	Source MAC address
Destination MAC address	Destination MAC address	Destination MAC address
Start time stamp	Time stamp	Start time stamp
End time stamp	–	End time stamp

features in both types of logs which can be considered for correlation are listed in Table 3. Anomalous logs are flow-based and hence have start and end time for the flow, however, signature-based logs have only time stamp of occurrence of event. This time stamp of signature-based logs only is considered as start and end time stamp to enable comparison and to maintain uniformity in correlated logs.

5.1 Correlation of Logs

The two types of logs (anomalous- and signature-based) are merged based on time stamp. Once the integrated logs are obtained, next step would be to analyze these logs for presence of potential APTs based on the common features. The main crux of this framework in Fig. 5 is to arrive at a potential APT chain based on correlation among all listed logs. For every record of integrated logs, if destination IP of any previous record is same as source IP of the current record, it indicates a continuity of attack and is considered to be part of one attack chain, a new attack chain is created. Further search is carried out to see if destination IP of the current record is the source IP of any future record. If so, it is further appended to the APT chain. This procedure is repeated for every record in the integrated log, to obtain the list of potential APT attack chains. The algorithm for verifying all these logs for potential APTs is given at Algorithm 1. The output of this algorithm, if implemented, would be list of all potential APT chains present in the dataset. This requires parallel processing to compare simultaneously every record in the log, with every other record, which is ahead in time, for presence of APT chain and is also computationally intensive, as the number of records in log files to be compared is very high. A code in Python has been written for the same to test the performance. However, it is pertinent to mention that, the challenge in testing this code is the availability of dataset which contains both the signature-based logs and anomalous logs. Also, it is difficult to generate a combined log from the existing datasets as different datasets would be having different IP addresses and different time stamps. Hence, generating the dataset

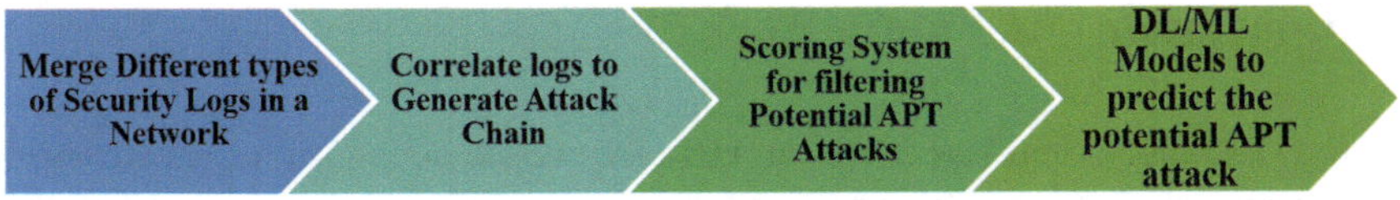

Fig. 6 APT detection with merged logs

in simulated environment is being explored to test the presented concept. Arriving at detection of potential APT attacks can be achieved by further applying effective ML/DL methods on the attack chain as shown in Fig. 6. After attack chains have been obtained from correlated logs, it is required to filter these attack chains further by providing a score to the attack chain. Each individual attack in the attack chain would be assigned a score. Signature-based attacks can be assigned score based on threat intelligence frameworks such as MITRE ATT&CK. Expert knowledge of cyber security analysts is required to assign scores to anomalous logs in attack chain. Each type of anomalous log is assigned score by security analysts based on severity of its impact. Threat rating of the entire attack chain is computed based on the individual event/attack score. The cumulative threat rating can be leveraged to identify potential APT attacks. Linked attacks which meet the threshold criteria need to be further passed on to ML/DL model for detection of potential APT attacks and alerting the cyber security team in real-time. In summary, presence of anomaly detection and signature-based security solutions in a network is quintessential and automated correlation of both types of logs through suggested framework would provide efficient solution for APT Detection.

Algorithm 1: Algorithm for correlation of divergent logs

```
df ← csvLogFile /* Integrated Log File */
N ← Total No. of Rows /* Total No. of records in the log file */
for i ← 1 to N − 1 do
    j ← i + 1
    APT ← df[i]
    while j > i and j ≤ N − 1 do
        if df[i].destIP = df[j].srcIP then
            /* Compare the Destination IP of prev record to the
               Source IP of the Current record */
            APT.append(df[j]) /* Add to the current APT Chain if the
               values matched */
            i ← j
            j ← j + 1
        end
        else
            j ← j + 1
        end
    end
end
```

6 Conclusion and Future Work

Tactics, Techniques and Procedures of APTs are ever-changing. Attackers are leveraging new techniques and trying to exploit unknown vulnerabilities. Existing solutions for detection of APTs have been elaborated in this paper, however, enhanced security strategies for detection and mitigation of complete APT attack chain, which leverage known and unknown attack patterns, adapting to dynamic changes of the APTs are required. Combined analysis of anomalies and signature-based logs will improve APT detection capabilities. Correlation of flow-based and packet-based logs is a challenge and needs to be considered by researchers for developing functional security solutions for APT detection, as individual solutions may not be effective. The concept of amalgamation of divergent logs is presented in this paper. To take forward this research, dataset needs to be generated and efficient algorithms are to be realized.

References

1. Abdullayeva F (2021) Advanced persistent threat attack detection method in cloud computing based on autoencoder and softmax regression algorithm. Array 10:100067
2. Bruce Schneier C (2022) Obligatory ChatGPT post. https://www.schneier.com/cryptogram/archives/2022/1215.html#cg22
3. Deloitte (2022) Advanced persistent threat: latest developments, potential impact and recommendations. https://www2.deloitte.com/ch/en/pages/risk/articles/advanced-persistent-threat.html
4. Parmar M, Domingo A (2019) On the use of cyber threat intelligence (CTI) in support of developing the commander's understanding of the adversary. In: MILCOM 2019—2019 IEEE military communications conference (MILCOM), pp 1–6
5. IDG Communications, Inc. (2023) APT groups use ransomware TTPs as cover for intelligence gathering and sabotage. https://www.csoonline.com/article/3686580/apt-groups-use-ransomware-ttps-as-cover-for-intelligence-gathering-and-sabotage.html
6. Print T (2022) Biggest cyberattack in recent years' hits Oil India HQ, hackers demand Rs 60 crore in Bitcoin. https://theprint.in/india/biggest-cyberattack-in-recent-years-hits-oil-india-hq-hackers-demand-rs-60-crore-in-bitcoin/914792/
7. Business Standard (2023) Hackers exploited discontinued web server at Tata Power: Microsoft. https://www.business-standard.com/article/companies/hackers-exploited-discontinued-web-server-at-tata-power-microsoft-122112400290_1.html
8. Rosenberg I, Sicard G, David E (2018) End-to-end deep neural networks and transfer learning for automatic analysis of nation-state malware. Entropy 20:390. https://doi.org/10.3390/e20050390
9. Naveen S, Puzis R, Angappan K (2020) Deep learning for threat actor attribution from threat reports. In: 2020 4th international conference on computer, communication and signal processing (ICCCSP), pp 1–6
10. Pajouh H, Azmoodeh A, Dehghantanha A, Parizi R (2020) MVFCC: a multi-view fuzzy consensus clustering model for malware threat attribution. IEEE Access 8:139188–139198. https://doi.org/10.1109/ACCESS.2020.3012907
11. Steffens T (2020) Attribution of advanced persistent threats: how to identify the actors behind cyber-Espionage. Springer Berlin Heidelberg. https://books.google.co.in/books?id=6FryDwAAQBAJ

12. Brandao P (2021) Advanced persistent threats (APT)-attribution-MICTIC framework extension. J Comput Sci 17:470–479
13. Mandiant (2019) Going ATOMIC: clustering and associating attacker activity at scale. https://www.mandiant.com/resources/clustering-and-associating-attacker-activity-at-scale
14. Huang L, Zhu Q (2020) A dynamic games approach to proactive defense strategies against advanced persistent threats in cyber-physical systems. Comput Secur 89:101660
15. Rass S, König S, Schauer S (2017) Defending against advanced persistent threats using game-theory. Publ Libr Sci (PLoS)
16. Kumar R, Singh S, Kela R (2022) Analyzing advanced persistent threats using game theory: a critical literature review
17. Moothedath S, Sahabandu D, Allen J, Clark A, Bushnell L, Lee W, Poovendran R (2020) A game-theoretic approach for dynamic information flow tracking to detect multistage advanced persistent threats. IEEE Trans Autom Control 65:5248–5263
18. Zhu T, Ye D, Cheng Z, Zhou W, Yu P (2023) Learning games for defending advanced persistent threats in cyber systems. IEEE Trans Syst Man Cybern Syst 53:2410–2422
19. Laurenza G, Lazzeretti R, Mazzotti L (2018) Malware triage for early identification of advanced persistent threat activities. CoRR. abs/1810.07321. http://arxiv.org/abs/1810.07321
20. Alrehaili M, Alshamrani A, Eshmawi A (2021) A hybrid deep learning approach for advanced persistent threat attack detection. In: The 5th international conference on future networks & distributed systems
21. Li J (2022) Detection of advanced persistent threat based on kill chain node mapping and LSTM. https://doi.org/10.1145/3501409.3501565
22. Shang L, Guo D, Ji Y, Li Q (2021) Discovering unknown advanced persistent threat using shared features mined by neural networks. Comput Netw 189:107937
23. Do Xuan C, Duong D (2022) Optimization of APT attack detection based on a model combining ATTENTION and deep learning. J Intell Fuzzy Syst 42. https://doi.org/10.3233/JIFS-212570
24. Cho D, Dao M (2021) A novel approach for APT attack detection based on combined deep learning model. Neural Comput Appl 33
25. Hofer-Schmitz K, Kleb U, Stojanovic B (2021) The influences of feature sets on the detection of advanced persistent threats. Electronics 10:704
26. Zimba A, Chen H, Wang Z, Chishimba M (2020) Modeling and detection of the multi-stages of advanced persistent threats attacks based on semi-supervised learning and complex networks characteristics. Future Gener Comput Syst 106:501–517
27. Alsanad A, Altuwaijri S (2022) Advanced persistent threat attack detection using clustering algorithms. https://doi.org/10.14569/IJACSA.2022.0130976
28. Anjum M, Iqbal S, Hamelin B (2021) ANUBIS: a provenance graph-based framework for advanced persistent threat detection
29. Han X, Pasquier T, Bates A, Mickens J, Seltzer M (2020) Unicorn: runtime provenance-based detector for advanced persistent threats. In: Proceedings 2020 network and distributed system security symposium
30. Cho D, Huong D (2022) A new approach for APT malware detection based on deep graph network for endpoint systems. Appl Intell 52:1–20
31. Ayoade G, Akbar K, Sahoo P, Gao Y, Agarwal A, Jee K, Khan L, Singhal A (2020) Evolving advanced persistent threat detection using provenance graph and metric learning. In: 2020 IEEE conference on communications and network security (CNS), pp 1–9
32. Oluoha O, Yange T, Okereke G, Bakpo F (2021) Cutting edge trends in deception based intrusion detection systems—a survey. J Inf Secur 12:250–269. https://doi.org/10.4236/jis.2021.124014
33. Baksi RP, Upadhyaya SJ (2020) Decepticon: a hidden Markov model approach to counter advanced persistent threats. Commun Comput Inf Sci 1186. https://doi.org/10.1007/978-981-15-3817-9_3
34. Amin MARA, Shetty S, Njilla L, Tosh DK, Kamhoua C. Hidden Markov model and cyber deception for the prevention of adversarial lateral movement. https://doi.org/10.1109/ACCESS.2021.3069105

35. Shu Z, Yan G (2018) Ensuring deception consistency for FTP services hardened against advanced persistent threats. In: Proceedings of the 5th ACM workshop on moving target defense
36. Myneni S, Chowdhary A, Sabur A, Sengupta S, Agrawal G, Huang D, Kang M (2020) DAPT 2020—constructing a benchmark dataset for advanced persistent threats
37. Coulter R, Zhang J, Pan L, Xiang Y (2021) Domain adaptation for windows advanced persistent threat detection. Comput Secur 112:102496
38. Medenou R, Mayo V, Balufo M, Castrillo M, Garrido F, Martinez A, Catalán D, Hu A, Rodriguez-Bermejo D, Vidal J, De Riquelme G, Berardi A, De Santis P, Torelli F, Sanchez S (2020) CYSAS-S3: a novel dataset for validating cyber situational awareness related tools for supporting military operations. In: Proceedings of the 15th international conference on availability, reliability and security. https://doi.org/10.1145/3407023.3409222
39. Al-Saraireh J, Masarweh A (2022) A novel approach for detecting advanced persistent threats. Egypt Inform J 23:45–55
40. Liu J, Shen Y, Simsek M, Kantarci B, Mouftah H, Bagheri M, Djukic P (2022) A new realistic benchmark for advanced persistent threats in network traffic. IEEE Netw Lett 4:162–166
41. Information Security (2021) HHS cyber security programme. https://www.hhs.gov/sites/default/files/zero-day-attacks-tlpwhite.pdf
42. Cynet (2022) Zero-day attacks, exploits, and vulnerabilities: a complete guide. https://www.cynet.com/zero-day-attacks/zero-day-vulnerabilities-exploits-and-attacks-a-complete-glossary/
43. Ramanathan RS, Panchal B, Saikrishna BVS (2022) Protection of national critical information infrastructure. Vivekananda International Foundations. https://www.vifindia.org/sites/default/files/Protection-of-National-Critical-Information-Infrastructure.pdf

Performance in Credit Risk Analysis Using Statistical and Classification Models

Yashika Jaiswal, Sudeep Sahu, Ziaurrahman, Shreya Baldua, Sudhanshu Gonge, Rahul Joshi, and Ketan Kotecha

Abstract The study of credit risk is a major concern for financial companies looking to make wise lending decisions and limit potential losses. In this study, the use of R, a potent open-source programming language, is examined in the context of credit risk analysis. This study offers a thorough framework to improve the accuracy and efficiency of credit risk assessment by utilizing the flexible data manipulation, statistical analysis, and machine learning capabilities of R. The importance of credit risk analysis in financial institutions is discussed in the paper's opening section, along with some of the difficulties it faces. The rich libraries, data processing capability, and data visualization features of R highlight how well-suited it is for this purpose. Data quality and consistency are stressed in the technique portion since it encompasses data collection, preprocessing, and feature engineering. To predict credit risk, a variety of statistical methods and machine learning models are used, which offers details on their benefits and interpretability. The research study also looks at model validation and evaluation, which ensures the stability and dependability of the credit risk models. Model accuracy, precision, and recall are evaluated

Y. Jaiswal · S. Sahu · Ziaurrahman · S. Baldua · S. Gonge (✉) · R. Joshi · K. Kotecha
Department of Computer Science and Engineering, Symbiosis Institute of Technology, Pune, Maharashtra, India
e-mail: sudhanshu.gonge@sitpune.edu.in

Y. Jaiswal
e-mail: yashika.jaiswal.btech2021@sitpune.edu.in

S. Sahu
e-mail: sudeep.sahu.btech2021@sitpune.edu.in

Ziaurrahman
e-mail: ziaurrahman.ziaurrahman.btech2021@sitpune.edu.in

S. Baldua
e-mail: shreya.baldua.btech2021@sitpune.edu.in

R. Joshi
e-mail: rahulj@sitpune.edu.in

K. Kotecha
e-mail: head@scaai.siu.edu.in

S. M. Thampi et al. (eds.), *Fifth International Conference on Computing and Network Communications*, Lecture Notes in Electrical Engineering 1219,
https://doi.org/10.1007/978-981-97-4540-1_36

using methods including exploratory data analysis (EDA), ROC analysis, and model performance indicators. This study concludes by highlighting how the use of R in credit risk analysis might enable financial institutions to make more knowledgeable lending decisions, lowering financial risks and promoting the stability of the financial sector. The purpose of the article is to offer a thorough methodology that financial institutions can use when performing credit risk analysis. This entails data gathering, preprocessing, feature engineering, statistical analysis, choosing a machine learning model, and model assessment. The goal of the paper is to provide practitioners with a detailed manual for implementing R-based data-driven credit risk analysis.

Keywords R-programming · Data processing · Financial institutions · Statistical analysis · Lending decisions · Model validation · Data-driven

1 Introduction

For well-informed lending decisions and risk reduction in the ever-changing financial environment, accurate credit risk analysis is essential [1]. This article investigates the field of credit risk assessment and provides a thorough framework that makes use of R, an open-source programming language renowned for its data manipulation, statistical analysis, and machine learning capabilities. Credit risk is the likelihood that a borrower would default on a loan, causing a financial loss. Credit risk essentially refers to the possibility that a lender will not get the principal and interest that is owed, which would disrupt cash flows and raise collection expenses. By examining a borrower's creditworthiness variables, such as their present debt burden and income, lenders can reduce credit risk [2]. Even if it is impossible to predict who will miss payments, credit risk can be effectively assessed and managed to decrease the impact of a loss. Lenders and investors can rank and have a brief idea on the uncertainty of not being repaid back by a particular client and can help with the decision-making of interest rate and eventually debt obligation issuers [3]. The importance of credit risk analysis within financial institutions and its difficulties are highlighted in the opening section. R is a good choice for these problems due to its powerful libraries and data processing abilities. High-quality data is essential for efficient credit risk analysis. The methodology builds a solid foundation by putting an emphasis on data gathering, preprocessing, and feature engineering. Credit risk is predicted using a variety of statistical and machine learning models, each of which provides a different set of insights [4, 5]. The research examines these models in depth and assesses their performance using parameters like accuracy and precision [6]. For credit risk models to remain stable, model validation and review are essential. Model performance indicators, ROC analysis, and exploratory data analysis are crucial [7, 8]. By allowing for better informed lending decisions, lowering financial risks, and boosting overall financial stability [9, 10], R has the potential to revolutionize credit risk analysis. This article intends to offer practitioners a thorough overview of data-driven credit risk analysis, from data processing to model evaluation.

A review of the literature will be conducted in Sect. 2 of this paper. An extensive and critical discussion on classification algorithms is followed by a presentation of earlier credit risk research. Section 3 contains the research methodology. Section 3.1, methodology, follows and continues the paper by describing various credit risk analysis approaches. We will talk about the outcomes in Sect. 3.3 based on the data analysis. The key conclusions, limits, and suggestions for additional research are all included in Sect. 4, along with a summary of the findings.

2 Literature Review

For financial organizations looking to make wise lending decisions while limiting possible losses, credit risk analysis is an essential concern [11]. The accuracy and effectiveness of credit risk assessment have been greatly improved by the incorporation of data-driven techniques and open-source programming languages. In keeping with the study in the paper, this literature review examines significant developments in this field.

To forward the objectives of the lenders, credit risk analysis seeks to assume an appropriate degree of risk while lending. These objectives may include things like business expansion, profitability, and qualitative elements. The management creates the rules that motivate the company to succeed [12]. Credit risk analysis has changed dramatically because of the use of open-source programming languages, particularly R. The sophisticated data manipulation, statistical analysis, and machine learning capabilities of R have become popular in the financial industry, as stated in the abstract and introduction of the study. The importance of R in credit risk analysis has been shown in several research. This model uses three classification algorithms, namely logistic regression, decision tree, and random forest. The models were implemented and verified using R-Studio. The best algorithm random forest was selected based on performance, flexibility, robustness, and interpretability [13].

2.1 Logistic Regression

Logistic regression, a modified variant of ordinary regression, is applied when the independent (input) variables are continuous, categorical, or both [14], and the dependent variable is binary (that is, it accepts only two values, typically signifying the occurrence or non-occurrence of some outcome event).

The model works on

Logistic Regression

$$\log\left(\frac{p}{1-p}\right) = \beta_0 + \beta_1 X_1 + \beta_2 X_2 + \cdots + \beta_k X_k \tag{1}$$

Here, p is the probability of the target variable being 1, and X_1, X_2, ..., X_k are the independent variables (predictors). $\beta_0, \beta_1, \beta_2, \ldots, \beta_k$ are known as the regression coefficients, which have to be estimated from the data. Logistic regression estimates the probability of a certain event occurring [15]. Credit risk analysis frequently uses the well-known and comprehensible algorithm of logistic regression. It works well when there is a predominantly linear relationship between the features and the target variable (such as creditworthiness) [16]. A logistic function is used in this approach to estimate the likelihood of a binary result, which often indicates good or unfavorable credit risk [17]. The influence of each attribute on the probability of a favorable or unfavorable credit risk outcome is shown by the logistic regression model's coefficients. This interpretability is important because it enables financial firms to understand the elements that go into a credit decision. The ease of use of logistic regression is one of its advantages. When the correlations between the variables are well understood, it is an excellent choice because it does not require complicated adjustment [18]. It might not, however, be able to capture complex or nonlinear correlations in the data.

2.2 *Decision Trees*

To analyze credit risk, decision trees provide a more adaptable method. Complex, nonlinear, and interactive interactions between characteristics and the goal variable can be accommodated by them [19]. In a decision tree, the dataset is divided into subsets based on the values of the features. This creates a hierarchy of decisions, which results in a classification at the leaf nodes [20, 21]. Because of this layout, decision trees are very easy to understand because the route from the root to a leaf node explains the decision-making process [15, 22]. As it offers clarity and insights into how a certain credit risk assessment was made, the interpretability of decision trees is a desirable quality. Decision trees can, however, be prone to overfitting [22] the training data, which leaves them less generalizable to novel, unforeseen data. To reduce this risk, proper regularization and pruning are required [23].

2.3 *Random Forest*

Decision trees serve as the cornerstone of the ensemble learning technique known as random forests. They are made up of a number of separate decision trees, each of which was trained using a different subset of the data [24] and has the option of taking into account a different subset of features. The overfitting risk connected to individual decision trees is decreased by the ensemble technique. The versatility of random forests allows them to handle both linear and nonlinear relationships in the data. They are excellent at identifying intricate patterns and interactions between characteristics and the target variable. This feature makes them ideal for credit risk assessments, where a variety of financial and nonfinancial elements may affect a person's or an entity's creditworthiness. The estimation of feature importance that random forests offer also enables analysts to pinpoint which factors are more important in the credit evaluation process. Random forests are a compromise between interpretability and performance, albeit they are not as easily interpreted as a single decision tree.

2.4 ROC Analysis

In order to evaluate the performance of binary classification models, such as those used to determine creditworthiness (e.g., good or bad credit risk), good or bad credit risk receiver operating characteristic analysis (ROC) [25] is a strong and well-known technique in the field of credit risk analysis and other areas. The best trade-off between sensitivity and specificity can be made using ROC analysis, which enables you to assess and visualize how well a model can discriminate between positive and negative classifications. Financial firms can choose the optimum trade-off between spotting potential positive and bad credit risks with the aid of ROC analysis in credit risk analysis. It is a useful tool for evaluating and optimizing the performance of credit scoring algorithms, ensuring that lending decisions are well-informed and sensible financially. In conclusion, the choice of the algorithm to be used in credit risk analysis is based on the complexity of the relationships within the data and the value of interpretability [26]. For simpler, linear relationships, logistic regression is appropriate. Random forests are a good option for many credit risk analysis situations because they offer a compromise between accuracy and interpretability while decision trees offer transparency but may need to be pruned carefully.

3 Methodology

The methodology used in this study, which makes use of the open-source programming language R, is intended to provide a thorough foundation for data-driven credit risk analysis. From data collection and preprocessing to model selection, validation, and evaluation, the research approach includes several crucial steps. The main goals are to increase financial sector stability, reduce financial risks, and give financial institutions the information they need to make wise lending decisions. The method are explained in the following flowcharts (Figs. 1, 2, and 3).

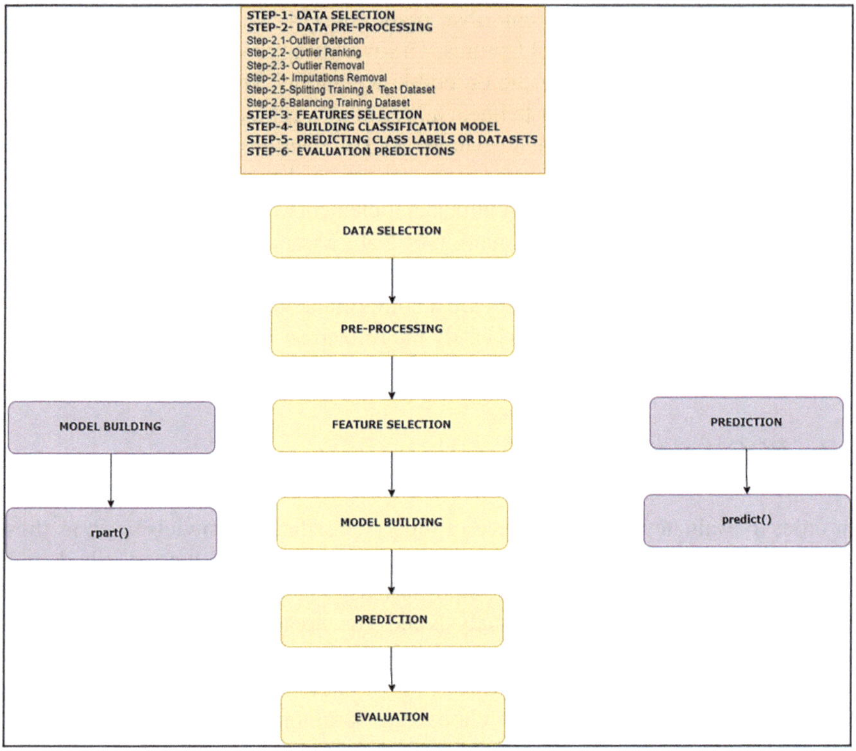

Fig. 1 Flowchart of credit risk analysis using R

3.1 Data Collection

Data collecting precedes credit risk analysis. The R environment is loaded with the dataset that was used in this investigation. It is anticipated that the data source will contain essential details including the credit limit, the credit history, and the applicants' personal data. The read.csv function is used to extract data from a specified dataset, as shown in the code. The dataset normally includes data on the amount and length of credit as well as the applicants' personal information. The process of data reading from CSV file using R-language is shown in following code (Fig. 4).

The dataset contains a variety of information, such as the amount and length of credit as well as the applicants' personal information for loans. Since the accuracy of later stages depends on the dataset's integrity, it is crucial that it be of the highest caliber and be as complete as possible. By laying the foundations for the entire credit risk analysis process and giving the information needed to make wise lending decisions, this phase creates the foundation.

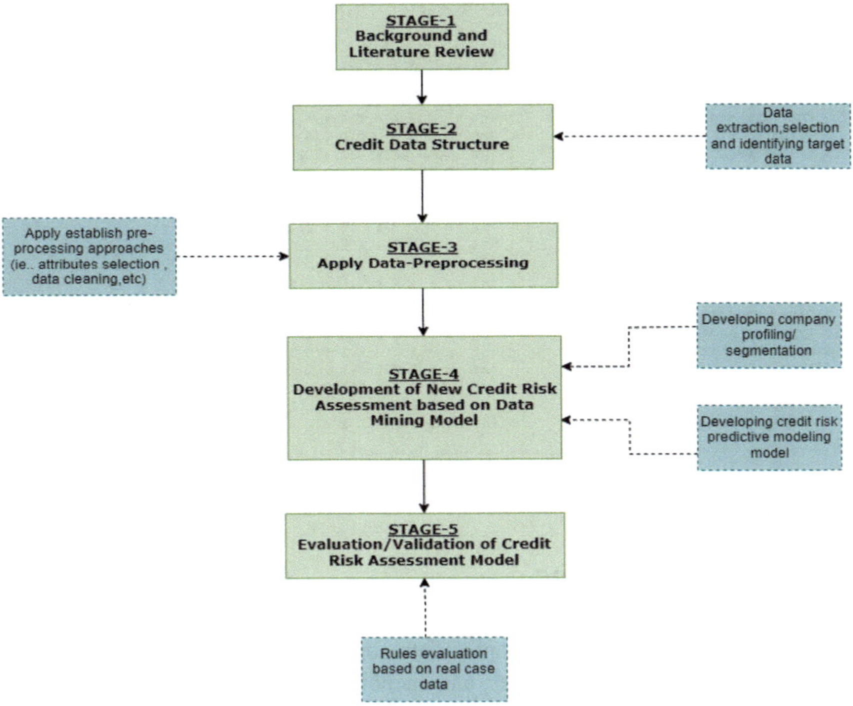

Fig. 2 Flowchart of credit risk analysis steps

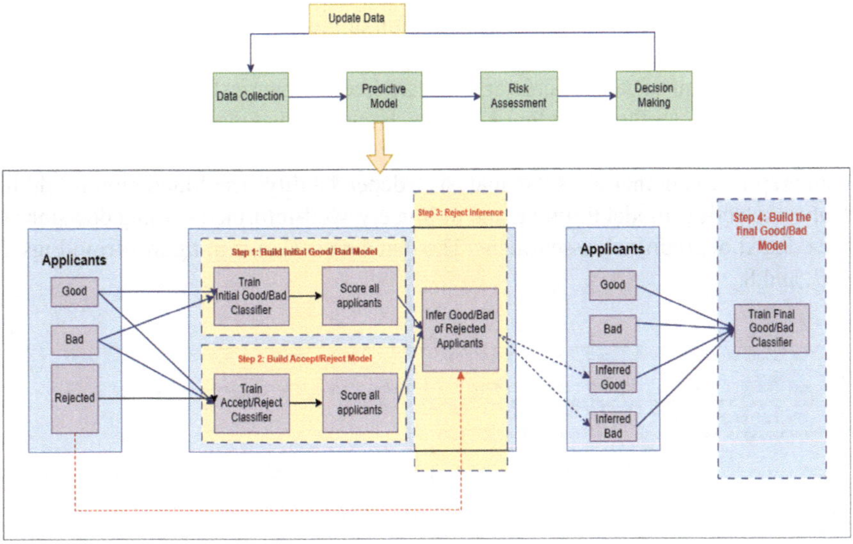

Fig. 3 Architecture of credit risk analysis

```
> credit <- read.csv('credit.csv', header = TRUE)
> str(credit)
'data.frame':	1000 obs. of  21 variables:
 $ Creditability                    : int  1 1 1 1 1 1 1 1 1 1 ...
 $ Status.of.exsiting.checking.amount: int  1 1 2 1 1 1 1 1 4 2 ...
 $ Duration.of.Credit..month.       : int  18 9 12 12 12 10 8 6 18 24 ...
 $ Credit.History                   : int  4 4 2 4 4 4 4 4 4 2 ...
 $ Purpose                          : int  2 0 9 0 0 0 0 0 3 3 ...
 $ Credit.Amount                    : int  1049 2799 841 2122 2171 2241 3398 1361 1098 3758 ...
 $ Savings.account.bonds            : int  1 1 2 1 1 1 1 1 1 3 ...
 $ Present.employement.since        : int  2 3 4 3 3 2 4 2 1 1 ...
 $ Installment.percent              : int  4 2 2 3 4 1 1 2 4 1 ...
 $ Sex...Marital.Status             : int  2 3 2 3 3 3 3 3 2 2 ...
 $ Guarantors                       : int  1 1 1 1 1 1 1 1 1 1 ...
 $ Present.Residence.since          : int  4 2 4 2 4 3 4 4 4 4 ...
 $ Property                         : int  2 1 1 1 2 1 1 1 3 4 ...
 $ Age..years.                      : int  21 36 23 39 38 48 39 40 65 23 ...
 $ Other.installment.plans          : int  3 3 3 3 1 3 3 3 3 3 ...
 $ Type.of.apartment                : int  1 1 1 1 2 1 2 2 2 1 ...
 $ No.of.Credits.at.this.Bank       : int  1 2 1 2 2 2 2 1 2 1 ...
 $ Occupation                       : int  3 3 2 2 2 2 2 2 1 1 ...
 $ No.of.dependents                 : int  1 2 1 2 1 2 1 2 1 1 ...
 $ Telephone                        : int  1 1 1 1 1 1 1 1 1 1 ...
 $ Foreign.Worker                   : int  1 1 1 2 2 2 2 2 1 1 ...
> 
```

Fig. 4 Importing and printing the imported dataset

3.2 Data Preprocessing

Data preparation is done to guarantee data consistency and quality. Dealing with outliers, addressing missing values, and encoding categorical variables are all part of this phase. Notably, the length of the credit, the amount of the credit, and the age are discretized into categories to improve their usefulness in the analysis. You can see how the cut function is used to discretize continuous variables into categories in the provided code, such as "duration of credit". For further modeling, this phase simplifies and improves the data. To work with various statistical and machine learning techniques, categorical variables are frequently transformed into numerical representation. One-hot encoding is one such method that may be used.

By correcting data anomalies and laying the groundwork for insightful inquiry, data preprocessing increases the analysis's dependability. The foundation is laid for feature selection, model training, and ultimately, well-informed lending decisions in the context of credit risk assessment. The outcome is shown in figures from Figs. 5, 6, 7, and 8.

```
> credit$Duration.of.Credit..month. <- cut(credit$Duration.of.Credit..month., c(0,12,18,24,Inf), labels = c(1:
4))
> credit$Credit.Amount <- cut(credit$Credit.Amount, c(0,1000,5000,10000,Inf), labels = c(1:4))
> credit$Age..years. <- cut(credit$Age..years., c(18,25,40,60,Inf), labels = c(1:4))
> head(credit[,c(3,6,14)],5)
  Duration.of.Credit..month. Credit.Amount Age..years.
1                          2             2           1
2                          1             2           2
3                          1             1           1
4                          1             2           2
5                          1             2           2
> 
```

Fig. 5 Printing the columns converted to factor variables

```
> for(i in 1:21) credit[, i] <- as.factor(credit[, i])
> null_values <- sapply(credit, function(x) sum(is.na(x)))
> print(null_values)
          Creditability Status.of.exsiting.checking.amount     Duration.of.Credit..month.
                      0                                  0                              0
         Credit.History                            Purpose                  Credit.Amount
                      0                                  0                              0
  Savings.account.bonds          Present.employement.since          Installment.percent
                      0                                  0                              0
    Sex...Marital.Status                        Guarantors       Present.Residence.since
                      0                                  0                              0
               Property                         Age..years.        Other.installment.plans
                      0                                  0                              0
      Type.of.apartment              No.of.Credits.at.this.Bank                 Occupation
                      0                                  0                              0
       No.of.dependents                          Telephone                 Foreign.Worker
                      0                                  0                              0
>
```

Fig. 6 Printing the columns with null values

Fig. 7 Printing the outliers

```
> print(outliers)
factor()
Levels: 1 2 3 4
>
```

Fig. 8 Boxplot of the credit amount

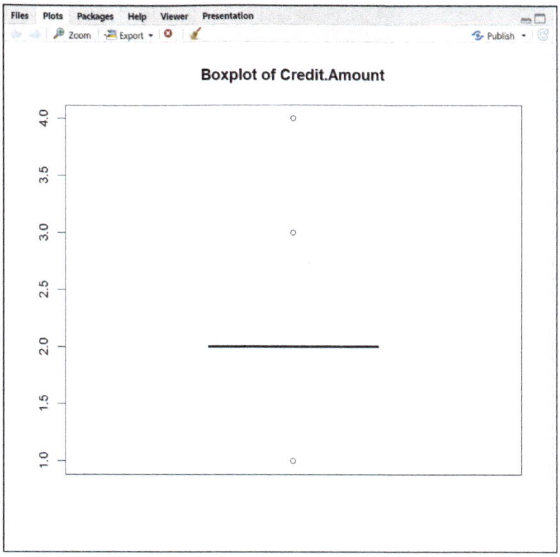

3.3 *Exploratory Data Analysis*

A crucial stage in comprehending the features of the dataset is exploratory data analysis. As seen in the code for the research paper, visualizations and summary statistics are produced to show data patterns and trends. Insights into the relationships between different credit risk categories and their distribution are provided by this phase. As shown in the code, methods like ggplot2 are used to build visualizations

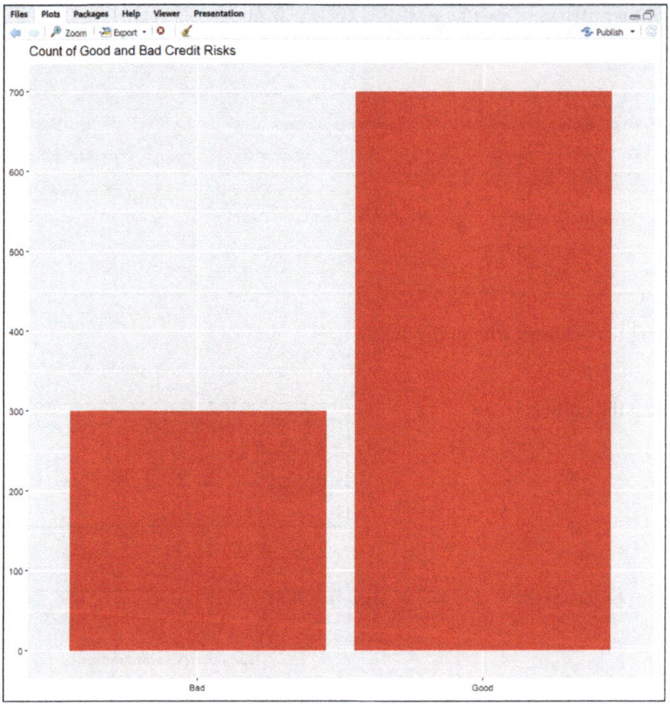

Fig. 9 Bar chart for count of bad and good credit risk

and summary statistics that indicate data patterns and trends. To summarize the key characteristics of the dataset, EDA also entails producing summary statistics, such as measures of central tendency, variance, and quartiles. With the use of these statistics, it is possible to spot data outliers or abnormalities that may need to be investigated further during the preprocessing stage. The exploratory data analysis is shown in Figs. 9, 10, 11, and 12 related to dataset.

Plot 1

The first figure, which was produced by the code, shows a bar chart showing the distribution of the "bad" and "good" credit risk categories. The message that this plot communicates is explained as follows:

Y-Axis (count): The credit risk count is represented by the y-axis. It denotes the number of occurrences connected to each credit risk category.
X-Axis (creditability): The "bad" and "good" labels on the x-axis, which measures creditability, indicate different credit risk categories.
Bars: The "bad" and "good" credit risks are represented by two bars, respectively. Each bar's height reflects the number of occurrences inside its relevant category.
Color: The bars have different hues for "bad" and "good" credit risks, respectively. The visual distinction between the categories is aided using color.

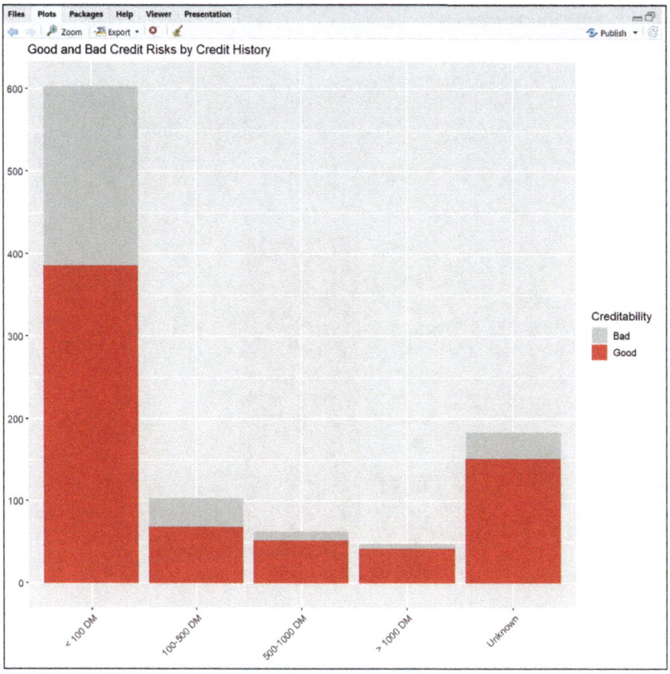

Fig. 10 Bar chart for bad and good credit risk by credit history

Title: The narrative is given the title "Count of Good and Bad Credit Risks," making it obvious what it is trying to accomplish.

Plot 2

Y-Axis (count): The number of credit risk occurrences is shown on the y-axis. It displays the number of candidates in each category.

X-Axis (saving account/bonds): The "savings account/bonds" variable, which indicates the applicants' financial history in relation to savings, is displayed on the x-axis in various subcategories. The following subcategories are available: "100 DM," "100–500 DM," "500–1000 DM," " > 1000 DM," and "Unknown."

Bars: For each subcategory, the chart shows two bars. One bar shows the number of "bad" credit risks (applicants judged to be dangerous), and the other shows the number of "good" credit risks (applicants judged to be creditworthy).

Color: "Bad" credit risks are depicted on the bars in one color, while "good" credit risks are depicted on the other. Visually distinguishing between the two credit risk groups is made easier using color.

Title: The title of the plot, "Good and Bad Credit Risks by Credit History," gives the visual depiction a clear context.

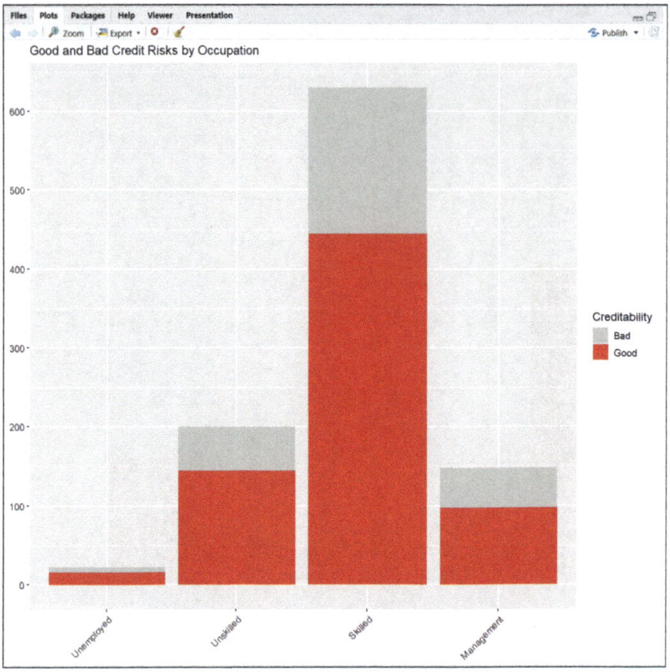

Fig. 11 Bar chart for bad and good credit risk by occupation

Plot 3

Y-Axis (count): The number of applicants falling into each group is represented on the y-axis by the count of credit risk instances.
X-Axis (occupation): The "occupation" variable's several subcategories, which indicate the various occupations of the applicants, are represented on the x-axis. The following subcategories are available: "unemployed," "unskilled," "skilled," and "management."
Bars: For each occupational subcategory, the chart shows two bars. Both bars show the number of "good" credit risks, or applicants who are thought to be creditworthy, and the number of "bad" credit risks, or applicants who are thought to be dangerous.
Color: The bars are color-coded, indicating "bad" credit risks with one color and "good" credit risks with a different color. This color differentiation makes it easier to distinguish between the two credit risk groups visually.
Title: The plot's title, "Good and Bad Credit Risks by Occupation," provides background information for the graphic.

Plot 4

Y-Axis (count): The number of applicants who fall into each category is displayed on the y-axis, which reflects the count of credit risk situations.

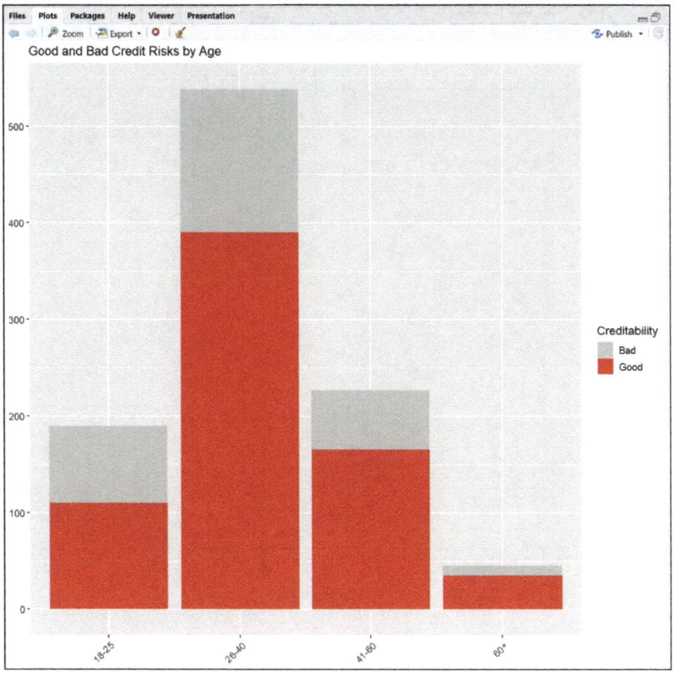

Fig. 12 Bar chart for bad and good credit risk by age

X-Axis (age groups): The x-axis shows the various age categories into which the candidates are divided. They are "18–25," "26–40," "41–60," and "60+."
Bars: Two bars, one for each age range, are shown. Both bars display the number of "good" credit risks, or applicants who are thought to be creditworthy, in comparison with the number of "bad" credit risks, or applicants who are thought to be dangerous.
Color: The bars have a color-coded system that shows "bad" credit risks in one color and "good" credit risks in a different color. The visual distinction between the two credit risk groups is aided using color.
Title: The narrative, "Good and Bad Credit Risks by Age," provides context for the graphic portrayal.

3.4 Logistic Regression

For the examination of credit risk, one of the most basic statistical modeling methods is logistic regression. You explained how to use logistic regression in your research report, and therefore, here is a more thorough description of this stage:

Building a logistic regression model in R requires the use of the GLM function (generalized linear model). You can use this function to specify the model's equation,

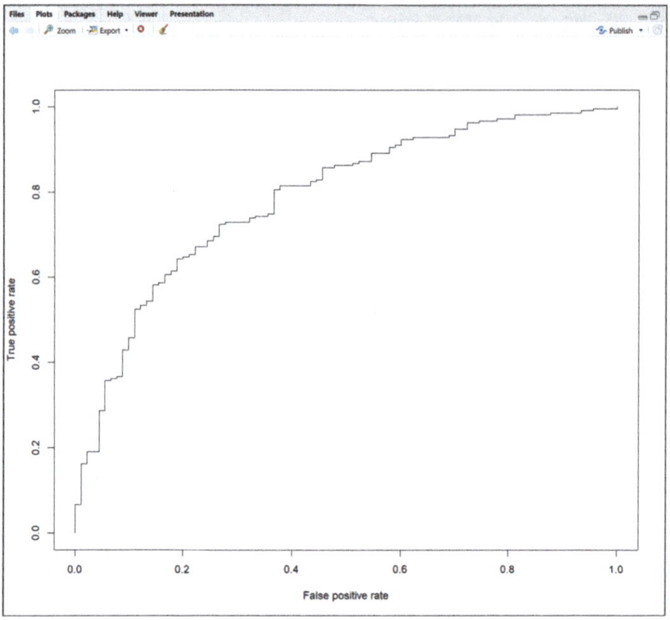

Fig. 13 Graph showing logistic regression (true positive rate vs. false positive rate)

which includes the predictors (features) you wish to utilize for the prediction as well as the target variable (creditability). Make predictions based on the testing dataset by using the predict function. These forecasts show how likely it is that a potential borrower will be rated as either a "good" or "bad" credit risk. Utilize the right measures to gauge the logistic regression model's effectiveness. You have used your code to show ROC analysis and to determine the area under the curve (AUC) as a model performance metric (Fig. 13).

For each predictor, logistic regression models offer comprehensible coefficients. The impact of each trait on the chance of having a "good" or "bad" credit risk is shown by these coefficients. Greater influences are suggested by coefficients with higher absolute values.

3.5 Decision Tree

The openness and simplicity of decision trees make them useful. It is obvious how each applicant's characteristics affect how they are classified as a credit risk because they constitute a sequence of binary decisions that result in a classification. Overfitting is a possibility, though. R's rpart function can be used to build a decision tree model. Provide details about the predictor variables you plan to utilize, as well as the goal variable (creditability). A tree-like structure is produced by the decision tree

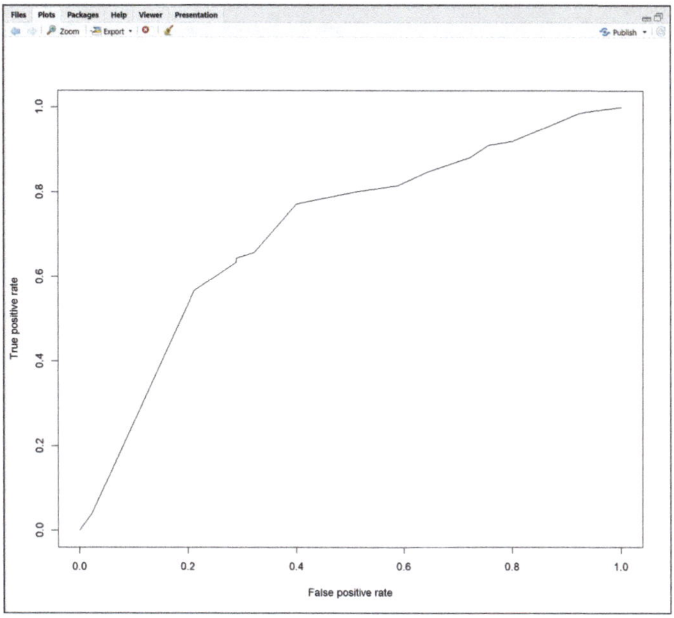

Fig. 14 Graph showing decision tree (true positive rate vs. false positive rate)

method by repeatedly dividing the data into subsets according to the chosen features. To comprehend the decision tree's structure and the standards that were applied, visualize it. You have implemented a useful visualization in your code by utilizing the fancyRpartPlot method (Figs. 14 and 15).

3.6 Random Forest

Comparatively speaking to individual decision trees, random forests offer better accuracy and resilience. The overfitting is decreased, and a more reliable forecast is given by combining many trees. As a result, financial institutions are better able to comprehend the elements that affect credit risk. They also reveal the significance of features.

Utilize R's random forest function to build a random forest model. Name the predictor variables and the target variable (creditability). A more reliable ensemble model is produced by random forest by combining several decision trees. To provide probability scores for each applicant's credit risk category, use the trained random forest model to generate predictions on the testing dataset. The feature importance scores provided by random forest aid in determining the factors that have the most impact on the evaluation of credit risk. Utilizing the varImpPlot function, you can see these scores (Figs. 16 and 17).

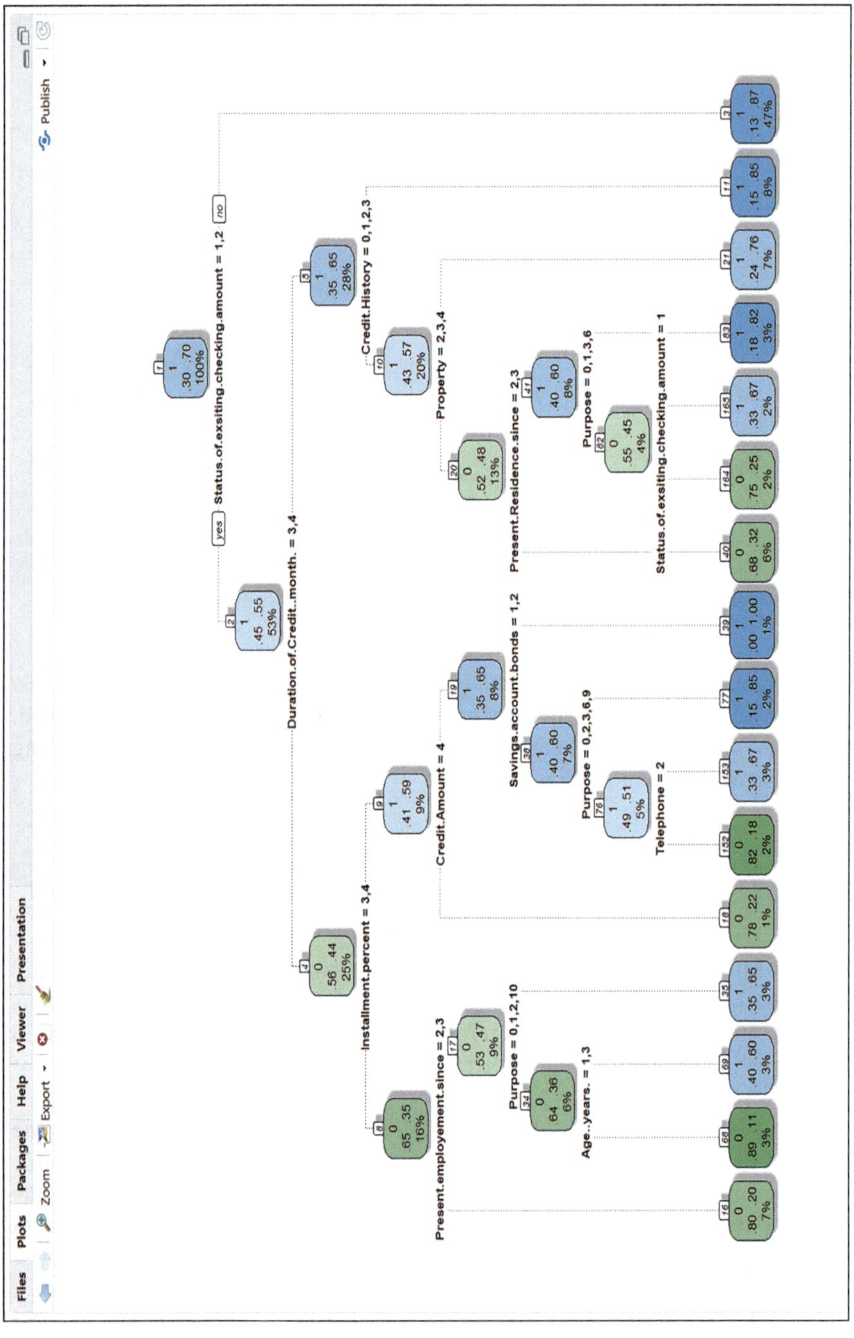

Fig. 15 Decision tree implementation for predicting credit risk analysis

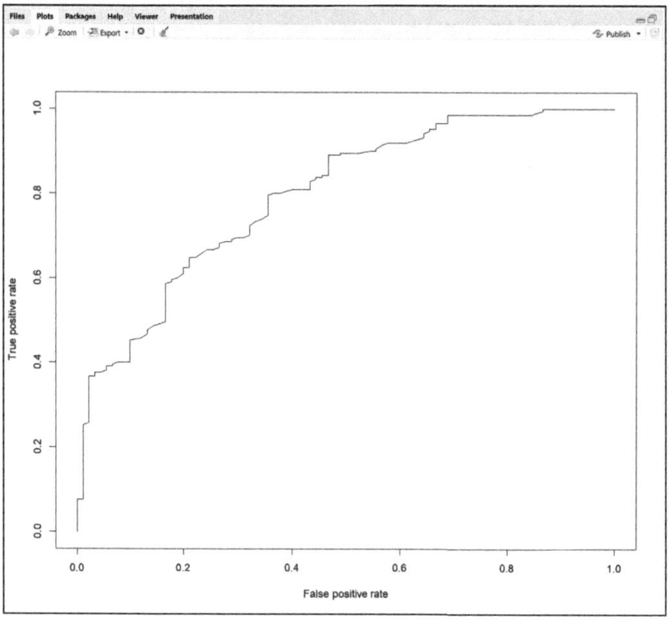

Fig. 16 Graph showing random forest (true positive rate vs. false positive rate)

3.7 *Confusion Matrix*

In R, generating a confusion matrix often entails evaluating the performance of a classification model, such as logistic regression or a machine learning technique. Accuracy, precision, recall, and other performance indicators of the model are assessed using the confusion matrix (Fig. 18).

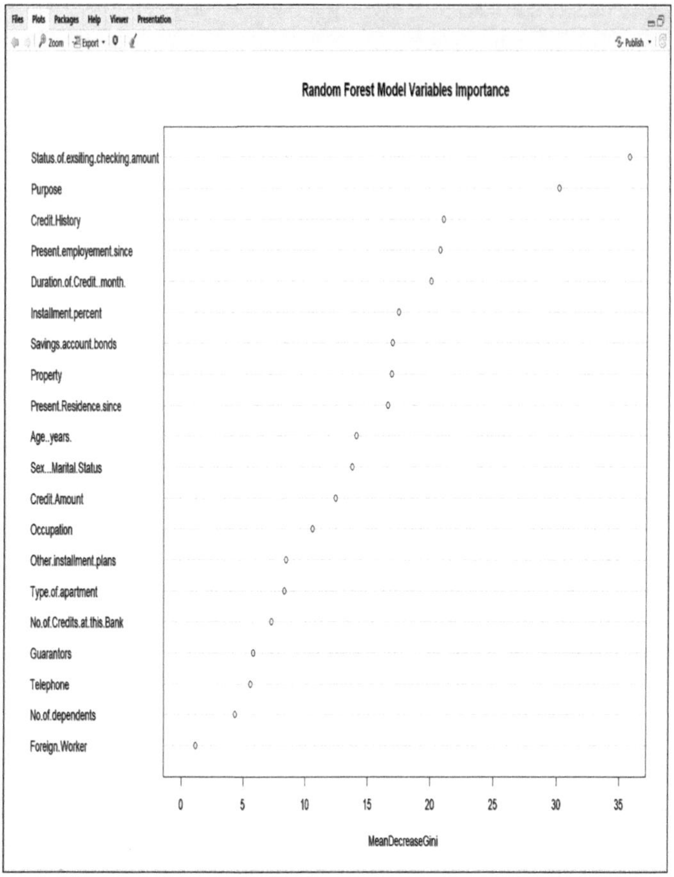

Fig. 17 Random forest plot

Fig. 18 Confusion matrix results

```
Confusion Matrix and Statistics

          Reference
Prediction   0   1
         0  40  50
         1  20 190

               Accuracy : 0.7667
                 95% CI : (0.7146, 0.8134)
    No Information Rate : 0.8
    P-Value [Acc > NIR] : 0.9330246

                  Kappa : 0.386

 Mcnemar's Test P-Value : 0.0005279

            Sensitivity : 0.6667
            Specificity : 0.7917
         Pos Pred Value : 0.4444
         Neg Pred Value : 0.9048
             Prevalence : 0.2000
         Detection Rate : 0.1333
   Detection Prevalence : 0.3000
      Balanced Accuracy : 0.7292

       'Positive' Class : 0

>
```

4 Conclusion

The results of our credit risk analysis, which used logistic regression, decision tree, and random forest models, are presented in this section. The results of our examination of model performance are informative. Its capacity to discern between "good" and "bad" credit risks is demonstrated by the logistic regression model's AUC for the decision tree model and AUC for the random forest model. The random forest model in particular highlights the importance of certain variables in credit risk assessment through feature importance analysis. Our investigation confirms that decision trees and logistic regression have an edge in terms of interpretability. These models offer useful insights into the relationship between application characteristics and credit risk categorization and offer unambiguous, step-by-step decision routes. Our models have broad applications in real life. These models can be used by financial organizations to make data-driven lending choices, improving their capacity to successfully manage credit risk. Institutions can lower financial risks and increase the general stability of the financial sector by implementing these approaches. A crucial characteristic of our models is robustness. We have rigorously evaluated them and found that they are stable in a range of scenarios, supporting their dependability for actual credit risk assessment. The results demonstrate the importance of our study in terms of real-world relevance and model interpretability. Our research indicates the possibility of data-driven credit risk analysis in the financial sector, to sum up. Our study highlights the significance of data-driven credit risk analysis in determining the direction of the financial industry. Financial institutions can better manage credit risk by utilizing predictive modeling and feature importance analysis, which will eventually promote economic stability and wise decision-making.

References

1. Shiv SJ, Murthy S, Challuru K (2018) Credit risk analysis using machine learning techniques. In: 2018 fourteenth international conference on information processing (ICINPRO), Bangalore, India, pp 1–5. https://doi.org/10.1109/ICINPRO43533.2018.9096854
2. Brock T, Schmitt KR (2023) Credit risk: definition, role of ratings, and examples. Investopedia. Available at: https://www.investopedia.com/terms/c/creditrisk.asp. Accessed: 25 Oct 2023
3. Board of Governors of the Federal Reserve System, Credit risk management. Accessed: 25 Oct 2023
4. Credit risk analysis applying logistic regression, neural networks and genetic algorithms models. Date of Issue: January 2021. Accessed on: 14 Oct 2023. https://www.researchgate.net/publication/228769399_Credit_Risk_Analysis_Applying_Logistic_Regression_Neural_Networks_and_Genetic_Algorithms_Models
5. A logistic regression model for consumer default risk. Date of Issue: May 2020. Accessed date: 14 Oct 2023. https://www.ncbi.nlm.nih.gov/pmc/articles/PMC9041570/
6. Credit risk analysis and prediction modelling of bank loans using R. Date of Issue: October 2016. Accessed date: 15 Oct 2023. https://www.researchgate.net/publication/309626126_Credit_Risk_Analysis_and_Prediction_Modelling_of_Bank_Loans_Using_R
7. Credit risk analysis and prediction modelling of bank loans using R0 Date of Issue: January 2019. Accessed date: 16 Oct 2023. https://www.enggjournals.com/ijet/docs/IJET16-08-05-414.pdf.
8. Credit risk analysis using machine learning. Date of Issue: May 2021. Accessed date: 16 Oct 2023. https://www.academia.edu/49558890/Credit_Risk_Analysis_using_Machine_Learning
9. Credit risk analysis using machine and deep learning models. Date of Issue: April 2018. Accessed date: 17 Oct 2023. https://www.mdpi.com/2227-9091/6/2/38
10. Applications of machine learning: consumer credit analysis. Date of Issue: May 2015. Accessed date: 17 Oct 2023. https://dspace.mit.edu/bitstream/handle/1721.1/100614/932622145-MIT.pdf?sequence=1
11. Röder J, Palmer M, Muntermann J (2022) Data-driven decision-making in credit risk management: the information value of analyst reports. Decis Support Syst 158:113770. https://doi.org/10.1016/j.dss.2022.113770
12. Lip G, Peterdy K (2023) Credit risk analysis. Corporate Finance Institute. Available at: https://corporatefinanceinstitute.com/resources/commercial-lending/credit-risk-analysis/. Accessed: 25 Oct 2023
13. Pandey TN, Jagadev AK, Mohapatra SK, Dehuri S (2017) Credit risk analysis using machine learning classifiers. In: 2017 international conference on energy, communication, data analytics and soft computing (ICECDS), Chennai, India, pp 1850–1854. https://doi.org/10.1109/ICECDS.2017.8389769
14. Dong G, Lai KK, Yen J (2010) Credit scorecard based on logistic regression with random coefficients. Procedia Comput Sci 1(1):2463–2468 (2010)
15. Satchidananda SS, Simha JB (2006) Comparing decision trees with logistic regression for credit risk analysis. International Institute of Information Technology, Bangalore, India
16. Li Y, Bellotti T, Adams N (2019) Issues using logistic regression with class imbalance, with a case study from credit risk modelling. Foundations of Data Science
17. Kruppa J et al (2013) Consumer credit risk: individual probability estimates using machine learning. Expert Syst Appl 40(13):5125–5131 (2013)

18. Cetiner E, Koçak T, Çağrı Güngör V (2018) Credit risk analysis based on hybrid classification: case studies on German and Turkish credit datasets. In: 2018 26th Signal processing and communications applications conference (SIU). IEEE (2018)
19. Swain PH, Hauska H (1977) The decision tree classifier: design and potential. IEEE Trans Geosci Electron 15(3):142–147
20. Quinlan JR (1996) Learning decision tree classifiers. ACM Comput Surv (CSUR) 28(1):71–72
21. Wang Y, Wang S, Lai KK (2005) A new fuzzy SVM to evaluate credit risk. IEEE Trans Fuzzy Syst 13:820–831
22. Ying X (2019) An overview of overfitting and its solutions. J Phys Conf Ser 1168
23. Leiva RG et al (2019) A novel hyperparameter-free approach to decision tree construction that avoids overfitting by design. IEEE Access 7(2019):99978–99987
24. Zhu L et al (2019) A study on predicting loan default based on the random forest algorithm. Procedia Comput Sci 162(2019):503–513
25. Zou KH, James O'Malley A, Mauri L, Receiver-operating characteristic analysis for evaluating diagnostic tests and predictive models
26. Martin D (2019) what is a ROC curve? A visualization with credit scores. Stacked Turtles. Available at: https://kiwidamien.github.io/what-is-a-roc-curve-a-visualization-with-credit-scores.html. Accessed: 25 Oct 2023

Enhancing Digital Forensic Readiness: Automated Detection of Missing and Null Log Values

Aparna Tripathi, Saurabh Shrivastava, and K. Praveen

Abstract The increase in cybercrime presents difficult problems in the digital sphere. The goal of this study is to improve incident response capabilities by examining Digital Forensic Readiness (DFR). In particular, the study presents an automated approach to identify missing log values, a crucial DFR component. This application monitors log files, spots irregularities, and generates automated security alerts. The study shows how crucial early cyber threat detection and prevention are for reducing operational disruptions. Prioritizing digital forensic capabilities is essential to fend off online threats and keep the workplace secure. With the potential for improvement and more research, this tool provides organizations looking to improve their Digital Forensic Readiness with some useful advantages.

Keywords Digital Forensic Readiness · Log analysis · Missing log values · Automated alerting · Incident response · Cybersecurity

1 Introduction

Organizations need to be Digital Forensic Readiness (DFR) to manage digital investigations effectively [1]. DFR's four primary building blocks are its people, processes, policy, and technology as shown in Fig. 1. While each element is significant, technology is especially critical to this endeavor. The main DFR operations are incident handling, evidence retention, and post-event analysis. Several techniques for improving digital forensic preparation were found in the literature review. It is crucial

A. Tripathi (✉) · S. Shrivastava · K. Praveen
TIFAC-CORE in Cyber Security, Amrita School of Engineering, Amrita Vishwa Vidyapeetham, Coimbatore, India
e-mail: cb.en.p2cys21002@cb.students.amrita.edu

S. Shrivastava
e-mail: s_saurabh@cb.amrita.edu

K. Praveen
e-mail: k_praveen@cb.amrita.edu

S. M. Thampi et al. (eds.), *Fifth International Conference on Computing and Network Communications*, Lecture Notes in Electrical Engineering 1219,
https://doi.org/10.1007/978-981-97-4540-1_37

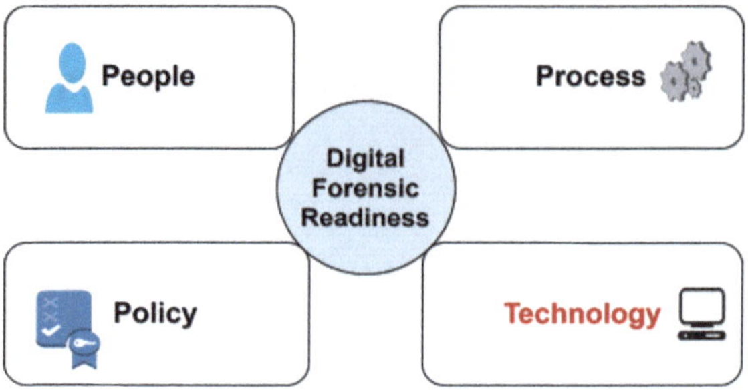

Fig. 1 Key factors of DFR

to remember that these solutions primarily concentrate on the safe preservation and recovery of prospective digital evidence rather than providing automated alerting for missing logs. Digital investigations require logs, but incomplete or missing records slow down the investigation and leave organizations exposed. Since manual analysis is time-consuming and prone to mistakes, automated technologies are required to find and report missing log entries.

To effectively identify and fix missing log information, enhance incident response capabilities, and lessen security threats, the study emphasizes the significance of automated solutions. Organizations should generally give top priority to the creation and adoption of technologies that streamline log analysis and guarantee the accuracy and comprehensiveness of digital evidence [2]. Investigations are significantly hampered by missing and null log values since it is more difficult to find important data and construct a timeline of events as a result. This study suggests an automated remedy to solve this issue and improve DFR.

2 Literature Survey

Several studies on digital forensics and software development logs are addressed in this review of the literature. According to Fred Cohen and colleagues, the multi-step process of digital forensics entails the identification, collection, storage, analysis, decryption, reconstruction, and presentation of digital evidence. Identification of the evidence, secure storage, analysis, and compelling presentation are among the difficulties this procedure offers. Himal Lalla and Stephen Flowerday have made a contribution by laying out a step-by-step procedure for obtaining and examining digital evidence from log files, with a special emphasis on network log mining in computer investigations.

Organizational "forensic readiness" is emphasized by Rowlingson, who also stresses the value of transparent policies, procedures, and instruments for assembling evidence. Gladyshev proposes a method for spotting data structure anomalies as signs of suspicious behavior in the context of intrusion investigations, utilizing formal software development methodologies for better forensic investigations and

Table 1 Proposed tool versus existing tools for Digital Forensic Readiness

Feature	Proposed tool	Secure RS	PDEF2S	W3RS
Missing log value alerts	Yes	No	No	No
Digital forensic preparation	Yes	Yes	Yes	Yes
Generic solution	Yes	Yes	Yes	Yes

intrusion detection. The importance of effective logging statement design was highlighted by Heng Li et al. in their study of the literature on log alterations in software development and their "just-in-time" method for counseling developers on log updates.

Overall, this review of the literature helps us better understand how to use digital forensics and logging techniques in software development and cybersecurity. The subjects covered range from organizational forensic readiness, recognition of mismatched data formats in intrusion investigations, and log modifications in software development through digital forensic methods and network log mining. The significance of accurate log file analysis, prompt log change recommendations, and adherence to forensic best practices are emphasized in these studies. Also created to aid with these efforts are the Digital Forensic Readiness (DFR) programs Windows Registry and RAM Readiness Storage (W3RS), Potential Digital Evidence Forensically Sound Storage (PDEF2S), and Secure RS.

A comparison table showcasing the proposed tool and existing tools is presented in Table 1.

3 Proposed Methodology

The following actions were taken in the suggested technique to address the issue of missing log entries and improve digital forensic investigations.

3.1 Selection of Services

To identify the vital services that produce logs essential for forensic investigations and security incident analyses, the IT architecture of the organization was evaluated. The following four crucial services were identified: SSHD, Apache, SYS Log, and Mail Log. Missing logs from these services can lead to various vulnerabilities and hinder the investigation and mitigation of potential attacks. For efficient forensic analysis and incident response, it's crucial to make sure that the logging configuration is correct and to find the logs in their intended locations. The log path is mentioned in Table 2. Below are the definitions of different services and the potential consequences that may occur if they lose their logs.

Table 2 Log path representation

Log type	Log path
Apache	/var/log/apache2/access.log
SSHD	/var/log/auth.log
Syslog	/var/log/syslog
Mail Server	/var/log/mail.log

Criteria for Log Selection

These essential services were chosen based on the following standards:

- **Operational Importance**: Services were selected based on how important they were to support regular operations and essential IT tasks.
- **Security Relevance**: With an emphasis on their responsibilities in recognizing and mitigating security threats, each service's importance in terms of security was a crucial consideration.
- **Compliance Alignment**: Compliance criteria were taken into account during the selection process to guarantee conformance to legal requirements and industry standards.
- **Incident Response Needs**: Services were chosen based on their applicability to incident response, providing crucial information for prompt security incident investigations.
- **Resource Efficiency**: In order to prevent waste, the effectiveness of monitoring and logging these services was considered.
- **Accessibility and Configurability**: Services were selected based on how simple it was to collect logs and how easily logging configurations could be managed.

SSHD Logs

A secure way to connect to a remote computer is through SSH (Secure Shell), and the tool that makes this connection possible is called "SSHD" [3]. Tracking and analyzing user activity, and identifying unauthorized access attempts, strange commands, and privilege escalation are all made possible by SSHD logs. Without SSHD logs, it is more difficult to spot harmful activity, such as unauthorized access by attackers or insider threats, and to take appropriate action.

Apache Logs

Internet-accessible webpages are hosted via the Apache program. Without Apache logs, it becomes difficult to recognize and assess web-based assaults, making it more difficult to track down attack sources, spot compromised systems, and gather data for forensic investigation [4].

SYS Logs

Sys logs are messages or recordings produced by the computer's operating system and software applications. They contain crucial data regarding system operations, problems, and occurrences [5]. They act as digital diaries that assist administrators in maintaining, securing, and improving system performance. Missing Sys logs make it challenging to track and investigate system-level security occurrences, which makes it more difficult to identify and analyze compromises and vulnerabilities.

Mail Log

Mail logs keep track of email-related activity and record important data that can be used to track, diagnose, and maintain email systems. Without them, it becomes difficult to investigate email threats and spot unusual activity, thereby providing security and data protection [6].

3.2 Log Identification and Location

Through careful examination, it was possible to identify and locate the log files for each service, including SSHD, Apache, Sys, and Mail [7]. To do this, one must look at each service's configuration files, system documentation, and logging procedures [8]. The SSHD configuration files and system settings were examined to determine the SSHD log locations. The location of the Apache web server logs was determined by examining the configuration files for Apache. The Sys log files were located through an analysis of system-level logging techniques such as syslog configuration files. Finally, the mail service logs were located by examining the mail server settings and related logging procedures. Establishing an efficient log-monitoring system is essential for identifying and examining security issues and unauthorized actions.

3.3 Log Parsing

Python scripts were created to extract crucial information, such as timestamps, events, IP addresses, and user actions, from SSHD, Apache, Sys, and Mail log files. To effectively analyze log entries, these programs employ log-parsing methods such as regular expressions and parsing algorithms [9]. This automation enhances the IT infrastructure's overall security and dependability, facilitating efficient forensic investigations and quick incident response.

Tables 3 and 4 provide a summary of various services and related attacks [10]. It draws attention to any security flaws and threats that businesses using these services

Table 3 Important log entries for forensic investigation

LOG type	Apache	SSHD	Sys log	Mail log
Timestamp	Yes	Yes	Yes	Yes
Message ID				Yes
IP address	Yes			
Request method	Yes			
URL	Yes			
HTTP status code	Yes			
User agent	Yes			
User information		Yes		
Session management		Yes		
Authentication		Yes		
Logins and user actions		Yes		
Service information			Yes	
Network related events			Yes	
Errors and failures			Yes	

Table 4 Services and types of attacks

Service	Types of attack
Apache	Cross-site scripting (XSS) Server-side request forgery (SSRF) SQL injection
SSHD	Brute force attack Man-in-the-middle attack
Sys Log	Log forging Log poisoning
Mail log	Spamming Email spoofing Phishing

may experience. The exact attack types frequently related to each service are listed in the table.

3.4 Development of an Alert Mechanism

For addressing missing log entries and enhancing digital forensic investigations, a Python-based alert mechanism script was created. The software compares expected and actual log entries methodically, looking for discrepancies and sending out prompt notifications. Organizations can use this to safeguard the integrity of log data, improve incident response, and increase the dependability of digital evidence. Investigation of

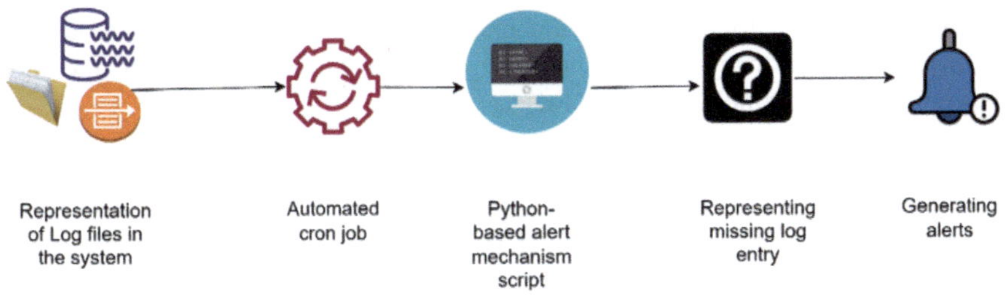

Fig. 2 Architecture of the tool

missing log entries in a variety of services turned up clear patterns, underscoring the difficulties they present in determining exact dates and identifying pertinent actions. The initiative aims to eliminate these gaps and improve digital evidence's overall integrity in Digital Forensic Readiness (DFR).

3.5 Automated Cron Job Setup

The alert system for detecting missing log values was automated by integrating a shell script with a Python script. A cron routine was set up to monitor log files continually and produce in-the-moment notifications [11]. By making sure no important log entries are missing, boosting digital forensic capabilities, and maintaining log data integrity, this automated system enhances incident response and forensic investigations. The tool architecture shown in Fig. 2 illustrates the overall structure and components of the tool implemented in this project.

3.6 Real-Time Notifications

The output file for missing log entries is updated on a regular basis by a cron job, and alerts generated by the alert mechanism script are logged there. These notifications can be analyzed by security analysts and system administrators to take the appropriate action. The output file offers continuous incident response and monitoring, ensuring prompt investigation and corrective measures for lost log entries.

3.7 Log Repository and Analysis

The organization saves all parsed logs, even those with missing entries, in a centralized database or SIEM system. This makes it possible to conduct in-depth audits for compliance and incident reconstruction [12]. Centralized storage improves the readiness of digital forensics, incident response, and log analysis capabilities by facilitating pattern recognition, abnormality detection, and adherence to retention requirements and compliance standards.

3.8 Customization and Scalability

Due to the high adaptability and scalability of the tool, the approach created for this study enables the integration of new services and log sources. To ensure continued monitoring, the alarm system and automated cron job can be expanded to include new services. Organizations can scale up or down their monitoring and forensic preparation in accordance with their specific demands, which improves incident response capabilities.

4 Experimental Results

Several trials were conducted to determine how well the proposed DFR tool would detect missing log values and issue alarms. These studies evaluated the tool's capacity to quickly create warnings, reliably identify inconsistencies in log data, and improve incident-response skills.

4.1 Log Parsing Accuracy

The methodology's log parsing scripts showed a high level of accuracy in retrieving pertinent data from log entries. The programs successfully analyzed the SSHD, Apache, Sys, and Mail services' log files to extract crucial information such as timestamps, events, IP addresses, and user activity. The effectiveness of the subsequent analysis and comparison of log entries was ensured by the correctness of the parsing procedure.

4.2 Detection of Missing Log Entries

By contrasting the expected log entries with the actual entries received, the alert mechanism script successfully identified missing log entries or entries with missing data. The script correctly identified instances during the tests in which log entries were absent or did not include anticipated data. Organizations were able to use this capacity to identify gaps in their log data which might indicate security issues or unauthorized activity.

Fig. 3 Missing alerts generated by proposed tool

4.3 *Timely Alert Generation*

When discrepancies or missing values were found, the automated cron job, which ran the alert mechanism script at regular intervals, reliably generated real-time notifications. The output file that contained the warnings served as a centralized area for tracking and keeping track of the discovered differences. The quick creation of alerts improves incident response capabilities by enabling security teams and system administrators to swiftly examine and address problems.

A screenshot of the tool generating missing alert is shown in Fig. 3.

4.4 *Scalability and Customization*

To handle more log sources and services, the exhibited methodologies and tools can be modified and scaled up. While the automated components of the tool regularly monitor log entries from new services, the technology's adaptability enables organizations to expand their log analysis and monitoring, as necessary. By spotting and filling up alarm system gaps and boosting issue response capabilities, the tool enhances incident detection.

4.5 *Overall Effectiveness*

By keeping an eye on log files, spotting missing log values, and promptly notifying users, the suggested DFR tool enhances incident reaction and investigation capabilities. It can be readily linked into current log management systems and is versatile and customizable. In order to support successful forensic and incident responses, the tool proactively addresses missing log entries, reduces security threats, and maintains the accuracy of log data.

5 Conclusion

The automated method for locating missing and null log values introduced in this study will make it easier for organizations to handle incidents and streamline log monitoring. Through log analysis techniques and easy integration with current log management systems, the suggested solution efficiently detects patterns and anticipated log values. The study shows the tool's dependability in identifying missing log information, sending security teams' quick notifications, and enhancing overall readiness for online threats. To ensure compliance with legal and regulatory norms, proactive methods like log monitoring can help with incident response, security breach detection, and preservation of digital evidence.

References

1. Tan J (2001) Forensic readiness, vol 1. @ Stake, Cambridge, MA
2. Elyas M et al (2014) Towards a systemic framework for digital forensic readiness. J Comput Inf Syst 54(3):97–105
3. Danda PK et al (2023) SSH-DAuth: secret sharing based decentralized OAuth using decentralized identifier
4. O'Leary M, O'Leary M (2019) Apache apache and modsecurity. In: Cyber operations: building, defending, and attacking modern computer networks, pp 721–788
5. Oliner A, Stearley J (2007) What supercomputers say: a study of five system logs. In: 37th annual IEEE/IFIP international conference on dependable systems and networks (DSN'07). IEEE
6. Mandela N, Mahmoud AAS, Agrawal AK (2023) A forensic analysis of the tor network in tails operating system. In: 2023 10th international conference on computing for sustainable global development (INDIACom). IEEE
7. Vinayakumar R, Soman KP, Poornachandran P (2017) Long short-term memory based operation log anomaly detection. In: 2017 international conference on advances in computing, communications and informatics (ICACCI). IEEE
8. Bellare M, Yee B (1997) Forward integrity for secure audit logs, vol 184. Technical report, Computer Science and Engineering Department, University of California at San Diego
9. Kauffman S (2023) Log analysis and system monitoring with nfer. Sci Comput Program 225:102909
10. Hemalatha T et al (2023) Secure and private data sharing in CPS e-health systems based on CB-SMO techniques. Meas Sens 27:100787
11. Joslin P (2023) Automated visualization pipeline for near real-time risk management system
12. Sekharan SS, Kandasamy K (2017) Profiling SIEM tools and correlation engines for security analytics. In: 2017 international conference on wireless communications, signal processing and networking (WiSPNET). IEEE
13. Li H et al (2017) Towards just-in-time suggestions for log changes. Empir Softw Eng 22:1831–1865
14. Lalla H et al (2012) A log file digital forensic model. In: Advances in digital forensics VIII: 8th IFIP WG 11.9 international conference on digital forensics, Pretoria, South Africa, January 3–5, 2012, Revised selected papers 8. Springer, Berlin, Heidelberg
15. Rowlingson R (2004) A ten step process for forensic readiness. Int J Digit Evid 2(3):1–28

Predicting and Evaluating Anomaly Detection and Traffic Analysis on Software Defined Networks Using a Hybrid Machine Learning Approach

S. Darshan, N. Radhika, and G. Radhika

Abstract In the rapidly evolving landscape of Software Defined Networks (SDNs), network security is of paramount importance. This research paper presents a comprehensive study on anomaly detection and traffic analysis in SDNs using machine learning algorithms. The study addresses a critical research gap by proposing an effective methodology for detecting network anomalies and analyzing traffic patterns. Leveraging a carefully selected dataset, we evaluate the performance of Logistic Regression, Decision Tree, Random Forest, Neural Networks and Hybrid Models. The results demonstrate the superior performance of the hybrid model built using Decision Tree and Neural Networks and the competitive accuracy of Neural Networks in SDN anomaly detection.

Keywords Software Defined Networks · Anomaly detection · Network traffic analysis · Machine learning · Hybrid model · Decision Tree · Neural Networks

1 Introduction

The advent of Software Defined Networks (SDNs) has ushered in a new era of network flexibility and centralized management. However, this evolution has also introduced complex security challenges. Ensuring the integrity and reliability of SDNs is crucial, and this paper delves into the realm of anomaly detection and traffic analysis, leveraging the power of machine learning algorithms to safeguard SDN environments. In today's interconnected world, where SDNs are used in telecommunications, data centers, and beyond, the need for robust network security solutions is more pressing than ever. Anomaly detection, the ability to identify deviations from

S. Darshan · N. Radhika (✉) · G. Radhika
Department of Computer Science and Engineering, Amrita School of Computing, Amrita Vishwa Vidyapeetham, Coimbatore, India
e-mail: n_radhika@cb.amrita.edu

G. Radhika
e-mail: g_radhika@cb.amrita.edu

S. M. Thampi et al. (eds.), *Fifth International Conference on Computing and Network Communications*, Lecture Notes in Electrical Engineering 1219,
https://doi.org/10.1007/978-981-97-4540-1_38

expected network behavior, is a cornerstone of network security. Moreover, the analysis of network traffic patterns can unveil valuable insights into the functioning and optimization of SDNs.

This research paper aims to address these critical aspects by presenting a comprehensive methodology that combines feature selection, model selection, and data balancing to enhance the accuracy and efficiency of anomaly detection in SDNs. We employ well-established ML algorithms, including Random Forests, Decision Trees, Logistic Regression, and Neural Networks, to evaluate their suitability for the task. Our study also sheds light on the superiority of the hybrid model and the competitive accuracy of Neural Networks, offering insightful information about Software Defined Networks to researchers, security experts, and network administrators. With the convergence of network security and machine learning, this research contributes to the ongoing efforts to fortify SDNs against malicious activities and ensure the seamless operation of these transformative network architectures.

2 Literature Review

In a recent study [1], a cutting-edge platform has been unveiled for the classification of encrypted network traffic, leveraging machine learning (ML) techniques. Engineered with a strong emphasis on practicality and performance, this platform introduces innovative feature extraction methods. Its workflow involves the stepwise processing of network traffic packets, encompassing feature extraction, data preparation, and ML modeling. Prominent ML algorithms, including GBTree, LightGBM, and XGBoost, are deployed in the ML stage. Notably, the XGBoost algorithm emerges as the top performer, achieving an exceptional $F1$ score of over 99%. These findings underscore the platform's viability in addressing real-world network traffic classification challenges with great efficiency and accuracy. A comparative study on online machine learning techniques for network traffic streams analysis. There are enormous amounts of traffic data streams produced by modern networks [2]. For a number of reasons, including network management of resources and cyber-security analysis, this data analysis is crucial. Data analytic techniques that can analyze network data continuously in response to the influx of fresh data are desperately needed. Techniques for online machine learning (OL) promise for facilitating these kinds of data analytics. This study examines and compares the OL approaches that enable data stream analytics in the networking space. In addition, the benefits of online learning are discussed, along with the drawbacks of using online machine learning-based stream analysis on network traffic, such as concept drift and imbalanced classes. The significance of traffic data analytics is also examined. The frameworks and tools for data stream processing that may be used to handle this kind of data online or in real time are evaluated, along with their benefits and drawbacks, and their compatibility with actual data processing paradigms is examined. An empirical investigation into the effectiveness of various ensemble and tree-based algorithms that can be used for network traffic categorization is carried out in order to investigate the efficacy of online machine learning approaches. Lastly, the remaining problems and potential paths for traffic data stream analysis are discussed. When addressing the needs and goals of online data streams predictive analytics and learning in the networking domain, this technical study offers the network research community insightful perspectives. In the analysis of attack detection in Software Defined Networks (SDN) using various machine learning methods [3], a more adaptable and efficiently run network design for networking of the future is provided by SDN. Thanks to the configurable centralized controller design, network managers may simply monitor and control the whole network. Because of its centralized nature, the central controller acts as the hub for several avenues of attack. Among them, distributed denial of service (DDoS) attacks that target SDN are the most well known. This project's goal is to classify SDN traffic as either attack or normal traffic by applying a machine learning technique. For finely granulated recognition, the Method of Feature Selection approach is then used, which includes the Fisher score method based on filtration systems, the wrapper-based method, and the analysis of variables (ANOVA) f-test. Then, to identify DDoS assaults on SDN controllers, a rule-based detection technique based on the Renyi joint entropy algorithm is applied. For this reason, a public dataset known as the "DDoS attack SDN Dataset" is maintained, consisting of 23 features in total. The dataset includes User Datagram Networking Protocol (UDP), Transmission Control Networking Protocol (TCP), and Internet Control Message Protocol (ICMP) for both normal and attack traffic. The collection includes over 100,000 recordings with statistical features including byte count, time sec, packet rate, and packet per flow, excluding variables that indicate the destination and source devices. Many classifiers are used in the classification process, such as k-Nearest Neighbor (k-NN), Support Vector Machine (SVM), XGBoost (XGB), and Artificial Neural Network (ANN). In the context of network attack detection in big data traffic, where big data scenarios are characterized by numerous devices, extensive traffic volumes, and high data dimensionality, a novel evolutionary computation-based machine learning approach has been proposed [4].

This approach addresses the crucial issue of feature selection from vast datasets, which directly impacts the effectiveness of attack detection. First, to solve the class disparity within the network attack traffic data acquired from large data traffic, a technique called Random Sample-SMOTE (Synthetic Minority Over-sampling Technique) has been created. Second, the LightGBM (Light Gradient Boosting Machine) model is used to evaluate and rank each feature relevance of the traffic which is being attacked inside the various categorical layers. Through retraining, the ideal feature values are chosen. Eventually, TuneGridSearchCV (Tune's Grid Search Cross-Validation) is used to optimize hyperparameters, yielding the best possible model once these chosen feature values are used for model training. Simulation studies show that the method described in this research is efficient in extracting characteristics from large-scale data flow. It successfully reduces feature dimensionality, substantially enhances detection accuracy, and results in approximately 40% reduction in computation time when compared to existing methods. The paper [5] employs a recursive network design to monitor traffic flow, ensuring precise anomaly identification and bolstering cyber-attack effectiveness within SDNs. The proposed model exhibits outstanding performance in detecting Distributed Denial-of-Service (DDoS) attacks, successfully stopping the deterioration of network forwarding performance. The approach is crafted to instruct users on optimizing traffic flow matching, therefore improving granularity, and proactively preventing congestion on the SDN data plane. The application of a learned control policy for matching traffic flow enables the acquisition of superior traffic data for real-time abnormality detection, ultimately enhancing cyber-attack detection performance. Interestingly, the suggested model's accuracy outperforms FMS, MMOS, Q-DATA, DATA, and DEEP-MC by margins of 19.23%, 16.25%, 47.61%, 16.25%, and 12.04%, respectively. In the face of escalating internet traffic volumes, Internet Service Providers (ISPs) and Network Service Providers (NSPs) are actively seeking methods to accurately predict the types of applications traversing their networks [6]. This prediction is vital for security and network monitoring, given the increasing use of dynamic port numbers, masquerading, and encryption techniques. Traditional approaches based on port or payload analysis are no longer sufficient. Consequently, machine learning has gained prominence in traffic classification. However, these algorithms grapple with imbalanced data issues. This paper employs network flow characteristics to identify traffic application types and introduces the cost-sensitive SVM (CMSVM) algorithm to address data imbalance. CMSVM combines a multi-class SVM algorithm with active learning, dynamically assigning weights to applications. Evaluations using MOORE_SET and NOC_SET datasets demonstrate that CMSVM reduces computational costs, enhances classification accuracy, and mitigates data imbalance compared to alternative machine learning techniques. A new method for predicting the traffic situation across a network is introduced [7], referred to as NN-ARIMA. It combines a statistical time series model, ARIMA, with a machine learning algorithm, MLP (Multilayer Perceptron). MLP captures network-scale co-movement patterns in traffic flows, while ARIMA processes residuals to extract location-specific traffic features. Experimental results demonstrate that post-processing MLP residuals with ARIMA significantly enhances traffic state prediction accuracy, reducing mean squared error by 8.9–13.4%. Comparative experiments involving Multidimensional Support Vector Regression (MSVR) also confirm the efficiency of ARIMA post-processing. This approach not only captures network-wide traffic patterns but also location-specific characteristics and nonlinearities in macroscopic traffic variables, leading to substantial prediction accuracy improvements when considering both aspects. The paper [8] delves into the assessment of machine learning-based categorization of protected network traffic, a crucial facet of network resource management in the context of evolving networks like 5G and beyond. The study focuses on traffic analysis systems for network monitoring and access control between Virtual Private Networks (VPNs) and non-VPNs. Various algorithms are explored for VPN traffic classification and detection, with a specific emphasis on enhancing precision, recall, and $F1$-score using ensemble classifiers. Optimizing the ensemble classifier parameters results in increased $F1$-score, precision, and recall. Comparing Bagging Decision Tree and Gradient Boosting methods to individual classifiers like k-Nearest Neighbors (k-NN), Multilayer Perceptron (MLP), and Decision Tree, the results look promising. With the suggested classifier, recognition accuracy that is attained is up to 93.80% on a test sample, surpassing the performance of all previously employed single algorithms [9].

3 Research Gap

Network security within Software Defined Networks (SDNs) has garnered significant attention in recent years due to the transformative nature of SDN technology. However, existing research in this domain often exhibits certain limitations and research gaps that have driven the need for our study. Many prior studies have primarily focused on traditional security methods or explored the individual efficacy of machine learning algorithms in SDN security. These studies often lack comprehensive methodologies for the integrated tasks of anomaly detection and traffic analysis in SDNs. Moreover, there is a scarcity of extensive comparative analyses among different machine learning approaches in this context. Our research paper aims to bridge these existing gaps by proposing a holistic methodology that integrates feature selection, model selection, and data-balancing techniques. This comprehensive approach acknowledges the intricate nature of network security in SDNs and provides a robust framework to address it.

4 Research Objectives

The primary objectives of this research encompass several key aspects of SDN network security:

1. Development of an Effective Methodology: Our foremost objective is to design and establish an effective methodology that combines various data preprocessing, feature selection, and machine learning techniques. This methodology should seamlessly integrate anomaly detection and traffic analysis, providing a holistic approach to enhancing network security in SDNs.
2. Evaluation of Machine Learning Algorithms: We aim to rigorously evaluate the performance of different machine learning algorithms in the context of SDN security. This assessment will offer insightful information on the advantages and disadvantages of various algorithms when applied to anomaly detection and traffic analysis tasks within SDN environments.

3. Identification of Suitable Algorithms: By conducting extensive comparative analyses, we seek to identify the most suitable machine learning algorithms for specific network security tasks within SDNs. This identification will facilitate the selection of the best-fitting algorithms for practical deployment in SDN networks.
4. Exploration of Hybrid Models: In addition to the immediate research objectives, our study sets the stage for research endeavors. We foresee the potential for building hybrid models that leverage the strengths of multiple machine learning algorithms. Such hybrid models could enhance the robustness and adaptability of SDN network security systems, which will be a focus of our future research efforts.

5 Problem Statement

Software Defined Networks (SDNs) represent a paradigm shift in network architecture, offering unprecedented flexibility and centralized control. However, this dynamic and centralized nature of SDNs introduces a set of unique security challenges. Among these challenges, two critical tasks stand out: detecting network anomalies and analyzing traffic patterns. The detection of network anomalies is crucial for identifying potentially malicious activities or deviations from normal network behavior. Such anomalies can be early indicators of security breaches, performance issues, or network misconfigurations. Additionally, analyzing traffic patterns helps network administrators optimize resource allocation and ensure efficient network operation [10].

Despite the importance of these tasks, there exists a significant gap in the field of SDN security. The current research landscape lacks comprehensive methodologies that effectively combine anomaly detection and traffic analysis within SDN environments. Many studies focus on traditional security approaches, while others explore individual machine learning algorithms in isolation. Our research aims to address this pressing problem by proposing a methodology that integrates anomaly detection and traffic analysis in SDNs. By doing so, we strive to enhance network reliability and security in these dynamic environments.

6 Motivation

The motivation behind this research is driven by the increasing adoption of Software Defined Networks (SDNs) across various industries, including telecommunications, data centers, and cloud computing. SDNs offer unparalleled advantages in terms of network flexibility, scalability, and centralized management [11]. As SDNs become integral components of modern infrastructure, the need for robust network security solutions becomes more critical than ever. The motivation for our research stems from several factors:

- Rising Adoption of SDNs: SDNs are becoming the backbone of modern network infrastructure, underpinning critical operations across diverse sectors. This widespread adoption underscores the urgency of fortifying SDNs against potential threats.
- Dynamic Nature of SDNs: The dynamic and software-driven nature of SDNs introduces unique vulnerabilities. Network security must evolve to address these new challenges effectively.
- Protection Against Malicious Activities: Safeguarding SDNs from malicious activities, such as cyberattacks and unauthorized access, is paramount to ensure the reliability and integrity of network operations.
- Network Efficiency: Analyzing traffic patterns and optimizing resource allocation within SDNs can lead to significant improvements in network efficiency, cost-effectiveness, and performance [12].

By conducting research in the field of anomaly detection and traffic analysis in SDNs, we seek to contribute to the development of advanced security solutions that align with the evolving nature of network architecture.

7 Methodology

This research study presents an approach that uses hybrid machine learning algorithms to solve the important features of anomaly detection and traffic analysis in Software Defined Networks (SDNs). This methodology is a thorough framework that has been thoughtfully designed to offer an all-encompassing approach to network security in SDN environments. The first step of the methodology involves the collection of network traffic data and system parameters. This data serves as the foundation for subsequent analysis. However, real-world data is often messy, and thus, data pre-processing becomes essential. Pre-processing techniques are applied to handle missing values and outliers, ensuring that the data is in a suitable format for analysis. To enhance the efficiency and effectiveness of machine learning models, feature selection is performed using SelectKBest, a method that selects the most relevant features for analysis. Additionally, categorical variables are encoded to make them compatible with machine learning algorithms, ensuring that they can contribute meaningfully to the models. The methodology incorporates various machine learning models to explore their performance in the context of SDN security. These models include Logistic Regression with L1 regularization, Decision Tree, Random Forest, and a Neural Network. Each model brings its unique strengths to the table, and their performances are rigorously evaluated. In addition to evaluating individual machine learning models, we explored the potential of hybrid models that combine the strengths of different algorithms. Specifically, we developed a hybrid model by sequentially combining a Decision Tree model with a Neural Network. This hybrid model leverages the decision-making capabilities of the Decision Tree to enhance the feature set for the Neural Network (Figs. 1 and 2).

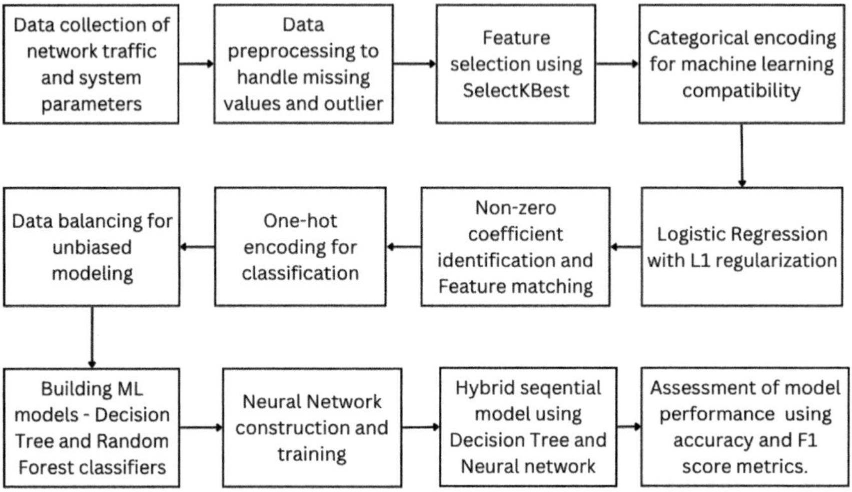

Fig. 1 Proposed methodology diagram for the machine learning and hybrid models

Fig. 2 Feed-forward neural network model with three hidden layers, using the ReLU activation function, and compiled with the Adam optimizer for binary classification

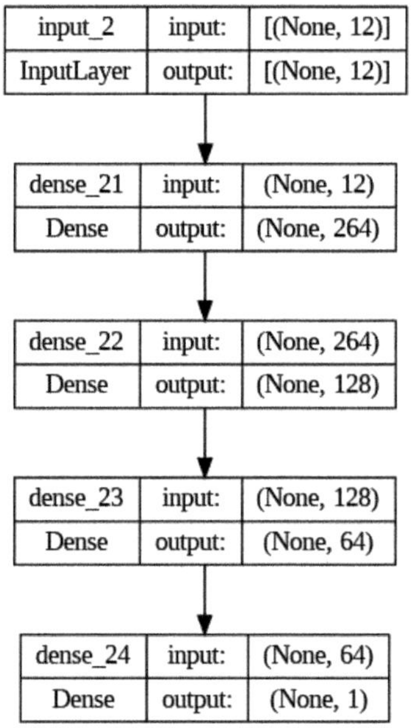

The hyperparameters used for building the machine learning models play a crucial role in their performance. These hyperparameters are carefully tuned to achieve optimal results. For example, in Logistic Regression, L1 regularization is applied to control overfitting, and the 'liblinear' solver is used [13]. In Decision Trees and Random Forest, parameters such as the maximum depth of the tree, the minimum number of samples required to split a node, and the number of estimators in the Random Forest are tuned. In the Neural Network, layers, activation functions, and the optimizer (Adam) are configured to achieve the best performance. Another vital aspect of the methodology is data balancing. Imbalanced datasets are a common challenge in anomaly detection tasks. To address this, the methodology includes techniques for data balancing, ensuring that the models are not biased toward the majority class. Overall, it combines data collection, pre-processing, feature selection, model selection, data balancing, and comprehensive evaluation to provide valuable insights into the best practices for securing SDNs. The hyperparameter tuning ensures that each machine learning model performs optimally, contributing to the overall success of the research [14].

8 Dataset Used

The dataset, which includes 25,192 samples with 41 characteristics, was acquired via Kaggle. The dataset used in this study comes from an extensive investigation into network intrusion detection. It was painstakingly assembled to mimic a military network, offering an authentic setting for testing intrusion detection methods. The process of acquiring the dataset comprised recording unprocessed TCP/IP dump data in a network environment that was created to mimic an ordinary US Air Force local area network. Within this LAN simulation, various intrusion attempts and network attacks were orchestrated to create a challenging and dynamic dataset for analysis.

- Network Environment Simulation: The LAN simulation closely mimics real-world network settings, making it a valuable resource for intrusion detection research.
- Attack Diversity: The dataset encompasses a wide variety of intrusions, including multiple attack types, to assess the efficacy of intrusion detection methods across different scenarios.
- Connection Records: Each entry in the dataset represents a network connection, which is a sequence of TCP packets exchanged between source and target IP addresses. These connections are conducted under well-defined protocols.
- Feature Set: For every TCP/IP connection, the dataset comprises a comprehensive set of 41 features, combining both quantitative (38 features) and qualitative (3 features) attributes. These features are extracted from both normal network traffic and attack data.
- The dataset classifies each connection into one of two categories:

  i. Normal: Represents network connections that adhere to standard behavior.
  ii. Anomalous: Corresponds to network connections that exhibit suspicious or attack-like behavior.

This dataset is instrumental in evaluating the performance of intrusion detection systems and machine learning algorithms in identifying network anomalies and attacks. The dataset's diverse attack scenarios and rich feature set make it an ideal choice for evaluating the effectiveness of machine learning algorithms in safeguarding network environments against malicious activities.

9 Results

The research outcomes underscore the effectiveness of machine learning algorithms in the domain of network security, particularly in the context of Software Defined Networks (SDNs). Logistic Regression achieved a commendable accuracy of 89.1%, providing a strong baseline for intrusion detection. However, Decision Tree and Random Forest models outperformed it, achieving testing accuracies of 92.2% and 92.5%, respectively. These tree-based models demonstrated robustness, with Decision Tree boasting a cross-validated mean training accuracy of 92.3%. Random Forest, with its ensemble approach, showcased even greater resilience, achieving a mean training accuracy of 92.6%. Furthermore, both Decision Tree and Random Forest models exhibited an impressive $F1$ score of around 0.92, highlighting their ability to balance precision and recall (Fig. 3).

Increasing training accuracy signifies the model's improved ability to fit the training data, while increasing validation accuracy reflects its capacity to generalize effectively to new, unseen data [15]. Concurrently, the decrease in both training and validation loss indicates the model's progress in making more accurate predictions and optimizing its performance. The efficacy of the model is compared with the existing models and the same is summarized in Table 1.

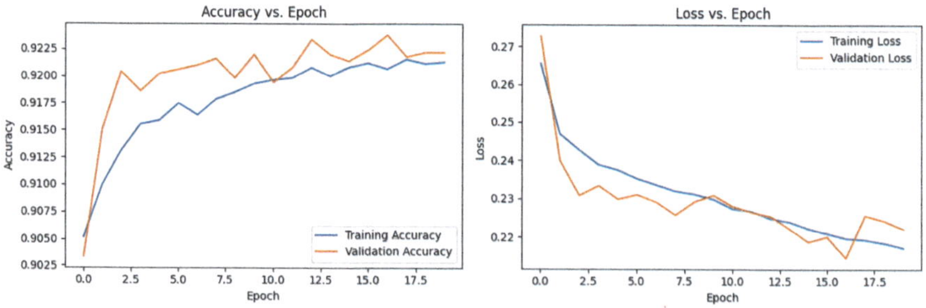

Fig. 3 Training accuracy versus validation accuracy and training loss versus validation loss graph for the neural network model

Table 1 Comparative analysis among other techniques

S. No.	Algorithm used	Accuracy (%)
1	Support Vector Classifiers	89.6
2	FF—Neural Networks	90.2
3	Logistic Regression	85.0
4	Decision Stump	88.5
5	Our hybrid approach	92.4

10 Conclusion

In Software Defined Networking (SDN), anomaly detection is essential for locating and addressing anomalous or perhaps harmful activity in a network. SDN offers a programmable and centralized method for managing networks, which makes the use of efficient anomaly detection methods possible. The hybrid model outperformed standalone models, like Neural Network and Logistic Regression, with an astounding testing accuracy of 92.88%. This outcome emphasizes how well anomaly detection and traffic analysis in SDNs can be enhanced by sequentially mixing machine learning techniques. In contrast, the Neural Network demonstrated competitive performance with a loss of 0.224 and an accuracy of 92.4%. All of these findings highlight how machine learning models may be used in SDNs for traffic analysis and anomaly detection, opening the door to more advanced network security solutions.

References

1. Bozkır R, Cicioğlu M, Çalhan A, Toğay C (2023) A new platform for machine-learning-based network traffic classification. Comput Commun 208:1–14. https://doi.org/10.1016/j.comcom.2023.05.010
2. Shahraki A, Abbasi M, Taherkordi A, Jurcut AD (2022) A comparative study on online machine learning techniques for network traffic streams analysis. Comput Netw 207:108836. https://doi.org/10.1016/j.comnet.2022.108836
3. Wang Y, Wang X, Ariffin MM, Abolfathi M, Alqhatani A, Almutairi L (2023) Attack detection analysis in software-defined networks using various machine learning method. Comput Electr Eng 108:108655. https://doi.org/10.1016/j.compeleceng.2023.108655
4. Wang Y, Zhang H, Wei Y, Wang H, Peng Y, Bin Z, Li W (2023) An evolutionary computation-based machine learning for network attack detection in big data traffic. Appl Soft Comput 138:110184. https://doi.org/10.1016/j.asoc.2023.110184
5. Shukla PK, Maheshwary P, Subramanian EK, Shilpa VJ, Varma PRK (2023) Traffic flow monitoring in software-defined network using modified recursive learning. Phys Commun 57:101997. https://doi.org/10.1016/j.phycom.2022.101997
6. Dong S (2021) Multi class SVM algorithm with active learning for network traffic classification. Expert Syst Appl 176:114885. https://doi.org/10.1016/j.eswa.2021.114885
7. Ma T, Antoniou C, Toledo T (2020) Hybrid machine learning algorithm and statistical time series model for network-wide traffic forecast. Transp Res Part C Emerg Technol 111:352–372. https://doi.org/10.1016/j.trc.2019.12.022

8. Sheik Abdullah A, Rajasekar D, Rahul A (2023) A machine learning approach for the identification and estimation of pulsars upon statistical analysis. In: 2023 international conference on network, multimedia and information technology (NMITCON), Bengaluru, India, pp 1–12. https://doi.org/10.1109/NMITCON58196.2023.10276169
9. Afuwape AA, Xu Y, Anajemba JH, Srivastava G (2021) Performance evaluation of secured network traffic classification using a machine learning approach. Comput Stand Interfaces 78:103545. https://doi.org/10.1016/j.csi.2021.103545
10. Karn I, Manikandan ST, Abdullah AS (2023) Disseminating the process of hurricane path prediction using multilayer perceptron and support vector machine upon varied kernel functions. In: 2023 international conference for advancement in technology (ICONAT), Goa, India, pp 1–5. https://doi.org/10.1109/ICONAT57137.2023.10080280
11. Radhika N, Abhilash V, Niranjan Kumar KS, Meduri P, Ram NCS (2020) Dynamic bandwidth allocation in an SDN environment using QoS queues and open flow. Int J Pharm Res 12(1):1060–1073
12. Vishnu Priya A, Radhika N (2019) Performance comparison of SDN OpenFlow controllers. Int J Comput Aided Eng Technol 11(4–5):467–479
13. Sreenivasa Chakravarthi S, Jagadeesh Kannan R, Anantha Natarajan V, Gao X-Z (2022) Deep learning based intrusion detection in cloud services for resilience management. Comput Mater Contin 71:5117–5133
14. Ganeshkumar M, Sowmya V, Gopalakrishnan EA, Soman KP (2022) Unsupervised deep learning approach for the identification of intracranial haemorrhage in CT images using PCA-Net and K-means algorithm. In: Intelligent vision in healthcare. Springer, Singapore, pp 23–31
15. Duraisamy P, Santhanakrishnan MN, Ramasamy S (2023) Real-time implementation of deep reinforcement learning controller for speed tracking of robotic fish through data-assisted modeling. Proc Inst Mech Eng Part C J Mech Eng Sci

Signal Processing, VLSI Design and Embedded Computing

Development of Tri-plate Structured Dipole Antenna Array with Incorporated Feed Network for S-Band Radar Application

Ankur Pandey, Tarlok Singh, Homen Borah, and Seema Doongarwal

Abstract This paper presents a vertically polarized linear antenna array of dipole elements with incorporated feed network, robustly designed for the application in S-Band surveillance and tracking radars. The dipole element used in the construction of array is designed with tri-plate structured metal grounds and balun feed. Elements in the array are fed with equal phase through two identical 1:24 incorporated feed networks. The array is designed to achieve the desired gain of more than 20 dBi with azimuth beamwidth of $2 \pm 0.3°$ in the required frequency band of 2.9–3.3 GHz. Designed antenna array has been realized and tested. Measured results are showing close agreement with simulation results.

Keywords Antenna array · Dipole antenna · Balun · Transmission line · Impedance transformer · Radome

1 Introduction

A linear antenna array is a part of phased array planar antenna aperture generally used for surveillance and tracking radar systems. The radiation pattern formed by antenna aperture has narrow beam in azimuth plane and co-secant squared pattern or constant height pattern in elevation plane making it suitable for air surveillance application [1]. A planar antenna aperture is synthesized using 40 numbers of linear antenna rows. These linear arrays are fed through amplifier modules like transmit/ receive modules (TRM). T/R modules provide amplification and electronic beam steering for electronically scanned antenna systems. Eighty numbers of T/R modules are employed to feed all the linear rows. So, antenna aperture works as semi-active phased array in which each antenna row is electronically controlled by two T/R modules to transmit high RF power, receive signal, and to provide phase for beam shaping as well as steering.

A. Pandey (✉) · T. Singh · H. Borah · S. Doongarwal
Development and Engineering-Antenna, Bharat Electronics Limited, Ghaziabad, UP, India
e-mail: ankurpandey@bel.co.in

S. M. Thampi et al. (eds.), *Fifth International Conference on Computing and Network Communications*, Lecture Notes in Electrical Engineering 1219,
https://doi.org/10.1007/978-981-97-4540-1_39

The radiating element used in the linear antenna array is a modified half wave dipole antenna fed through tri-plate transmission line. Considering the manufacturability, light weight aperture and beamwidth requirement, the proposed element is well suited for designing antenna array. The proposed design has the advantage of effective impedance matching, simplified balun structure, transmission topology with air substrate, and easy realization with respect to present state of art designs. Another considerable advantage of the low level of mutual coupling in the array design is achieved by using the proposed dipole element due to its unique structure. For feeding dipole elements with required amplitude distribution and equal phase, inline power divider has been implemented in the form of 1:24 way networks. Two similar networks are employed to excite all the elements with two separate input feed through RF connector. The design of the feed network is based on stripline topology with two metal grounds and one center conductive transmission strip sandwiched between the grounds. Extreme radiating dipoles at each end of the array of 50 elements are used for the application of side lobe blanking/cancelation in full planar aperture. These elements are excited through a 50-Ω conductive strip with separate RF inputs provided by two RF connectors which are terminated during testing of the array.

One sample of the linear antenna row is manufactured for validation. The structure of all dipoles is fabricated by machining two running conductive sheets. The feed network is fabricated through wire cutting process forming 1.5 m long strip of 1 mm thickness. Mounting brackets are designed for handling and installation of antenna rows on planar aperture frame. The antenna row is enclosed with radome cover for structural protection from harsh environment and water ingress.

2 Design Considerations

Conventionally, it is known that dipole antenna is a balanced radiating element with two poles/arms having equal length and extending in opposite directions from the feed point [2]. It should ideally be fed through a balanced transmission line. The conventional method of feeding dipole element is through coaxial balun which comprises of a coaxial line and a short circuited wire of length approximately equal to quarter operating wavelength of the dipole [3]. The short circuited wire and the coaxial line's outer conductor combine to form a two-conductor shorted stub which is fixed at the top of the balun called feed point [4] (Fig. 1).

Due to this shorted stub, infinite impedance is presented at the feed point which prevents current flow or coupling from radiating dipole to the surface of outer conductor of the coaxial line [5]. Similar functionality is achieved in the presented design of tri-plate structured dipole element with ease in manufacturing of large arrays and feed network. Radiating dipole is called tri-plate structured as the arms of the dipole are two co-planar conductive ground plates and third plate lying between these plates forms the stripline feed and balun to the element. Ensuring proper impedance matching is crucial in order to avoid unintended energy loss during

Fig. 1 Dipole element with coaxial balun feed [4]

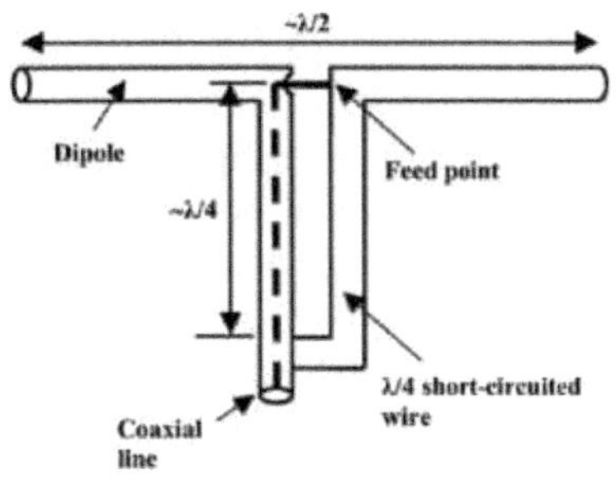

both the antenna radiation and reception function. To obtain substantially constant impedance over the required bandwidth, two ground planes are tapered up to the arms of dipole to form extended aperture. This extended aperture, in other words, acts as an impedance transformer between the dipole element and tri-plate structured transmission line. Longitudinal slots are introduced at the center of these extended top and bottom ground planes which are aligned with each other [6]. The radiation from dipole antenna is plane polarized perpendicular to dipole arms. And element shares the same plane as the tri-plate feed network. The required vertical polarization of antenna array is therefore accomplished by placing the tri-plate structure containing arms of dipole parallel to the Earth surface.

Single dipole element is fed through center conductor of tri-plate transmission line which is simply a 50-Ω conductive strip extended up to dipoles to form a balun. The application of balun lies in the concept that the energy is fed to dipoles in such a manner that current always flows in the same direction in both the arms of the half wave dipole to constructively radiate. Also, it matches the balanced dipole element with unbalanced transmission line. Balun is designed as L-shaped conductive strip terminated at one of the arms of dipole and connected with top and bottom arms through conductive pin [7]. The initial length of the balun runs along one side of the longitudinal dipole slot which is nearly quarter wavelength long. This length is then extended across the slot between the arms running with one of the arms up to an optimized length. The strip is then shorted to the top and bottom arms of the dipole through conductive pin whose diameter is optimized for RF performance.

The novelty of the design lies in impedance matching technique and balun design. The concept is implemented in S-Band, and linear array of 50 elements is designed in the frequency band of 2.9–3.3 GHz with incorporated feed network.

3 Constituent Parts

The design of main constituent parts of linear antenna array is described as follows with their functions and significance to achieve optimum RF performance.

3.1 Radiating Element

Tri-plate structured dipole antenna used as radiating element is basically a half wave dipole fed through the extended center conductor of tri-plate transmission line. The critical dimension of dipole antenna is length and width of dipole arms and the length of longitudinal slot in between the arms which decides the effective impedance of the element and hence determines the working frequency and bandwidth of the antenna. Length of dipole arms is nominally taken as half wavelength corresponding to operating frequency, and depth of slot is optimized at around quarter wavelength. These dimensions of dipole element are optimized for array environment fulfilling the requirement of active return loss in the array.

A Z-shaped projection plate is introduced in the design which is fixed with top and bottom ground plate near the end of longitudinal slot. It absorbs a fraction of the power from dipole and re-radiates it at low angle with respect to elevation plane. This makes the radiated beam of individual element broader and further helps in broad beam scanning of the array.

3.2 Feed Network

In order to feed linear antenna array of 48 dipole elements, feed network consists of two symmetrical 1:24 inline power divider networks placed sideways. Transmission line structure used in the design of power dividers is basically a stripline topology of transmission with air substrate in which top and bottom plate works as ground and conductive strip sandwiched between the plates at constant air gap acts as center transmission path (Fig. 2).

Equal spacing of the center conductor from top and bottom ground plate is essentially maintained across the length of the structure with the support of Teflon spacers which are placed at different location containing center conductor in between and are screwed along with top and bottom ground plates within the air gap (Fig. 3).

Input to each 1:24 way power divider network is provided through coaxial RF connector soldered to the strip parallel to tri-plate plane. Each output of the network

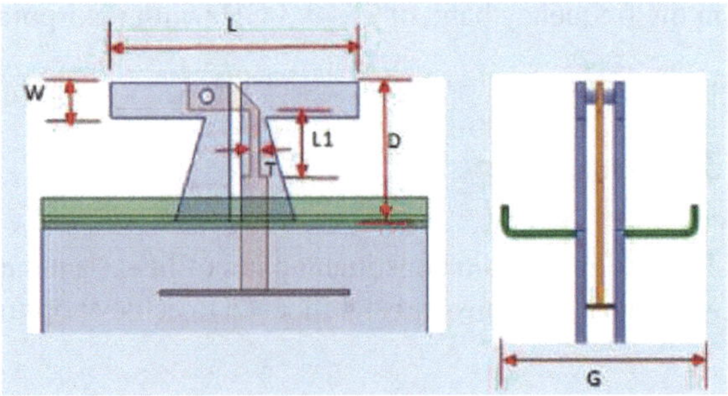

Fig. 2 Tri-plate structured dipole antenna element

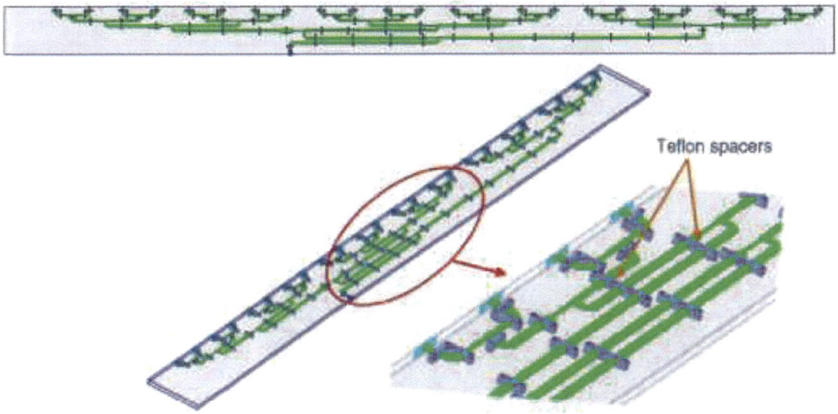

Fig. 3 1:24 inline power divider network with Teflon spacers

is designed with balun to feed RF energy to individual dipole element. Amplitude tapering for the array construction is chosen as per Taylor distribution for side lobe level of 35 dB and N-bar of 4 with uniform phase excitation.

3.3 Linear Array

The array of tri-plate structured dipole elements is designed by repeating the element with optimized inter element spacing of 0.7 times operating wavelength. All the elements lie in the same plane and share the common ground plates in the array.

While designing this type of array, strong coupling between neighboring elements is an undesirable feature that needs to be addressed. It adversely affects the returns loss of the element and finally the required beam shape. To cater this effect, Z projection plate is purposely designed with bend plate portions that helps to minimize the effect of mutual coupling among the elements in vertical plane as it prevents the power radiated from one element to neighboring element. The dimension of projection plate is required to be tuned for the frequency of operation. The elements are made to constructively combine to generate a narrow beam in azimuth plane through amplitude tapering as per suitable Taylor distribution with equal phase excitation fed through incorporated feed network. Ground plates are electrically connected to form a common ground with the help of metal spacers fixed at suitable positions which also help in maintaining air gap between the ground plates (Fig. 4).

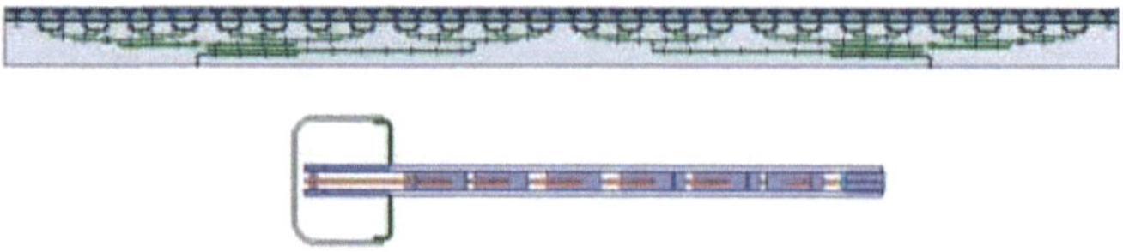

Fig. 4 Linear antenna array of 50 dipole elements with feed network

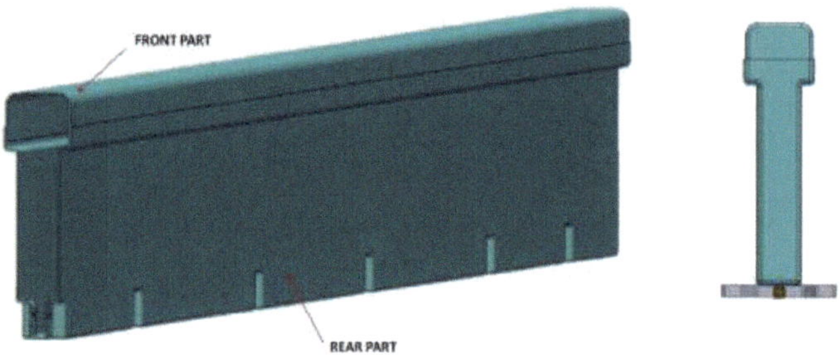

Fig. 5 Structural model of radome with front and rear part combined

3.4 Radome Cover

Radome is a structural protective enclosure used to protect the surface of antenna row from harsh weather and environmental conditions. It is designed with two separate parts glued together through appropriate adhesive and sealant. The front part of radome is critical for RF performance and is designed as C shaped structure with rounded corners resting on Z projection plate to cover all the dipole elements. The rear part of radome is designed to cover ground plates and open spacing between them. Glass Epoxy Prepeg (300GSM) with layer thickness of nearly 0.25 mm is used for fabricating radome. Thickness of both the parts is kept 1 mm to obtain optimum structural strength along with minimal RF loss. The effect of radome material on RF performance is catered during unit cell simulation and array synthesis by using structural model of the radome with optimized distance form dipole arms (Fig. 5).

4 Design Validation

4.1 3D Modeling and RF Simulation

The conceptual design of tri-plate structured dipole element is modeled and simulated using advanced RF simulation tool for required frequency band. To optimize single element with respect to planar antenna array, unit cell simulation has been performed for active S parameters. Single element exhibits the simulated gain of more than 6 dBi and elevation beam of 123° with minimum 16 dB active return loss at boresight and 11 dB while 30° scanning in array environment (Figs. 6, 7; Table 1).

The layout of 1:24 way power divider network is generated as per Track width calculation of conductive strip for impedance of each distribution line based on stripline transmission topology. The model of feed network is then created and simulated to achieve required amplitude and phase distribution. Amplitude variation of ± 0.5 dB and phase variation of ± 5° could be achieved among the output ports across the frequency band in the simulation.

A linear array of 50 dipole elements is synthesized with tri-plate feed network and equivalent radome structure using Ansys Electronics Desktop for analysis of active S parameter and electromagnetic radiation. The complete row of elements is fed through four coaxial RF connectors; two are used to provide RF input to two separate 1:24 feed networks and the other two are used to feed two extreme SLB/SLC elements. Considering the end application of dipole antenna row in planar antenna aperture, row is simulated with lattice boundary in vertical plane to optimize active return loss. This simulation is equivalent to place at least 9 numbers of antenna rows parallel to each other on either side of center row in vertical plane with inter row spacing of 50 mm (Figs. 8 and 9).

The realized gain of the antenna row achieved in the simulation is more than 20 dBi at required frequency spots with elevation beamwidth of nearly 130° and average side lobe level of 30 dB. Azimuth bandwidth of 2.0 ± 0.3°, derived from

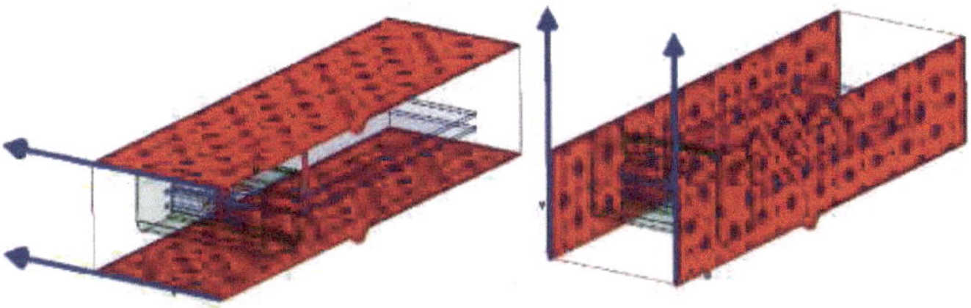

Fig. 6 Lattice boundary in horizontal and vertical plane of single radiating element for unit cell simulation

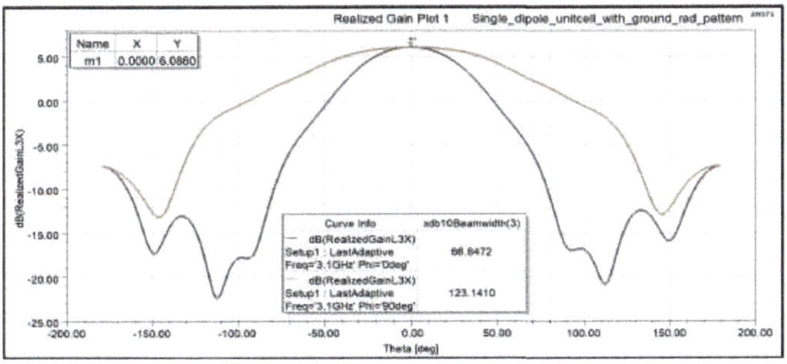

Fig. 7 Simulated radiation pattern of single dipole element at 3.1 GHz

Table 1 Optimized design parameters of radiating element

S. No.	Design parameters	Initial calculated (mm)	Final opt. value (mm)
1	Dipole length (*L*)	48	45
2	Dipole depth (*D*)	24	24
3	Dipole width (*W*)	5	6
4	Balun track length	12	11.5
5	Balun track width	2.2	1.7
6	Cover plate gap (*G*)	28	31.2

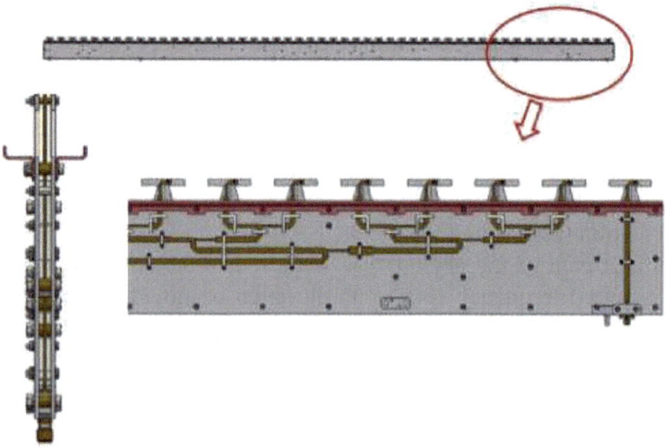

Fig. 8 3D model of linear antenna array of 50 elements with incorporated feed network

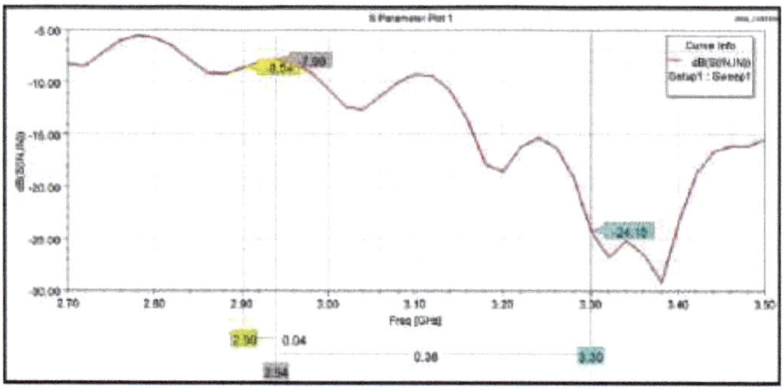

Fig. 9 Simulated active return loss of linear antenna row while using in the planar aperture

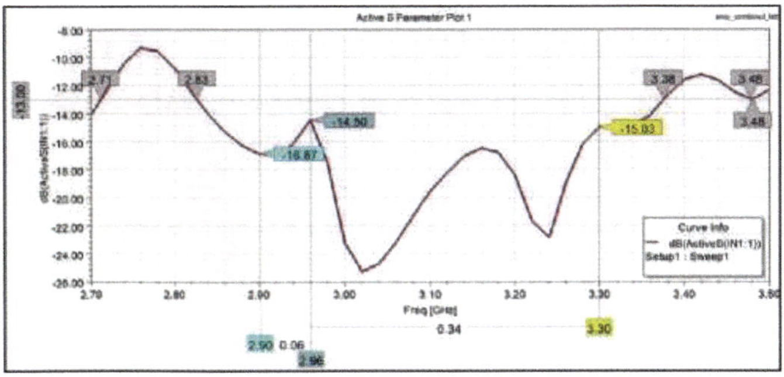

Fig. 10 Simulated return loss of single linear antenna row

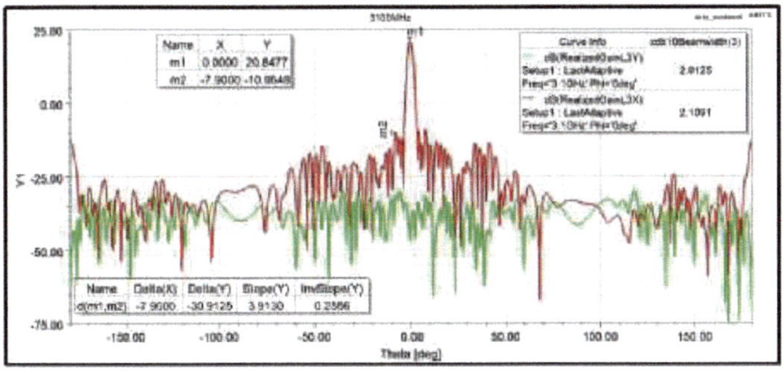

Fig. 11 Azimuthal radiation pattern of linear antenna row at 3.1 GHz

system calculation is achieved across the frequency band using 48 dipole elements in the designed antenna row. The antenna array exhibits a simulated active return loss of better than 14 dB across the band in array environment, whereas in standalone condition, the simulated return loss is better than 8 dB. 3D CAD Tool is used for modeling the final structure of linear antenna row with radome, mounting brackets, and RF connectors. The dimensions of the designed antenna row are 3500 mm × 200 mm × 35 mm (Figs. 10, 11, 12; Table 2).

4.2 *Fabrication and Measurements*

After validation using RF simulation tool, one linear antenna row has been manufactured, assembled, and tested without radome at bench level for return loss measurement in standalone condition. Return loss better than 8.5 dB is achieved across the required frequency band matching with the simulated results.

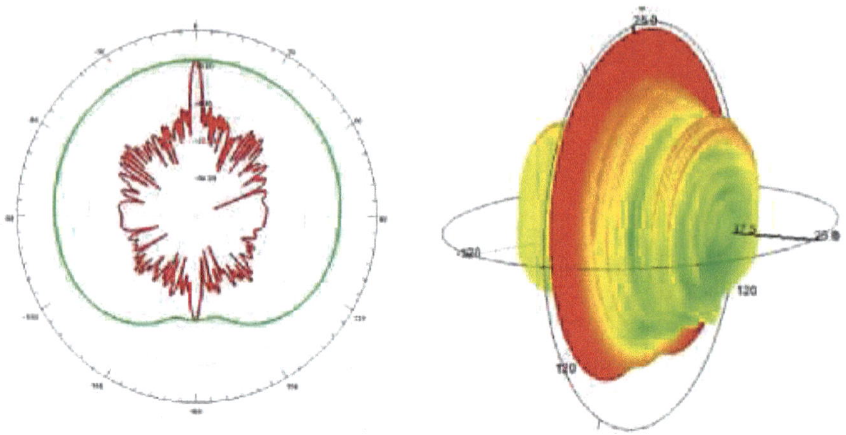

Fig. 12 2D and 3D polar radiation plot of linear antenna row showing azimuth and elevation patterns at center frequency, 3.1 GHz

Table 2 Summarized results of simulated radiation pattern of linear antenna array

Freq. (GHz)	Gain (dB)	SLL (dB)	AZ BW (°)	EL BW (°)
2.9	20.3	− 25.1	2.32	129.3
3.0	21.2	− 32.7	2.09	130.2
3.1	20.8	− 30.9	2.10	131.4
3.2	21.6	− 35.2	2.00	132.6
3.3	21.7	− 28.0	1.95	133.9

To check air tightness of antenna array after assembly of radome covers, inlet and outlet ports are provided at either ends of the row to conduct air leak test (Figs. 13, 14, 15, 16; Table 3).

Fig. 13 Fabricated antenna row without radome along with air leak test ports and customized connector

Fig. 14 Return loss measurement set-up for fabricated antenna row with radome

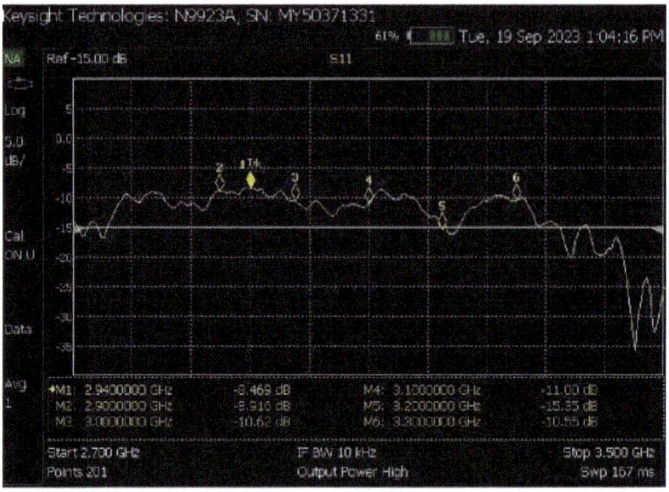

Fig. 15 Snapshot of measured standalone S parameter of antenna port 1

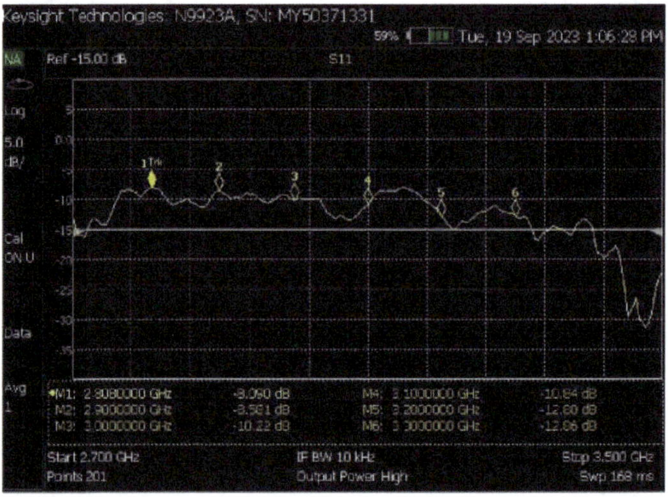

Fig. 16 Snapshot of measured standalone S parameter of antenna port 2

Table 3 Summarized results of standalone return loss for antenna ports of linear array

Standalone return loss (dB)						
Freq. (GHz)	2.9	3.0	3.1	3.2	3.3	Worst
ANT port-1	– 8.9	– 10.6	– 11	– 15.3	10.5	– 8.5 (2.94 GHz)
ANT port-2	– 8.6	– 10.2	– 10.8	– 12.8	– 12.8	– 8.6 (2.9 GHz)

5 Conclusion

A linear antenna array of tri-plate structured dipole element with incorporated feed network is designed, realized, and tested at 2.9–3.3 GHz. Measured S parameter is showing close agreement with simulation result. Radiation parameters and active return loss are yet to be validated after fabrication of 4 more numbers of similar antenna rows along with radome covers which are already under manufacturing. For successful operation of radar, 40 numbers of designed antenna array are required to create planar antenna aperture for scanning in azimuth and elevation planes. Easy manufacturability, light weight structure and broad beam scanning of the antenna row makes it suitable for land-based surveillance radar applications.

References

1. Skolnik M (2008) Radar handbook, 3rd edn. McGraw Hill, New York, pp 2.98–2.100
2. Balanis CA (2005) Antenna theory: analysis and design. Wiley, New York, pp 538–541
3. Palson CL, Lakshmy EKM, Sunny AE, Krishna DD (2016) Analysis of coaxial stub type balun fed dipole antenna. In: IEEE 2016 sixth international symposium on embedded computing and system design (ISED)
4. Li RL, Wu T, Pan B, Lim K, Laskar J, Tentzeris MM (2009) Equivalent-circuit analysis of a broadband printed dipole with adjusted integrated balun and an array for base station applications. IEEE Trans Antennas Propag 57(7)
5. Pozar DM (1998) Microwave engineering, 2nd edn. Wiley, New York, pp 246–248
6. Wu D, Yin Y, Guo M, Shen R (2005) Wideband dipole antenna for 3G base stations. In: IEEE international symposium on microwave, antenna, propagation & EMC technologies for wireless communications, vol 1
7. Ávila-Navarro E, Antón J, Blanes JM, Reig C (2010) Broadband printed dipole with integrated via-hole balun for WiMAX applications. Microw Opt Technol Lett 53:52–55

Scattering Control and RCS Reduction by Polarization Conversion Metasurface

Krunal Patel and Manjusha Joshi

Abstract In this paper, the concept of Coding Metasurfaces is presented for RCS reduction. RCS reduction is the predominant parameter in the Defence Applications. Metamaterials have the ability to change the properties of the electromagnetic wave. The two-dimensional versions of the metamaterials are known as metasurfaces. The Polarization Conversion Metasurface has the ability to achieve RCS reduction over a wide bandwidth. The shape of the reflected beam can be changed by different coding schemes of the metasurfaces. By gradually rotating the unit cell, the phase gradient can be achieved which is used to obtain wide RCS reduction bandwidth. In this paper, various Polarization Conversion Metasurfaces are discussed with different coding schemes in order to demonstrate their RCS reduction capabilities. The main novelty of the paper is that it introduces the Coding Metasurfaces and their RCS reduction performance for a plane wave incident at the normal direction and oblique incidence.

Keywords Radar cross section reduction · Polarization Conversion Metasurface · Coding Metasurface · CST Microwave Studio

1 Introduction

The common techniques for RCS reduction are target shaping and the use of Radar Absorbing Materials. The target shaping methods may not be useful due to the aerodynamic shape of the object, and Radar Absorbing Materials are narrow band in nature. The use of Frequency Selective Surfaces and Electromagnetic Bandgap structures is also prevalent. In Ref. [1], the target shaping method was discussed to redirect the backscattered radiation. In Ref. [2], a metamaterial absorber was designed using

K. Patel (✉)
MPSTME, NMIMS University, Mumbai, Maharashtra, India
e-mail: krunalmailbox@rediffmail.com

M. Joshi
Amity University, Mumbai, Mumbai, Maharashtra, India

S. M. Thampi et al. (eds.), *Fifth International Conference on Computing and Network Communications*, Lecture Notes in Electrical Engineering 1219,
https://doi.org/10.1007/978-981-97-4540-1_40

circular sectors which offered peak absorptivity of 91% at 10.44 GHz. The design exhibited greater than 90% absorption at 70° angle of incidence. Dang et al. [3] designed High Impedance Surface-based absorbers which used two stacked layers of resistive Frequency Selective Surfaces. In Ref. [4], MTM absorber was used to achieve RCS reduction, but the device exhibited narrow band performance. It showed RCS reduction over the frequency range of 7–12.8 GHz with fractional bandwidth of 58.59%. Tennant and Chambers [5] designed an absorber using active Frequency Selective Surface controlled by PIN diodes. The performance of the absorber could be controlled over the frequency range of 9–13 GHz. The design was also complex due to the incorporation of PIN diodes. Huang et al. [6] proposed a metamaterial absorber using six close rings designed in two metallic layers. The metallic layers were separated by FR-4 substrates. The absorber achieved peak absorptivity of 93.3%, 93.6% and 94.9% at the frequencies of 4.76 GHz, 8.65 GHz and 13.55 GHz, respectively. The RCS reduction of a microstrip antenna over a wide band was achieved in Ref. [7]. The overall antenna RCS reduction bandwidth was 112.06%. The self-repeated fractal geometries can also be utilized to design Polarization Conversion Metasurfaces. One such approach is followed in Ref. [8]. A single CSRR structure was reported in Ref. [9] for beam steering applications. This CSRR structure can be modified by the fractal curves to obtain RCS reduction. From the literature review, it is observed that metamaterial absorbers are bandlimited in nature. Recently, the Coding Metasurfaces [10] are being developed to obtain wideband RCS reduction. These Polarization Conversion Metasurfaces (PCM) can be designed by rotating the unit cells gradually and obtaining the required phase gradient for RCS reduction. The Coding Metasurface can also be designed by arranging the PCM unit cells in the different directions. In Ref. [11], a Coding Metasurface was investigated using polarization rotators. The designed metasurface showed 10 dB RCS reduction over the frequency range of 1.9–5.5 THz (97.29%). The overall RCS reduction was obtained for a narrow frequency band. Reference [12] used two single-band Artificial Magnetic Conductors (AMCs) in checkerboard configuration to obtain the phase difference of $180° \pm 37°$. RCS reduction of 10 dB was obtained from 4.1 to 7.9 GHz (63%). The reported overall RCS reduction was not obtained over a wide frequency band. Reference [13] reported the checkerboard configuration with single-band AMCs and double-band AMCs. The overall RCS reduction band was from 12.5 to 24 GHz (63.01%). A metamaterial absorber structure was reported in Ref. [14] which showed overall absorption bandwidth of 76.92% over the frequency range of 8–18 GHz.

The reported literature shows that the RCS reduction is obtained over a narrow frequency band. The published literature also doesn't include scattering control and beam shaping of the reflected wave by changing the directions of the unit cells. In this paper, a new type of PCM unit cell is designed and various Coding Metasurfaces are investigated. The paper also shows how the scattering can be controlled by changing the coding scheme. This also helps to select the best configuration for RCS reduction.

2 Development of the Polarization Conversion Metasurface Unit Cell

Figure 1 shows the evolution of Polarization Converter Metasurface from a ring-shaped structure. Figure 1a shows the circular ring with single split. It is the Step I in the design of the unit cell. Figure 1b shows the circular ring with two splits. It shows the Step II of the design. The dimensions of the unit cells are 7 mm × 7 mm. The radius of the outer ring is 3 mm, and the radius of the inner ring is 2.5 mm. The thickness of the substrate is 2.4 mm. The type of the substrate is FR-4 (glass–epoxy) with relative permittivity of 4.3. The unit cell simulations are done in the Frequency Domain Solver with Periodic Boundary Conditions.

As shown in Step I of the design if only one split is created then the co-polar reflection coefficient is still higher. The minima are located at 5.1, 12.18, 21.18 and 24.39 GHz for TE mode as shown in Fig. 2a. The values of the reflection coefficients are − 5.03 dB, − 1.32 dB, − 4.25 dB and − 7.7 dB, respectively. The cross-polar reflection coefficient observes maximum values at 5.04, 12.3, 20.24 and 24.63 GHz for the TE mode. The values of the cross-polar reflection coefficient are − 65.86 dB, − 59.36 dB, − 49.08 dB and − 51.32 dB, respectively. As shown in Fig. 2b, the minima of the co-polar reflection coefficient are observed at 7.65, 21.28 and 24.59 GHz for the TM mode. The values of the reflection coefficient are − 1.4 dB, − 3 dB and − 8.75 dB, respectively. The maxima of the cross-polar reflection coefficient are observed at 5.04, 12.3, 20.27, 22.78 and 24.56 GHz for this mode. The values of the cross-polar reflection coefficient are − 65.8 dB, − 59.29 dB, − 49.8 dB, − 62 dB and − 52.27 dB, respectively, at these frequencies.

For the ring structure with two splits in the orthogonal directions, the reflection coefficient is minimized at the frequencies of 6.39, 10.38, 14.98, 20.7 and 24.74 GHz for the TE mode as shown in Fig. 3a. The values of the reflection coefficients are − 17.58 dB, − 26.19 dB, − 26.2 dB, − 24.03 dB and − 10.57 dB at these frequencies,

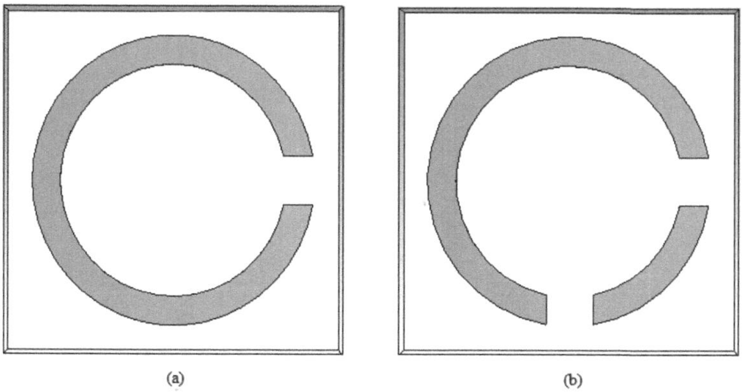

Fig. 1 Steps for designing the Polarization Converters **a** Step I, **b** Step II

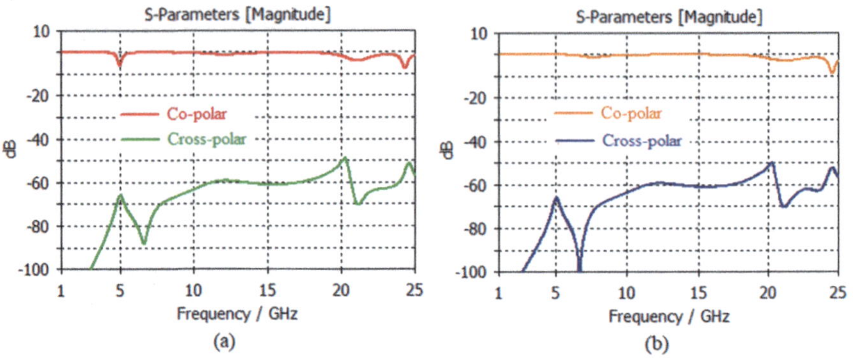

Fig. 2 Co-polar and cross-polar reflection coefficients for Step I **a** TE mode, **b** TM mode

respectively. It is observed that if an additional split is created then the number of resonances increases and the polarization conversion efficiency of the structure increases. Figure 3b shows the co-polar and cross-polar reflection coefficients for the TM mode. The co-polar and cross-polar reflection coefficient values are almost the same as those of the TE mode.

To observe the effects of the substrate height on the bandwidth, the parametric analysis is done for various heights of the substrate as shown in Fig. 4a. The first resonance is observed almost at the frequency of 6.3924 GHz for the various heights. The second and third resonance frequencies are increased with the reduction in the height of the substrate. The fourth resonance values are obtained at 20.696 GHz, 21.044 GHz, 21.55 GHz, 22.089 GHz and 22.63 GHz for the substrate heights of 2.4 mm, 2.3 mm, 2.2 mm, 2.1 mm and 2 mm, respectively. The simulations results show that if the height of the substrate decreases then the bandwidth of polarization conversion increases. Figure 4b shows the variations in the cross-polar reflection coefficient with respect to the height of the substrate. Figure 5a shows the parametric analysis of the co-polar reflection coefficient with respect to the height of the substrate

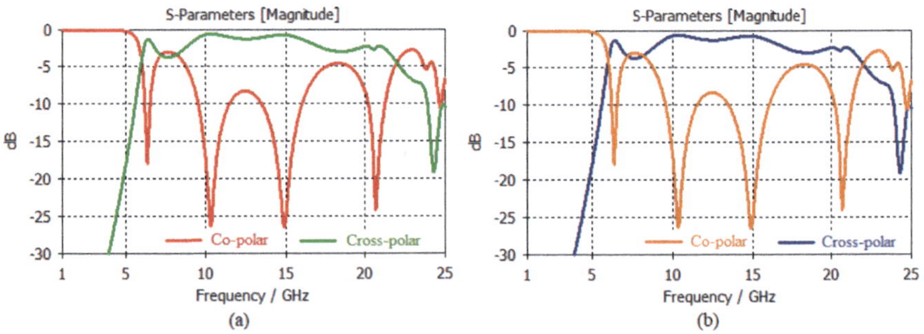

Fig. 3 **a** Co-polar and cross-polar reflection coefficients for Step II (TE mode), **b** co-polar and cross-polar reflection coefficients for Step II (TM mode)

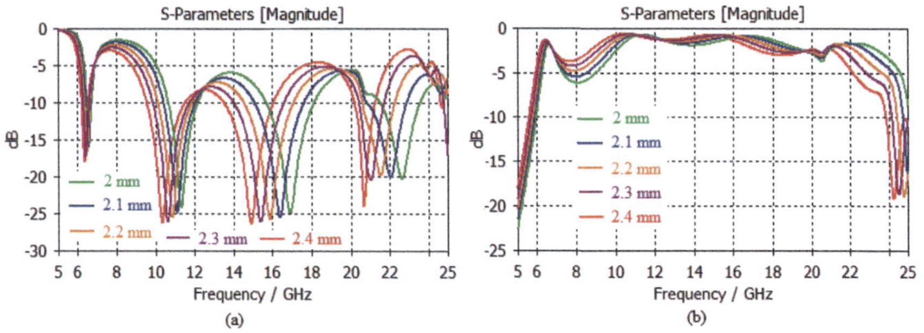

Fig. 4 **a** Parametric analysis w.r.t. the height of the substrate for TE mode (co-polar reflection coefficient), **b** parametric analysis w.r.t. the height of the substrate for TE mode (cross-polar reflection coefficient)

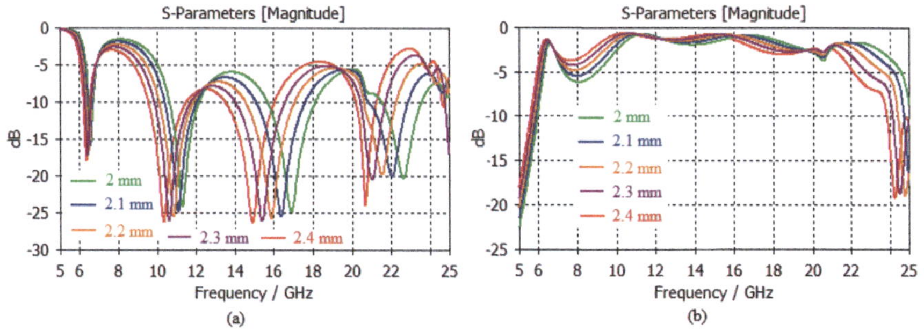

Fig. 5 **a** Parametric analysis w.r.t. the height of the substrate for TM mode (co-polar reflection coefficient), **b** parametric analysis w.r.t. the height of the substrate for TM mode (cross-polar reflection coefficient)

for the TM mode. Figure 5b shows the cross-polar reflection coefficients for the different heights of the substrate in the case of TM mode.

3 Design of the Coding Metasurfaces

Figure 6 shows the designed metasurfaces using the Polarization Conversion Metasurface unit cells. The design of Fig. 6a shows that the directions of all the unit cells are the same. This design is designated as Metasurface 1. In Fig. 6b the unit cells in the adjacent regions are oppositely directed. This design is designated as Metasurface 2. It has 10/01 coding scheme. Figure 7a shows the design of Metasurface 3, and Fig. 7b shows the design of Metasurface 4. Metasurface 3 indicates the metasurface with the coding scheme 1–0–1. Whereas, Metasurface 4 shows the metasurface with the coding scheme 10/10.

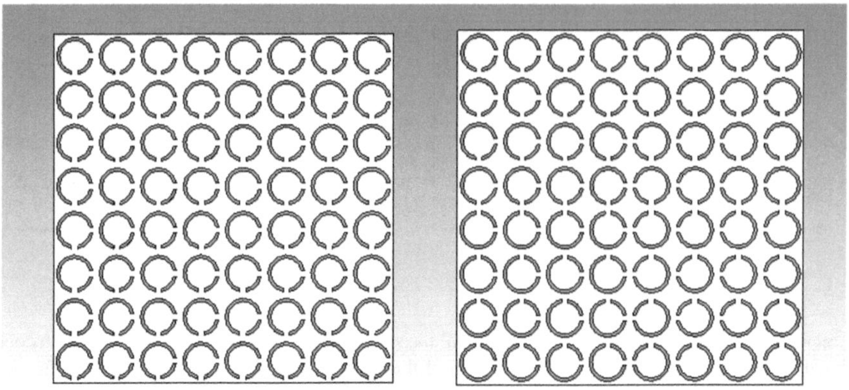

(a) (b)

Fig. 6 Designs of metasurfaces **a** Metasurface 1, **b** Metasurface 2

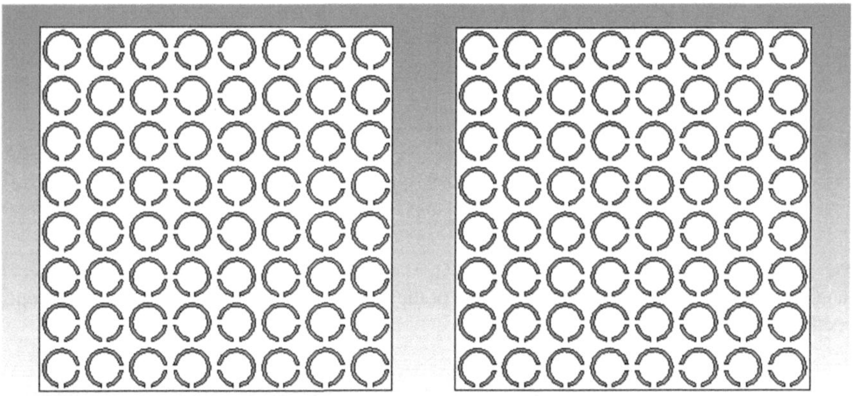

(a) (b)

Fig. 7 Designs of metasurfaces **a** Metasurface 3, **b** Metasurface 4

Figure 8 compares the RCS reduction performance of the Metal Sheet, Metasurface 1 and Metasurface 2 in the case of normal incidence. The RCS simulations are done in the Time Domain Solver of CST Microwave Studio software. It is observed that Metasurface 2 shows RCS reduction from 5 to 25 GHz frequency range due to the oppositely directed unit cells in the adjacent regions of the metasurface. The overall RCS reduction bandwidth is 133.33%. In the case of Metasurface 1, no RCS reduction is observed due to the arrangement of the unit cells in the same directions.

Figure 9 shows the RCS comparison of various coding schemes for the TM mode. From the comparison, it is evident that Metasurface 4 shows better RCS reduction for the TM mode. For the Metasurface 3, the minimum values of RCS are observed at 6 GHz (− 19.49 dBsm), 10.69 GHz (− 16.43 dBsm), 14.97 GHz (− 11.85 dBsm),

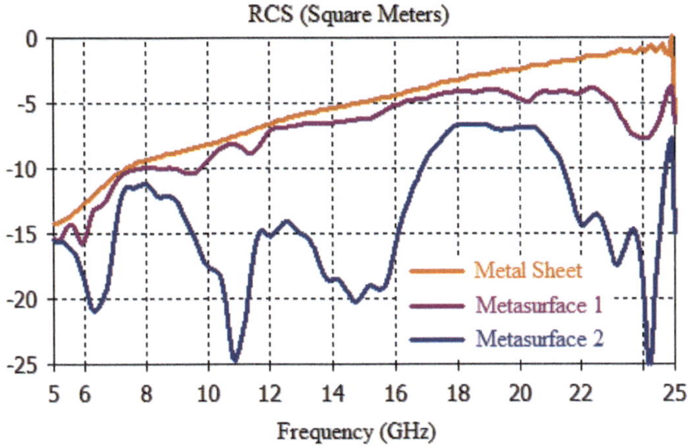

Fig. 8 RCS comparison at the normal incidence

21.38 GHz (− 9.89 dBsm) and 24.09 GHz (− 11.06 dBsm). In the case of Metasurface 4, the RCS reduction peaks are observed at 6.18 GHz (− 24.07 dBsm), 10.52 GHz (− 39.89 dBsm), 14.71 GHz (− 27.29 dBsm), 21.5 GHz (− 16.09 dBsm) and 23.98 dBsm (− 16.58 dBsm). Metasurface 2 shows the RCS reduction peaks at 6.45 GHz (− 21.33 dBsm), 10.92 GHz (− 24.6 dBsm), 14.72 GHz (− 20.24 dBsm), 22.05 GHz (− 14.36 dBsm), 23.15 GHz (− 17.35 dBsm) and 24.18 GHz (− 25.88 dBsm).

Figure 10a–c shows the comparison of 3D scattering from the Metasurface 3, Metasurface 4 and Metasurface 2, respectively. Metasurface 2 with PCM unit cells in the mirror configuration generates 4-beam pattern. It is observed that the Metasurface

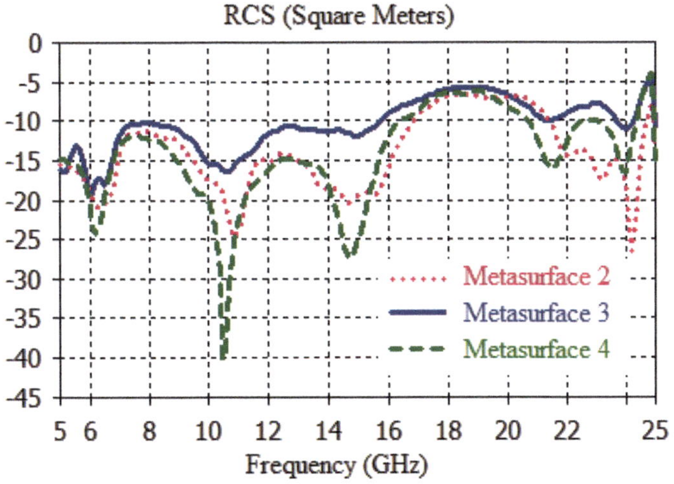

Fig. 9 RCS comparison of various coding schemes (TM mode)

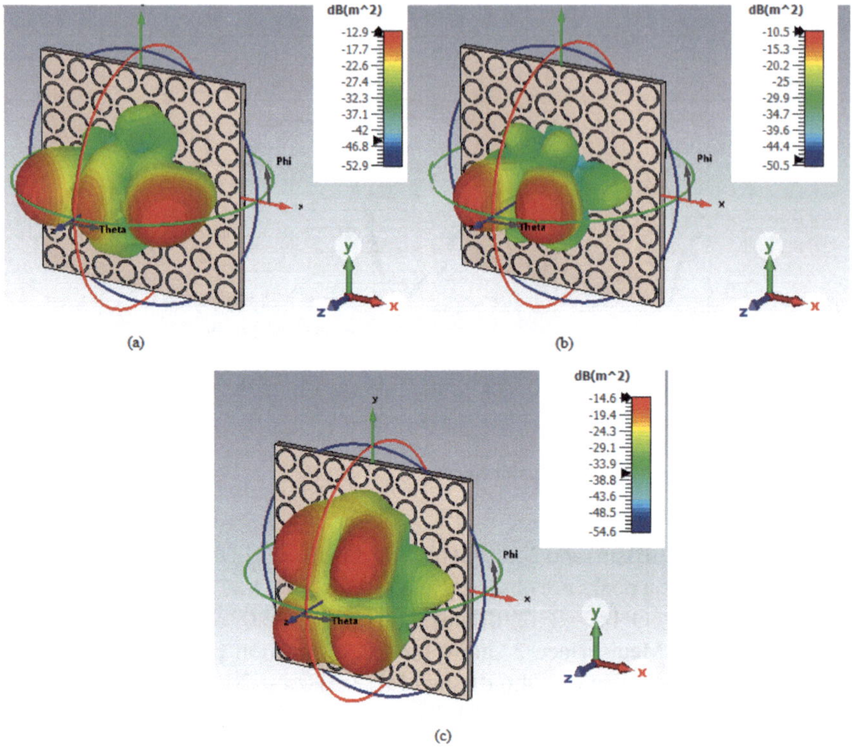

Fig. 10 3D scattering from the metasurfaces **a** Metasurface 3, **b** Metasurface 4, **c** Metasurface 2

3 with 1–0–1 coding scheme generates 3-beam pattern. Metasurface 4 generates 2-beam pattern.

Figure 11 shows the RCS comparison of the samples for the TE mode. For Metasurface 2, the RCS reduction peaks are observed at 6.45 GHz (− 21.33 dBsm), 10.92 GHz (− 24.6 dBsm), 14.72 GHz (− 20.24 dBsm), 22.05 GHz (− 14.36 dBsm), 23.15 GHz (− 17.35 dBsm) and 24.18 GHz (− 25.88 dBsm). Metasurface 3 shows RCS reduction peaks at 5.99 GHz (− 18.71 dBsm), 11.09 GHz (− 15.67 dBsm), 21.83 GHz (− 9.7 dBsm) and 24.13 GHz (− 11.76 GHz). In the case of Metasurface 4, the minimum values of RCS are observed at 6.54 GHz (− 23.45 dBsm), 10.8 GHz (− 32.42 dBsm), 13.69 GHz (− 24.11 dBsm) and 21.68 dBsm (− 17.69 dBsm).

Figure 12 compares the Bistatic RCS performance of the metasurfaces in the $\phi = 0°$ plane for the normal incidence at 10.69 GHz frequency. The RCS values of the Metal Sheet, Metasurface 2, Metasurface 3 and Metasurface 4 are − 6.2 dBsm, − 24.14 dBsm, − 16.45 dBsm and − 31.48 dBsm, respectively. The Metasurface 3 generates 3 beams at 0 and ± 40°. The Metasurface 4 generates 2 beams at ± 19°. As far as Bistatic RCS reduction is concerned, Metasurface 2 shows superior Bistatic RCS reduction performance as compared with the other configurations. It shows RCS reduction over the angular range of ± 45°.

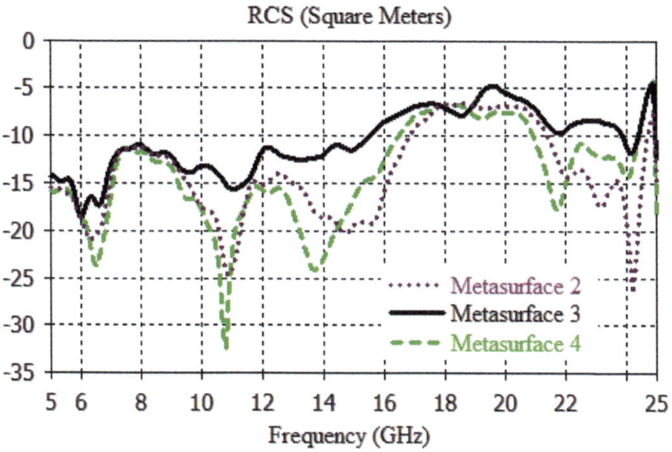

Fig. 11 RCS comparison of various coding schemes (TE mode)

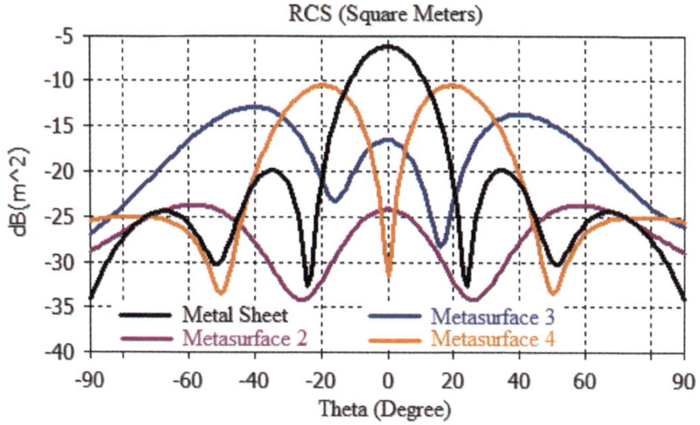

Fig. 12 Bistatic RCS comparison for the metasurfaces at $\theta = 0°$, $\phi = 0°$

Figure 13 compares the Bistatic scattering RCS of the metasurfaces in the $\theta = 0°$, $\phi = 90°$ plane at 10.69 GHz frequency. The RCS values of the Metallic Plate, Metasurface 2, Metasurface 3 and Metasurface 4 are − 6.2 dBsm, − 24.14 dBsm, − 16.45 dBsm and − 31.48 dBsm, respectively, in the boresight direction. The Metasurface 2, Metasurface 3 and Metasurface 4 show RCS reduction of 17.94 dBsm, 10.25 dBsm and 25.28 dBsm, respectively. The Metasurface 4 shows good RCS reduction as compared to the other designs in the boresight direction in the $\phi = 90°$ plane. In the other directions also, Metasurface 4 maintains RCS reduction over a wide angular range. Whereas, the Metasurface 2 shows Bistatic Scattering RCS reduction almost over ± 45°. The RCS reduction performance of Metasurface 4 is different in both the principal planes. So, the Metasurface 2 is preferred for the design of low RCS platforms.

Table 1 shows the comparison of the proposed work with the relevant literature. The reported literature in the comparison table showed narrow RCS reduction bandwidth. The absorber-based designs are not capable to obtain scattering control and wide RCS reduction bandwidth. The comparison shows that the proposed designs are capable to obtain wide RCS reduction bandwidth along with the ability of scattering control.

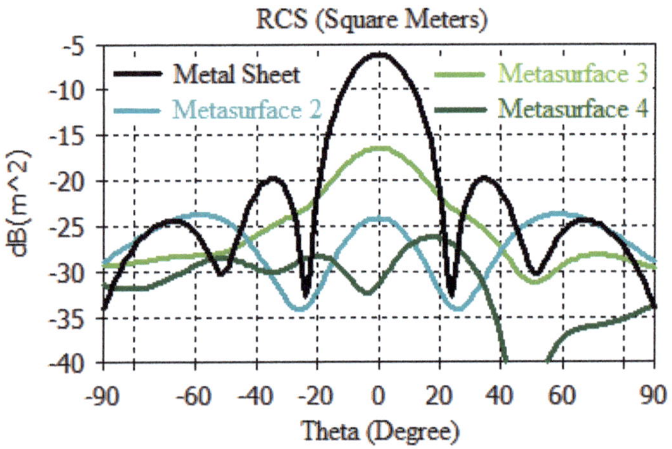

Fig. 13 Bistatic RCS comparison for the metasurfaces at $\theta = 0°$, $\phi = 90°$

Table 1 Comparison with relevant work

References	Methodology used	Percentage bandwidth	Remarks
[11], 2020	Coding Metasurface	1.9–5.5 THz (97.29%), 10 dB RCS reduction BW	Relatively narrow bandwidth
[12], 2015	Single-band AMCs	4.1–7.9 GHz (63%) (10 dB RCS reduction BW)	Narrow bandwidth
[13], 2013	Checkerboard configuration	12.5–24 GHz (63.01%) (overall)	Relatively narrow bandwidth
[14], 2019	Metamaterial absorber	8–18 GHz (76.92%) (overall absorption BW)	Beam shaping is not possible
Proposed work	Coding Metasurface	5–25 GHz (133.33%) (overall)	Beam shaping is possible, wide RCS reduction bandwidth

4 Conclusion

A C-shaped Polarization Conversion Metasurface unit cell is designed, and RCS reducer metasurfaces are investigated based on this design. In terms of RCS reduction peaks, the Metasurface 4 with 10/10 coding scheme shows better performance for TE as well as TM modes. But it has no four-fold symmetry. So its RCS reduction performance is different for TE mode and TM mode in the case of normal incidence. It generates 2-beams pattern. The Metasurface 3 with 1–0–1 coding scheme has 3-beams pattern, and it shows less RCS reduction as compared to Metasurface 2 and Metasurface 4. Metasurface 2 has same RCS reduction performance for TE mode as well as TM mode due to four-fold symmetry for the normal incidence. As far as Bistatic RCS reduction performance is considered, then Metasurface 2 shows better RCS reduction performance over a wide angular range. Metasurface 2 gives wide RCS reduction bandwidth of 133.33% starting from 5 to 25 GHz. This can also be used to design antennas with low observability. The use of Polarization Conversion Metasurface has noteworthy advantages over metamaterial absorbers and traditional frequency selective surfaces.

References

1. Knott EF, Tuley MT, Shaeffer JF (1993) Radar cross section, 2nd edn. Artech House, Norwood, MA
2. Lee D, Hwang JG, Lim D, Hara T, Lim S (2016) Incident angle and polarization insensitive metamaterial absorber using circular sectors. Sci Rep 6:27155
3. Dang K-Z, Shi J-M, Wang J-C, Lin Z-D, Wang Q-C (2015) Tunable wideband absorber based on resistively loaded lossy high-impedance surface. Chin Phys B 24(10):104104
4. Nguyen TT, Lim S (2018) Angle and polarization insensitive broadband metamaterial absorber using resistive fan-shaped resonators. Appl Phys Lett 112(2):021605
5. Tennant A, Chambers B (2004) A single-layer tuneable microwave absorber using an active FSS. IEEE Microw Wirel Compon Lett 14(1):46–47
6. Huang L, Chen H (2011) Multi-band and polarization insensitive metamaterial absorber. Prog Electromagn Res 113:103–110
7. Patel K, Joshi M (2023) Wideband RCS reduction of microstrip antenna using artificial magnetic conductor metasurface. Sādhanā 48(88)
8. Hong T, Peng K, Gu Y, Jiang W, Gong S (2018) Wideband polarization conversion metasurface for RCS reduction of antennas. Int J Antennas Propag 2018:1–8
9. Cao W, Xiang Y, Zhang B, Liu A, Yu T, Guo D (2011) A low-cost compact patch antenna with beam steering based on CSRR-loaded ground. IEEE Antennas Wirel Propag Lett 10:1520–1523
10. Ma Q, Xiao Q, Hong QR, Gao X, Galdi V, Cui TJ (2022) Digital coding metasurfaces: from theory to applications. IEEE Antennas Propag Mag 64(4):96–109
11. Qi Y, Zhang B, Liu C, Deng X (2020) Ultra-broadband polarization conversion metasurface and its application in polarization converter and RCS reduction. IEEE Access 8:116675–116684
12. Chen W, Balanis CA, Birtcher CR (2015) Checkerboard EBG surfaces for wideband radar cross section reduction. IEEE Trans Antennas Propag 63(6):2636–2645

13. Iriarte Galarregui JC, Tellechea Pereda A, de Falcón JLM, Ederra I, Gonzalo R, de Maagt P (2013) Broadband radar cross-section reduction using AMC technology. IEEE Trans Antennas Propag 61(12):6136–6143
14. Sun L, Sun J, Yang B, Gao X, Long H, Shao Y (2019) A simplified design of broadband metamaterial absorber covering X- and Ku-band. Mater Res Express 6:125805

Performance of SOVA Decoded SCCPM with SNR Mismatch

Zhijie Guo, Lei Cao, and Ramanarayanan Viswanathan

Abstract Concatenated coding schemes with iterative decoding structures require the exchange of soft information between component decoders to improve the decoding results, which are heavily dependent on the accuracy of statistical properties of the channel, i.e., the signal to noise ratio (SNR). Incorrect estimation of the channel SNR is known as SNR mismatch (SNRM) and can lead to serious degradation in iterative decoding performance. We study the relationship between the number of pilot symbols and the performance of a Maximum Likelihood Estimation (MLE) channel estimator operating over a Binary Input Additive White Gaussian Noise (BIAWGN) channel. The simulation shows that using a larger number of pilot symbols leads to more precise estimations of both the noise variance and symbol energy. Nonetheless, the accuracy comes at the expense of bandwidth efficiency. There exists a substantial tradeoff between bandwidth efficiency and estimation accuracy. Furthermore, We investigate the impact of the SNRM on the performance of the Soft Output Viterbi Algorithm (SOVA) decoded serially concatenated continuous phase modulation (SCCPM) system. Our simulation shows that under-estimation of the channel SNR has a more detrimental effect on the SCCPM system compared to over-estimation.

Keywords SOVA · SCCPM · SNR mismatch

Z. Guo · L. Cao (✉) · R. Viswanathan
Department of Electrical and Computer Engineering, University of Mississippi, University, MS 38677, USA
e-mail: lcao@olemiss.edu

Z. Guo
e-mail: zguo2@go.olemiss.edu

R. Viswanathan
e-mail: viswa@olemiss.edu

S. M. Thampi et al. (eds.), *Fifth International Conference on Computing and Network Communications*, Lecture Notes in Electrical Engineering 1219,
https://doi.org/10.1007/978-981-97-4540-1_41

1 Introduction

Continuous Phase Modulation (CPM) [1] is a class of digital phase modulation schemes known for the constant envelope and inherent memory properties. CPM has been widely used in various application areas, including satellite communication and Bluetooth technology. Its popularity arises from its excellent bandwidth and energy efficiency, attributed to the continuous phase property, which effectively narrows the spectrum's main lobe. T. Aulin and Sundberg gave a comprehensive description and analysis of the full-response CPM signal, power spectrum properties, and error performances in [2]. Another important feature of CPM that attracted research interest is the memory of phase. The memory property lets the signal phase possess a form of finite-state trellis. Therefore, trellis-based decoding algorithms can be implemented [3–5].

A CPM signal can be decomposed into two separate parts: a linear encoder called the Continuous Phase Encoder (CPE) and a Memoryless Modulator (MM). This approach allows the coding (decoding) process independent from the modulation (demodulation) process. Rimoldi [6] proved that the CPE is in a form of convolutional code (CC) with code rate 1 and proposed a form of tilted-phase trellis that simplifies the structure of the finite-state trellis of CPM with a reduced number of states in the trellis. Various types of channel encoders such as convolutional codes and LDPC codes can be cascaded with the CPE to constitute serially concatenated continuous phase modulation (SCCPM) systems. An iterative decoding process can be performed in the SCCPM to obtain more coding gain. Much work on SCCPM has been done based on the CPM decomposition model [5, 7].

There are two algorithms to decode the channel codes with a structure of finite-state trellis. One is the Viterbi algorithm (VA) that is a maximum likelihood sequence estimation (MLSE) decoder to search and select the optimum path in the trellis [8, 9]. The other is the BCJR algorithm that is a maximum a posteriori (MAP) decoder [10]. The BCJR MAP algorithm obtains the optimum decode results bit-wise and it can provide the reliability information for each decoded bit. VA can also be modified to generate the reliability information for each decoded symbol. The modified VA is referred to as the Soft Output Viterbi Algorithm (SOVA), which was proposed by Hagenauer [11, 12]. With the rapid development of turbo coding technology, to achieve a better bit error rate (BER) performance and a lower decoding delay, a large number of works have been carried out to modify the BCJR algorithm to reduce the complexity, such as the Max-Log-MAP algorithm proposed by Koch and Baier [13] and Erfanian et al. [14] with a sub-optimum performance by using an approximation in the log domain. Robertson et al. [15] further improved the approximation in the previous Max-Log-Map algorithm and the improved algorithm is referred to as Log-Map algorithm. The Log-Map can approach the performance of the conventional BCJR algorithm by less than 0.1 dB at BER of 10^{-4} and outperform the Max-log-MAP by 0.4 dB but with a significant reduction in the complexity [16].

Although the (Max-)Log-MAP algorithm significantly reduces the complexity, the SOVA is still the least complex among these soft-in soft-out (SISO) decoders.

Generally, the SOVA performs about 0.6 dB worse than the Max-Log-MAP decoder using the iterative decoding but with only about half of the complexity at BER of 10^{-4} [3, 5, 11]. But in [15], it has been shown that the complexity of Max-Log-MAP compared with SOVA is not always significantly large under a condition of small constraint length. The simplicity of the SOVA still raises the interest to modify it to get better performance. For example, a Bi-Directional SOVA was proposed in [2] which uses the backward SOVA decoding simultaneously with the forward SOVA in the iterative decoding and it has been shown that it outperforms the Max-Log-Map for around 0.1 dB at BER of 10^{-4} and underperforms the BCJR algorithm for about 0.4 dB. An improvement has been made in [17] by using a scaling factors scheme, which can provide 0.2 dB gain on the Bi-SOVA at BER of 10^{-4}. It can be noticed that the complexity of Bi-SOVA is about twice of SOVA and thus the Bi-SOVA has a slightly larger complexity than Max-Log-MAP when a small constraint length is considered. However, the requirement of memory used to store information of branches in Bi-SOVA is less than Max-Log-Map [17].

Concatenated coding schemes and other decoding schemes such as belief propagation (BP) decoding require the exchange of soft information between component decoders to improve the decoding results, which is heavily dependent on the accuracy of statistical properties of the channel. Specifically, these SISO decoders require the accurate knowledge of the signal-to-noise ratio (SNR) of the channel [18–20]. Generally, pilot symbols can be used to estimate the channel SNR at the cost of a reduced data rate. In this paper, we show the relationship between the number of pilot symbols and the accuracy of the estimation of channel SNR. We also study the impact of SNR mismatch (SNRM) for a SCCPM system designed in [1]. The SCCPM is a serial concatenation of a convolutional code (CC) and a M-ary full-response CPM. SOVA is used as the SISO decoder for each component module in SCCPM.

The rest of the paper is organized as follows. Section 2 describes the CPM scheme, its decomposition model, and the SCCPM system. Section 3 presents the details of the Viterbi decoding algorithm and the SOVA. Section 4 discusses the effect of the number of pilot symbols on the performance of a BIAWGN channel estimator and discusses SNRM and its impact on the SOVA decoded SCCPM system. Finally, Sect. 5 concludes the paper.

2 SCCPM System Model

The conventional CPM signal can be expressed as [2]:

$$s(t, \boldsymbol{\alpha}) = \sqrt{\frac{2E_s}{T}} cos(2\pi f_c t + \phi(t; \boldsymbol{\alpha}) + \phi_0) \tag{1}$$

where E_s is the symbol energy, T is the symbol time and the ϕ_0 is initial phase which is assumed to be *zero* without loss of generality. The $\phi(t; \boldsymbol{\alpha})$ is the information-carrying phase which is defined by:

$$\phi(t;\boldsymbol{\alpha}) = 2\pi h \sum_{k=0}^{n} \alpha_k q(t - kT), t > 0 \tag{2}$$

where the $\boldsymbol{\alpha}$ is the sequence of M-ary data symbols, α_k is a symbol selected from the alphabet $\pm 1, \pm 3, \ldots, \pm(M-1)$, and h is the modulation index of CPM signal which affects the number of states of the signal phases. The modulation index h is defined by

$$h = \frac{K}{P} \tag{3}$$

where K and P are relatively prime positive integers. The $q(t)$ is the phase pulse which is defined by:

$$q(t) = \int_{0}^{t} g(\tau)d\tau \tag{4}$$

where $g(t)$ is called frequency pulse which controls the shape of the phase pulse, for example, a rectangular frequency pulse commonly referred to as LREC is defined as:

$$g(t) = \begin{cases} \frac{1}{2LT} & 0 \le t \le LT \\ 0 & \text{otherwise} \end{cases} \tag{5}$$

where L is referred to as the memory of the signal [6], LT is the duration time of the pulse. When $L = 1$, the signal is called full-response CPM. When $L > 1$, the signal is called partial-response CPM. For simplicity, we only consider the full response case, that is, $L = 1$.

Rimoldi's CPM decomposition approach [6] allows a conventional CPM signal to be decomposed into one linear encoder with memory and one memoryless modulator (MM). The structure of the phase trellis is modified into a form of time-invariant trellis, which is called a physical tilted-phase. The physical tilted-phase trellis simplifies the structure of the phase trellis and reduces the number of states. The CPE is a trellis encoder with rate 1, as stated by Rimoldi [6], the CPE itself would be a convolutional encoder.

Consider a $h = K/P$ LREC M-ary CPM signal, for the sake of convenience, the CPM symbol sequence $\boldsymbol{\alpha}$ is modified as data sequence $\mathbf{u}$ which is defined by $u_k = \alpha_k + (M-1)/2$ and we can have $u_k \in \{0, 1, 2, ...M-1\}$.

Let

$$U_n = R_P \left[2\pi h \sum_{k=0}^{n-L} u_k \right] \tag{6}$$

where $R_P[\cdot]$ is the "modulo P operator".

Let $t = \tau + nT$, the tilted-phase $\psi(t)$ during the n-th symbol interval is given by [6]:

$$\psi(\tau + nT, \mathbf{u}) = R_{2\pi}\left[2\pi h U_n + 4\pi h \sum_{k=0}^{L-1} U_{n-k} q(\tau + kT) + \omega(\tau)\right] \tag{7}$$

where $R_{2\pi}[\cdot]$ is the "modulo 2π operator" and

$$\begin{aligned}\omega(\tau) &= \pi h(M-1)\frac{\tau}{T} \\ &\quad - 2\pi h(M-1)\sum_{k=0}^{L-1} q(\tau + kT) + \pi h(L-1)(M-1), \\ 0 &\le \tau \le T\end{aligned} \tag{8}$$

is the data-independent term. It has been shown in [6] that the number of states for $\psi(t, \mathbf{u})$ will decrease to PM^{L-1} in the phase trellis while the number of states is $2PM^{L-1}$ for the original M-ary CPM when K is odd.

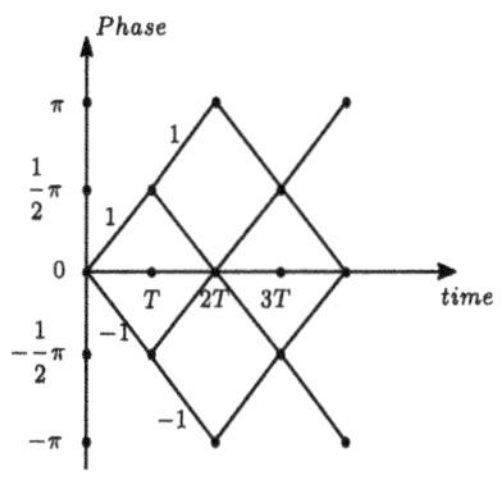

Fig. 1 Conventional MSK phase trajectory diagram

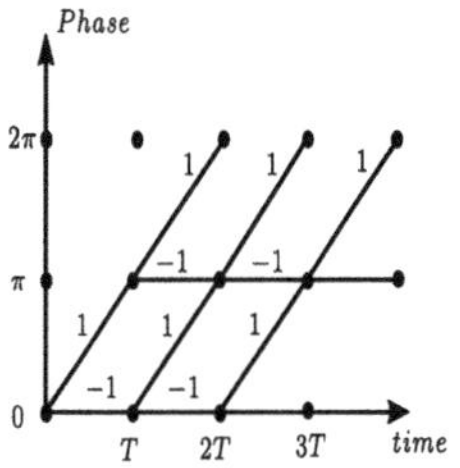

Fig. 2 Physical tilted-phase MSK phase trajectory diagram

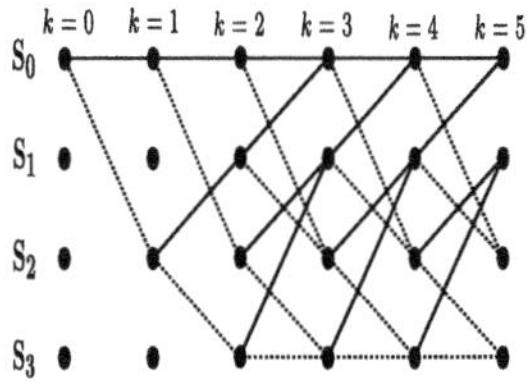

Fig. 3 Trellis diagram for a (2, 1, 3) convolutional code

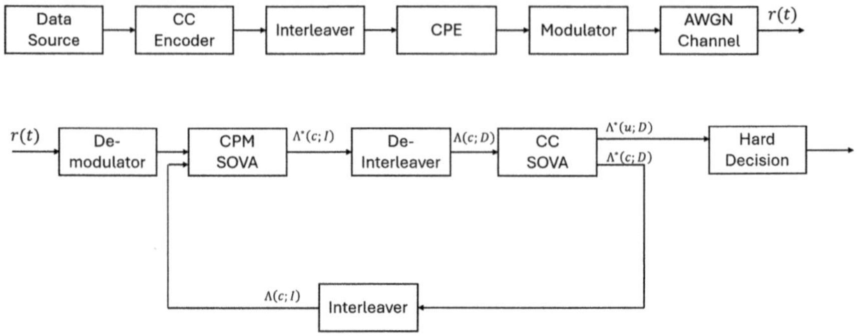

Fig. 4 Block diagram of SOVA decoded SCCPM system

Figure 1 is the conventional phase trellis for the MSK signal and Fig. 2 shows the physical tilted-phase trellis for the Minimum-Shift Keying (MSK) signal. MSK is the simplest signal of CPM, which is a binary, 1REC CPM with $h = 1/2$. It has 2 states in the physical tilted-trellis, whereas the original MSK has $2 \times 2 \times 2^{1-1} = 4$ states in the phase trellis (Fig. 3).

Figure 4 shows the structure of the SCCPM system. The CC has a rate of 1/2. There is a pair of bit-wise pseudo-random interleaver and deinterleaver placed between CC and continuous phase encoder/decoder. The interleaver/deinterleaver pair is used to permute the code sequence to disperse the burst errors that could fail the CC decoding. Assume that the encoder will reset the state to 0 when each data frame is transmitted. In Fig. 4, the soft input and output of SOVA blocks are denoted by $\Lambda(\cdot)$ and $\Lambda^*(\cdot)$. $\Lambda(\cdot, I)$ and $\Lambda(\cdot, D)$ denote interleaved and deinterleaved sequences respectively. The $\Lambda(c, \cdot)$, which includes both information bits and parity bits, denotes the soft information of CC coded bit sequence, and $\Lambda(u, \cdot)$, which only contains the information bits, denotes the soft information of VA decoded bit sequence. The soft output of SOVA $\Lambda^*(\cdot)$ gives the log-likelihood ratio (LLR) of selecting a correct branch for decoding bit u_k at stage k at the state of the survival branch [11]. The soft information from the SOVA decoder of CPE will be the *a-priori* information for the SOVA decoder of CC. The soft information outputs from the SOVA decoder of CPE will be fed back to the SOVA decoder of CC as the *a-priori* information. The soft information exchange between the CPM SOVA and the CC SOVA will proceed with the iterative decoding until the stop criteria are reached [1].

3 Viterbi Algorithm and SOVA

3.1 Viterbi Algorithm

The classic Viterbi decoding consists of two parts: a forward process and a trace-back (backward) process. For simplicity, a trellis with two branches entering and leaving at each state node is assumed in this section, i.e., two paths merging at each state node at each stage except the first and the second stage, and two branches heading to two different states at the next stage (See Fig. 3).

In the procedure of decoding, each node metric consists of the branch metric and node metric of the previous state of the incident branch. The branch metric is the cost from the possible state of the previous stage to the current state at the current stage. The forward process operates "add", "compare", and "select" steps at each stage. The "add" step adds the branch metric and the previous state node metric and the "compare" step compares the two node state metrics of two paths converging at the same node at the current stage. The "select" step selects the path with the best node metric as the survival path and then discards the other one. When the forward process reaches a certain depth in the trellis, the VA traces back the trellis along the survival path and outputs the decoded results.

3.2 Soft Information

Assuming that BPSK modulation is used with the transmitted symbols to be $+1$ or -1. We use u_k to denote the k-th symbol. The soft outputs from both components decoders are in term of LLR that is a natural logarithm of two probability values.

At receiver, the LLR for the value of the decoded bit u_k conditioned on the observed signal sequence $\mathbf{r}$ is represented by:

$$L(u_k \mid \mathbf{r}) = \ln\left(\frac{P(u_k = +1 \mid \mathbf{r})}{P(u_k = -1 \mid \mathbf{r})}\right) \tag{9}$$

where $P(u_k = +1 \mid \mathbf{r})$ and $P(u_k = -1 \mid \mathbf{r})$ denote the probabilities of $u_k = +1$ and $u_k = -1$, respectively, when the observed signal sequence is given. Equation (9) is called the *a posteriori* LLR of decoded bit u_k. The sign of the LLR gives the estimated bit. For simplicity, we write $L(u_k \mid \mathbf{r})$ as $L(u_k)$. Using Bayes' theorem, the *a posterior* LLR (9) can be rewritten as [21]:

$$L(u_k) = \ln\frac{P(\mathbf{r}, u_k = +1)/P(\mathbf{r})}{P(\mathbf{r}, u_k = -1)/P(\mathbf{r})} = \ln\frac{P(\mathbf{r}, u_k = +1)}{P(\mathbf{r}, u_k = -1)} \tag{10}$$

The probability $P(u_k)$ can be written as:

$$\begin{aligned} P(u_k = \pm 1) =& \frac{e^{-L(u_k)/2}}{1 + e^{-L(u_k)}} e^{\pm L(u_k)/2} \\ =& A_k \times e^{u_k L(u_k)/2} \end{aligned} \tag{11}$$

where A_k is a common factor for both $P(u_k = +1)$ and $P(u_k = -1)$:

$$A_k = \frac{e^{-L(u_k)/2}}{1 + e^{-L(u_k)}} \tag{12}$$

and the nature logarithm is

$$\ln P(u_k) = \ln A_k + \frac{1}{2} u_k L(u_k) \tag{13}$$

3.3 Mathematical Description of the SOVA

Consider a state sequence $\mathbf{s}_s^k$ as a survival path along the trellis from the starting stage to stage k, and the state at stage k is $S^k = s$ and the state at stage $k-1$ is $S^{k-1} = s'$. Let $\mathbf{y}$ be the observed sequence, then the *a posteriori* probability of the state sequence given $\mathbf{y}$ is [3, 16]:

$$P(\mathbf{s}_s^k \mid \mathbf{y}_{j \le k}) = \frac{P(\mathbf{s}_s^k, \mathbf{y}_{j \le k})}{P(\mathbf{y}_{j \le k})} \tag{14}$$

It can be noticed that the *a posteriori* probability is proportional to $P(\mathbf{s}_s^k, \mathbf{y}_{j \le k})$. Therefore, the metric $P(\mathbf{s}_s^k, \mathbf{y}_{j \le k})$ should be maximized to obtain the optimal result. The metric can be calculated in a recursive form. That is, it can be obtained by multiplying the previous stage metric by the probability of transition from the state at stage $k-1$ to the state at stage k. This probability is denoted as $\gamma_k(s', s)$:

$$\begin{aligned} \gamma_k(s', s) \triangleq& P(y_k, S^k = s \mid S^{k-1} = s') \\ =& P(y_k, s \mid s') \end{aligned} \tag{15}$$

Then the metric $P(\mathbf{s}_s^k, \mathbf{y}_{j \le k})$ is:

$$\begin{aligned} P(\mathbf{s}_s^k, \mathbf{y}_{j \le k}) =& P(\mathbf{s}_{s'}^{k-1}, \mathbf{y}_{j \le k-1}) P(y_k, s | s') \\ =& P(\mathbf{s}_{s'}^{k-1}, \mathbf{y}_{j \le k-1}) \gamma_k(s', s) \end{aligned} \tag{16}$$

The path metric M_p for the survival path $\mathbf{s}_s^k$ is given by:

$$M_p(\mathbf{s}_s^k) \triangleq \ln\left(P\left(\mathbf{s}_s^k, \mathbf{y}_{j\leq k}\right)\right) = M_p(\mathbf{s}_{s'}^{k-1}) + \ln\left(\gamma_k(s', s)\right) \tag{17}$$

We can notice that the $\ln(\gamma_k(s', s))$ is the branch metric from state s' to state s during the internal $[k-1, k]$. The $\gamma_k(s', s)$ is calculated by using the definition (15) and using Bayes' rule:

$$\begin{aligned}\gamma_k(s', s) &= P\left(\mathbf{y}_k, s \mid s'\right) \\ &= P\left(\mathbf{y}_k \mid s, s'\right) P\left(s \mid s'\right) \\ &= P\left(\mathbf{y}_k \mid \mathbf{x}_k\right) P(u_k)\end{aligned} \tag{18}$$

where u_k is the symbol input in the trellis during the interval $[k-1, k]$ that causes the state transition from s' to s, and $\mathbf{x}_k$ is the output of transition from state s' to s along the trellis, which is an M-ary symbol and $M = 2^n$, where n is the number of bits per symbol. The $\mathbf{y}_k$ is the observed M-ary codeword symbol at the receiver. In an AWGN channel, the probability $P\left(\mathbf{y}_k \mid \mathbf{x}_k\right)$ is given by:

$$P\left(\mathbf{y}_k \mid \mathbf{x}_k\right) = \frac{1}{\sqrt{2\pi}\sigma}\exp\left(\frac{D(\mathbf{y}_k, \mathbf{x}_k)}{2\sigma^2}\right) \tag{19}$$

where $D(\mathbf{y}_k, \mathbf{x}_k)$ is the Euclidean distance of two symbols $\mathbf{y}_k$ and $\mathbf{x}_k$.

The nature logarithm of $P\left(\mathbf{y}_k \mid \mathbf{x}_k\right)$ is represented as:

$$\begin{aligned}\ln\left(P\left(\mathbf{y}_k \mid \mathbf{x}_k\right)\right) &= n \times \ln\left(\frac{1}{\sqrt{2\pi\sigma^2}}\right) - \frac{1}{2\sigma^2}\left\{\sum_{i=1}^{n} y_i^2 + \sum_{i=1}^{n} x_i^2\right\} + \frac{2}{2\sigma^2}\sum_{i=1}^{n}(y_i x_i) \\ &= C_k + \frac{1}{2}L_c\sum_{i=1}^{n}(y_i x_i)\end{aligned} \tag{20}$$

where

$$C_k = n \times \ln\left(\frac{1}{\sqrt{2\pi\sigma^2}}\right) - \frac{1}{2\sigma^2}\left\{\sum_{i=1}^{n} y_i^2 + \sum_{i=1}^{n} x_i^2\right\} \tag{21}$$

is independent of u_k or x_k so that it can be considered as a constant and omitted. x_i and y_i are individual bits within the codewords $\mathbf{x_k}$ and $\mathbf{y_k}$. L_c is referred to as channel reliability measure [12, 16]:

$$L_c = 2a/\sigma^2 \tag{22}$$

where a is fading amplitude of channel and $a = 1$ when AWGN is assumed.

We substitute (13) and (20) into (18) and omit the constant term C_k and the common factor A_k due to the fact that they do not affect the metrics comparison. The nature logarithm of the branch metric $\gamma_k(s', s)$ is given by:

$$\ln \gamma_k(s', s) = \frac{1}{2} L_c \sum_{i=1}^{n} (y_i x_i) + \frac{1}{2} u_k L(u_k) \tag{23}$$

The path metric M_p (17) can be represented as:

$$\begin{aligned} M_p(\mathbf{s}_s^k) &= M_p(\mathbf{s}_{s'}^{k-1}) + \ln\left(\gamma(s', s)\right) \\ &= M_p(\mathbf{s}_{s'}^{k-1}) + \frac{1}{2} L_c \sum_{i=1}^{n} (y_i x_i) + \frac{1}{2} u_k L(u_k) \end{aligned} \tag{24}$$

which is the recursive equation for path metric of SOVA. To avoid the output of SOVA overflow, the output LLRs are normalized by:

$$\alpha = 4 \times D_{free} \times \frac{E_s}{N_o} \tag{25}$$

where D_{free} is the free distance of the convolutional code. After calculation, we can get that the free distance of the CPM shown in Fig. 2 is 2 and the free distance of the CC shown in Fig. 3 is 4.

In SOVA algorithm, at stage k of the trellis, there are two paths that enter the state $S^k = s$ and by using the "select" unit in VA, one path is selected as the survival path $\mathbf{s}_s^k$ and another one path $\mathbf{s}_s^{k'}$ is discarded. The path metric difference between the survival path and the discarded path, which is denoted as Δ_s^k, is calculated and stored:

$$\Delta_k^s = M_p(\mathbf{s}_s^k) - M_p(\mathbf{s}_s^{k'}) \geq 0 \tag{26}$$

The probability of the survival path is correctly selected is given by:

$$P(\text{correct}) = \frac{P(\mathbf{s}_s^k)}{P(\mathbf{s}_s^k) + P(\mathbf{s}_s^{k'})} \tag{27}$$

Using the definition of path metric (17), (27) can be rewritten as:

$$P(\text{correct}) = \frac{e^{M_p(\mathbf{s}_s^k)}}{e^{M_p(\mathbf{s}_s^k)} + e^{M_p(\mathbf{s}_s^{k'})}} = \frac{e^{\Delta_s^k}}{e^{\Delta_s^k} + 1} \tag{28}$$

then the LLR of selecting the correct survival path at stage k is given by:

$$L(u_k) = \ln \frac{P(\text{correct})}{1 - P(\text{correct})} = \Delta_s^k \tag{29}$$

where u_k is the decoded result corresponding to the survival path $\mathbf{s}_s^k$ at stage k. It can be noted that the reliability of selecting a correct path is simply given by Δ_s^k. It has been shown [11] that at stage l, all of the surviving paths in the trellis will have the same starting state node at stage $l - \sigma$, where usually the σ is at most five times the constraint length of the convolutional code. At the interval from stage $k = l - \sigma$ to stage $k = l$, the decoding bit u_k may be affected by the selection of survival paths at each stage. Therefore, to get the soft information of decoded result u_k at stage l, the differences of metrics for all states along the survival path from stage $l - \sigma$ to l need to be taken into account. Hagenauer in [2] showed that the LLR of selection of a correct path can be approximated by:

$$L(u_k \mid \mathbf{y}_{j \leq k}) \approx \min_{\substack{i=k-\sigma,\ldots,k-1,k \\ u_s^i \neq u_d^i}} \Delta_{s_i}^i \tag{30}$$

where u_s^i and u_d^i denote the bits decoded at stage i based on the survival path and the discarded path, respectively.

4 SNR Mismatch and Impact

4.1 *Impact of the Number of Pilot Symbols on SNR Estimation*

From (24) and the normalization factor (25), it can be noticed that the channel information is used to calculate the path metric in the SOVA and to normalize the SOVA outputs. When the channel information is not accurate, the decoding performance must be degraded. In general, the SNRM is from the estimation errors of the channel parameters. To simplify the discussion, we assume binary phase shift keying (BPSK) modulation. Given a Binary-input AWGN channel with input $x = \pm 1$, the output of the channel can be described as $y = x\sqrt{E_s} + n$, where E_s is the symbol energy and $n \sim N(0, \sigma^2)$ is the channel noise with mean zero and noise variance σ^2. We can have $y \sim N(x, \sigma^2)$. The output of the receiver at time k is:

$$r_k = x_k\sqrt{E_s} + n_k \tag{31}$$

where noise component n_k is identically independent distributed (iid) zero-mean Gaussian random variable with variance σ^2.

We use a pilot-based channel estimation method introduced in [22] to study the relationship between the number of pilot symbols used by the estimator and the SNRM. This estimator is derived from the probability density function (pdf) of the received block of data conditioned on the known symbols (pilot symbols) and the Maximum Likelihood Estimation (MLE). Let $A = \frac{\sqrt{E_s}}{\sigma^2}$ and $B = \sqrt{E_s}$, these two factors can be estimated by:

$$\hat{B} = \frac{1}{N_p}\left(\sum_{i\in\{x_i=+1\}} r_i - \sum_{i\in\{x_i=-1\}} r_i\right) \tag{32}$$

$$\hat{A} = \frac{\hat{B}}{\frac{1}{N_p}\sum_{i=1}^{N_p} r_i^2 - \hat{B}^2} \tag{33}$$

where N_p is the number of known pilot symbols at the receiver.

The estimated symbol energy $E_{s,e}$ and the estimated noise variance σ_e^2 is obtained by:

$$E_{s,e} = \hat{B}^2 \tag{34}$$

$$\sigma_e^2 = \frac{\hat{B}}{\hat{A}} \tag{35}$$

The signal to noise ratio (SNR) per symbol is:

$$SNR = \frac{E_s}{N_o} = \frac{E_s}{2\sigma^2} \tag{36}$$

The SNR for estimated symbol energy $E_{s,e}$ and noise variance σ_e^2 are:

$$SNR_e = \frac{E_{s,e}}{N_{o,e}} = \frac{E_{s,e}}{2\sigma_e^2} = \frac{\hat{A}\hat{B}}{2} \tag{37}$$

When SNR_e is erroneously estimated, we define the SNRM ratio Υ:

$$\Upsilon = \frac{SNR_e}{SNR} \tag{38}$$

We define two types of the SNRM:

1. SNR over-estimation occurs when $SNR_e > SNR$, i.e., $\Upsilon > 0$ dB
2. SNR under-estimation occurs when $SNR_e < SNR$, i.e., $\Upsilon < 0$ dB

Figure 5 displays the performance of the pilot-based ML estimator for different numbers of pilot symbols used. The μ_{N_p} and σ_{N_p} represent the mean values and standard deviations of the estimator results, respectively. We can observe that the larger N_p is, the greater the probability of getting an accurate estimate, i.e., the SNRM is closer to 0 dB. From Fig. 5, we can get that when $N_p = 512$, the probability of -1 dB $\leq \Upsilon \leq 1$ dB is 89.93% and the probability of $\Upsilon \leq -1$ dB is 5.22%, which means it still has some possibility to get a harmful estimation to degrade the iterative decoding performance. When $N_p = 64$, the probability of $\Upsilon \leq -1$ dB is 23.8%, which means it has a large probability of getting a harmful estimation compared with a larger N_p. However, in a practical system, one cannot afford to use too many pilot symbols that could significantly reduce the bandwidth efficiency.

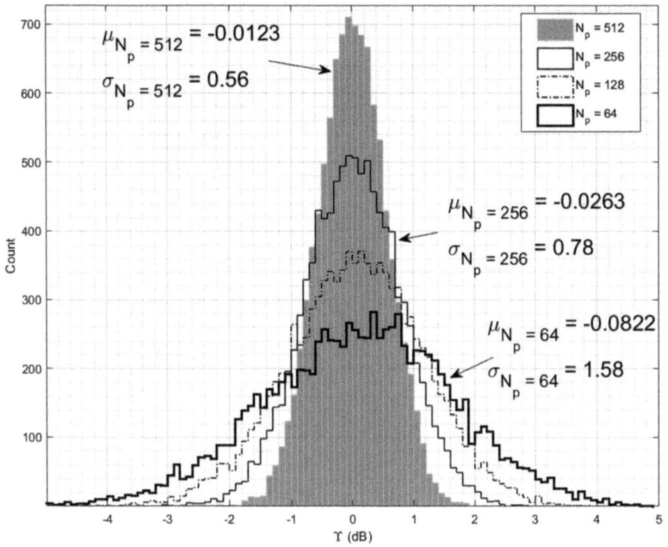

Fig. 5 Histogram for the MLE channel estimator. Number of trails $= 10^4$. Numbers of pilot symbols N_p = 64, 128, 256, and 512. True $E_s/N_0 = 1$ dB

4.2 Impact of SNRM on the Performance of SCCPM

Some simulation results are displayed to show the sensitivity of the SOVA decoded SCCPM to the SNRM issue. In this experiment, the inner code is a binary 1REC CPE with $h = 1/2$ and the outer code is a (2, 1, 3) binary CC. Each CC data frame contains $N = 2048$ bits of data. Therefore, there are 1024 bits of source data per coded data frame. For simplicity, we only consider the noise variance to be erroneously estimated and assume the estimated symbol energy is equal to the true value. We use Bit Error Rate (BER) after decoding as the performance measure.

Figure 6 shows the BER performance of the SCCPM after 8 iterations across a range of SNRM $\Upsilon = [-3, 3]$ dB. Figure 7 shows how the BER performance changes for varying iteration numbers with different SNRM ($\Upsilon = [-2, 2]$ dB). We can observe that the underestimate (the estimated SNR to be lower than the true SNR) of the channel SNR is more detrimental than over-estimation when the channel SNR is $E_b/N_0 = 1$ dB.

From Fig. 7, we can find that when the channel SNR is $E_b/N_0 = 1$ dB, the BER performance of under-estimation of the channel SNR may get worse with the increase of iterations, and the over-estimation of the channel SNR makes the iterative decoding system lose some coding gain but it does not introduce more errors during the decoding process.

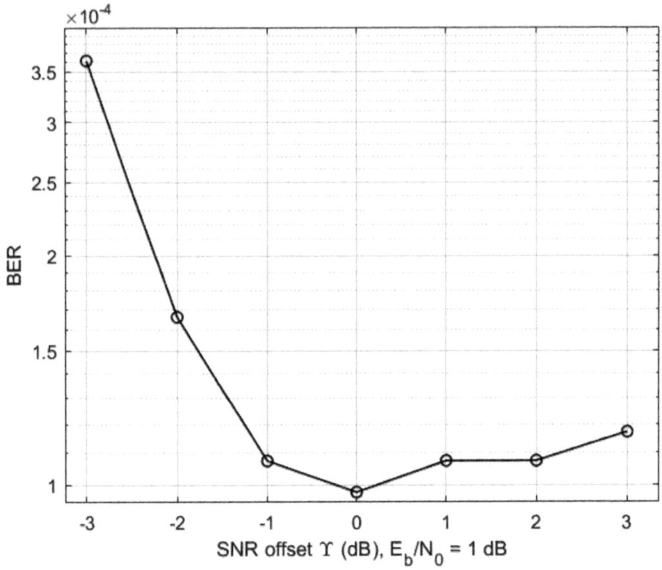

Fig. 6 BER performance after 8 iterations with SNRM = [−3, 3] dB

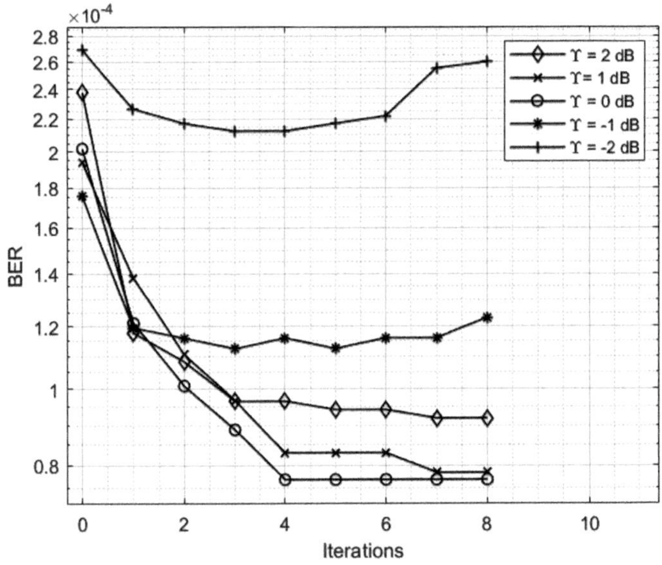

Fig. 7 BER performance of SCCPM when iterations increase from 0 to 8 with different SNRM, the true channel SNR $E_b/N_0 = 1$ dB

5 Conclusion

In this study, we have studied the effect of the number of pilot symbols and the performance of a MLE channel estimator in a BIAWGN channel. Our simulations have shown that a larger number of pilot symbols allow the estimator to obtain more accurate estimations of the noise variance and symbol energy. This accuracy in estimation, however, comes at the cost of bandwidth efficiency, presenting a significant tradeoff between bandwidth efficiency and estimation accuracy. Additionally, we investigate the impact of SNRM on the BER performance of a SOVA decoded SCCPM iterative decoding system. The simulation shows that under-estimation of the channel SNR has a more detrimental effect on the SCCPM system compared to over-estimation. In future work, we will explore strategies to improve the robustness of the SCCPM system against SNRM and investigate more accurate estimation algorithms.

References

1. Guo Z, Cao L (2022) Serially concatenated continuous phase modulation with SOVA turbo decoding. In: SoutheastCon 2022, pp 605–610. https://doi.org/10.1109/SoutheastCon48659.2022.9764064
2. Aulin T, Sundberg C (1981) Continuous phase modulation—part I: full response signaling. IEEE Trans Commun 29(3):196–209. https://doi.org/10.1109/TCOM.1981.1095001
3. Cao L, Chen CW (2004) Constrained SOVA decoding in concatenated codes. In: 2004 IEEE wireless communications and networking conference (IEEE Cat. No. 04TH8733), vol 4, pp 2280–2284. https://doi.org/10.1109/WCNC.2004.1311443
4. Cao L, Chen CW, Orlik P, Zhang J, Gu D (2004) A trellis-based technique for blind channel estimation and equalization. J Commun Network 6(1):19–25. https://doi.org/10.1109/JCN.2004.6596983
5. Moqvist P, Aulin T (2001) Serially concatenated continuous phase modulation with iterative decoding. IEEE Trans Commun 49(11):1901–1915. https://doi.org/10.1109/26.966054
6. Rimoldi B (1988) A decomposition approach to CPM. IEEE Trans Inf Theory 34(2):260–270. https://doi.org/10.1109/18.2634
7. Rimoldi B, Li Q (1995) Coded continuous phase modulation using ring convolutional codes. IEEE Trans Commun 43(11):2714–2720. https://doi.org/10.1109/26.481222
8. Forney G (1973) The Viterbi algorithm. Proc IEEE 61(3):268–278. https://doi.org/10.1109/PROC.1973.9030
9. Viterbi A (1967) Error bounds for convolutional codes and an asymptotically optimum decoding algorithm. IEEE Trans Inf Theory 13(2):260–269. https://doi.org/10.1109/TIT.1967.1054010
10. Bahl L, Cocke J, Jelinek F, Raviv J (1974) Optimal decoding of linear codes for minimizing symbol error rate (corresp). IEEE Trans Inf Theory 20(2):284–287. https://doi.org/10.1109/TIT.1974.1055186
11. Hagenauer J, Hoeher P (1989) A Viterbi algorithm with soft-decision outputs and its applications. In: 1989 IEEE global telecommunications conference and exhibition communications technology for the 1990s and beyond, vol 3, pp 1680–1686. https://doi.org/10.1109/GLOCOM.1989.64230
12. Hagenauer J, Papke L (1994) Decoding Turbo-codes with the soft output Viterbi algorithm (SOVA). In: Proceedings of 1994 IEEE international symposium on information theory, p 164. https://doi.org/10.1109/ISIT.1994.394808

13. Koch W, Baier A (1990) Optimum and sub-optimum detection of coded data disturbed by time-varying intersymbol interference (applicable to digital mobile radio receivers). In: [Proceedings] GLOBECOM 90: IEEE global telecommunications conference and exhibition, vol 3, pp 1679–1684. https://doi.org/10.1109/GLOCOM.1990.116774
14. Erfanian J, Pasupathy S, Gulak G (1994) Reduced complexity symbol detectors with parallel structure for ISI channels. IEEE Trans Commun 42(234):1661–1671. https://doi.org/10.1109/TCOMM.1994.582868
15. Robertson P, Villebrun E, Hoeher P (1995) A comparison of optimal and sub-optimal MAP decoding algorithms operating in the log domain. In: Proceedings IEEE international conference on communications ICC 95, vol 2, pp 1009–1013. https://doi.org/10.1109/ICC.1995.524253
16. Woodard J, Hanzo L (2000) Comparative study of Turbo decoding techniques: an overview. IEEE Trans Veh Technol 49(6):2208–2233. https://doi.org/10.1109/25.901892
17. Chang YC, Lain JK (2005) Improved decoding with the bi-directional SOVA for turbo codes. In: 2005 IEEE 61st vehicular technology conference, vol 1, pp 673–677. https://doi.org/10.1109/VETECS.2005.1543377
18. El-Khamy M, Wu J, Lee J, Roh H, Kang I (2012) Near-optimal Turbo decoding in presence of SNR estimation error. In: 2012 IEEE global communications conference (GLOBECOM), pp 3737–3742. https://doi.org/10.1109/GLOCOM.2012.6503698
19. Fadhel H, Kharel A, Cao L (2020) Performance of raptor codes on the BIAWGN channel in the presence of SNR mismatch. In: 2020 IEEE wireless communications and networking conference (WCNC), pp 1–6. https://doi.org/10.1109/WCNC45663.2020.9120480
20. Summers T, Wilson S (1998) SNR mismatch and online estimation in Turbo decoding. IEEE Trans Commun 46(4):421–423. https://doi.org/10.1109/26.664291
21. Langton C. Tutorial 24A—turbo coding and MAP decoding—part 1. https://complextoreal.com/tutorials/tutorial-24a-turbo-coding-and-map-decoding-part-1/
22. Moghari M, Shahrrava B (2004) Channel estimation for Turbo decoding on AWGN channels. In: IEEE 5th workshop on signal processing advances in wireless communications, 2004, pp 46–49. https://doi.org/10.1109/SPAWC.2004.1439200

Compact Broadband Waveguide Polarizer for X-Band Satellite Communication (SATCOM) Applications

Neeresh Kumar, Manoj Dwivedi, and Bal Mukund Jha

Abstract This paper presents design and development of a compact wideband waveguide polarizer for X-band satellite communication (SATCOM) applications. The polarizer has been designed based on the antiphase posts separated by finite distance in a circular waveguide. The posts have variable height in the waveguide. The polarizer has been analyzed using 3D EM software, and a prototype is fabricated and measured. The measured axial ratio (AR) of less than 1.15 dB in the required frequency of operation is 7.25–8.4 GHz; however, it has 3-dB AR bandwidth of 37.2% over the frequency band of 6.26–9.16 GHz, and return loss is better than − 27 dB required frequency band of operation while it has more than 38% return loss bandwidth (6.25–9.20 GHz).

Keywords Circular waveguide · Polarizer · Axial ratio · SATCOM applications

1 Introduction

Polarization is an important characteristic of the radiation field, and it is primarily of two types: linear and elliptical/circular polarization. In linear polarization, if electric field vector is not aligned between the transmitting and receiving antennas, then polarization mismatch is occurred which degrades the signal power in the required polarization. That is why, in few SATCOM frequency bands such as S, X, and Ka band, circular polarization is chosen to eliminate the effect of polarization mismatch.

In satellite communication, generally reflector-based antenna systems are employed for communication which utilized a feed assembly which includes feed horn, ortho-mode transducer, and waveguide polarizer [1]. The purity of circular polarization primarily depends on the waveguide polarizer, and feed horn does not contribute to it [2]. Thus, polarizer is main element for converting linear to circular polarization or vice versa. Few authors [3, 4] have presented septum inside

N. Kumar (✉) · M. Dwivedi · B. M. Jha
Development and Engineering, Antenna Bharat Electronics Ltd., Ghaziabad, UP, India
e-mail: neereshkumar@bel.co.in

S. M. Thampi et al. (eds.), *Fifth International Conference on Computing and Network Communications*, Lecture Notes in Electrical Engineering 1219,
https://doi.org/10.1007/978-981-97-4540-1_42

waveguides-based polarizers. These types of polarizers include two ports, each one for left hand circularly polarized (LHCP) wave and right hand circularly polarized (RHCP) but inherently suffer from low cross-port isolation between the two signals. In [4], it is discussed that authors were able to achieve maximum of 20% the relative bandwidth with the use of a septum polarizer while keeping the isolation between two orthogonal ports higher than 25 dB.

In [5], a segment of square waveguide with diagonal ridges with adjustment iris has been used as phase shifting medium to achieve required phase difference in circular polarization, but this has fabrication challenges and required tight tolerances. Instead of using square waveguide, use of a circular waveguide can be done to make the feed assembly compact, light, and simple structure. The required phase of 90° in circular waveguide is generally done by either septums [5, 6], corrugations/iris [7], or posts [8, 9]. The above designs either need complex manufacturing or tuning after fabrication to match the simulation results or have very small relative axial ratio (AR) bandwidth if the required isolation between orthogonal ports is more than 25 dB. The issue is addressed by using multiple cylindrical posts inserted in circular waveguide.

In this paper, design method and analysis of compact waveguide polarizer at X-band for large relative axial ratio (AR) bandwidth with isolation between orthogonal ports is discussed. Further, the research work provides parametric analysis of polarizer with respect to AR bandwidth. A prototype of polarizer is fabricated and measured for return loss and axial ratio. The measured results are found to be in close agreement with simulated results.

2 Theory and Design Procedure

In this type of polarizers, the right hand circularly polarized (RHCP) and left hand circularly polarized (LHCP) waves at first physical port converted into linearly polarized dominant modes TE_{11} and TE_{11}*, respectively, at the second physical port. The polarizer is electrically 4-port devices with the following scattering matrix [10]:

$$S = 1/\sqrt{2}\begin{bmatrix} 0 & 0 & e^{j\phi} & e^{j(\phi-\pi/2)} \\ 0 & 0 & e^{j(\phi-\pi/2)} & e^{j\phi} \\ e^{j\phi} & e^{j(\phi-\pi/2)} & 0 & 0 \\ e^{j(\phi-\pi/2)} & e^{j\phi} & 0 & 0 \end{bmatrix} \quad (1)$$

There are two basic requirements for the design of polarizers. Firstly, the linearly polarized modes must be equally divided (or combined) into orthogonal components with identical magnitude.

$$|S_{31}| = |S_{32}| \quad (2)$$

$$|S_{41}| = |S_{42}| \quad (3)$$

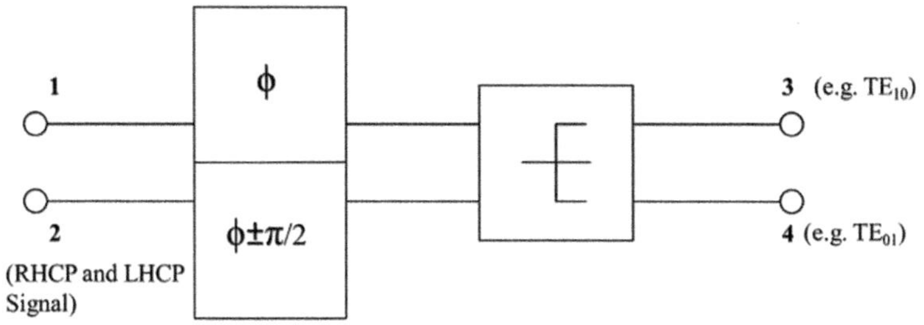

Fig. 1 Circuit diagram of polarizer

Secondly, these components should have a differential phase shift of $\pm\ \pi/2$.

$$\Delta\phi = \measuredangle S_{31} - \measuredangle S_{32} = +(-)\,\pi/2 \tag{4}$$

Hence, polarizer is a device with an equal power divider consisting two independent phase shifters. Therefore, a physical alignment of 45° is required to fulfill the above conditions. Circuit diagram of the polarizer is given in Fig. 1.

When phase shifting medium is kept at 45° orientation (as shown in Fig. 2), the incident TE_{10} is divided into two 3 dB components in which one TE_{10} is parallel and TE_{01} is perpendicular to the phase shifting medium as shown in figure. The axial ratio of polarizer depends only on these two parameter: the amplitude of these orthogonal components and phase difference between them. The axial ratio (AR) formula is given below [11]:

$$\mathrm{AR} = \sqrt{\frac{(E_1/E_2)^2 + 1 + \sqrt{(E_1/E_2)^4 + 1 + 2(E_1/E_2)^2\cos 2\delta}}{(E_1/E_2)^2 + 1 - \sqrt{(E_1/E_2)^4 + 1 + 2(E_1/E_2)^2\cos 2\delta}}} \tag{5}$$

where

E_1, E_2 are magnitude of orthogonal components
δ is the phase difference.

So if $E_1 = E_2$ then

$$\mathrm{AR} = \tan(\delta/2) \tag{6}$$

And

$$\mathrm{AR\,(dB)} = 20\log * \mathrm{AR} \tag{7}$$

From Eqs. (6) and (7), it can be deduced that if power in both components is equally divided and phase difference $\delta = 90°$ then

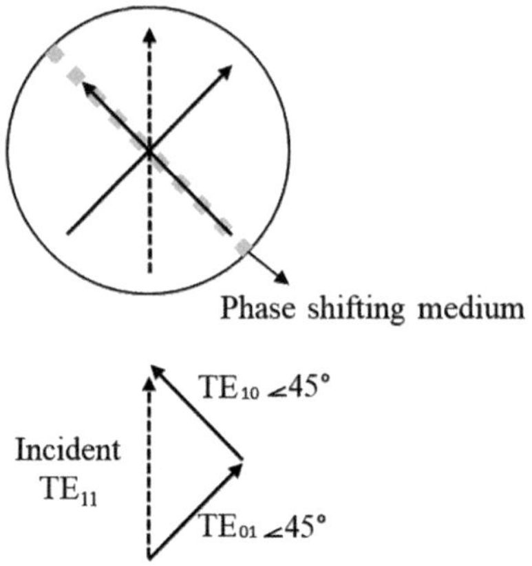

Fig. 2 Alignment of differential phase shifting medium

$$\text{AR} = 1 \text{ or}$$
$$\text{AR (dB)} = 0\,\text{dB}$$

Which means the wave is perfect circularly polarized. But this is not the general case. The linear signal is converted into elliptically polarized signal with AR (dB) less than 3 dB, and then, this is sufficient for practical applications.

As discussed above, the phase shifting region can be realized by proper taper/ transformer or posts design. Here, design with antiphase posts as phase shifting medium has been shown.

Figure 3 shows cross-sectional view of polarizer with post parameters.

The design of circular waveguide polarizer starts with the determination of radius (r), which is selected as per cut off frequency of dominant mode TE_{11} and is given

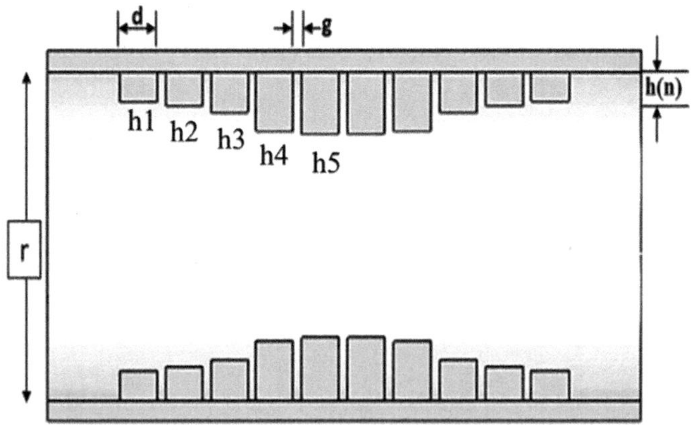

Fig. 3 Cross-sectional view of polarizer

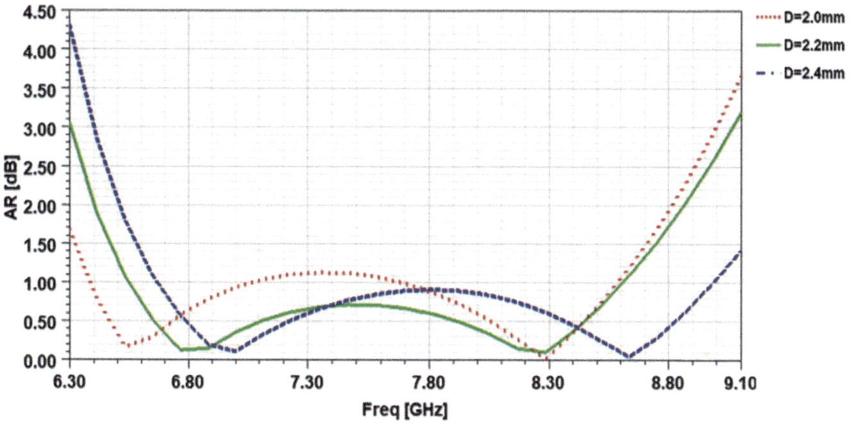

Fig. 4 Post diameter (d) variation

below [12]:

$$f_c = \frac{1.841}{2\pi r\sqrt{\mu\varepsilon}} \tag{8}$$

Here, the diameter (d) of the cylindrical posts affects the AR bandwidth. Variation of d with respect to AR is shown in Fig. 4.

As, it can be seen from Fig. 4, increasing the post diameter shifts the AR to lower frequency, and decreasing the post diameter shifts the AR to higher frequency. In our design, d has been kept as $\lambda g/20$ for desired frequency band of operation. Gap (g) between the post affects the level of axial ratio matching, and its variation with respect to AR is shown in Fig. 5.

As, it can be seen from Fig. 4, the post gap (g) controls AR bandwidth of the polarizer, and decreasing the gap (g) increases the AR bandwidth, but it poses two issues, one is gap cannot be reduced beyond a certain value; otherwise, posts will overlap which is practically not feasible, and other is AR value at center frequency tends to go high. In this design, post gap (g) has been kept $\lambda g/10$ at the center frequency of operation for optimum results. Height $h(n)$ of the posts is varied and has been optimized to achieve acceptable level of matching.

An important parameter of polarizer is height of posts. The height profile affects the level of matching while individual post heights affect the axial ratio also. As a variation, the height of center post ($h5$) was varied with respect to return loss and axial ratio, and results are shown in Figs. 6 and 7, respectively.

As it can be seen from Fig. 6, the return loss bandwidth of polarizer does not change with center post height ($h5$) variation rather value of return loss is varied in which we have improved return loss at $h5 = 6.2$ mm.

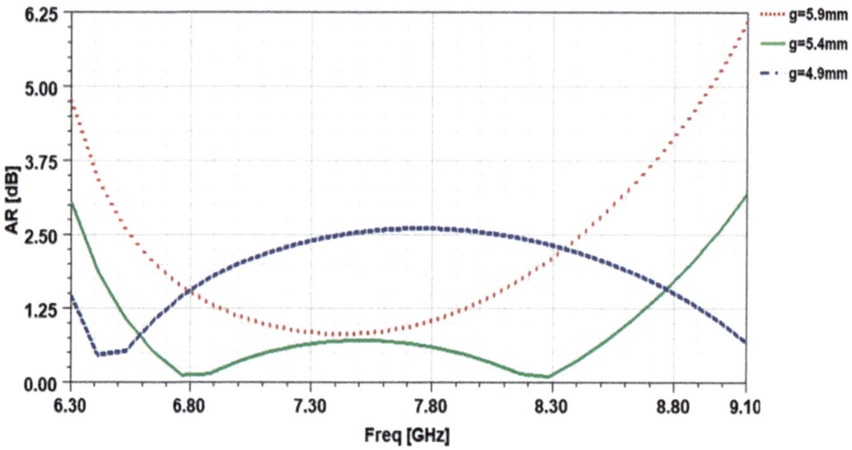

Fig. 5 Post gap (g) variation

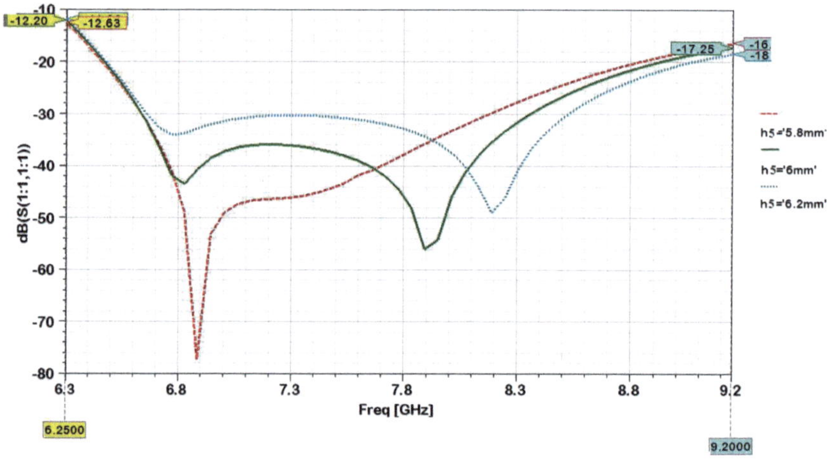

Fig. 6 RL versus center post height ($h5$) variation

3 Analysis, Results, and Discussion

In this design, a polarizer with a circular waveguide and five pair of posts is designed as shown in Fig. 3. The polarizer designed based on the above design method was simulated in Ansys Electronic Desktop [3].

There are two degenerative modes of TE_{11} at each port, and hence electrically it is a 4-port device. As discussed in the previous section, design objective is to achieve that the incident TE_{11} shall be divided into two components with equal power and orthogonal phase. Plot of power delivered to two orthogonal modes is shown in Fig. 8.

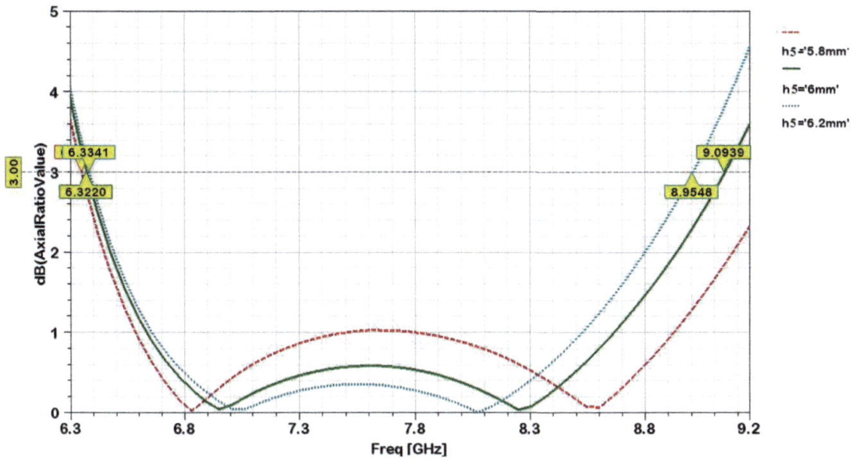

Fig. 7 AR versus center post height ($h5$) variation

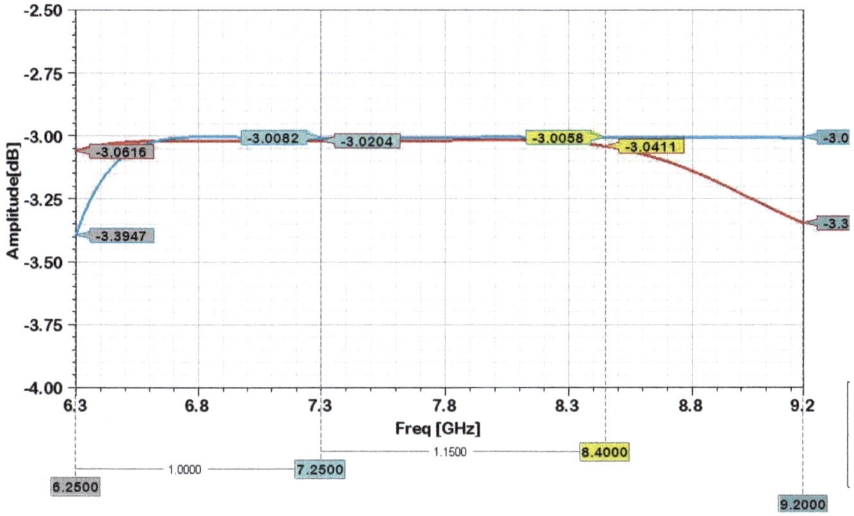

Fig. 8 Power delivered to two orthogonal modes

As it can be seen from Fig. 8, the output power in both modes is equal to − 3 dB for desired frequency range of 7.25–8.4 GHz while it is up to − 3.4 dB for extended range of frequency 6.25–9.2 GHz which means the input is being divided with equal amplitudes. The phase difference of the two modes is shown in Fig. 9.

The phase difference of the two modes is within $90° \pm 4°$ in frequency range 7.25–8.4 GHz which shows high orthogonality of the generated output signals, while its 65° for extended range of frequency 6.25–9.2 GHz. As given in Eq. (6), the axial

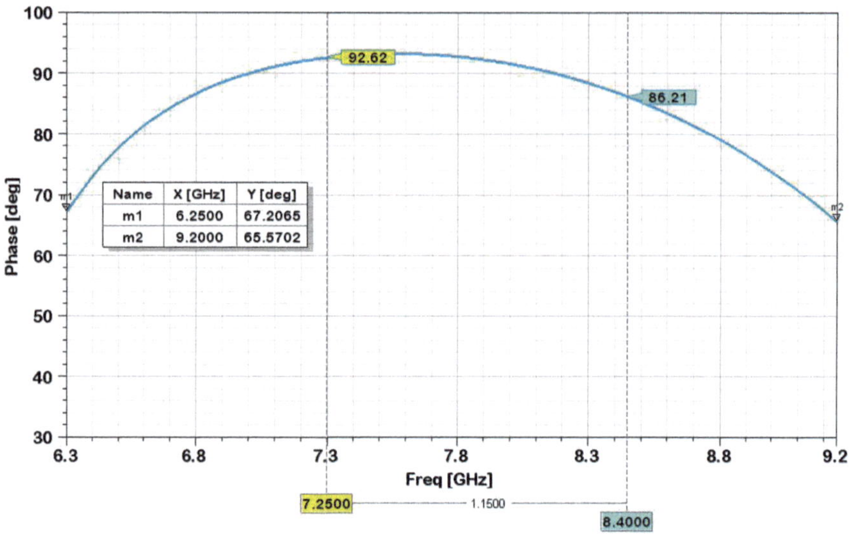

Fig. 9 Phase difference between orthogonal modes w.r. to frequency

ratio primarily depends on the phase difference if the amplitude is equal, and hence, the AR is 0.6 dB within frequency range 7.25–8.4 GHz, while it is gone up to 3 dB in the frequency range 6.25–9.2 GHz because phase difference at corner frequencies 6.25 and 9.2 GHz is about 65°.

Plot of AR with respect to frequency is shown in Fig. 10.

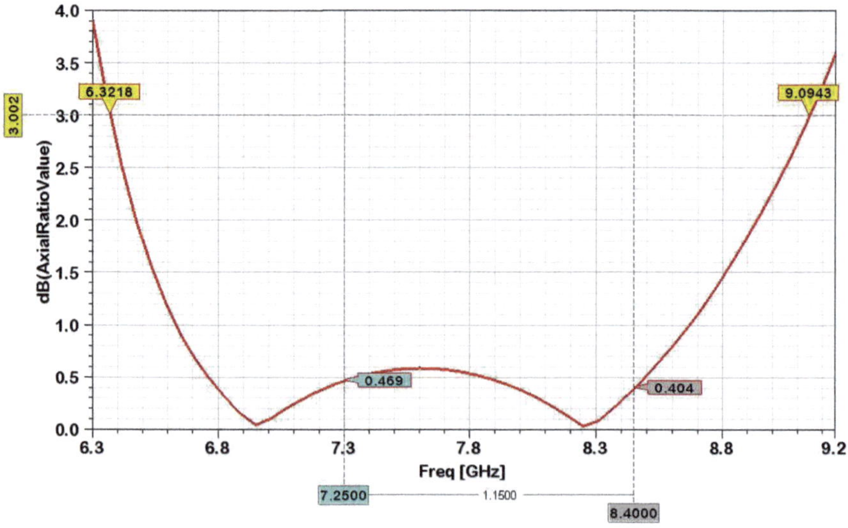

Fig. 10 Axial ratio plot

Fig. 11 Fabricated polarizer

So, same design has been used as the achieved axial ratio is less than 0.6 dB in the operating frequency band of 7.25–8.4 GHz; however, this can be optimized in the full band based on the requirement of axial ratio value and bandwidth. The designed polarizer was fabricated. The fabricated polarizer has circular flanges matching to feed horn and locating holes at 45°. The prototype is shown in Fig. 11.

The prototype was tested at bench level for return loss. The test setup included two circular to rectangular transition at the both ends of the polarizer and finally connected to waveguide to coaxial adapter as shown in Fig. 12.

The return loss of the polarizer is shown in Fig. 13.

The return loss is better than − 27.5 dB for desired frequency band 7.25–8.4 GHz while it is better than − 11 dB for 6.25–9.20 GHz. The fabricated polarizer was connected with feed horn and measured in near field test range (NFTR) as shown in Fig. 14.

With respect to Fig. 15 which depicts RHCP and LHCP patterns of feed with polarizer at 7.25 GHz, both the simulated and measured results are in close agreement, and cross-polarization at 0° is also similar.

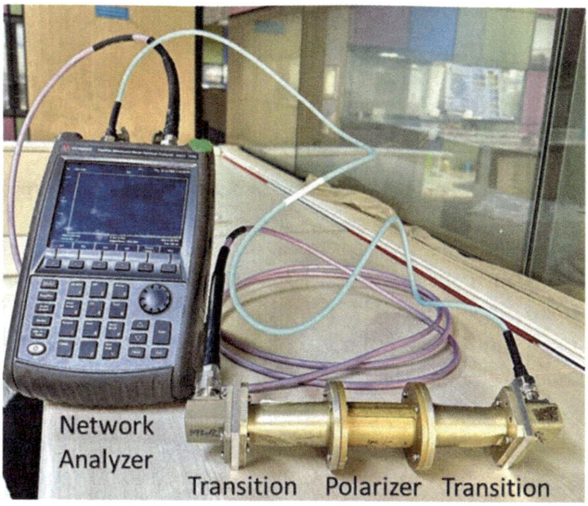

Fig. 12 Polarizer test setup

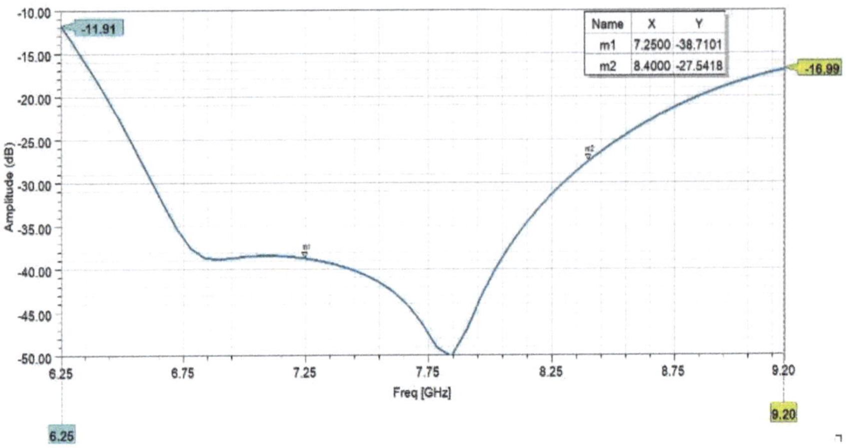

Fig. 13 Return loss

Fig. 14 Near field measurement of polarizer with feed horn

With respect to Fig. 16 which depicts RHCP and LHCP patterns of feed with polarizer at 8.4 GHz, both the simulated and measured results are in close agreement, and cross-polarization at 0° is − 48.4 dB in simulation while − 43.7 dB in measurements. The radiation pattern of feed with polarizer was measured in the wideband frequency range of 6.25–9.20 GHz to validated axial ratio bandwidth. The measured axial ratio (AR) versus frequency is shown in Fig. 17.

As seen from Fig. 13, the measured axial ratio (AR) is in close agreement with simulated results. The overall achieved results are summarized in Table 1.

As seen from Fig. 13, the measured axial ratio (AR) is in close agreement with simulated results. The AR in frequency range 7.25–8.4 GHz is better than 1.15 dB

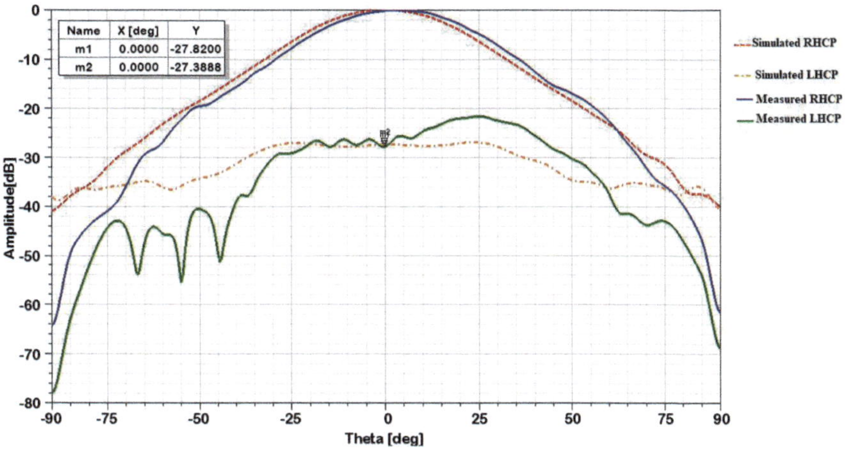

Fig. 15 Simulated versus measured radiation pattern at 7.25 GHz

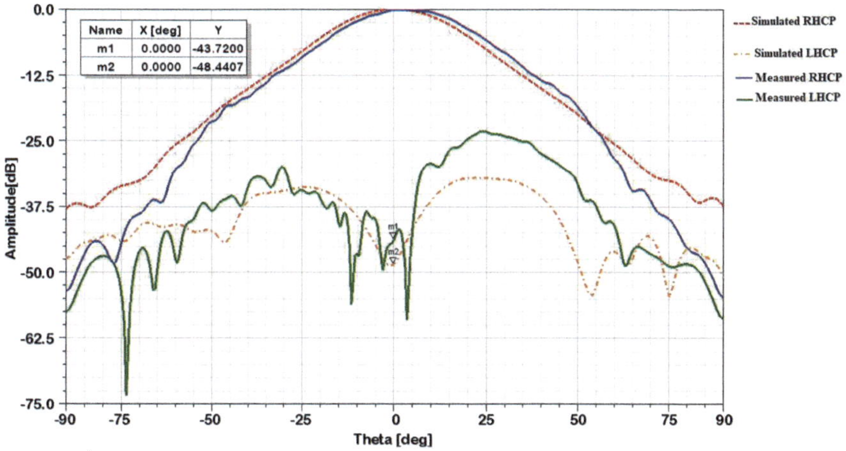

Fig. 16 Simulated versus measured radiation pattern at 8.4 GHz

against 1.0 dB simulated value while 3 dB axial ratio bandwidth is 6.26–9.12 GHz which is approximately 37.2%.

4 Conclusion

In this article, the detailed design procedure has been discussed, the same has been simulated, and measured results of compact wideband waveguide polarizer at X-band have been discussed. The achieved axial ratio for 7.25–8.4 GHz is better than 1.15 dB

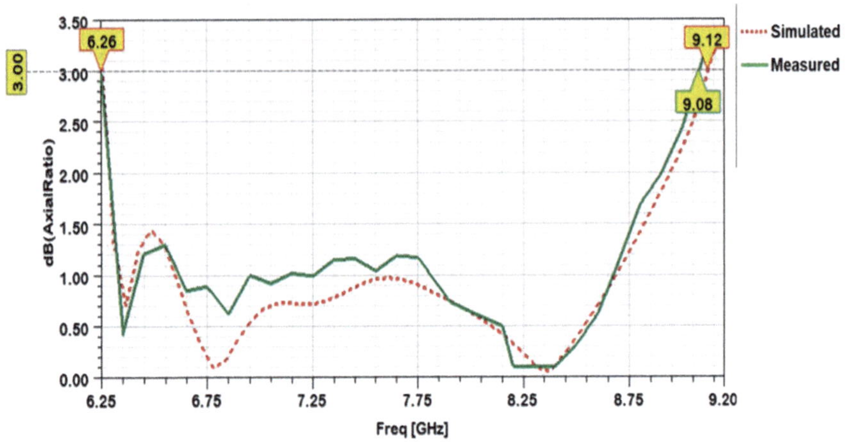

Fig. 17 Simulated versus measured boresight axial ratio

Table 1 Results summary

S. No.	Parameter	Achieved results (%)	Remarks
1	VSWR bandwidth	38	RL ≤ − 11 dB (6.26–9.12 GHz)
2	3-dB AR bandwidth	37.2	AR ≤ 1.15 [7.25–8.4 GHz] AR ≤ 3 [6.26–9.12 GHz]

against 1.0 dB simulated value while 3 dB axial ratio bandwidth is 6.26–9.12 GHz which is approximately 37.2%, and it has more than 38% return loss bandwidth (6.25–9.20 GHz). The length of fabricated polarizer is only 7 cm which is compact for the required application.

Acknowledgements This work was supported by Bharat Electronics Ltd., A Government of India Enterprise under Ministry of Defense.

References

1. Olver AD, Clarricoats PJB et al (1994) Microwave horns and feeds. Institution of Electrical Engineers, London
2. Stutzman WL (2018) Polarization in electromagnetic systems. Artech House, Norwood, MA
3. Deutschmann B, Jacob AF (2020) Broadband septum polarizer with triangular common port. IEEE Trans Microw Theory Techn 68(2):693–700
4. Nikolic N et al (2018) A septum polarizer with integrated square to circular tapered waveguide transition. In: IEEE international symposium on antennas and propagation & USNC/URSI national radio science meeting, Boston, USA, July 2018, pp 725–726. https://doi.org/10.1109/APUSNCURSINRSM.2018.8608909

5. Rud LA, Shpachenko KS (2012) Polarizers on a segment of square waveguide with diagonal ridges and adjustment iris. Radioelectron Commun Syst 55(10):458–463
6. Leal-Sevillano CA et al (2013) A 225 GHz circular polarization waveguide duplexer based on a septum orthomode transducer polarizer. IEEE Trans Terahertz Sci Technol 3(5):574–583
7. Dubrovka F et al (2020) Compact X-band stepped-thickness septum polarizer. In: IEEE Ukrainian microwave week (UkrMW), Kharkiv, Ukraine, Sept 2020, pp 135–138
8. Chung M-H, Je D-H, Han S-T, Kim SR (2014) Development of a 85~115 GHz 90-deg phase shifter using corrugated square waveguide. In: European microwave conference (EuMC), Rome, Italy, Oct 2014, pp 1146–1149
9. Zhu QC, Williamson AG, Neve MJ (2007) Reactance of posts in circular waveguide. IEEE Trans Microw Theory Techn 55(8):1685–1688
10. Sharma SB, Singh VK, Dey R, Chakrabarty S (2009) Analysis of a post discontinuity in an oversized circular waveguide. IEEE Trans Microw Theory Techn 57(8):1989–1995
11. Uher J, Bornemann J, Rosenberg U (1993) Waveguide components for antenna feed system: theory and CAD. Artech House, London
12. Sun L (2013) Axial ratio bandwidth of a circularly polarized microstrip antenna, chap 10. INTECH, pp 229–246

Parallel Digital VLSI Architecture for Combined SVM Training and Classification

Sidhant Priyadarshi and Saroja V. Siddamal

Abstract The paper presents a Parallel Digital Hardware architecture for Integrated Support Vector Machine Training and Classification. The architecture is designed with three layers of parallelism. As use case, 1000 data sets of diabetic patients are considered to classify diabetic and non-diabetic based on the Support Vector Machine (SVM) Algorithm. The design is synthesized using Cadence Genus with TSMC 180 nm technology. Design is much faster than software tools which are used to execute behavioral codes of AI/ML or faster than general-purpose CPU's. The design takes 1024×1024 clock cycles of period 7.147 ns which is in total 7.494 ms to train raw data sets which is attractively much better than software.

Keywords Support Vector Machine · Hyper plane · Parallel architecture · Karush–Kuhn–Tucker conditions

1 Introduction

Parallel architecture adds a new dimension to the development of the system by the usage of more processors. All the processes are working concurrently to solve the same problem. By the utilization of a large number of processors, performance is higher than the performance of a single processor at a given point in time. The actual purpose of parallel processing is to increase the processing capability and increase throughput. Usage of AI/ML applications in real-time is increasing in an attractive manner but to achieve efficiency it requires training of huge data sets which consumes much time. This is what motivates today's world to work on hardware models that can train and classify data with higher efficiency and achieve a speed of almost 80 times that of training using software.

S. Priyadarshi · S. V. Siddamal (✉)
KLE Technological University, Hubballi, India
e-mail: sarojavs@kletech.ac.in

S. M. Thampi et al. (eds.), *Fifth International Conference on Computing and Network Communications*, Lecture Notes in Electrical Engineering 1219,
https://doi.org/10.1007/978-981-97-4540-1_43

2 Literature Survey

The research literature available shows the work of various researchers in developing accelerators using ML algorithms. The work in [1] presents, PROMISE the first end-to-end design of a programmable mixed-signal accelerator from Instruction Set Architecture (ISA) to a high-level language compiler for acceleration of diverse ML algorithms. The authors first identify prevalent operations in widely used ML algorithms and key constraints in supporting these operations for a programmable mixed-signal accelerator. Experiments show that PROMISE can accelerate diverse ML algorithms with energy efficiency competitive even with fixed-function digital ASICs for specific ML algorithms. Cascaded Support Vector Machine (SVM), a powerful training algorithm, is leveraged to significantly improve the scalability of hardware-based SVM training and develop an efficient parallel VLSI architecture. The parallel SVM processors provide a significant training time speedup and energy reduction compared with the software SVM algorithm running on a general-purpose CPU [2]. Improvement in speed with efficiency in training and classifying data is implemented using algorithms like gradient ascent and processing data in parallel with other units [3]. SODA Synthesizer, an automated open-source high-level ML framework to Verilog modular compiler targeting AI/ML Application-Specific Integrated Circuits accelerators. SODA tightly couples the Multi-Level Intermediate Representation (MLIR) compiler infrastructure and open-source HLS approaches [4]. A parallel architecture designed for a complex pattern classification from high-energy physics involving thousands of patterns is reported and discussed, comparing the performance obtained by this architectural solution to that of a simpler sequential architecture [5]. The MAPLE architecture in [6], explores its design space with a simulator, and illustrates how to automatically map application kernels to the hardware. The implementation of a multiclass classifier based on SVMs which maps its inherent degrees of parallelism to the GPU programming model to improve the through and efficiency is presented in [7]. A parallel FPGA-based coprocessor for Support Vector Machines (SVMs), a machine learning algorithm whose applications include recognition tasks such as learning scenes, situations, and concepts, and reasoning tasks such as analyzing the recognized scenes and semantics is presented in [8].

3 Mathematical Model of SVM

To achieve classification of datasets with higher accuracy, AI/ML algorithms are needed to be transformed into hardware architectures but to do so it requires mathematical models. Similarly to perform SVM on hardware it has certain algorithms like Gradient Ascent which works on the principle of Lagrange's multiplier which again requires some conditions called Karush–Kuhn–Tucker (KKT's) conditions to minimize the distance between margins and support vectors which is done by maximizing weights (W).

$$\alpha_i^{\text{new}} = \alpha_i - \frac{1}{K(\overrightarrow{x_i}, \overrightarrow{x_j})} y_i \left(\sum_{j=1}^{N} \alpha_i y_i K(\overrightarrow{x_i}, \overrightarrow{x_j}) + b \right) \tag{1}$$

$$\alpha_i^{\text{new}} = 1 - y_i \left(\sum_{j=1}^{N} \alpha_i y_i K(\overrightarrow{x_i}, \overrightarrow{x_j}) + b \right) - 1 \tag{2}$$

Lagrange's multiplier discriminates support vectors and non-support vectors. Equations 1 and 2 are derived from KKT conditions shown in Eqs. 3, 4, and 5. The weight (W) is maximizing or minimizing the distance between margins and support vectors by minimizing the half of square of weight.

$$\alpha_i = 0 \Rightarrow y_i \left(\sum_{j=1}^{N} \alpha_i y_i K(\overrightarrow{x_i}, \overrightarrow{x_j}) + b \right) \geq 1 \tag{3}$$

$$0 < \alpha_i < C \Rightarrow y_i \left(\sum_{j=1}^{N} \alpha_i y_i K(\overrightarrow{x_i}, \overrightarrow{x_j}) + b \right) = 1 \tag{4}$$

$$\alpha_i = C \Rightarrow y_i \left(\sum_{j=1}^{N} \alpha_i y_i K(\overrightarrow{x_i}, \overrightarrow{x_j}) + b \right) \leq 1 \tag{5}$$

By partially differentiating Eq. 2 with respect to weight (W), bias (B) and equating both with zero as shown in Eq. 6 following conclusion is made as shown in Eq. 7. Equation 7 shows the weight obtained which is equal to the summation of α_i, y_i, $K(x_i, x_j)$. The bias obtained after partial differential is equal to zero.

$$\frac{\delta\alpha}{\delta W} = \frac{\delta\alpha}{\delta B} = 0 \tag{6}$$

$$W = \sum_{j=1}^{N} \alpha_i y_i \phi(x) \tag{7}$$

4 Proposed Hardware Architecture for SVM

To achieve parallelism the proposed novel approach works on three levels of training. The training modules are arranged in the ratio of 4:2:1. Each trainer module is designed to consist of Input memory and a Gradient ascent module. The first level of training data is stored in the 1st stage output memory unit. Figure 1 shows the proposed SVM Architecture. The architecture has 15 sub-top modules in which 4 Trainer units and 4 output memories are cascaded in a parallel in the first layer. In the second layer, 2 Trainer units and output memories are cascaded in parallel and finally, one trainer unit and one output memory along with the classifier are cascaded in series with the output from the previous layer as shown in Fig. 1. The design has to perform the Support Vector Machine (SVM) algorithm being synthesizeable in nature which can be done by optimal technique to find the best hyperplane between label + 1 and − 1 that is diabetic and non-diabetic class as referenced in this design. The Gradient Ascent (GA) module shown in Fig. 2 is the main module of design which performs training by differentiating between support vectors and non-support vectors which are performed by the gradient ascent algorithm. The Gradient Ascent algorithm works on a principle of Lagrange's theorem to get minimum margins between label + 1 and − 1 by maximizing the half of the square of weight (W) which needs to satisfy 3 conditions given by Karush–Kuhn–Tucker (KKT) until convergence. Training is done on data sets that are stored in the Input Memory of the trainer which sends the alpha value on each clock until the complete alpha values are sent for each X and Y value. After collecting support vectors and their respective Lagrange's multiplier value from GA, these are then sent to the output memory module which is stored in parallel until training is completed. Once training is completed, the data from the memory module is sent to the next stage trainer module on the active enable signal. The classifier is shown in Fig. 3 is activated when it receives an enable signal generated once the training is completed. Now, the classifier takes input from the user to check whether the data is diabetic or non-diabetic and also inputs from output memory to perform Eq. 8 when $f(X)$ is greater then 0 it is labeled + 1 as non-diabetic and is labeled − 1 and diabetic respectively.

$$f(X) = \sum_{i=1}^{N_{sv}} \alpha_{sv} \times Y_{sv} \times K(X, X_{sv}) \tag{8}$$

5 Hardware Implementation and Results

5.1 FPGA Implementation

The designed architecture is simulated using ISIM simulator, Fig. 4 shows the simulation results. The simulation is done for the design having data sets of blood glucose levels to classify between diabetic and non-diabetic persons. A value greater than 180 is considered as diabetic which is − 1 whereas less than 180 is considered to be non-diabetic which is label + 1. As shown in Fig. 4 test data of 011001000 gives out Class (marked in blue) output as − 1 as 200 is considered to be diabetic. As

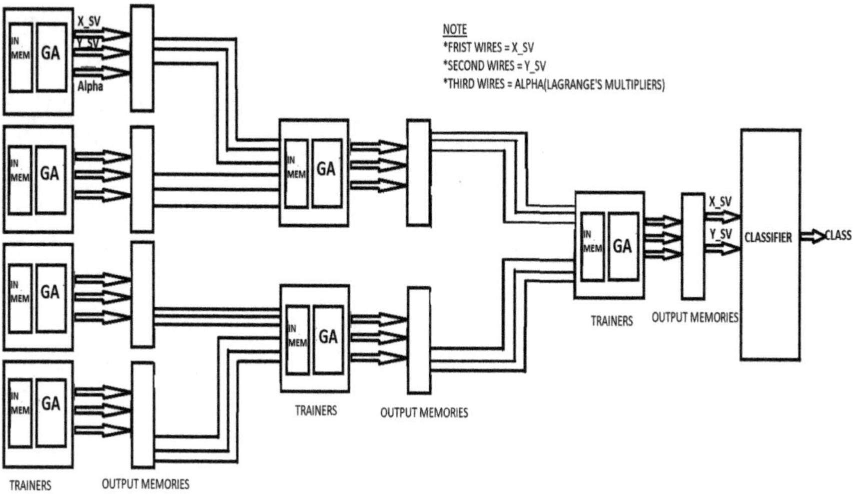

Fig. 1 Architecture of cascaded SVM model

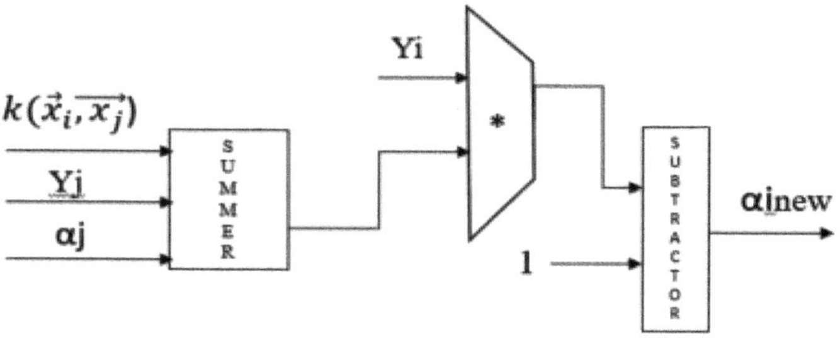

Fig. 2 Architecture of gradient ascent module

Fig. 3 Architecture of classifier module

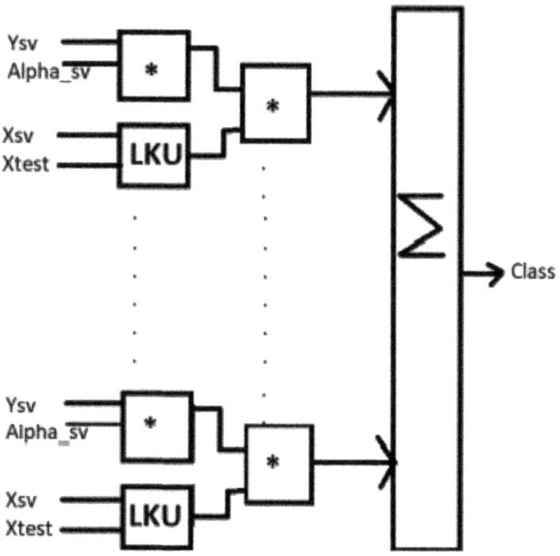

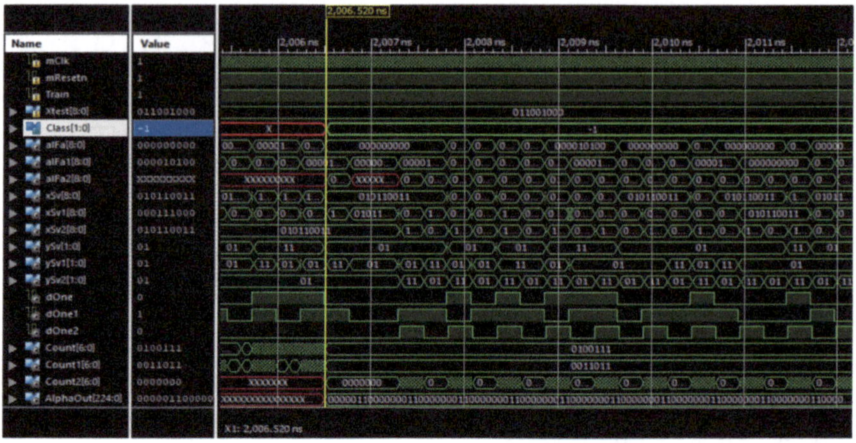

Fig. 4 Classification output of diabetic data

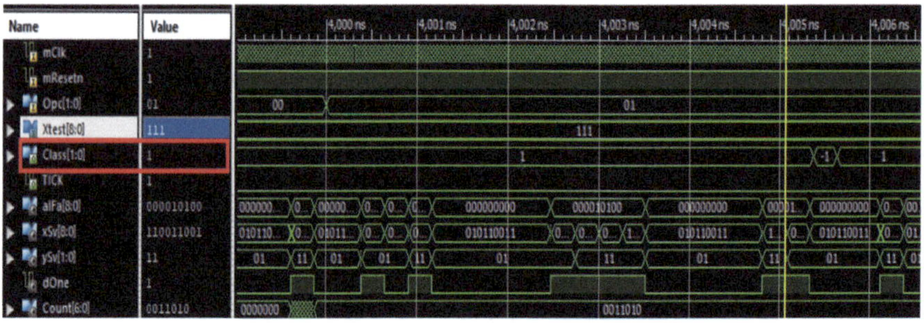

Fig. 5 Classification output of non-diabetic data

Table 1 Hardware utilization summary

Mnemonics	Opcodes	Available
Slice registers	943	4800
Slice LUT's	1348	2400
LUT-FF pairs	114	114
Bonded IOBs	21	132
BUFG	3	16
DSP48A1s	8	8

shown in Fig. 5 test data of 1101111 gives Class (marked in red) output as + 1 as 111 is considered to be non-diabetic. The design is synthesized on Xilinx ISE 14.7 and ported to Spartan6 XC6SLX4. Table 1 shows the hardware resource utilized.

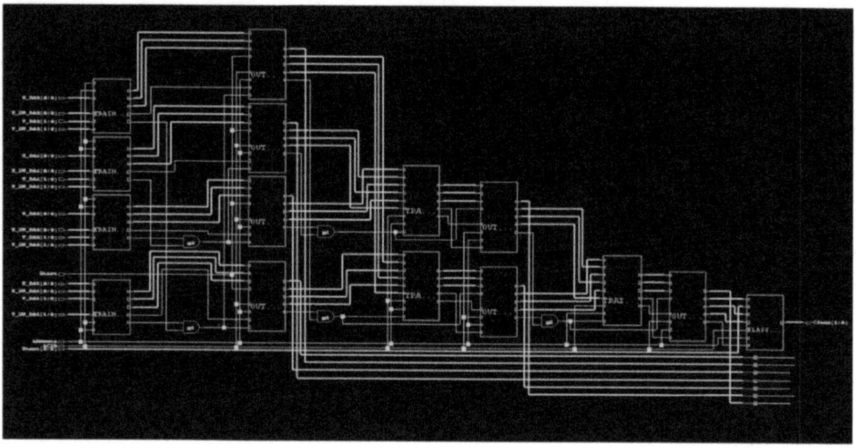

Fig. 6 RTL view after HDL elaboration

Table 2 Timing report

Parameters	Values (ns)
Clock period	7.147 (139.917 MHz)
Setup time	18.626
Propagation delay	3.597
Slack	3.55

5.2 ASIC Implementation

The proposed architecture is implemented in CMOS 180 nm technology. The design has to perform classification based on the blood glucose levels of diabetic and non-diabetic persons which is + 1 and − 1 respectively. 1000 data set is used to train the module. Figure 6 shows the RTL view after HDL elaboration taken using the cadence genus tool.

5.3 Timing Analysis

The timing analysis of the NETLIST is performed using the Tempus tool. As shown in Table 2 it is observed that the design works on a maximum operating frequency of 139.917 MHz and has an output propagation delay of 3.597 ns which gives a positive slack of 3.55 ns.

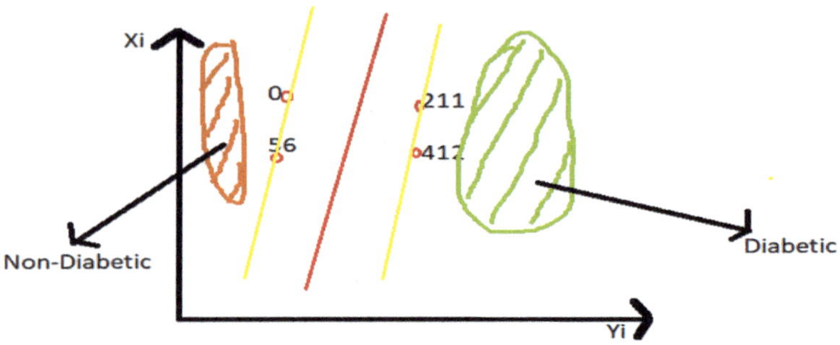

Fig. 7 Plot of support vector

5.4 *Support Vectors Plot*

Figure 7 is a hand-plotted graph of hyperplane and support vectors obtained by training datasets ranging from 0 to 511. As shown, support vectors 0 and 56 are obtained after training 25 sets of data from the non-diabetic class and 211 and 412 are obtained from the diabetic class. The circle marked in orange represents non-diabetic sets of data ranging from 0 to 179 and the green circle represents diabetic regions ranging from 180 to 511.

6 Conclusion

The authors have the design of parallel digital VLSI architecture for combined SVM training and classification which is developed using Xilinx ISE 14.7 and synthesized using Cadence Genus tool with TSMC 180 nm library. Design is much faster than software tools which are used to execute behavioral codes of AI/ML or faster than general-purpose CPU's. The design takes 1024×1024 clock cycles of period 7.147 ns which is in total 7.494 ms to train raw data sets which is attractively much better than software.

References

1. Wang Q (2016) Architectures and design of VLSI machine learning systems. IEEE Trans Very Large Scale Integr (VLSI) Syst
2. Wang Q, Li P, Kim Y (2014) A parallel digital VLSI architecture for integrated support vector machine training and classification. IEEE Trans Very Large Scale Integr (VLSI) Syst 1471–1484
3. Zhang JJ, Agostini NB, Song S, Tan C, Limaye A, Amatya V, Manzano J, Minutoli M, Castellana VG, Tumeo A et al (2021) Towards automatic and agile AI/ML accelerator design with end-to-end synthesis. In: IEEE 32nd international conference on application-specific systems, architectures, and processors (ASAP), pp 218–225

4. Srivastava P, Kang M, Gonugondla SK, Lim S, Choi J, Adve V, Kim NS, Shanbhag N (2018) PROMISE: an end-to-end design of a programmable mixed-signal accelerator for machine-learning algorithms. In: ACM/IEEE 45th annual international symposium on computer architecture (ISCA), pp 43–56
5. Biasi I, Boni A, Zorat A (2005) A reconfigurable parallel architecture for SVM classification. In: Proceedings. 2005 IEEE international joint conference on neural networks, pp 2867–2872
6. Cadambi S, Majumdar A, Becchi M, Chakradhar S, Graf HP (2010) A programmable parallel accelerator for learning and classification. In: Proceedings of the 19th international conference on parallel architectures and compilation techniques, pp 273–284
7. Herrero-Lopez S, Williams JR, Sanchez A (2010) Parallel multiclass classification using SVMs on GPUs. In: Proceedings of the 3rd workshop on general-purpose computation on graphics processing units, pp 2–11
8. Cadambi S, Durdanovic I, Jakkula V, Sankaradass M, Cosatto E, Chakradhar S, Graf HP (2009) A massively parallel FPGA-based coprocessor for support vector machines. In: 2009 17th IEEE symposium on field-programmable custom computing machines, pp 115–122

Design and Verification of Pipelined Trivium Cipher

M. S. Praveen Kumar, K. Shailaja, and K. Sudeendra Kumar

Abstract Cryptography and ciphers are essential for secure communication and data security. In order to ensure secrecy and integrity, plain text is converted into ciphertext using mathematical procedures and techniques. A type of symmetric encryption known as stream ciphers encrypts data bit by bit, usually in real time. Through this research, we have tried to design a new cipher technique that uses the pipelining concept to consume less power while taking up less space than conventional ciphers. Pipelined trivium ciphers are more efficient for ASIC implementation and can be placed in IP formats with ASIC implementation at online security services like blockchain technology. They are implemented based on the structure of conventional trivium ciphers and focused on reducing propagation delay to operate at a higher speed than conventional trivium ciphers. In order to attain excellent performance while consuming less energy, brand-new synchronous stream ciphers called pipelined trivium ciphers are developed by the team through this research work for security systems and online services. The designed pipelined trivium cipher is 1.18 GHz faster than the conventional trivium cipher with its slack value greater by a difference of 55 ps than the conventional trivium cipher.

Keywords Trivium cipher · Low power · Light weight cipher

M. S. Praveen Kumar (✉) · K. Shailaja
JSS Science and Technology University, Mysuru, India
e-mail: ms_praveenkumar@sjce.ac.in

K. Shailaja
e-mail: kshaila_15@sjce.ac.in

K. Sudeendra Kumar
PES University, Bangalore, India

S. M. Thampi et al. (eds.), *Fifth International Conference on Computing and Network Communications*, Lecture Notes in Electrical Engineering 1219,
https://doi.org/10.1007/978-981-97-4540-1_44

1 Introduction

A cipher is a set of predetermined procedures that can be used as a technique in cryptography to perform encryption or decryption. Encipherment is a different, less frequently used term. Information is transformed into a code or cipher when it is encoded or enciphered. In everyday speech, the terms "cipher" and "code" are interchangeable [1] since they both refer to a series of operations that encrypt a message. However, in cryptography, especially classical cryptography, the ideas are different. Algorithms with low computational and spatial complexity are known as lightweight ciphers. Lightweight encryption is employed in restricted devices like RFID tags, fire and security detectors, wireless communication devices, and other IoT devices in the present era of downsizing. Stream ciphers are symmetric ciphers that produce cipher text by encrypting the bitstream of plain text with a corresponding key stream. Therefore, a stream cipher with a high level of security but minimal computational complexity might be referred to as a lightweight stream cipher. The pipelined trivium cipher is a variant of the trivium stream cipher that introduces pipeline stages to enhance its performance. It is designed to provide secure and efficient encryption for various applications. The cipher follows the principles of synchronous stream ciphers, generating a keystream to encrypt plaintext data. By incorporating pipeline stages, the pipelined trivium cipher enables parallel processing and improves throughput. This enhancement allows for faster encryption and decryption operations, making it suitable for high-speed data transmission and resource-constrained devices.

2 Motivation

The crucial criteria for cryptographic hardware applications include privacy, data protection, and information security. Several billion electronic gadgets exchange vast volumes of protected data thanks to modern technology. The main ideology behind designing pipelined trivium cipher is to create security solutions that can operate on devices with restricted resources by utilizing less memory, less computational power, and less energy. Comparing the trivium ciphers to traditional encryption, it is anticipated that they will be quicker and easier. The major limitations we face while designing lightweight ciphers often have to do with power consumption, gate equivalents, and timing. On devices with limited resources, the lightweight cryptography compromises implementation costs, execution speed, security, performance, and energy consumption.

3 Objectives

The objectives considered to be achieved while designing the pipelined trivium cipher are pointed out below.

- Implementation of conventional trivium cipher and an analyzing the encrypted results
- Pipelined high-performance implementation of trivium stream cipher.
- To analyze and verify the reduction of power consumption and switching action in pipelined trivium cipher.

4 Related Work

The literature review has been painstakingly crafted by a thorough examination including numerous important factors. Came across various authors who have shown interest to include information on the various stream cipher implementations and algorithms, including ZUC, Snow3g, Grain V1, Mickey V2, and A5/1 [2]. We also learned about the crucial role the trivium cipher plays in embedded systems [3] and IoT [4] due to its superior performance in terms of speed and switching frequency. The amount of data shared through plainly visible communication channels like wireless broadcasts has increased as a result of the Internet of Things widespread adoption.

A common method for ensuring the confidentiality of the data is encryption. We also discovered an energy-efficient trivium cipher implementation [5] based on a threshold implementation masking scheme that is demonstrably shielded against such leakages. Additionally, the review covered the trivium cipher implementations' susceptibility to fault attacks [6] and suggested a defense against bitstream modification attacks [7].

Trivium cipher multi radix architecture [1, 8] versions generate 16 bit/clock cycles; hence the measurements show that the MPLP version has the lowest power usage per bit (0.69 pJ/bit at 50 MHz) out of all the other stream ciphers. The notion of lightweight cryptography, which refers to cryptographic algorithms designed for usage in devices with minimal or extremely low resources, is described clearly by author Panasenko et al. [9]. A bitstream modification attack on the trivium stream cipher is also been taken under consideration, a global standard under ISO/IEC 29192-3 [7]. Valencia-Barrero et al. provide a shift register parallelization-based [10] low-power hardware implementation of the trivium stream cipher [11]. The purpose of this article is to assess the applicability of this technique and to compare the low-power and standard implementations' core areas and power usage. Meena et al. [12] have clearly tried to explain that in numerous applications, including the Internet of Things, the depletion of power is a crucial essential requirement. In the Internet of Things, data transport is one of the most important and crucial tasks that consume the most power when performing cryptography. The research's ultimate goal is to create a pipelined trivium cipher implementation that uses less power and increases throughput, which improves the performance of IoT-based systems.

5 Design and Implementation

The methodology is divided into three sections, one explaining the proposed flow diagram development and the other two explaining the methodology for the development of conventional as well as pipelined trivium cipher.

5.1 Proposed Flow Diagram

The proposed flow diagram is illustrated in Fig. 1. The steps involved in illustration Fig. 1 are:

(1) Design and implementation of the algorithm for trivium low-power ciphers to optimize the basic design.
(2) Simulate the behavioral model of the designed algorithm using Cadence NC.
(3) Verilog Simulation Tool: Validation or testing of the designed algorithm for ciphers using test bench.
(4) Measure and model the performance metrics like power consumption, time and space complexity for each step using the RC—Encounter tool in cadence.
(5) Comparing results of both conventional and pipelined trivium ciphers.

5.2 Conventional Trivium Cipher

Trivium is a synchronous stream cipher with an 80-bit initialization vector and an 80-bit secret key that can generate a pseudorandom key stream of up to 264 bits. The cipher's construction is based on an internal state register with a 288-bit cyclic shift, and its nonlinear feedback is given by combinational logic AND and XOR gates. Three shift registers of different lengths are used to implement the trivium technique. Each internal shift register has a different length; the first register contains 93 bits, the second 84 bits, and the third 111 bits. Each register is added *m* bits in serial order, depending on the *m* output bits. Then, key stream bits are generated each clock cycle. The bits added to trivium's shift registers are produced via XOR gates. The control signals are applied to the different blocks of conventional trivium cipher as shown in Fig. 2. The majority of the power used in this architecture and mode of operation is consumed by the flip flops in the cyclic shift registers. In order to reduce switching activity while maintaining a consistent input frequency and supply voltage, we focused on applying parallelization techniques. The shift register is effectively

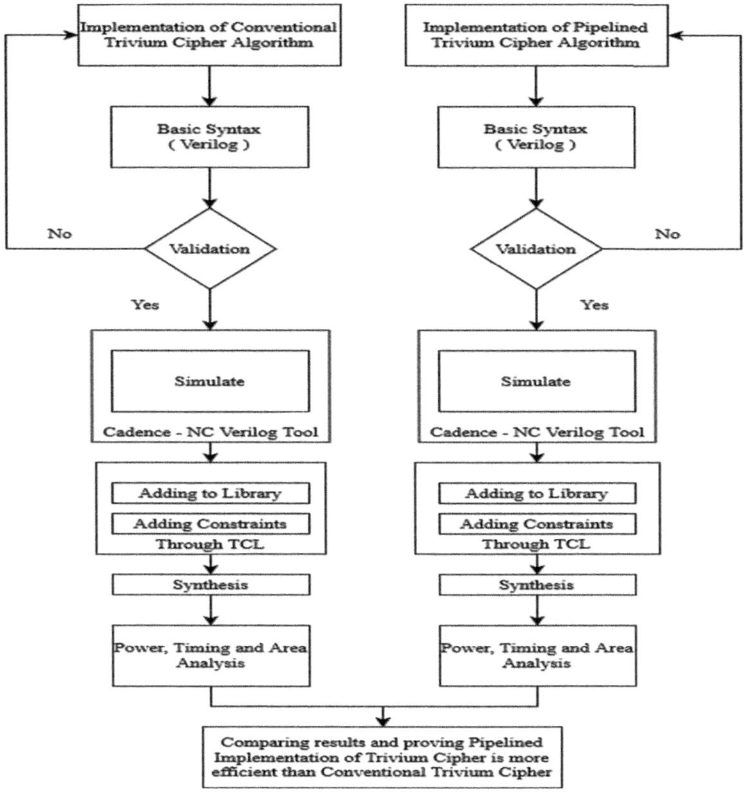

Fig. 1 Overall proposed block diagram

split into odd and even shift registers with half-bits using this technique. Dynamic power consumption is directly influenced by clock frequency, supply voltage, output capacitance, and switching activity factor, which quantifies the average proportion of clock cycles during which a signal transfer takes place. When the number of logic transitions in the cell output increases, correspondingly the switching power also increases.

5.3 Pipelined Trivium Cipher

The trivium cipher with pipelining will make use of different control signals for proper operation of the cipher and in order to get desired outputs. The control signals are applied to the different blocks of pipelined trivium cipher as shown in Fig. 3. The cipher engine is the heart of the system, which contains shift registers of three different lengths, the total length of three shift registers will be 288 bits. It's divided

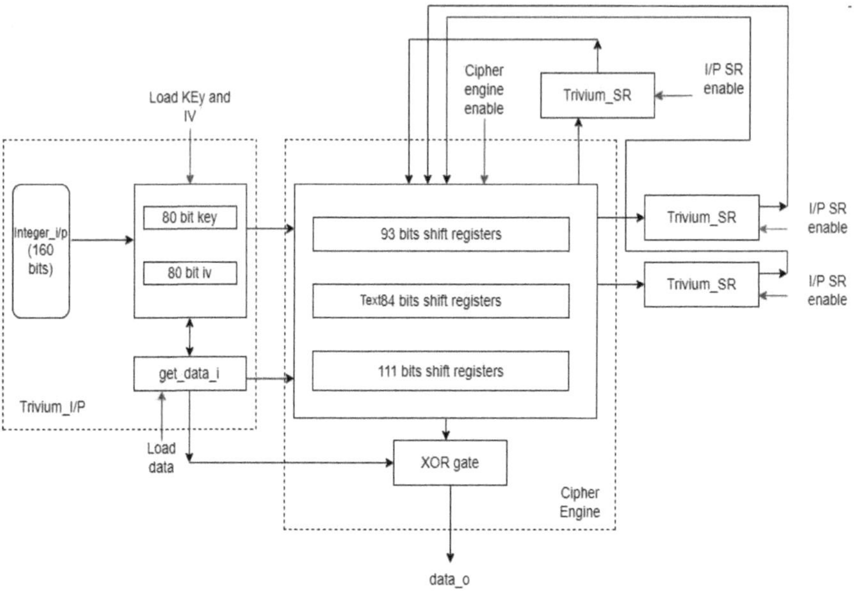

Fig. 2 Conventional trivium cipher—block-level representation

into 3 different lengths in order to achieve strong encryption with the help of nonlinear feedback, where nonlinear feedback is nothing but involving combinational logic.

The nonlinear feedback involving combinational logic is mainly XOR and AND gates. The nonlinear feedback circuits with the help of control signals will wake up all the 3 shift registers. Shift registers used are basically serial in serial out shift registers in order to achieve the cyclic operation and proper shifting of the data toward the nonlinear feedback in turn to get the desired encryption.

Inputs are applied to the shift registers through integer input block which consists of 2 shift registers of the same length, i.e., key and IV of 80 bits needed along with data. Apply this huge amount of data into the pipelined cipher's integer input block that would act like a buffer because it will accept the data from the user and applies into the system's integer input block which in turn would help in reducing the path delay.

The path delay has to be reduced in order to increase the speed of operation by keeping latch time positive for the desired operation of pipelined trivium cipher. Serial input connections are necessary to be connected with respect to the cipher engine block in order to generate the proper encryption, which can be achieved through trivium input block having proper control signal enabled at different states of pipelined trivium cipher independent of feedback signals. This path also involves D-flip flops in order to reduce the path delay by keeping latch time positive for the desired operation of pipelined trivium cipher. The cipher engine consists of three shift registers of different lengths where register 1 is 93 bits, register 2 is 84 bits, and register 3 is 111 bits.

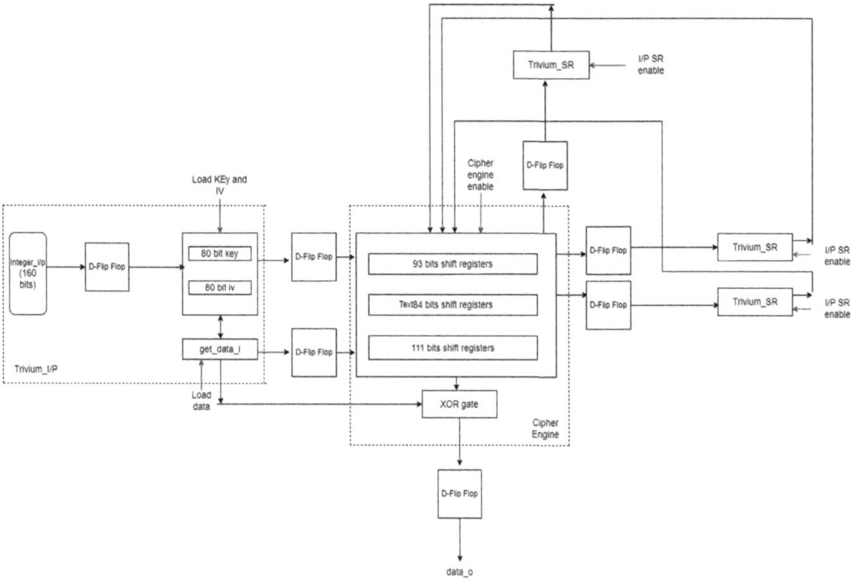

Fig. 3 Pipelined trivium cipher—block-level representation

A total of 288 bits of shift registers are available in the cipher engine block at the first stage, register 1 is loaded in the next stage, register 2 is loaded in third stage and finally, register 3 is loaded at the end stage.

Data shifter over shift register in each clock and nonlinear feed backs together work for achieving proper encryption through feedback formed by AND and XOR gate which in turn involves combinational logic plus D-flip flops in its structure to reduce the path delay by keeping the latch time positive. The nonlinear feedback circuit present in the trivium SR block is connected to the trivium input block and in turn, involves the XOR operation to be carried out with the input data to achieve strong and efficient encryption. Here D-flip flops are contributed to reducing the path delay by keeping latch time positive and thereby increasing the operating speed to achieve good performance.

The process of transforming a logical design description into a physically realizable netlist that can be constructed, on the other hand, is known as physical synthesis. RC synthesis is a tool utilized in this process. RC synthesis optimizes the design while taking into account physical restrictions like timing, routing, and placement.

Library Used for Synthesis: The Standard Cell library created for the 90 nm semiconductor process technology is known as the "saed90nm" library. It is a set of flip flops, logic gates, and other digital circuit elements that have been predefined for use in the design and production of integrated circuits (ICs) made with the 90 nm technology.

For the purpose of simulating digital designs created using Verilog and System Verilog HDLs, NC-Verilog is a simulator included in the cadence tool set.

6 Results and Analysis

6.1 Schematic and Waveforms

The implemented conventional trivium cipher (Fig. 4) and pipelined trivium cipher (Fig. 5) are stream ciphers that consist of 7 blocks plus a trivium top block, with block 7 being the cipher engine. The cipher engine combines the initialization vector (IV) and the key to generate the keystream used for encryption and decryption. In the trivium cipher, the IV and key are loaded into the cipher engine during the initialization phase. The cipher engine then processes the IV and key to generate a long keystream, which is used to encrypt or decrypt the plaintext or ciphertext.

Out of 7 blocks, block 0 (data *i*) helps in loading the initialization vectors and the key toward block 6 which acts as a backbone to trivium top in controlling the clock inputs from blocks 5, 4, and 2, respectively. Ultimately, the key and IV get into cipher engine block 7 through the trivium top acting as channelizer. Finally, after applying the trivium algorithm inside the cipher engine, the encrypted text passes through

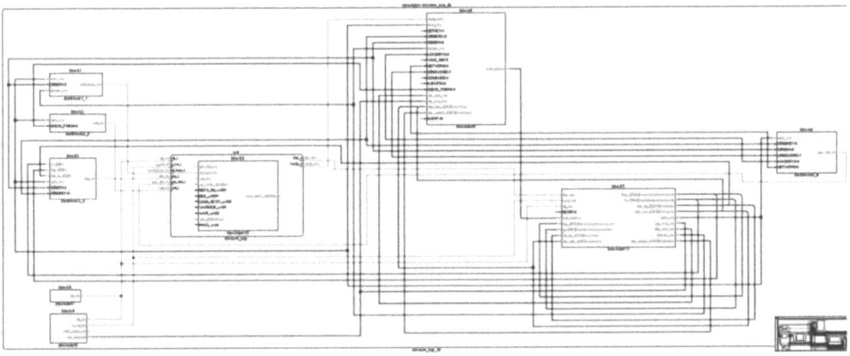

Fig. 4 Conventional trivium cipher schematic—NC-Verilog

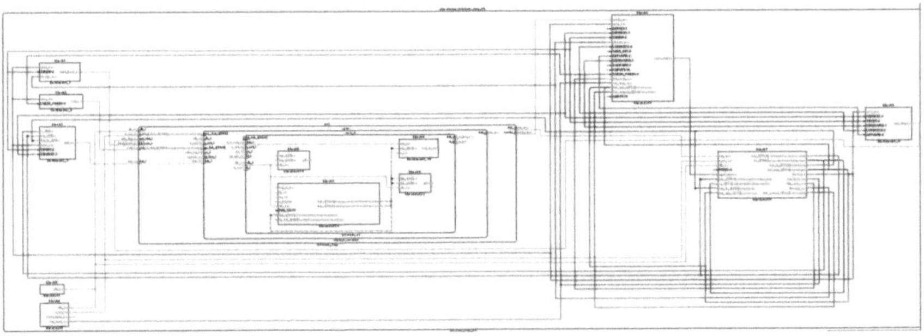

Fig. 5 Pipelined trivium cipher schematic—NC-Verilog

trivium top data *o* port to provide encrypted output. The waveform clearly depicts the output encrypted result through data *o* for text input data *i*. Suppose the input is an 8-bit numerical "00000000". When it comes to the matter of final schematic and output waveforms both pipelined and conventional trivium ciphers generate similar results through waveforms and their schematic superficially looks the same since the pipelining concept has been applied only inside the cipher engine and input sr blocks. After analysis we are able to conclude that waveform-based results obtained from both conventional and pipelined versions of trivium cipher are the same but vary in terms of operating speed, area utilized, and power efficiency.

After synthesizing both the conventional trivium cipher design and pipelined trivium cipher using the RC synthesis tool in the cadence environment, an analysis has been performed to evaluate the area, power, and timing characteristics of the design.

Area Analysis: The hardware resources needed to implement the synthesized trivium cipher design are revealed by the area analysis. The area analysis with respect to both conventional as well as pipelined ciphers have been tabulated thoroughly in Tables 1 and 2.

The effectiveness of the design with regard to resource usage can be evaluated by looking at metrics like gate count, cell count, and area utilization. After area analysis, the total design occupied area with respect to the entire design top block—trivium top of pipeline trivium cipher is less and is equal to 210.84 mm^2. Whereas on the

Table 1 Area analysis—conventional trivium cipher

Instance	Cells	Cell area (μm^2)	Net area (μm^2)	Total area (μm^2)	Wire load (S)
trivium top	837	20,501	1040	21,541	35,000
cphr	610	13,502	590	14,093	16,000
reg c	233	5003	210	5213	8000
reg a	197	4426	234	4659	8000
reg b	178	4038	145	4183	8000
key iv	160	6046	0	6046	8000

Table 2 Area analysis—pipelined trivium cipher

Instance	Cells	Cell area (μm^2)	Net area (μm^2)	Total area (μm^2)	Wire load (S)
trivium top	831	20,174	911	21,084	35,000
cphr	604	13,156	515	13,671	16,000
reg c	234	5045	210	5255	8000
reg a	194	4226	196	4422	8000
reg b	173	3807	107	3914	8000
key iv	161	6071	51	6121	8000

other hand, the conventional trivium cipher total design occupied area with respect to its trivium top base module is slightly more and equal to 215.41 mm^2.

Power Analysis: Understanding the synthesized trivium cipher design's power consumption profile is the main goal of power analysis. The power characteristics of the design can be examined.

For a thorough knowledge of the design's power consumption, it is crucial to assess both dynamic power (power used during switching activities) and static power (leakage power). The power analysis with respect to both conventional as well as pipelined ciphers have been tabulated thoroughly in Tables 3 and 4.

After power analysis the total power consumption with respect to the entire design top block—trivium top of pipeline trivium cipher is less and is equal to 512.043 μW whereas, on the other hand, conventional trivium cipher total power with respect to its trivium top base module is slightly more and equal to 554.974 μW.

Timing Analysis: The performance of the synthesized trivium cipher design is assessed in terms of its timing requirements through timing analysis. To make that the design complies with timing requirements and performs within the intended frequency range, metrics including critical path delay, setup time, hold time, and maximum operating frequency are checked.

For the cipher to function properly and operate reliably, timing analysis aids in identifying potential timing violations or performance bottlenecks that need to be resolved. The slack value, used in timing analysis, quantifies the amount of time a

Table 3 Power analysis: conventional trivium cipher

Instance	Cells	Leakage power (nW)	Dynamic power (nW)	Total power (nW)
trivium_top	837	81333.761	473640.997	554974.758
cphr	610	55747.729	351976.596	407724.325
reg_c	233	20944.477	115987.159	136931.636
reg_a	197	18189.282	116062.433	134251.716
reg_b	178	16412.739	111295.606	127708.345
key_iv	160	21001.715	35448.720	56450.435

Table 4 Power analysis: pipelined trivium cipher

Instance	Cells	Leakage power (nW)	Dynamic power (nW)	Total power (nW)
trivium_top	831	86040.989	426002.292	512043.281
cphr	604	55640.075	323571.138	379211.212
reg_c	234	21287.640	124845.522	146133.162
reg_a	194	17894.785	101519.210	119413.995
reg_b	173	16055.121	94795.294	110850.415
key_iv	161	25867.688	14597.295	40464.984

signal path can be delayed without going against the timing restrictions of a digital circuit.

The pipelined trivium cipher has a slack of 7074 ps compared to the traditional trivium cipher's 7019 ps. (Clearly visible in the conventional and pipelined trivium ciphers' timing analysis snippets in Figs. 6 and 7.) Accordingly, the pipelined trivium cipher outperforms the conventional trivium cipher by 1.18 GHz. These positive slack numbers suggest that there is some room for error or extra time for both trivium cipher variants. They imply that, taking into account potential variables such process variations, temperature changes, or age effects, the timing requirements of the circuit are easily met. Essentially, even if there are delays within the specified range indicated by the slack values, the circuits can still function properly.

```
           Pin                    Type      Fanout   Load Slew Delay Arrival
                                                     (fF) (ps)  (ps)   (ps)
------------------------------------------------------------------------------
(clock clk_i)                    launch                                  0 R
cur_state_r_reg[1]/CLK                                        0          0 R
cur_state_r_reg[1]/Q             DFFARX1         5   10.0    54  +201    201 F
g3/IN1                                                           +50     251
g3/QN                            NAND2X0         1    2.8    58   +37    288 R
g2/IN1                                                           +52     340
g2/Q                             OR2X1           7   18.7    74   +77    418 R
g557/IN1                                                         +185    603
g557/QN                          NAND2X2        88  170.0   369  +203    806 F
cphr/ce_i
  reg_c/ce_i
    g1205/INP                                                    +279   1085
    g1205/ZN                     INVX0           5   14.5   185  +106   1191 R
    g3815/INP                                                    +219   1410
    g3815/ZN                     INVX2         112  205.8   291  +192   1602 F
    g3814/IN2                                                    +299   1901
    g3814/QN                     NOR2X0         28   48.5   436  +226   2127 R
    drc3828/INP                                                  +176   2304
    drc3828/ZN                   INVX0           3    5.7   137   +71   2374 F
    drc3827/INP                                                   +51   2426
    drc3827/ZN                   INVX0          27   46.4   306  +168   2594 R
    g3726/IN2                                                    +154   2749
    g3726/Q                      AO22X1          1    1.5    42  +112   2861 R
    dat_r_reg[9]/D               DFFARX1                          +33   2894
    dat_r_reg[9]/CLK             setup                        0   +88   2981 R
- - - - - - - - - - - - - - - - - - - - - - - - - - - - - - - - - - - - - - -
(clock clk_i)                    capture                               10000 R
------------------------------------------------------------------------------
Cost Group   : 'clk_i' (path_group 'clk_i')
Timing slack :     7019ps
Start-point  : cur_state_r_reg[1]/CLK
End-point    : cphr/reg_c/dat_r_reg[9]/D
```

Fig. 6 Conventional trivium cipher—timing analysis

```
           Pin                  Type      Fanout  Load  Slew Delay Arrival
                                                  (fF)  (ps)  (ps)   (ps)
--------------------------------------------------------------------------
(clock clk_i)                   launch                                0 R
cur_state_r_reg[1]/CLK                                    0           0 R
cur_state_r_reg[1]/Q            DFFARX1        5  10.8   56  +203   203 F
g562/IN2                                                      +77   280
g562/QN                         NAND3X0        7  18.1  141   +79   359 R
g557/IN1                                                     +185   544
g557/QN                         NAND2X2       87 161.0  358  +209   753 F
cphr/ce_i
  reg_c/ce_i
    g1214/INP                                                +279  1032
    g1214/ZN                    INVX0          5  14.5  183  +105  1137 R
    g3828/INP                                                +219  1356
    g3828/ZN                    INVX2        112 205.8  291  +191  1547 F
    g3827/IN2                                                +299  1846
    g3827/QN                    NOR2X0        28  48.5  436  +226  2072 R
    drc3841/INP                                              +176  2248
    drc3841/ZN                  INVX0          3   5.7  137   +71  2319 F
    drc3840/INP                                               +51  2371
    drc3840/ZN                  INVX0         27  46.4  306  +168  2539 R
    g3739/IN2                                                +154  2694
    g3739/Q                     AO22X1         1   1.5   42  +112  2805 R
    dat_r_reg[9]/D              DFFARX1                       +33  2838
    dat_r_reg[9]/CLK            setup                     0   +88  2926 R
- - - - - - - - - - - - - - - - - - - - - - - - - - - - - - - - - - - - -
(clock clk_i)                   capture                           10000 R
--------------------------------------------------------------------------
Cost Group    : 'clk_i' (path_group 'clk_i')
Timing slack  :    7074ps
Start-point   : cur_state_r_reg[1]/CLK
End-point     : cphr/reg_c/dat_r_reg[9]/D
```

Fig. 7 Pipelined trivium cipher—timing analysis

7 Conclusion and Future Work

Pipelined trivium cipher has unquestionably outperformed conventional trivium cipher in all areas, including power efficiency, lower space, and higher frequency of operation. The traditional version is more straightforward and frequently used in software-based encryption settings, whereas the pipelined version offers performance gains and is ideally suited for hardware implementations. The designed pipelined trivium cipher is 1.18 GHz faster than conventional trivium cipher in terms of operating frequency. Area-wise the pipelined trivium cipher occupies 2.32% less area when compared with conventional design. In terms of power efficiency pipelined trivium cipher consumes 8.40% less than conventional trivium cipher.

Despite the pipelined trivium cipher's enhanced power efficiency, additional decrease can be made by using clock gating techniques. Power wastage can be reduced by selectively turning off clock signals for inactive circuit components. Stronger and more effective encryption can be produced by improving the cipher's nonlinear feedback mechanism, ensuring increased protection for sensitive data. Application-specific integrated circuit (ASIC) implementation of pipelined trivium ciphers and its incorporation into security systems as Intellectual Property (IP) formats also offer a promising future. These developments could displace Advanced Encryption Standard (AES) systems and offer a workable substitute with better performance and security. Pipelined trivium ciphers can be improved upon and used in a variety of security and encryption fields with ongoing research and development in these areas.

References

1. Mora-Gutiérrez JM, Jiménez-Fernández CJ, Valencia-Barrero M (2020) ASIC design and power characterization of standard and low power multi-radix trivium. IEEE Trans Circuits Syst II Express Briefs 67(11):2682–2686
2. Islam S, Haq IU (2016) Cube attack on trivium and A5/1 stream ciphers. In: 2016 13th international Bhurban conference on applied sciences and technology (IBCAST)
3. Manifavas C, Hatzivasilis G, Fysarakis K, Papaefstathiou Y (2016) A survey of lightweight stream ciphers for embedded systems. Secur Commun Netw
4. Philip MA, Vaithiyanathan (2017) A survey on lightweight ciphers for IoT devices. In: 2017 international conference on technological advancements in power and energy (TAP energy)
5. Montoya M, Hiscock T, Bacles-Min S, Molnos A, Fournier JJA (2018) Energy-efficient masking of the trivium stream cipher. In: 2018 25th IEEE international conference on electronics, circuits and systems (ICECS)
6. Potestad-Ordóñez FE, Jiménez-Fernández CJ, Valencia-Barrero M (2016) Fault attack on FPGA implementations of trivium stream cipher. In: 2016 IEEE international symposium on circuits and systems (ISCAS)
7. Ngo K, Dubrova E, Moraitis M (2020) Attacking trivium at the bitstream level. In: 2020 IEEE 38th international conference on computer design (ICCD)
8. Mora-Gutiérrez JM, Jiménez-Fernández CJ, Valencia-Barrero M (2017) Multiradix trivium implementations for low-power IoT hardware. IEEE Trans Very Large Scale Integr (VLSI) Syst 25(12):3401–3405. Buchanan WJ, Li S, Asif R (2017) Lightweight cryptography methods. J Cyber Secur Technol
9. Panasenko S, Smagin S (2011) Lightweight cryptography: underlying principles and approaches. Int J Comput Theory Eng 3(4)
10. Mora-Gutiérrez JM, Jiménez-Fernández CJ, Valencia-Barrero M (2013) Low power implementation of trivium stream cipher. In: Integrated circuit and system design. Power and timing modeling, optimization and simulation. Springer
11. Jafarpour A, Mahdlo A, Akbari A, Kianfar K (2011) Grain and trivium ciphers implementation algorithm in FPGA chip and AVR micro controller. In: 2011 IEEE international conference on computer applications and industrial electronics (ICCAIE)
12. Meena A, Shiyamala S, Kumar SV, Saleemnawaz MA, Uthirasamy R (2022) Low power enhanced trivium implementation using parallel pipeline technique. In: 2022 international conference on smart technologies and systems for next generation computing (ICSTSN)
13. Potestad-Ordóñez FE, Jiménez-Fernández CJ, Valencia-Barrero M (2016) Fault attack on FPGA implementations of trivium stream cipher. In: IEEE international symposium on circuits and systems (ISCAS)
14. Mora-Gutiérrez JM, Jiménez-Fernández CJ, Valencia-Barrero M (2016) Trivium hardware implementations for power reduction. Int J Circuit Theory Appl
15. Potestad-Ordóñez FE, Tena-Sánchez E, Mora-Gutiérrez JM, Valencia-Barrero M, Jiménez-Fernández CJ (2021) Trivium stream cipher countermeasures against fault injection attacks and DFA. IEEE Access 9:168444–168454
16. Wang ZC, Dai ZB (2015) High-speed realization of trivium based on multi-core cryptographic processor. In: 2015 IEEE 11th international conference on ASIC (ASICON), pp 1–4

17. Beckwith L, Diehl W (2020) New directions for newhope: improving performance of post-quantum cryptography through algorithm-level pipelining. In: 2020 international conference on field-programmable technology (ICFPT)
18. De Cannière C (2006) TRIVIUM: a stream cipher construction inspired by block cipher design principles. In: Katsikas SK, López J, Backes M, Gritzalis S, Preneel B (eds) Information security. Lecture notes in computer science, vol 4176. Springer, Berlin, Heidelberg
19. Kitsos P, Sklavos N, Provelengios G, Skodras AN (2013) FPGA-based performance analysis of stream ciphers ZUC, Snow3g, Grain V1, Mickey V2, Trivium and E0. Microprocess Microsyst

GPSR Compliance

The European Union's (EU) General Product Safety Regulation (GPSR) is a set of rules that requires consumer products to be safe and our obligations to ensure this.

If you have any concerns about our products, you can contact us on ProductSafety@springernature.com

In case Publisher is established outside the EU, the EU authorized representative is:

Springer Nature Customer Service Center GmbH
Europaplatz 3
69115 Heidelberg, Germany

Batch number: 10725962

Printed by Printforce, the Netherlands